Approximate Iterative Algorithms

Approximate Iterative Algorithms

Anthony Almudevar
Department of Biostatistics and Computational Biology,
University of Rochester, Rochester, NY, USA

CRC Press
Taylor & Francis Group
Boca Raton London New York

CRC Press is an imprint of the
Taylor & Francis Group, an **informa** business
A BALKEMA BOOK

CRC Press
Taylor & Francis Group
6000 Broken Sound Parkway NW, Suite 300
Boca Raton, FL 33487-2742

First issued in paperback 2019

Typeset by MPS Limited, Chennai, India

ISBN-13: 978-0-415-62154-0 (hbk)
ISBN-13: 978-0-367-37888-2 (pbk)

Library of Congress Cataloging-in-Publication Data

Almudevar, Anthony, author.
Approximate iterative algorithms / Anthony Almudevar, Department of Biostatistics and Computational Biology, University of Rochester, Rochester, NY, USA.
pages cm
Includes bibliographical references and index.
ISBN 978-0-415-62154-0 (hardback) — ISBN 978-0-203-50341-6 (eBook PDF)
1. Approximation algorithms. 2. Functional analysis. 3. Probabilities.
4. Markov processes. I. Title.
QA76.9.A43A46 2014
519.2′33—dc23

2013041800

Visit the Taylor & Francis Web site at
http://www.taylorandfrancis.com

and the CRC Press Web site at
http://www.crcpress.com

Table of contents

PART III
Application to Markov decision processes 247

Chapter 1

Introduction

The scope of this volume is quite specific. Suppose we wish to determine the solution V^* to a fixed point equation $V = TV$ for some operator T. Under suitable conditions, V^* will be the limit of an iterative algorithm

$$\begin{aligned} V_0 &= v_0 \\ V_k &= TV_{k-1}, \quad k = 1, 2, \ldots, \end{aligned} \tag{1.1}$$

where v_0 is some initial solution. Such algorithms are ubiquitous in applied mathematics, and their properties well known.

Then suppose (1.1) is replaced with an approximation

$$\begin{aligned} V_0 &= v_0 \\ V_k &= \hat{T}_k V_{k-1}, \quad k = 1, 2, \ldots, \end{aligned} \tag{1.2}$$

where each $\hat{T}_k$ is close to T in some sense. The subject of this book is the analysis of algorithms of the form (1.2). The material in this book is organized around three questions:

(Q1) If (1.1) converges to V^*, under what conditions does (1.2) also converge to V^*?
(Q2) How does the approximation affect the limiting properties of (1.2)? How close is the limit of (1.2) to V^*, and what is the rate of convergence (particularly in comparison to that of (1.1))?
(Q3) If (1.2) is subject to design, in the sense that an approximation parameter, such as grid size, can be selected for each $\hat{T}_k$, can an approximation schedule be determined which minimizes approximation error as a function of computation time?

From a theoretical point of view, the purpose of this book is to show how quite straightforward principles of functional analysis can be used to resolve these questions with a high degree of generality. From the point of view of applications, the primary interest is in dynamic programming and Markov decision processes (MDP), with emphasis on approximation methods and computational efficiency. The emphasis is less on the construction of specific algorithms then with the development of theoretical tools with which broad classes of algorithms can be defined, and hence analyzed with a common theory.

The book is divided into three parts. Chapters 2–8 cover background material in real analysis, linear algebra, measure theory, probability theory and functional analysis. This section is fairly extensive in comparison to other volumes dealing specifically with MDPs. The intention is that the language of functional analysis be used to express concepts from the other disciplines, in as general but concise a manner as possible. By necessity, many proofs are omitted in these chapters, but suitable references are given when appropriate.

Chapters 9–11 form the core of the volume, in the sense that the questions (Q1)–(Q3) are largely considered here. Although a number of examples are considered (most notable, an analysis of the Robbins-Monro algorithm), the main purpose is to deduce properties of general classes of approximate iterative algorithms on Banach and Hilbert spaces.

The remaining chapters deal with Markov decision processes (MDPs), which forms the principal motivation for the theory presented here. A foundation theory of MDPs is given in Chapters 12 and 13, from the point of view of functional analysis, while the remain chapters discuss approximation methods.

Finally, I would like to acknowledge the patience and support of colleagues and family, especially Cynthia, Benjamin and Jacob.

Part I

Mathematical background

Chapter 2

Real analysis and linear algebra

In this chapter we first define notation, then review a number of important results in real analysis and linear algebra of which use will be made in later chapters. Most readers will be familiar with the material, but in a number of cases it will be important to establish which of several commonly used conventions will be used. It will also prove convenient from time to time to have a reference close at hand. This may be especially true of the section on spectral decomposition.

2.1 DEFINITIONS AND NOTATION

In this section we describe the notational conventions and basic definitions to be used throughout the book.

2.1.1 Numbers, sets and vectors

A *set* is a collection of distinct objects of any kind. Each member of a set is referred to as an *element*, and is represented once. A set E may be *indexed*. That is, given an index set $\mathcal{T}$, each element may be assigned a unique index $t \in \mathcal{T}$, and all indices in $\mathcal{T}$ are assigned to exactly one element of E, denoted x_t. We may then write $E = \{x_t;\ t \in \mathcal{T}\}$.

The set of (finite) real numbers is denoted $\mathbb{R}$, and the set of extended real numbers is denoted $\bar{\mathbb{R}} = \mathbb{R} \cup \{-\infty, \infty\}$. The restriction to nonegative real numbers is written $\mathbb{R}_+ = [0, \infty)$ and $\bar{\mathbb{R}}_+ = \mathbb{R}_+ \cup \{\infty\}$. We use standard notation for open, closed, left closed and right closed intervals (a, b), $[a, b]$, $[a, b)$, $(a, b]$. A reference to a interval I on $\bar{\mathbb{R}}$ may be any of these types.

The set of (finite) integers will be denoted $\mathbb{I}$, while the extended integers will be $\mathbb{I}_\infty = \mathbb{I} \cup \{-\infty, \infty\}$. The set of natural numbers $\mathbb{N}$ is taken to be the set of positive integers, while $\mathbb{N}_0$ is the set of nonnegative integers. A rational number is any real number expressible as a ratio of integers.

Then $\mathbb{C}$ denotes the complex numbers $z = a + bi \in \mathbb{C}$, where $i = \sqrt{-1}$ is the imaginary number and $a, b \in \mathbb{R}$. Note that i is added and multiplied as though it were a real number, in particular $i^2 = -1$. Multiplication is defined by $z_1 z_2 = (a_1 + b_1 i)(a_2 + b_2 i) = a_1 a_2 - b_1 b_2 + (a_1 b_2 + a_2 b_1)i$. The *conjugate* of $z = a + bi \in \mathbb{C}$ is written $\bar{z} = a - bi$, so that $z\bar{z} = a^2 + b^2 \in \mathbb{R}$. Together, z and $\bar{z}$, without reference to their order, form a *conjugate pair*.

The absolute value of $a \in \mathbb{R}$ is denoted $|a| = \sqrt{a^2}$, while $|z| = (z\bar{z})^{1/2} = (a^2 + b^2)^{1/2} \in \mathbb{R}$ is also known as the magnitude or modulus of $z \in \mathbb{C}$.

If $\mathcal{S}$ is a set of any type of number, $\mathcal{S}^d$, $d \in \mathbb{N}$, denotes the set of d-dimensional vectors $\tilde{s} = (s_1, \ldots, s_d)$, which are ordered collections of numbers $s_i \in \mathcal{S}$. In particular, the set of d-dimensional real vectors is written $\mathbb{R}^d$. When $0, 1 \in \mathcal{S}$, we may write the zero or one vector $\vec{0} = (0, \ldots, 0)$, $\vec{1} = (1, \ldots, 1)$, so that $c\vec{1} = (c, \ldots, c)$.

A collection of d numbers from $\mathcal{S}$ is *unordered* if no reference is made to the order (they are unlabeled). Otherwise the collection is *ordered*, that is, it is a vector. An unordered collection from $\mathcal{S}$ differs from a set in that a number $s \in \mathcal{S}$ may be represented more than once. Braces $\{\ldots\}$ enclose a set while parentheses $(\ldots)$ enclose a vector (braces will also be used to denote indexed sequences, when the context is clear).

2.1.2 Logical notation

We will make use of conventional logical notation. We write $S_1 \Rightarrow S_2$ if statement S_1 implies statement S_2, and $S_1 \Leftrightarrow S_2$ whenever $S_1 \Rightarrow S_2$ and $S_2 \Rightarrow S_1$ both hold. In addition, 'for all' is written $\forall$, 'there exists' is written $\exists$ and 'such that' is written $\ni$.

2.1.3 Set algebra

If x is, or is not, an element of E, we write $x \in E$ or $x \notin E$. If all elements in A are also in B then A is a *subset* of B, that is $A \subset B$. If $A \subset B$ and $B \subset A$ then $A = B$. If $A \subset B$ but $A \neq B$, then A is a *strict subset* of B. Define the empty set, or null set, $\emptyset$, which contains no elements. We may write $\emptyset \subset A$ for any set A.

Set algebra is defined for the class of all subsets of a nonempty set Ω, commonly known as a *universe*. Any set we consider may only contain elements of Ω. This always includes both $\emptyset$ and Ω. Set operations include union $(A \cup B) = (A \text{ or } B) = (A \vee B)$ (all elements in either A or B), intersection $(A \cap B) = (A \text{ and } B) = (A \wedge B)$ (all elements in both A and B), complementation $(\sim A) = (\text{not } A) = (A^c)$ (all elements in Ω not in A), relative complementation, or set difference, $(B \sim A) = (B - A) = (B \text{ not } A) = (BA^c)$ (all elements in B not in A). For any indexed collection of subsets $A_t \subset \Omega$, $t \in \mathcal{T}$, the union is $\cup_{t \in \mathcal{T}} A_t$, the set of all elements in at least one A_t, and the intersection is $\cap_{t \in \mathcal{T}} A_t$, the set of all elements in all A_t. *De Morgan's Law* applies to any index set $\mathcal{T}$ (finite or infinite), that is,

$$\cup_{t \in \mathcal{T}} A_t^c = (\cap_{t \in \mathcal{T}} A_t)^c \quad \text{and} \quad \cap_{t \in \mathcal{T}} A_t^c = (\cup_{t \in \mathcal{T}} A_t)^c.$$

The cardinality of a set E is the number of elements it contains, and is denoted $|E|$. If $|E| < \infty$ then E is a finite set. We have $|\emptyset| = 0$. If $|E| = \infty$, this statement does not suffice to characterize the cardinality of E. Two sets A, B are in a *1-1 correspondence* if a collection of pairs (a, b), $a \in A$, $b \in B$ can be constructed such that each element of A and of B is in exactly one pair. In this case, A and B are of equal cardinality. The pairing is known as a *bijection*.

If the elements of A can be placed in a 1-1 correspondence with $\mathbb{N}$ we say A is *countable* (is *denumerable*). We also adopt the convention of referring to any subset of a countable set as countable. This means all finite sets are countable. If for countable A we have $|A| = \infty$ then A is *infinitely countable*. Note that by some conventions, the term countable is reserved for infinitely countable sets. For our purposes, it is more natural to consider the finite sets as countable.

All infinitely countable sets are of equal cardinality with $\mathbb{N}$, and so are mutually of equal cardinality. informally, a set is countable if it can be written as a list, finite or infinte. The set $\mathbb{N}^d$ is countable since, for example, $\mathbb{N}^2 = \{(1,1),(1,2),(2,1),(1,3),(2,2),(3,1),\ldots\}$. The set of rational numbers is countable, since the pairing of numerator and denominator, in any canonical representation, is a subset of $\mathbb{N}^2$.

A set A is *uncountable* (is *nondenumerable*) if $|A| = \infty$ but A is not countable. The set of real numbers, or any nonempty interval of real numbers, is uncountable.

If $A_1, \ldots, A_d$ are d sets, then $A_1 \times A_2 \times \cdots \times A_d = \times_{i=1}^d A_i$ is a product set, consisting of the set of all ordered selections of one element from each set $a_i \in A_i$. A vector is an element of a product set, but a product set is more general, since the sets A_i need not be equal, or even contain the same type of element. The definition may be extended to arbitrary forms of index sets.

2.1.4 The supremum and infimum

For any set $E \subset \mathbb{R}$, $x = \max E$ if $x \in E$ and $y \leq x\ \forall y \in E$. Similarly $x = \min E$ if $x \in E$ and $y \geq x\ \forall y \in E$. The quantities $\min E$ or $\max E$ need not exist (consider $E = (0,1)$).

The *supremum* of E, denoted $\sup E$ is the *least upper bound* of E. Similarly, the *infimum* of E, denoted $\inf E$ is the *greatest lower bound* of E. In contrast with the min, max operations, the supremum and infimum always exist, possibly equalling $-\infty$ or ∞. For example, if $E = (0,1)$, then $\inf E = 0$ and $\sup E = 1$. That is, $\inf E$ or $\sup E$ need not be elements of E. All numbers in $\bar{\mathbb{R}}$ are both upper and lower bounds of the empty set $\emptyset$, which means

$$\inf \emptyset = \infty \quad \text{and} \quad \sup \emptyset = -\infty.$$

If $E = \{x_t;\ t \in \mathcal{T}\}$ is an indexed set we write, when possible,

$$\max E = \max_{t \in \mathcal{T}} x_t, \quad \min E = \min_{t \in \mathcal{T}} x_t, \quad \sup E = \sup_{t \in \mathcal{T}} x_t, \quad \inf E = \inf_{t \in \mathcal{T}} x_t.$$

For two numbers $a, b \in \bar{\mathbb{R}}$, we may use the notations $\max\{a,b\} = x \vee y = \max(a,b)$ and $\min\{a,b\} = x \wedge y = \min(a,b)$.

2.1.5 Rounding off

Rounding off will proceed by the floor and ceiling conventions $\lfloor 1.99 \rfloor = 1 = \lfloor 1 \rfloor$ and $\lceil 1.01 \rceil = 2 = \lceil 2 \rceil$.

When we write $x \approx 3.45$, we mean $x \in [3.445, 3.455)$. This convention is adopted throughout.

2.1.6 Functions

If X, Y are two sets, then a function $f : X \to Y$ assigns a unique element of Y to each element of X, in particular $y = f(x)$. We refer to X and Y as the *domain* and *range* (or *codomain*) of f. The *image* of a subset $A \subset X$ is $f(A) = \{f(x) \in Y \mid x \in A\}$, and the *preimage* (or *inverse image*) of a subset $B \in Y$ is $f^{-1}(B) = \{x \in X \mid f(x) \in B\}$. We say f is *injective* (or *one-to-one*) if $f(x_1) \neq f(x_2)$ whenever $x_1 \neq x_2$, f is *surjective* (alternatively, *many-to-one* or *onto*) if $Y = f(X)$, and f is *bijective* if it is both injective and surjective. An injective, surjective or bijective function is also referred to as an *injection*, *surjection* or *bijection*. A bijective function f is *invertible*, and possesses a unique *inverse* function $f^{-1}: Y \to X$ which is also bijective, and satisfies $x = f^{-1}(f(x))$. Only bijective functions are invertible. Note that a preimage may be defined for any function, despite what is suggested by the notation.

An indicator function f maps a domain $\mathcal{X}$ to $\{0,1\}$ by specifying a set $E \subset \mathcal{X}$ and setting $f(x) = 1$ if $x \in E$ and $f(x) = 0$ otherwise. This may be written explicitly as $f(x) = I\{x \in E\}$, or I_E when the context is clear.

For real valued functions f, g, $(f \vee g)(x) = f(x) \vee g(x)$, $(f \wedge g)(x) = f(x) \wedge g(x)$. We write $f \equiv c$ for constant c if $f(x) = c\ \forall x$. A function f on $\mathbb{R}$ satisfying $f(x) = -f(-x)$ or $f(x) = f(-x)$ is an *odd* or *even* function. A real valued function f will sometimes be decomposed into positive and negative components $f = f^+ - f^-$ where $f^+ = f(x)I\{f(x) > 0\}$ and $f^- = |f(x)|I\{f(x) < 0\}$.

For mappings $f: X \to Y$ and $g: Y \to Z$, where f is surjective, we denote the composition $(g \circ f): X \to Z$, evaluated by $g(f(x)) \in Z\ \forall x \in X$.

2.1.7 Sequences and limits

A sequence of real numbers $a_0, a_1, a_2, \ldots$ will be written $\{a_k\}$. Depending on the context, a_0 may or may not be defined. For any sequence of real numbers, by $\lim_{k\to\infty} a_k = a \in \mathbb{R}$ is always meant that $\forall \epsilon > 0\ \exists K \ni k > K \Rightarrow |a - a_k| < \epsilon$. A reference to $\lim_{k\to\infty} a_k$ implies an assertion that a limit exists. This will sometimes be written $a_k \to a$ or $a_k \to_k a$ when the context makes the meaning clear.

When a limit exists, a sequence is *convergent*. If a sequence does not converge it is *divergent*. This excludes the possibility of a limit ∞ or $-\infty$ for a convergent sequence. However, it is sometimes natural to think of a sequence with a 'limit' in $\{-\infty, \infty\}$. We can therefore write $\lim_{k\to\infty} a_k = \infty$ if $\forall M\ \exists K \ni k > K \Rightarrow a_k > M$, and $\lim_{k\to\infty} a_k = -\infty$ if $\forall M\ \exists K \ni k > K \Rightarrow a_k < M$. Either sequence is *properly divergent*.

If $a_{k+1} \geq a_k$, the sequence must possess a limit a, possibly ∞. This is written $a_k \uparrow a$. Similarly, if $a_{k+1} \leq a_k$, there exists a limit $a_k \downarrow a$, possibly $-\infty$. Then $\{a_k\}$ is an *nondecreasing* or *nonincreasing* sequence (or *increasing*, *decreasing* when the defining inequalities are strict).

Then $\limsup_{k\to\infty} a_k = \lim_{k\to\infty} \sup_{i\geq k} a_i$. This quantity is always defined since $a'_k = \sup_{i\geq k} a_i$ defines an nonincreasing sequence. Similarly $\liminf_{k\to\infty} a_k = \lim_{k\to\infty} \inf_{i\geq k} a_i$ always exists. We always have $\liminf_{k\to\infty} a_k \leq \limsup_{k\to\infty} a_k$ and $\lim_{k\to\infty} a_k$ exists if and only if $a = \liminf_{k\to\infty} a_k = \limsup_{k\to\infty} a_k$, in which case $\lim_{k\to\infty} a_k = a$.

When limit operations are applied to sequences of real values functions, the limits are assumed to be evaluated pointwise. Thus, if we write $f_n \uparrow f$, this means that $f_n(x) \uparrow f(x)$ for all x, and therefore f_n is a *nondecreasing sequence* of functions, with analagous conventions used for the remaining types of limits.

Note that pointwise convergence of a function $\lim_{n\to\infty} f_n = f$ is distinct from *uniform convergence* of a sequence of functions, which is equivalent to $\lim_{n\to\infty} \sup_x |f_n(x) - f(x)| = 0$. Of course, uniform convergence implies pointwise convergence, but the converse does not hold. Unless uniform convergence is explicitly stated, pointwise convergence is intended.

When the context is clear, we may use the more compact notation $\tilde{d}=(d_1,d_2,\ldots)$ to represent a sequence $\{d_k\}$. If $\tilde{a}=\{a_k\}$ and $\tilde{b}=\{b_k\}$ then we write $\tilde{a}\leq\tilde{b}$ if $a_k\leq b_k$ for all k.

Let $\mathcal{S}$ be the class of all sequences of finite positive real numbers which converge to zero, and let $\mathcal{S}^-$ be those sequences in $\mathcal{S}$ which are nonincreasing. If $\{a_k\}\in\mathcal{S}$ we define the *lower* and *upper convergence rates* $\lambda^l\{a_k\}=\liminf_{k\rightarrow\infty}a_{k+1}/a_k$ and $\lambda^u\{a_k\}=\limsup_{k\rightarrow\infty}a_{k+1}/a_k$. If $0<\lambda^l\{a_k\}\leq\lambda^u\{a_k\}<1$ then $\{a_k\}$ *converges linearly*. If $\lambda^u\{a_k\}=0$ or $\lambda^l\{a_k\}=1$ then $\{a_k\}$ *converges superlinearly* or *sublinearly*, respectively. We also define a weaker characterization of linear convergence by setting $\hat{\lambda}^l\{a_k\}=\liminf_{k\rightarrow\infty}a_k^{1/k}$ and $\hat{\lambda}^u\{a_k\}=\limsup_{k\rightarrow\infty}a_k^{1/k}$.

When $\lambda^l\{a_k\}=\lambda^u\{a_k\}=\rho$ we write $\lambda\{a_k\}=\rho$. Similarly $\hat{\lambda}^l\{a_k\}=\hat{\lambda}^u\{a_k\}=\rho$ is written $\hat{\lambda}\{a_k\}=\rho$.

A sequence $\{a_k\}$ is *of order* $\{b_k\}$ if $\limsup_k a_k/b_k<\infty$, and may be written $a_k=O(b_k)$. If $a_k=O(b_k)$ and $b_k=O(a_k)$ we write $a_k=\Omega(b_k)$. Similarly, for two real valued mappings f_t,g_t on $(0,\infty)$ we write $f_t=O(g_t)$ if $\limsup_{t\rightarrow\infty}f_t/g_t<\infty$, and $f_t=\Omega(g_t)$ if $f_t=O(g_t)$ and $g_t=O(f_t)$.

A sequence $\{b_k\}$ *dominates* $\{a_k\}$ if $\lim_k a_k/b_k=0$, which may be written $a_k=o(b_k)$. A stronger condition holds if $\lambda^u\{a_k\}<\lambda^l\{b_k\}$, in which case we say $\{b_k\}$ *linearly dominates* $\{a_k\}$, which may be written $a_k=o_\ell(b_k)$. Similarly, for two real valued mappings f_t,g_t on $(0,\infty)$ we write $f_t=o(g_t)$ if $\lim_{t\rightarrow\infty}f_t/g_t=0$, that is, g_t *dominates* f_t.

2.1.8 Infinite series

Suppose we are given sequence $\{a_k\}$. The corresponding *series* (or *infinite series*) is denoted

$$\sum_{k=1}^{\infty}a_k=\sum_k a_k=a_1+a_2+\cdots.$$

Some care is needed in defining a sum of an infinite collection of numbers. First, define *partial sums*

$$S_n=\sum_{k=1}^{n}a_k=a_1+a_2+\cdots+a_n,\quad n\geq 1.$$

We may set $S_0=0$. It is natural to think of evaluating a series by sequentially adding each a_n to a cumulative total S_{n-1}. In this case, the total sum equals $\lim_n S_n$, assuming the limit exists. We say that the series (or simply, the sum) exists if the limit exists (including $-\infty$ or ∞). The series is *convergent* if the sum exists and is finite. A series is *divergent* if it is not convergent, and is *properly divergent* if the sum exists but is not finite.

It is important to establish whether or not the value of the series depends on the order of the sequence. Precisely, suppose $\sigma : \mathbb{N} \mapsto \mathbb{N}$ is a bijective mapping (essentially, an infinite permutation). If the series $\sum_k a_k$ exists, we would like to know if

$$\sum_k a_k = \sum_k a_{\sigma(k)}. \tag{2.1}$$

Since these two quantities are limits of distinct partial sums, equality need not hold. This question has a quite definite resolution. A series $\sum_k a_k$ is called *absolutely convergent* if $\sum_k |a_k|$ is convergent (so that all convergent series of nonnegative sequences are absolutely convergent). A convergent sequence is *unconditionally convergent* if (2.1) holds for all permutations σ. It may be shown that a series is absolutely convergent if and only if it is unconditionally convergent. Therefore, a convergent series may be defined as *conditionally convergent* if either it is not absolutely convergent, or if (2.1) does not hold for at least one σ. Interestingly, by the *Riemann series theorem*, if $\sum_k a_k$ is conditionally convergent then for any $L \in \mathbb{R}$ there exists permutation σ_L for which $\sum_k a_{\sigma_L(k)} = L$.

There exist many well known tests for series convergence, and can be found in most calculus textbooks.

Let $E = \{a_t;\ t \in \mathcal{T}\}$ be a infinitely countable indexed set of extended real numbers. For example, we may have $\mathcal{T} = \mathbb{N}^d$. When there is no ambiguity, we can take $\sum_t a_t$ to be the sum of all elements of E. Of course, in this case the implication is that the sum does not depend on the summation order. This is the case if and only if there is a bijective mapping $\sigma : \mathbb{N} \mapsto \mathcal{T}$ for which $\sum_k a_{\sigma(k)}$ is absolutely convergent. If this holds, it holds for all such bijective mappings. All that is needed is to verify that the cumulative sum of the elements $|a_t|$, taken in any order, remains bounded. This is written, when possible

$$\sum_{t \in \mathcal{T}} a_t = \sum_t a_t.$$

We also define for a sequence $\{a_k\}$ the product $\prod_{k=1}^{\infty} a_k$. We will usually be interested in products of positive sequences, so this may be converted to a series by the log transformation:

$$\log\left(\prod_{k=1}^{\infty} a_k\right) = \sum_{k=1}^{\infty} \log(a_k)$$

so that the issues are largely the same as for series. Similarly, for indexed set $E = \{a_t;\ t \in \mathcal{T}\}$, we may define $\prod_{t \in \mathcal{T}} a_t = \prod_t a_t$ when no ambiguity arises. This will be the case when, for example, either $a_t \in (0, 1]$ for all t or $a_t \in [1, \infty)$ for all t.

Finally, we make note of the following convention. We will sometimes be interested in summing over a strict subset of the index set $\mathcal{T}' \subset \mathcal{T}$. This poses no particular problem if the series $\sum_t a_t$ is well defined. If it happens that $\mathcal{T}' = \emptyset$, we will take

$$\sum_{t \in \emptyset} a_t = 0 \quad \text{and} \quad \prod_{t \in \emptyset} a_t = 1. \tag{2.2}$$

2.1.9 Geometric series

We will make use of the following geometric series:

$$\sum_{i=0}^{\infty} \frac{(i+m)!}{i!} r^i = \frac{m!}{(1-r)^{m+1}} \quad \text{for } r^2 < 1, \ m = 0, 1, 2, \ldots$$

$$\sum_{i=0}^{n} r^i = \frac{1-r^{n+1}}{1-r} \quad \text{for } r \neq 1. \tag{2.3}$$

2.1.10 Classes of real valued functions

Suppose $\mathcal{X}$ is a subset of $\bar{\mathbb{R}}$. The real valued function $f : \mathcal{X} \to \bar{\mathbb{R}}$ is a *bounded function* if $\sup_{x\in\mathcal{X}} |f(x)| < \infty$. In addition f is *bounded below* or *bounded above* if $\inf_{x\in\mathcal{X}} f(x) > -\infty$ or $\sup_{x\in\mathcal{X}} f(x) < \infty$.

A real valued function $f : \mathcal{X} \to \bar{\mathbb{R}}$ is *lower semicontinuous* at x_0 if $x_n \to_n x_0$ implies $\liminf_n f(x_n) \geq f(x_0)$, or *upper semicontinuous* at x_0 if $x_n \to x_0$ implies $\limsup_n f(x_n) \leq f(x_0)$. We use the abbreviations *lsc* and *usc*. A function is, in general, *lsc* (*usc*) if it is *lsc* (*usc*) at all $x_0 \in \mathcal{X}$. Equivalently, f is *lsc* if $\{x \in \mathcal{X} \mid f(x) \leq \lambda\}$ is closed for all $\lambda \in \mathbb{R}$, and is *usc* if $\{x \in \mathcal{X} \mid f(x) \geq \lambda\}$ is closed for all $\lambda \in \mathbb{R}$. A function is continous (at x_0) if and only if it is both *lsc* and *usc* (at x_0). Note that only sequences in $\mathcal{X}$ are required for the definition, so that if f is *lsc* or *usc* on $\mathcal{X}$, it is also *lsc* or *usc* on $\mathcal{X}' \subset \mathcal{X}$.

A set $\mathcal{X} \subset \mathbb{R}^d$ is *convex* if for any $p \in [0,1]$ and any $x_1, x_2 \in \mathcal{X}$ we also have $px_1 + (1-p)x_2 \in \mathcal{X}$. A real valued function $f : \mathcal{X} \to \mathbb{R}$ on a convex set $\mathcal{X}$ is *convex* if for any $p \in [0,1]$ and any $x_1, x_2 \in \mathcal{X}$ we have $pf(x_1) + (1-p)f(x_2) \geq f(px_1 + (1-p)x_2)$. Additionally, f is *strictly convex* if $pf(x_1) + (1-p)f(x_2) > f(px_1 + (1-p)x_2)$ whenever $p \in (0,1)$ and $x_1 \neq x_2$. If $-f$ is (strictly) convex then f is *(strictly) concave*.

The usual kth order partial derivatives, when they exist, are written $\partial^k f / \partial x_{i_1} \ldots \partial x_{i_k}$, and if $d = 1$ the kth total derivative is written $d^k f / dx^k = f^{(k)}(x)$. A derivative is a function on $\mathcal{X}$, unless evaluation at a specific value of $x \in \mathcal{X}$ is indicated, as in $d^k f / dx^k |_{x=x_0} = f^{(k)}(x_0)$. The first and second total derivative will also be written $f'(x)$ and $f''(x)$ when the context is clear.

The following function spaces are commonly defined: $C(\mathcal{X})$ is the set of all continuous real valued functions on $\mathcal{X}$, while $C_b(\mathcal{X}) \subset C(\mathcal{X})$ denotes all bounded continuous functions on $\mathcal{X}$. In addition, $C^k(\mathcal{X}) \subset C(\mathcal{X})$ is the set of all continuous functions on $\mathcal{X}$ for which all order $1 \leq j \leq k$ derivatives exist and are continuous on $\mathcal{X}$, with $C^\infty(\mathcal{X}) \subset C(\mathcal{X})$ denoting the class of functions with continuous derivatives of all orders (the infinitely divisible functions). Note that a function on $\mathbb{R}$ may possess derivatives $f'(x)$ everywhere (which are consistent in direction), without $f'(x)$ being continuous.

When defining a function space, the convention that $\mathcal{X}$ is open, with $\bar{\mathcal{X}}$ representing the closure of $\mathcal{X}$ when needed, is sometimes adopted. This ensures that the convential definitions of continuity and differentiability apply (formally, any bounded function defined on a finite set $\mathcal{X}$ is continuous, since the only convergent sequences in $\mathcal{X}$ are constant ones).

2.1.11 Graphs

A *graph* is a collection of *nodes* and *edges*. Most commonly, there are m nodes uniquely *labeled* by elements of set $V=\{1,\ldots,m\}$. We may identify the set of nodes as V (although sometimes *unlabeled graphs* are studied). An *edge* is a connection between two nodes, of which there are two types. A *directed edge* is any ordered pair from V, and an *undirected edge* is any unordered pair from V. Possibly, the two nodes defining an edge are the same, which yields a *self edge*. If E is any set of edges, then $G=(V,E)$ defines a *graph*. If all edges are directed (undirected), the graph is described as directed (undirected), but a graph may contain both types.

It is natural to imagine a dynamic process on a graph defined by node occupancy. A directed edge (v_1,v_2) denotes the possibly of a transition from v_1 to v_2. Accordingly, a *path* within a directed graph $G=(V,E)$ is any sequence of nodes $v_0,v_1,\ldots,v_n$ for which $(v_{i-1},v_i)\in E$ for $1\leq i\leq n$. This describes a path from v_0 to v_n of length n (the number of edges needed to construct the path).

It will be instructive to borrow some of the terminology associated with the theory of Markov chains (Section 5.2). For example, if there exists a path starting at i and ending at j we say that j is *accessible* from i, which is written $i\to j$. If $i\to j$ and $j\to i$ then i and j *communicate*, which is written $i\leftrightarrow j$. The connectivity properties of a directed graph are concerned with statements of this kind, as well as lengths of the relevant paths.

The *adjacency matrix* $adj(G)$ of graph G is an $m\times m$ 0-1 matrix with element $g_{i,j}=1$ if and only if the graph contains directed edge (i,j). The path properties of G can be deduced directly from the iterates $adj(G)^n$ (conventions for matrices are given in Section 2.3.1).

Theorem 2.1 *For any directed graph G with adjacency matrix $A_G=adj(G)$ there exists a path of length n from node i to node j if and only if element i,j of A_G^n is positive.*

Proof Let $g[k]_{i,j}$ be element i,j of A_G^k. All such elements are nonnegative. Suppose, as an induction hypothesis, the theorem holds for all paths of length n', for any $n'<n$. We may write

$$g[n]_{i,j}=\sum_{k=1}^{m} g[n']_{i,k}g[n-n']_{k,j},$$

from which we conclude that $g[n]_{i,j}>0$ if and only if for some k we have $g[n']_{i,k}>0$ and $g[n-n']_{k,j}>0$. Under the induction hypothesis, the latter statement is equivalent to the claim that for all $n'<n$ there is a node k for which there exists a path of length n' from i to k and a path of length $n-n'$ from k to j. In turn, this claim is equivalent to the claim that there exists a path of length n from i to j. The induction hypothesis clearly holds for $n=1$, which completes the proof. ///

It is interesting to compare Theorem 2.1 to the *Chapman-Kolmogorov* equations (5.4) associated with the theory of Markov chains. It turns out that many important properties of a Markov chain can be understood as the path properties of a directed graph. It is especially important to note that in Theorem 2.1 we can, without loss of generality, replace the '1' elements in A_G with any positive number. Accordingly, we give an alternative version of Theorem 2.2 for nonnegative matrices.

Theorem 2.2 *Let A be an $n \times n$ matrix of nonnegative elements $a_{i,j}$. Let $a[k]_{i,j}$ be element i,j of A^k. Then $a[n]_{i,j} > 0$ if and only if there exists a finite sequence of $n+1$ indices $v_0, v_1, \ldots, v_n$, with $v_0 = i$, $v_n = j$, for which $a_{v_{k-1},v_k} > 0$ for $1 \leq k \leq n$.*

Proof The proof follows that of Theorem 2.1. ///

The implications of this type of path structure are discussed further in Sections 2.3.4 and 5.2.

2.1.12 The binomial coefficient

For any $n \in \mathbb{N}_0$ the factorial is written $n! = \prod_{i=1}^{n} i$. By convention, $0! = 1$ (compare to (2.2)). The *binomial coefficient* is

$$\binom{n}{k} = \frac{n!}{k!(n-k)!}, \quad n \geq k, \;\; n, k \in \mathbb{N}_0.$$

Given $m \geq 2$, if $n_i \in \mathbb{N}_0$, $i = 1, \ldots, m$, and $n = n_1 + \cdots + n_m$, then the *multinomial coefficient* is

$$\binom{n}{n_1, \ldots, n_m} = \frac{n!}{\prod_{i=1}^{m} n_i!}.$$

The *Binomial Theorem* states that for $a, b \in \mathbb{R}$ and $n \in \mathbb{N}$ the following equality holds

$$(a+b)^n = \sum_{i=0}^{n} \binom{n}{i} a^i b^{n-i}. \tag{2.4}$$

2.1.13 Stirling's approximation of the factorial

The factorial $n!$ can be approximated accurately using series expansions. See, for example, Feller (1968) (Chapter 2, Volume 1). Stirling's approximation for the factorial is given by

$$s_n = (2\pi)^{1/2} n^{n+1/2} e^{-n}, \quad n \geq 1,$$

and if we set $n! = s_n \rho_n$, we have

$$e^{1/(12n+1)} < \rho_n < e^{1/(12n)}. \tag{2.5}$$

The approximation is quite sharp, guaranteeing that (*a*) $\lim_{n\to\infty} n!/s_n = 1$; (*b*) $1 < n!/s_n < e^{1/12} < 1.087$ for all $n \geq 1$; (*c*) $(12n+1)^{-1} < \log(n!) - \log(s_n) < (12n)^{-1}$ for all $n \geq 1$.

2.1.14 L'Hôpital's rule

Suppose $f, g \in C(\mathcal{X})$ for open interval $\mathcal{X}$, and for $x_0 \in \mathcal{X}$ we have $\lim_{x \to x_0} f(x) = \lim_{x \to x_0} g(x) = b$, where $b \in \{-\infty, 0, \infty\}$. The ratio $f(x_0)/g(x_0)$ is not defined, but the limit $\lim_{x \to x_0} f(x)/g(x)$ may be. If $f, g \in C^1(\mathcal{X} - \{x_0\})$, and $g'(x) \neq 0$ for $x \in \mathcal{X} - \{x_0\}$ then *l'Hôpital's Rule* states that

$$\lim_{x \to x_0} f(x)/g(x) = \lim_{x \to x_0} f'(x)/g'(x),$$

provided the right hand limit exists.

2.1.15 Taylor's theorem

Suppose f is n times differentiable at x_0. The nth order *Taylor's polynomial* about x_0 is defined as

$$P_n(x; x_0) = \sum_{i=1}^{n} \frac{f^{(i)}(x_0)}{i!}(x - x_0)^i, \tag{2.6}$$

and the *remainder term* is given by

$$R_n(x; x_0) = f(x) - P_n(x; x_0). \tag{2.7}$$

The use of $P_n(x; x_0)$ to approximate $f(x)$ is made precise by *Taylor's Theorem*:

Theorem 2.3 *Suppose f is $n+1$ times differentiable on $[a, b]$, $f \in C^n([a, b])$, and $x_0 \in [a, b]$. Then for each $x \in [a, b]$ there exists $\eta(x)$, satisfying* $\min(x, x_0) \leq \eta(x) \leq \max(x, x_0)$ *for which*

$$R_n(x; x_0) = \frac{f^{(n+1)}(\eta(x))}{(n+1)!}(x - x_0)^{n+1}, \text{ (Lagrange form)} \tag{2.8}$$

as well as $\eta'(x)$, also satisfying $\min(x, x_0) \leq \eta'(x) \leq \max(x, x_0)$, *for which*

$$R_n(x; x_0) = \frac{f^{(n+1)}(\eta'(x))}{(n+1)!}(x - \eta'(x))^n(x - x_0). \text{ (Cauchy form)} \tag{2.9}$$

The Lagrange form of the remainder term is the one commonly intended, and we adopt that convention here, although it is worth noting that alternative forms are also used.

2.1.16 The l^p norm

The *l^p norm* for $p \geq 0$ is defined for any $x = (x_1, \ldots, x_n) \in \mathbb{R}^n$ by

$$\|x\|_p = \left(\sum_{i=1}^{n} |x_i|^p \right)^{1/p}.$$

for $p < \infty$, and

$$\|x\|_\infty = \max_i |x_i| .$$

when $p = \infty$.

2.1.17 Power means

For a collection of positive numbers $\tilde{a} = (a_1, \ldots, a_n)$ the *power mean* is defined as $M_p[\tilde{a}] = \left(n^{-1} \sum_{i=1}^n a_i^p\right)^{1/p}$ for finite nonzero p. The definition is extended to $p = 0, -\infty, \infty$ by the existence of well defined limits, yielding $M_{-\infty}[\tilde{a}] = \min_i\{a_i\}$, $M_0[\tilde{a}] = \left(\prod_{i=1}^n a_i\right)^{1/n}$ and $M_\infty[\tilde{a}] = \max_i\{a_i\}$.

Theorem 2.4 *Suppose for positive numbers $\tilde{a} = (a_1, \ldots, a_n)$ and real number $p \in (-\infty, 0) \cup (0, \infty)$ we define power mean $M_p[\tilde{a}] = \left(n^{-1} \sum_{i=1}^n a_i^p\right)^{1/p}$. Then*

$$\lim_{p \to 0} M_p[\tilde{a}] = \left(\prod_{i=1}^n a_i\right)^{1/n} = M_0[\tilde{a}], \tag{2.10}$$

$$\lim_{p \to \infty} M_p[\tilde{a}] = \max_i\{a_i\} = M_\infty[\tilde{a}] \text{ and} \tag{2.11}$$

$$\lim_{p \to -\infty} M_p[\tilde{a}] = \min_i\{a_i\} = M_{-\infty}[\tilde{a}], \tag{2.12}$$

which justifies the conventional definitions of $M_{-\infty}[\tilde{a}]$, $M_0[\tilde{a}]$ and $M_\infty[\tilde{a}]$. In addition, $-\infty \leq p < q \leq \infty$ implies $M_p[\tilde{a}] \leq M_q[\tilde{a}]$, with equality if and only if all elements of $\tilde{a}$ are equal.

Proof By l'Hôpital's Rule,

$$\lim_{p \to 0} \log(M_p[\tilde{a}]) = \lim_{p \to 0} \frac{n^{-1} \sum_{i=1}^n \log(a_i) a_i^p}{n^{-1} \sum_{i=1}^n a_i^p} = n^{-1} \sum_{i=1}^n \log(a_i) = \log(M_0[\tilde{a}]).$$

Relabel $\tilde{a}$ so that $a_1 = \max_i\{a_i\}$. Then

$$\lim_{p \to \infty} M_p[\tilde{a}] = \lim_{p \to \infty} n^{-1} a_1 \left(\sum_{i=1}^n (a_i/a_1)^p\right)^{1/p} = a_1 = \max_i\{a_i\} = M_\infty[\tilde{a}].$$

The final limit of (2.12) can be obtained by replacing a_i with $1/a_i$.

That the final statement of the theorem holds for $0 < p < q < \infty$ follows from Jensen's inequality (Theorem 4.13), and the extension to $0 \leq p < q \leq \infty$ follows from the limits in (2.12). It then follows that the statement holds for $-\infty \leq p < q \leq 0$ after replacing a_i with $1/a_i$, and therefore it holds for $-\infty \leq p < q \leq \infty$. ///

The cases $p = 1, 0, -1$ correspond to the *arithmetic mean*, *geometric mean* and *harmonic mean* which will be denoted $AM[\tilde{a}] \geq GM[\tilde{a}] \geq HM[\tilde{a}]$, respectively.

2.2 EQUIVALENCE RELATIONSHIPS

The notion of *equivalence relationships and classes* will play an important role in our analysis. Suppose $\mathcal{X}$ is a set of objects, and $\sim$ defines a *binary relation* between two objects $x, y \in \mathcal{X}$.

Definition 2.1 *A binary relation* $\sim$ *on a set* $\mathcal{X}$ *is an* equivalence relation *if it satisfies the following three properties for any* $x, y, z \in \mathcal{X}$*:*

> ***Reflexivity*** $x \sim x$.
> ***Symmetry*** *If* $x \sim y$ *then* $y \sim x$.
> ***Transitivity*** *If* $x \sim y$ *and* $y \sim z$ *then* $x \sim z$.

Given an equivalence relation, an *equivalence class* is any set of the form $E_x = \{y \in \mathcal{X} \mid y \sim x\}$. If $y \in E_x$ then $E_y = E_x$. Each element $x \in \mathcal{X}$ is in exactly one equivalence class, so $\sim$ induces a partition of $\mathcal{X}$ into equivalence classes.

In Euclidean space, 'is parallel to' is an equivalence relation, while 'is perpendicular to' is not.

For finite sets, cardinality is a property of a specific set, while for infinite sets, cardinality must be understood as an equivalence relation.

2.3 LINEAR ALGEBRA

Formal definitions of both a *field* and a *vector space* are given in Section 6.3. For the moment we simply note that the notion of real numbers can be generalized to that of a *field* $\mathbb{K}$, which is a set of *scalars* that is closed under the rules of addition and multiplication comparable to those available for $\mathbb{R}$. Both $\mathbb{R}$ and $\mathbb{C}$ are fields.

A *vector space* $\mathcal{V} \subset \mathbb{K}^n$ is any set of vectors $x \in \mathbb{K}^n$ which is closed under linear and scalar composition, that is, if $x, y \in \mathcal{V}$ then $ax + by \in \mathcal{V}$ for all scalars a, b. This means the zero vector $\vec{0}$ must be in $\mathcal{V}$, and that $x \in \mathcal{V}$ implies $-x \in \mathcal{V}$.

Elements $x_1, \ldots, x_m$ of $\mathbb{K}^n$ are *linearly independent* if $\sum_{i=1}^m a_i x_i = 0$ implies $a_i = 0$ for all i. Equivalently, no x_i is a linear combination of the remaining vectors. The *span* of a set of vectors $\tilde{x} = (x_1, \ldots, x_n)$, denoted $span(\tilde{x})$, is the set of all linear combinations of vectors in $\tilde{x}$, which must be a vector space. Suppose the vectors in $\tilde{x}$ are not linearly independent. This means that, say, x_m is a linear combination of the remaining vectors, and so any linear combination in $span(\tilde{x})$ including x_m may be replaced with one including only the remaining vectors, so that $span(\tilde{x}) = span(x_1, \ldots, x_{m-1})$. The *dimension* of a vector space $\mathcal{V}$ is the minimum number of vectors whose span equals $\mathcal{V}$. Clearly, this equals the number in any set of linearly independent vectors which span $\mathcal{V}$. Any such set of vectors forms a *basis* for $\mathcal{V}$. Any vector space has a basis.

2.3.1 Matrices

Let $M_{m,n}(\mathbb{K})$ be the set of $m \times n$ matrices A, for which $A_{i,j} \in \mathbb{K}$ (or, when required for clarity, $[A]_{i,j} \in \mathbb{K}$) is the element of the ith row and jth column. When the field need not be given, we will write $M_{m,n} = M_{m,n}(\mathbb{K})$. We will generally be interested in $M_{m,n}(\mathbb{C})$, noting that the real matrices $M_{m,n}(\mathbb{R}) \subset M_{m,n}(\mathbb{C})$ can be considered a special case of complex matrices, so that any resulting theory holds for both types. This is important to note, since even when interest is confined to real valued matrices, complex numbers enter the analysis in a natural way, so it is ultimately necessary to consider complex vectors and matrices. Definitions associated with real matrices (transpose, symmetric, and so on) have analgous definitions for complex matrices, which reduce to the more familiar definitions when the matrix is real.

The *square matrices* are denoted as $M_m = M_{m,m}$. Elements of $M_{m,1}$ are *column vectors* and elements of $M_{1,m}$ are *row vectors*. A matrix in $M_{m,n}$ is equivalently an ordered set of m row vectors or n column vectors. The transpose $A^T \in M_{n,m}$ of a matrix $A \in M_{m,n}$ has elements $A'_{j,i} = A_{i,j}$. For $A \in M_{n,k}$, $B \in M_{k,m}$ we always understand matrix multiplication to mean that $C = AB \in M_{n,m}$ possesses elements $C_{i,j} = \sum_{k'=1}^{k} A_{i,k'} B_{k',j}$, so that matrix multiplication is generally not commutative. Then $(A^T)^T = A$ and $(AB)^T = B^T A^T$ where the product is permitted.

In the context of matrix algrebra, a vector $x \in \mathbb{K}^n$ is usually assumed to be a column vector in $M_{n,1}$. Therefore, if $A \in M_{m,n}$ then the expression Ax is understood to be evaluated by matrix multiplication. Similarly, if $x \in \mathbb{K}^m$ we may use the expression $x^T A$, understanding that $x \in M_{m,1}$.

When $A \in M_{m,n}(\mathbb{C})$, the *conjugate matrix* is written $\bar{A}$, and is the component-wise conjugate of A. The identity $\bar{A}\bar{B} = \overline{AB}$ holds. The *conjugate transpose* (or *Hermitian adjoint*) of A is $A^* = \bar{A}^T$. As with the transpose operation, $(A^*)^* = A$ and $(AB)^* = B^* A^*$ where the product is permitted. This generally holds for arbitrary products, that is $(ABC)^* = (BC)^* A^* = C^* B^* A^*$, and so on. For $A \in M_{m,n}(\mathbb{R})$, we have $A = \bar{A}$ and $A^* = A^T$, so the conjugate transpose may be used in place of the transpose operation when matrices are real valued. We always may write $(A + B)^* = A^* + B^*$ and $(A + B)^T = A^T + B^T$ where dimensions permit.

A matrix $A \in M_n(\mathbb{C})$ is *diagonal* if the only nonzero elements are on the diagonal, and can therefore be referred to by the diagonal elements $diag(a_1, \ldots, a_n) = diag(A_{1,1}, \ldots, A_{n,n})$. A diagonal matrix is *positive diagonal* or *nonnegative diagonal* if all diagonal elements are positive or nonegative.

The identity matrix $I \in M_m$ is the matrix uniquely possessing the property that $A = IA = AI$ for all $A \in M_m$. For $M_m(\mathbb{C})$, I is diagonal, with diagonal entries equal to 1. For any matrix $A \in M_m$ there exists at most one matrix $A^{-1} \in M_m$ for which $AA^{-1} = I$, referred to as the *inverse* of A. An inverse need not exist (for example, if the elements of A are constant).

The *inner product* (or *scalar product*) of two vectors $x, y \in \mathbb{C}^n$ is defined as $\langle x, y \rangle = y^* x$ (a more general definition of the inner product is given in Definition 6.13). For any $x \in \mathbb{C}^n$ we have $\langle x, x \rangle = \sum_i \bar{x}_i x_i = \sum_i |x_i|^2$, so that $\langle x, x \rangle$ is a nonnegative real number, and $\langle x, x \rangle = 0$ if and only if $x = \vec{0}$. The magnitude, or *norm*, of a vector may be taken as $\|x\| = (\langle x, x \rangle)^{1/2}$ (a formal definition of a norm is given in Definition 6.6).

Two vectors $x, y \in \mathbb{C}^n$ are *orthogonal* if $\langle x, y \rangle = 0$. A set of vectors $x_1, \ldots, x_m$ is orthogonal if $\langle x_i, x_j \rangle = 0$ when $i \neq j$. A set of m orthogonal vectors are linearly independent, and so form the basis for an m dimensional vector space. If in addition $\|x_i\| = 1$ for all i, the vectors are *orthonormal*.

A matrix $Q \in M_n(\mathbb{C})$ is *unitary* if $Q^* Q = QQ^* = I$. Equivalently, Q is unitary if and only (*i*) its column vectors are orthonormal; (*ii*) its row vectors are orthonormal; (*iii*) it possesses inverse $Q^{-1} = Q^*$. The more familiar term *orthogonal matrix* is usually reserved for a real valued unitary matrix (otherwise the definition need not be changed).

A unitary matrix preserves magnitude, since $\langle Qx, Qx\rangle = (Qx)^*(Qx) = x^*Q^*Qx = x^*Ix = x^*x = \|x\|^2$.

A matrix $Q \in M_n(\mathbb{C})$ is a *permutation* matrix if each row and column contains exactly one 1 entry, with all other elements equal to 0. Then $y = Qx$ is a permutation of the elements of $x \in \mathbb{C}^n$. A permutation matrix is always orthogonal.

Suppose $A \in M_{m,n}$ and let $\alpha \subset \{1,\ldots,m\}$, $\beta \subset \{1,\ldots,n\}$ be any two nonempty subsets of indices. Then $A[\alpha,\beta] \in M_{|\alpha|,|\beta|}$ is the *submatrix* of A obtained by deleting all elements except for $A_{i,j}$, $i \in \alpha$, $j \in \beta$. If $A \in M_n$, and $\alpha = \beta$, then $A[\alpha,\alpha]$ is a *principal submatrix*.

The determinant associates a scalar with $A \in M_m(\mathbb{C})$ through the recursive formula

$$\det(A) = \sum_{i=1} (-1)^{i+j} A_{i,j} \det(A^{i,j}) = \sum_{j=1} (-1)^{i+j} A_{i,j} \det(A^{i,j})$$

where $A^{i,j} \in M_{m-1}(\mathbb{C})$ is the matrix obtained by deleting the ith row and jth column of A. Note that in the respective expressions any j or i may be chosen, yielding the same number, although the choice may have implications for computational efficiency. As is well known, for $A \in M_1(\mathbb{C})$ we have $\det(A) = A_{1,1}$ and for $A \in M_2$ we have $\det(A) = A_{1,1}A_{2,2} - A_{1,2}A_{2,1}$. In general, $\det(A^T) = \det(A)$, $\det(A^*) = \overline{\det(A)}$, $\det(AB) = \det(A)\det(B)$, $\det(I) = 1$ which implies $\det(A^{-1}) = \det(A)^{-1}$ when the inverse exists.

A large class of algorithms are associated with the problem of determining a solution $x \in \mathbb{K}^m$ to the *linear systems of equations* $Ax = b$ for some fixed $A \in M_m$ and $b \in \mathbb{K}^m$.

Theorem 2.5 *The following statements are equivalent for $A \in M_m(\mathbb{C})$, and a matrix satisfying any one is referred to as* nonsingular, *any other matrix in $M_m(\mathbb{C})$* singular:

(i) *The columns vectors of A are linearly independent.*
(ii) *The row vectors of A are linearly independent.*
(iii) $\det(A) \neq 0$.
(iv) *$Ax = b$ possesses a unique solution for any $b \in \mathbb{K}^m$.*
(v) *$x = \vec{0}$ is the only solution of $Ax = \vec{0}$.*

Matrices $A, B \in M_n$ are *similar*, if there exists a nonsingular matrix S for which $B = S^{-1}AS$. Simlarity is an equivalence relation (Definition 2.1). A matrix is *diagonalizable* if it is similar to a diagonal matrix. Diagonalization offers a number of advantages. We always have $B^k = S^{-1}A^kS$, so that if A is diagonal, this expression is particularly easy to evaluate. More generally, diagonalization can make apparent the behavior of a matrix interpreted as a transformation. Suppose in the diagonalization $B = S^{-1}AS$ we know that S is orthogonal, and that A is real. Then the action of B on a vector is decomposed into S (a change in coordinates), A (elementwise scalar multiplication) and S^{-1} (the inverse change in coordinates).

2.3.2 Eigenvalues and spectral decomposition

For $A \in M_n(\mathbb{C})$, $x \in \mathbb{C}^n$, and $\lambda \in \mathbb{C}$ we may define the *eigenvalue equation*

$$Ax = \lambda x, \tag{2.13}$$

and if the pair (λ, x) is a solution to this equation for which $x \neq \vec{0}$, then λ is an *eigenvalue* of A and x is an associated *eigenvector* of λ. Any such solution (λ, x) may be called an *eigenpair*. Clearly, if x is an eigenvector, so is any nonzero scalar multiple. Let R_λ be the set of all eigenvectors x associated with λ. If $x, y \in R_\lambda$ then $ax + by \in R_\lambda$, so that R_λ is a vector space. The dimension of R_λ is known as the *geometric multiplicity* of λ. We may refer to R_λ as an *eigenspace* (or *eigenmanifold*). In general, the *spectral properties* of a matrix are those pertaining to the set of eigenvalues and eigenvectors.

If $A \in M_n(\mathbb{R})$, and λ is an eigenvalue, then so is $\bar{\lambda}$, with associated eigenvectors $R_{\bar{\lambda}} = \bar{R}_\lambda$. Thus, in this case eigenvalues and eigenvectors occur in conjugate pairs. Simlarly, if λ is real there exists a real associated eigenvector.

The eigenvalue equation may be written $(A - \lambda I)x = 0$. However, by Theorem 2.5 this has a nonzero solution if and only if $A - \lambda I$ is singular, which occurs if and only if $p_A(\lambda) = \det(A - \lambda I) = 0$. By construction of a determinant, $p_A(\lambda)$ is an order n polynomial in λ, known as the *characteristic polynomial* of A. The set of all eigenvalues of A is equivalent to the set of solutions to the *characteristic equation* $p_A(\lambda) = 0$ (including complex roots). The multiplicity of an eigenvalue λ as a root of $p_A(\lambda)$ is referred to as its *algebraic multiplicity*. A *simple eigenvalue* has algebraic multiplicity 1. The geometric multiplicity of an eigenvalue can be less, but never more, than the algebraic multiplicity. A matrix with equal algebraic and geometric multiplicities for each eigenvalue is a *nondefective matrix*, and is otherwise a *defective matrix*.

We therefore denote the set of all eigenvalues as $\sigma(A)$. An important fact is that $\sigma(A^k)$ consists exactly of the eigenvalues $\sigma(A)$ raised to the kth power, since if (λ, x) solves $Ax = \lambda x$, then $A^2 x = A\lambda x = \lambda A x = \lambda^2 x$, and so on. A quantity of particular importance is the *spectral radius* $\rho(A) = \max\{|\lambda| \mid \lambda \in \sigma(A)\}$. There is sometimes interest in ordering the eigenvalues by magnitude. If there exists an eigenvalue $\lambda_1 = \rho(A)$, this is sometimes referred to as the *principal eigenvalue*, and any associated eigenvector is a *principal eigenvector*.

In addition we have the following theorem:

Theorem 2.6 *Suppose $A, B \in M_n$, and $|A| \leq B$, where $|A|$ is the element-wise absolute value of A. Then $\rho(A) \leq \rho(|A|) \leq \rho(B)$.*

In addition, if all elements of $A \in M_n(\mathbb{R})$ are nonnegative, then $\rho(A') \leq \rho(A)$ for any principal submatrix A'.

Proof See Theorem 8.1.18 of Horn and Johnson (1985). ///

Suppose we may construct n eigenvalues $\lambda_1, \ldots, \lambda_n$, with associated eigenvectors $\nu_1, \ldots, \nu_n$. Then let $\Lambda \in M_n$ be the diagonal matrix with ith diagonal element λ_i, and let $V \in M_n$ be the matrix with ith column vector ν_i. By virtue of (2.13) we can write

$$AV = V\Lambda. \tag{2.14}$$

If V is invertable (equivalently, there exist n linearly independent eigenvectors, by Theorem 2.5), then

$$A = V\Lambda V^{-1}, \tag{2.15}$$

so that A is diagonalizable. Alternatively, if A is diagonalizable, then (2.14) can be obtained from (2.15) and, since V is invertable, there must be n independent eigenvectors. The following theorem expresses the essential relationship between diagonalization and spectral properties.

Theorem 2.7 *For square matrix $A \in M_n(\mathbb{C})$:*

(i) *Any set of $k \leq n$ eigenvectors $\nu_1, \ldots, \nu_k$ associated with distinct eigenvalues $\lambda_1, \ldots, \lambda_k$ are linearly independent,*
(ii) *A is diagonalizable if and only if there exist n linearly independent eigenvectors,*
(iii) *If A has n distinct eigenvalues, it is diagonalizable (this follows from (i) and (ii)),*
(iv) *A is diagonalizable if and only if it is nondefective.*

Right and Left Eigenvectors

The eigenvectors defined by (2.13) may be referred to as *right eigenvectors*, while *left eigenvectors* are nonzero solutions to

$$x^*A = \lambda x^*, \tag{2.16}$$

(note that some conventions do not explicitly refer to complex conjugates x^* in (2.16)). This similarly leads to the equation $x^*(A - \lambda I) = 0$, which by an argument identical to that used for right eigenvectors, has nonzero solutions if and only if $p_A(\lambda) = 0$, giving the same set of eigenvalues as those defined by (2.13). There is therefore no need to distinguish between 'right' and 'left' eigenvalues. Then, fixing eigenvalue λ we may refer to the *left eigenspace* L_λ as the set of solution x to (2.16) (in which case, R_λ now becomes the *right eigenspace* of λ).

The essential relationship between the eigenspaces is summarized in the following theorem:

Theorem 2.8 *Suppose $A \in M_n(\mathbb{C})$.*

(i) *For any $\lambda \in \sigma(A)$ L_λ and R_λ have the same dimension.*
(ii) *For any distinct eigenvalues $\lambda_1, \ldots, \lambda_m$ from $\sigma(A)$, any selection of vectors $x_i \in R_{\lambda_i}$ for $i = 1, \ldots, m$ are linearly independent. The same holds for selections from distinct L_λ.*
(iii) *Right and left eigenvectors associated with distinct eigenvalues are orthogonal.*

Proof Proofs may be found in, for example, Chapter 1 of Horn and Johnson (1985). ///

Next, if V is invertible, multiply both sides of (2.15) by V^{-1} yielding

$$V^{-1}A = \Lambda V^{-1}.$$

Just as the column vectors of V are right eigenvectors, we can set $U^* = V^{-1}$, in which case the ith column vector υ_i of U is a solution x to the left eigenvector equation (2.16) corresponding to eigenvalue λ_i (the ith element on the diagonal of Λ). This gives the diagonalization

$$A = V\Lambda U^*.$$

Since $U^*V = I$, indefinite multiplication of A yields the *spectral decomposition*:

$$A^m = V\Lambda^m U^* = \sum_{i=1}^{n} \lambda_i^m v_i v_i^*. \tag{2.17}$$

The apparent recipe for a spectral decomposition is to first determine the roots of the characteristic polynomial, and then to solve each resulting eigenvalue equation (2.13) after substituting an eigenvalue. This seemingly straightforward procedure proves to be of little practical use in all but the simplest cases, and spectral decompositions are often difficult to construct using any method. However, a complete spectral decomposition need not be the objective. First, it may not even exist for many otherwise interesting models. Second, there are many important problems related to A that can be solved using spectral theory, but without the need for a complete spectral decomposition. For example:

(*i*) Determining bounds $\|Ax\| \leq a\,\|x\|$ or $\|Ax\| \geq b\,\|x\|$,
(*ii*) Determining the convergence rate of the limit $\lim_{k\to\infty} A^k = A^\infty$,
(*iii*) Verifying the existence of a scalar λ and vector v for which $Av = \lambda v$, and guaranteeing that (for example) λ and v are both real and positive.

Basic spectral theory relies on the identification of special matrix forms which impose specific properties on a the spectrum. We next discuss two cases.

2.3.3 Symmetric, Hermitian and positive definite matrices

A matrix $A \in M_n(\mathbb{C})$ is *Hermitian* if $A = A^*$. A Hermitian real valued matrix is *symmetric*, that is, $A = A^T$. The spectral properties of Hermitian matrices are quite definitive (see, for example, Chapter 4, Horn and Johnson (1985)).

Theorem 2.9 *A matrix $A \in M_n(\mathbb{C})$ is Hermitian if and only if there exists a unitary matrix U and real diagonal matrix Λ for which $A = U\Lambda U^*$.*

A matrix $A \in M_n(\mathbb{R})$ is symmetric if and only if there exists a real orthogonal Q and real diagonal matrix Λ for which $A = Q\Lambda Q^T$.

Clearly, the matrices Λ and U may be identified with the eigenvalues and eigenvectors of A, with n eignevalue equation solutions given by the respect columns of $AU = U\Lambda U^* U = U\Lambda$. An important implication of this is that all eigenvalues of a Hermitian matrix are real, and eigenvectors may be selected to be orthonormal.

If we interpet $x \in \mathbb{C}^n$ as a column vector $x \in M_{n,1}$ we have *quadratic form* x^*Ax, which is interpretable either as a 1×1 complex matrix, or as a scalar in $\mathbb{C}$, as is convenient.

If A is Hermitian, then $(x^*Ax)^* = x^*A^*x = x^*Ax$. This means if $z = x^*Ax \in \mathbb{C}$, then $z = \bar{z}$, equivalently $x^*Ax \in \mathbb{R}$. A Hermitian matrix A is *positive definite* if and only if $x^*Ax > 0$ for all $x \neq \vec{0}$. If instead $x^*Ax \geq 0$ then A is *positive semidefinite*. A non-symmetric matrix satisfying $x^TAx > 0$ can be replaced by $A' = (A + A^T)/2$, which is symmetric, and also satisfies $x^TA'x > 0$.

Theorem 2.10 *If $A \in M_n(\mathbb{C})$ is Hermitian then x^*Ax is real. If, in addition, A is positive definite then all of its eigenvalues are positive. If it is positive semidefinite then all of its eigenvalues are nonnegative.*

If A is positive semidefinite, and we let λ_{min} and λ_{max} be the smallest and largest eigenvalies in $\sigma(A)$ (all of which are nonnegative real numbers) then it can be shown that

$$\lambda_{min} = \min_{\|x\|=1} x^*Ax \quad \text{and} \quad \lambda_{max} = \max_{\|x\|=1} x^*Ax.$$

If A is positive definite then $\lambda_{min} > 0$. In addition, since the eigenvalues of A^2 are the squares of the eigenvalues of A, and since for a Hermitian matrix $A^* = A$, we may also conclude

$$\lambda_{min} = \min_{\|x\|=1} \|Ax\| \quad \text{and} \quad \lambda_{max} = \max_{\|x\|=1} \|Ax\|,$$

for any positive semidefinite matrix A.

Any diagonalizable matrix A possesses a kth root, $A^{1/k}$, meaning $A = \left(A^{1/k}\right)^k$. Given diagonalization $A = Q^{-1}\Lambda Q$, this is easily seen to be $A^{1/k} = Q^{-1}\Lambda^{1/k}Q$, where $[\Lambda^{1/k}]_{i,j} = \left(\Lambda_{i,j}\right)^{1/k}$. If A is a real symmetric positive definite matrix then $A^{1/2}$ is real, symmetric and nonsingular.

2.3.4 Positive matrices

A real valued matrix $A \in M_{m,n}(\mathbb{R})$ is *positive* or *nonnegative* if all elements are positive or nonnegative, respectively. This may be conveniently written $A > 0$ or $A \geq 0$ as appropriate.

The spectral properties of $A \geq 0$ are quite precisely characterized by the *Perron-Frobenius Theorem* which is discussed below.

If $P \in M_n$ is a *permutation matrix* then the matrix P^TAP is obtained from A by a common permutation of the row and column indices.

Definition 2.2 *A matrix $A \in M_n(\mathbb{R})$ is* reducible *if $n = 1$ and $A = 0$, or there exists a permutation matrix P for which*

$$P^TAP = \begin{bmatrix} B & C \\ 0 & D \end{bmatrix} \tag{2.18}$$

where B and D are square matrices. Otherwise, A is irreducible.

The essential feature of a matrix of the form (2.18) is that the block of zeros is of dimension $a \times b$ where $a + b = n$. It can be seen that this same block remains 0 in any power $(P^TAP)^k$. The same will be true for A, subject to a label permutation. Clearly, this structure will not change under any relabeling, which is the essence of the permutation transformation. The following property of irreducible matrices should be noted:

Theorem 2.11 *If $A \in M_n(\mathbb{R})$ is irreducible, then each column and row must contain at least 1 nondiagonal nonzero element.*

Proof Suppose all nondiagonal elements of row i of matrix $A \in M_n(\mathbb{R})$ are 0. After relabeling i as n, there exists a $1 \times (n-1)$ block of 0's conforming to (2.18). Similarly, if all nondiagonal elements of column j are 0, relabeling j as 1 yields a similar block of 0's. ///

Irreducibility may be characterized in the following way:

Theorem 2.12 *For a nonnegative matrix $A \in M_n(\mathbb{R})$ the following statements are equivalent:*

- *(i)* *A is irreducible,*
- *(ii)* *The matrix $(I + A)^{n-1}$ is positive.*
- *(iii)* *For each pair i, j there exists k for which $[A^k]_{i,j} > 0$.*

Condition *(iii)* is often strengthened:

Definition 2.3 *A nonnegative matrix $A \in M_n$ is* primitive *if there exists k for which A^k is positive.*

Clearly, Definition 2.3 implies statement *(iii)* of Theorem 2.12, so that a primitive matrix is also irreducible.

The main theorem follows (see, for example, Horn and Johnson (1985)):

Theorem 2.13 (Perron-Frobenius Theorem) *For any primitive matrix $A \in M_n$, the following hold:*

- *(i)* $\rho(A) > 0$,
- *(ii)* *There exists a simple eigenvalue $\lambda_1 = \rho(A)$,*
- *(iii)* *There is a positive eigenvector v_1 associated with λ_1,*
- *(iv)* *$|\lambda| < \lambda_1$ for any other eigenvalue λ.*
- *(v)* *Any nonnegative eigenvector is a scalar multiple of v_1.*

If A is nonnegative and irreducible, then (i)–(iii) hold.

If A is nonnegative, then $\rho(A)$ is an eigenvalue, which possesses a nonnegative eigenvector. Furthermore, if v is a positive eigenvector of A, then its associated eigenvalue is $\rho(A)$.

One of the important consequences of Theorem 2.13 is that an irreducible matrix A possesses a unique principal eigenvalue $\rho(A)$, which is real and positive, with a positive principal eigenvector. Noting that A^T is also irreducible, we may conclude that the left principal eigenvector is also positive.

We cannot rule out $\rho(A) = 0$ for $A \geq 0$ ($A \equiv 0$, among other examples). However, a convenient lower bound for $\rho(A)$ exists, a consequence of Theorem 2.6, which implies that $\max_i A_{i,i} \leq \rho(A)$.

Suppose a nonnegative matrix $A \in M_n$ is diagonalizable, and $\rho(A) > 0$. A normalized spectral decomposition follows from (2.17):

$$\left[\rho(A)^{-1}A\right]^m = \sum_{i=1}^{n} \left[\rho(A)^{-1}\lambda_i\right]^m \nu_i \upsilon_i^*.$$

To fix ideas, suppose A is primitive. By Theorem 2.13 there exists a unique principal eigenvalue, say $\lambda_1 = \rho(A)$, and any other eigenvalue satisfies $|\lambda_j| < \rho(A)$. Then

$$\left[\rho(A)^{-1}A\right]^m = \nu_1\upsilon_1^* + O\left(m^{m_2-1}\left[\rho(A)^{-1}|\lambda_{SLEM}|\right]^m\right), \tag{2.19}$$

where λ_{SLEM} is the *second largest eigenvalue in magnitude* and m_2 is the algebraic multiplicity of λ_{SLEM}, that is, any eigenvalue other than λ_1 (not necessarily unique) maximizing $|\lambda_j|$. Since $|\lambda_{SLEM}| < \rho(A)$ we have limit

$$\lim_{m\to\infty} \left[\rho(A)^{-1}A\right]^m = \nu_1\upsilon_1^*, \tag{2.20}$$

where ν_1, υ_1 are the principal right and left eigenvectors, with convergence at a geometric rate $O\left(\left[\rho(A)^{-1}|\lambda_{SLEM}|\right]^m\right)$. For this reason, the quantity $|\lambda_{SLEM}|$ is often of considerable interest. Note that in this representation, the normalization $\langle \nu_i, \upsilon_i \rangle = 1$ is implicit.

However, existence of the limit (2.20) for primitive matrices does not depend on the diagonalizability of A, and is a direct consequence of Theorem 2.13. When A is irreducible, the limit (2.20) need not exist, but a weaker statement involving asymptotic averages will hold. These conclusions are summarized in the following theorem:

Theorem 2.14 *Suppose nonegative matrix $A \in M_n(\mathbb{R})$ is irreducibile. Let ν_1, υ_1 be the principal right and left eigenvectors, normalized so that $\langle \nu_1, \upsilon_1 \rangle = 1$. Then*

$$\lim_{N\to\infty} N^{-1} \sum_{m=1}^{N} \left[\rho(A)^{-1}A\right]^m = \nu_1\upsilon_1^*. \tag{2.21}$$

If A is primitive, then (2.20) *also holds.*

Proof See, for example, Theorems 8.5.1 and 8.6.1 of Horn and Johnson (1985). ///

A version of (2.21) is available for nonnegative matrices which are not necessarily irreducible, but which satisfy certain other regularity conditions (Theorem 8.6.2, Horn and Johnson (1985)).

2.3.5 Stochastic matrices

We say $A \in M_n$ is a *stochastic matrix* if $A \geq 0$, and each row sums to 1. It is easily seen that $A\vec{1} = \vec{1}$, and so $\lambda = 1$ and $v = \vec{1}$ form an eigenpair. Since $\vec{1} > 0$, by Theorem 2.13 we must have $\rho(A) = 1$.

In addition, for a general stochastic matrix, any positive eigenvector v satisfies $Av = v$.

If A is also irreducible then $\lambda = 1$ is a simple eigenvalue, so any solution to $Av = v$ must be a multiple of $\vec{1}$ (in particular, any positive eigenvector must be a multiple of $\vec{1}$).

If A is primitive, any nonnegative eigenvector v must be a multiple of $\vec{1}$. In addition, all eigenvalues other than the principal have modulus $|\lambda_j| < 1$.

We will see that is can be very advantageous to verify the existence of a principal eigenpair (λ_1, v_1) where $\lambda_1 = \rho(A)$ and $v_1 > 0$. This holds for any stochastic matrix.

2.3.6 Nonnegative matrices and graph structure

The theory of nonnegative matrices can be clarified by associating with a square matrix $A \geq 0$ a graph $G(A)$ possessing directed edge (i,j) if and only if $A_{i,j} > 0$. Following Theorems 2.1–2.2 of Section 2.1.11, we know that $A^n_{i,j} > 0$ if and only if there is a path of length n from i to j within $G(A)$.

By (*iii*) of Theorem 2.12 we may conclude that A is irreducible if and only if all pairs of nodes in $G(A)$ communicate (see the definitions of Section 2.1.11).

Some important properties associated with primitive matrices are summarized in the following theorems.

Theorem 2.15 *If $A \in M_n(\mathbb{R})$ is a primitive matrix then for some finite k' we have $A^k > 0$ for all $k \geq k'$.*

Proof By Definition 2.3 there exists finite k' for which $A^{k'} > 0$. Let i,j be any ordered pair of nodes in $G(A)$. Since a primitive matrix is irreducible, we may conclude from Theorem 2.11 that there exists node k such that (k,j) is an edge in $G(A)$. By Theorem 2.2 there exists a path of length k' from i to k, and therefore also a path of length k' from i to j. This holds for any i,j, therefore by Theorem 2.2 $A^{k'+1} > 0$. The proof is completed by successively incrementing k'. ///

Thus, for a primitive matrix A all pairs of nodes in $G(A)$ communicate, and in addition there exists k' such that for any ordered pair of nodes i,j there exists a path from i to j of any length $k \geq k'$.

Any irreducible matrix with positive diagonal elements is also primitive:

Theorem 2.16 *If $A \in M_n(\mathbb{R})$ is an irreducible matrix with positive diagonal elements, then A is also a primitve matrix.*

Proof Let i,j be any ordered pair of nodes in $G(A)$. There exists at least one path from i to j. Suppose one of these paths has length k. Since, by hypothesis, $A_{j,j} > 0$ the edge (j,j) in included in $G(A)$, and can be appended to any path ending at j. This means there also exists a path of length $k+1$ from i to j. The proof is completed by noting that there must be some finite k' such that any two nodes may be joined by a path of length no greater than k', in which case $A^{k'} > 0$. ///

A matrix can be irreducible but not primitive. For example, if the nodes of $G(A)$ can be partitioned into subsets V_1, V_2 such that all edges (i,j) are formed by nodes from distinct subsets, then A cannot be primitive. To see this, suppose $i,j \in V_1$. Then any path from i to j must be of even length, so that the conclusion of Theorem 2.15 cannot hold. However, if $G(A)$ includes all edges not ruled out by this restriction, it is easily seen that A is irreducible.

Finally, we characterize the conectivity properties of a reducible nonnegative matrix. Consider the representation (2.18). Without loss of generality we may take the identity permutation $P = I$. Then the nodes of $G(A)$ may be partitioned into V_1 and V_2 in such a way that there can be no edge (i, j) for which $i \in V_1$ and $j \in V_2$. This means that no node in V_2 is accessible from any node in V_1, that is, there cannot be any path beginning in V_1 and ending in V_2.

We will consider this issue further in Section 5.2, where it has quite intuitive interpretations.

Chapter 3

Background – measure theory

Measure theory provides a rigorous mathematical foundation for the study of, among other things, integration and probability theory. The study of stochastic processes, and of related control problems, can proceed some distance without reference to measure theoretic ideas. However, certain issues cannot be resolved fully without it, for example, the very existence of an optimal control in general models. In addition, if we wish to develop models which do not assume that all random quantities are stochastically independent, which we sooner or later must, the theory of martingale processes becomes indepensible, an understanding of which is greatly aided by a familiarity with measure theoretic ideas. Above all, foundational ideas of measure theory will be required for the function analytic construction of iterative algorithms.

3.1 TOPOLOGICAL SPACES

Suppose we are given a set Ω, and a sequence $x_k \in \Omega$, $k \geq 1$. It is important to have a precise definition of the convergence of x_k to a limit. If $\Omega \subset \mathbb{R}^n$ the definition is standard, but if Ω is a collection of, for example, functions or sets, more than one useful definition can be offered. We may consider pointwise convergence, or uniform convergence, of a sequence of real-valued functions, each being the more appropriate for one or another application.

One approach to this problem is to state an explicit definition for convergence ($x_n \rightarrow_n x \in \mathbb{R}$ iff $\forall \epsilon > 0 \exists N_\epsilon \ni \sup_{n \geq N_\epsilon} |x_n - x| < \epsilon$). The much more comprehensive approach is to endow Ω with additional structure which induces a notion of proximity. This is achieved through the notion of a *neighborhood* of any $x \in \Omega$, a type of subset which includes x. If x_k remains in any neighborhood of x for all large enough k then we can say that x_k converges to x.

This idea is formalized by the *topology*:

Definition 3.1 *Let $\mathcal{O}$ be a collection of subsets of a set Ω. Then $(\Omega, \mathcal{O})$ is a* topological space *if the following conditions hold:*

(i) $\Omega \in \mathcal{O}$ *and* $\emptyset \in \mathcal{O}$,
(ii) *if* $A, B \in \mathcal{O}$ *then* $A \cap B \in \mathcal{O}$,
(iii) *for any collection of sets* $\{A_t\}$ *in* $\mathcal{O}$ *(countable or uncountable) we have* $\cup_t A_t \in \mathcal{O}$.

In this case $\mathcal{O}$ is referred to a topology *on Ω. If $\omega \in O \in \mathcal{O}$ then O is a* neighborhood *of ω.*

The sets $\mathcal{O}$ are called *open sets*. Any complement of an open set is a *closed set*. They need not conform to the common understanding of an open set, since the *power set* $\mathcal{P}(\Omega)$ (that is, the set of all possible subsets) satisfies the definition of a topological space. However, the class of open sets in $(-\infty, \infty)$ as usually understood does satisfy the definition of a topological space, so the term 'open' is a useful analogy.

A certain flexibility of notation is possible. We may explicitly write the topological space as $(\Omega, \mathcal{O})$. When it is not necessary to refer to specific properties of the topology $\mathcal{O}$, we can simply refer to Ω alone as a topological space. In this case an open set $O \subset \Omega$ is understood to be an element of some topology $\mathcal{O}$ on Ω.

Topological spaces allow a definition of convergence and continuity:

Definition 3.2 *If $(\Omega, \mathcal{O})$ is a topological space, and ω_k is a sequence in Ω, then ω_k* converges *to $\omega \in \Omega$ if and only if for every neighborhood O of ω there exists K such that $\omega_k \in O$ for all $k \geq K$.*

A mapping $f : X \to Y$ between topological spaces X, Y is continuous *if for any open set E in Y the preimage $f^{-1}(E)$ is an open set in X.*

A continuous bijective mapping $f : X \to Y$ between topological spaces X, Y is a homeomorphism *if the inverse mapping $f^{-1} : Y \to X$ is also continuous. Two topological spaces are* homeomorphic *if there exists a homeomorphism $f : X \to Y$.*

We may have more than one topology on Ω. In particular, if $\mathcal{O}$ and $\mathcal{O}'$ are topologies on Ω then if $\mathcal{O}' \subset \mathcal{O}$ we say $\mathcal{O}'$ is a *weaker* topology than $\mathcal{O}$, which is a *stronger* topology than $\mathcal{O}'$. Since convergence is defined as a condition imposed on a class of open sets, a weaker topology necessarily has a less stringent definition of convergence. The weakest topology is $\mathcal{O} = \{\Omega, \emptyset\}$, in which case all sequences converge to all elements of Ω. The strongest topology is the set of all subsets of Ω. Since the topology includes all singletons, the only convergent sequences are constant ones, which essentially summarizes the notion of convergence on sets of countable cardinality.

We can see that the definition of continuity for a mapping between topological spaces $f : X \to Y$ requires that Y is small enough, and that X is large enough. Thus, if f is continuous, it will remain continuous if Y is replaced by a weaker topology, or X is replaced by a stronger topology. In fact, any f is continuous if Y is the weakest topology, or X is the strongest topology. We also note that the definitions of semicontinuity of Section 2.1.10 apply directly to real-valued functions on topologies.

The study of topology is especially concerned with those properties which are unaltered by homeomorphisms. From this point of view, two homeomorphic topological spaces are essentially the same.

If $\Omega' \subset \Omega$ and $\mathcal{O}' = \{U \cap \Omega' \mid U \in \mathcal{O}\}$, then $(\Omega', \mathcal{O}')$ is also a topology, sometimes referred to as the *subspace topology*. Note that Ω' need not be an element of $\mathcal{O}$.

An *open cover* of a subset E of a topological space X is any collection U_α, $\alpha \in \mathcal{I}$ of open sets containing E in its union. We say E is a *compact set* if any open covering of E contains a finite subcovering of E (the definition may be applied to X itself). This idea is a generalization of the notion of bounded closure (see Theorem 3.3). Similarly, a set E is a *countably compact set* if any countable open covering of E contains a finite subcovering of E. Clearly, countable compactness is a strictly weaker property than compactness.

3.1.1 Bases of topologies

We say $B(\mathcal{O}) \subset \mathcal{O}$ is a *base* for $\mathcal{O}$ if all open sets are unions of sets in $B(\mathcal{O})$. This suggests that a topology may be constructed from a suitable class of subsets $\mathcal{G}$ of Ω by taking all unions of members of $\mathcal{G}$ and then including Ω and $\emptyset$. As might be expected, not all classes $\mathcal{G}$ yield a topology in this manner, but conditions under which this is the case are well known:

Theorem 3.1 *A class of subsets $\mathcal{G}$ of Ω is a base for some topology if and only if the following two conditions hold (i) every point $x \in \Omega$ is in at least one $G \in \mathcal{G}$; (ii) if $x \in G_1 \cap G_2$ for $G_1, G_2 \in \mathcal{G}$ then there exists $G_3 \in \mathcal{G}$ for which $x \in G_3 \subset G_1 \cap G_2$.*

The proof of Theorem 3.1 can be found in, for example, Kolmogorov and Fomin (1970) (Chapter 3 of this reference can be recommended for this topic).

3.1.2 Metric space topologies

Definition 3.3 *For any set X a mapping $d: X \times X \to [0, \infty)$ is called a* metric, *and (X, d) is a* metric space, *if the following axioms hold:*

> ***Identifiability*** *For any $x, y \in X$ we have $d(x, y) = 0$ if and only if $x = y$,*
> ***Symmetry*** *For any $x, y \in X$ we have $d(x, y) = d(y, x)$,*
> ***Triangle inequality*** *For any $x, y, z \in X$ we have $d(x, z) \leq d(x, y) + d(y, z)$.*

Convergence in a metric space follows from the metric, so that we write $x_n \to_n x$ if $\lim_n d(x_n, x) = 0$. Of course, this formulation assumes that $x \in X$, and we may have sequences exhibiting 'convergent like' behavior even is it has no limit in X.

Definition 3.4 *A sequence $\{x_n\}$ in a metric space (X, d) is a* Cauchy sequence *if for any $\epsilon > 0$ there exists N such that $d(x_n, x_m) < \epsilon$ for all $n, m \geq N$. A metric space is* complete *if all Cauchy sequences converge to a limit in X.*

Generally any metric space can always be *completed* by extending X to include all limits of Cauchy sequences (see Royden (1968), Section 5.4).

Definition 3.5 *Given metric space (X, d), we say $x \in X$ is a* point of closure *of $E \subset X$ if it is a limit of a sequence contained entirely in E. In addition, the* closure *$\bar{E}$ of E is set of all points of closure of E. We say A is a* dense subset *of B if $A \subset B$ and $\bar{A} = B$.*

Clearly, any point in E is a point of closure of E, so that $E \subset \bar{E}$. A metric space is *separable* if there is a countable dense subset of X. The real numbers are separable, since the rational numbers are a dense subset of $\mathbb{R}$.

A metric space also has natural topological properties. We may define an *open ball* $B_\delta(x) = \{y \mid d(y, x) < \delta\}$.

Theorem 3.2 *The class of all open balls of a metric space (X, d) is the base of a topology.*

Proof We make use of Theorem 3.1. We always have $x \in B_\delta(x)$, so condition (i) holds. Next, suppose $x \in B_{\delta_1}(y_1) \cap B_{\delta_2}(y_2)$. The for some $\epsilon > 0$ we have $d(x, y_1) < \delta_1 - \epsilon$ and $d(x, y_2) < \delta_2 - \epsilon$. Then by the triangle inequality $x \in B_\epsilon(x) \subset B_{\delta_1}(y_1) \cap B_{\delta_2}(y_2)$, which completes the proof. ///

A topology on a metric space generated by the open balls is referred to as the *metric topology*, which always exists by Theorem 3.2. For this reason, every metric space can be regarded as a topological space. We adopt this convention, with the understanding that the topology being referred to is the metric topology. We then say a topological space $(\Omega, \mathcal{O})$ is *metrizable* (*completely metrizable*) if it is homeomorphic to a metric space (complete metric space), in which case there exists a metric which induces the topology $\mathcal{O}$. This generalizes the notion of a metric space. Homeomorphisms form an equivalence class, and metrics are equivalent if they induce the same topolgy.

Additional concepts of continuity exist for mappings $f : \mathcal{X} \to \mathcal{Y}$ between metric spaces $(\mathcal{X}, d_x)$ and $(\mathcal{Y}, d_y)$. We say f is *uniformly continuous* if for every $\epsilon > 0$ there exists $\delta > 0$ such that $d_x(x_1, x_2) < \delta$ implies $d_y(f(x_1), f(x_2)) < \epsilon$. A family of functions F mapping $\mathcal{X}$ to $\mathcal{Y}$ is *equicontinuous* at $x_0 \in \mathcal{X}$ if for every $\epsilon > 0$ there exists $\delta > 0$ such that for any $x \in \mathcal{X}$ satisfying $d_x(x_0, x) < \delta$ we have $\sup_{f \in F} d_y(f(x_0), f(x)) < \epsilon$. We say F is equicontinuous if it is equicontinuous at all $x_0 \in \mathcal{X}$.

Theorem 3.3 (Heine-Borel Theorem) *In the metric topology of $\mathbb{R}^m$ a set S is compact if and only if it is closed and bounded.*

3.2 MEASURE SPACES

In elementary probability, we have a set of possible outcomes Ω, and the ability to assign a probability $P(A)$ to any subset of outcomes $A \subset \Omega$. If we ignore the interpretation of $P(A)$ as a probability, then P becomes simply a *set function*, which, as we expect of a function, maps a set of objects to a number. Formally, we write, or would like to write, $P : \mathcal{P}(\Omega) \to [0, 1]$, where $\mathcal{P}(E)$ is the power set of E, or the class of all subsets of E. It is easy enough to write a rule $y = x^2 + x + 1$ which maps any number x to a number y, but this can become more difficult when the function domain is a power set. If $\Omega = \{1, 2, \ldots\}$ is countable, we can use the following process. We first choose a probability for each singleton in Ω, say $P(\{i\}) = p_i$, then extend the definition by setting $P(E) = \sum_{i \in E} p_i$. Of course, there is nothing preventing us from defining an alternative set function, say $P^*(E) = \max_{i \in E} p_i$, which would possess at least some of the properties expected of a probability function. We would therefore like to know if we may devise a precise enough definition of a probability function so that any choice of p_i yields exactly one extension, since definitions of random variables on countable spaces are usually given as probabilities of singletons.

The situation is made somewhat more complicated when Ω is uncountable. It is univerally accepted that a random variable X on $\mathbb{R}$ can be completely defined by the cumulative distribution function $F(x) = P\{X \leq x\}$, which provides a rule for calculating only a very small range of elements of $\mathcal{P}(\mathbb{R})$. We can, of course, obtain probabilities of intervals though subtraction, that is $P\{X \in (a, b]\} = F(b) - F(a)$, and so on, eventually for open and closed intervals, and unions of intervals. We achieve the same effect if we use a density $f(x)$ to calculate probabilities $P\{X \in E\} = \int_E f(x)dx$, since our methods of calculating an integral almost always assume E is a collection of intervals. We are therefore confronted with the same problem, that is, we would like to define probabilities for a simple class of events $E \in \mathcal{E} \subset \mathcal{P}(\Omega)$, for example, singletons or half intervals $(-\infty, x]$, and extend the probability set function P to power sets $\mathcal{P}(\Omega)$ in such a way that P satisfies a set of axioms we regard as essential to our understanding of a probability calculus.

The mathematical issues underlying such a construction relate to the concept of *measurability*, and it is important to realize that it affects both countable and uncountable sets Ω. Suppose we propose a random experiment consisting of the selection of any positive integer at random. We should have no difficulty deciding that the set E_k consisting of all integers divisible by k should have probability $1/k$. To construct such a probability rule, we may set $\Omega = \mathbb{I}_+$, and define a class of subsets $\mathcal{F}_0$ as those $E \subset \mathbb{I}_+$ for which the limit

$$P^*(E) = \lim_{n \to \infty} n^{-1} |E \cap \{1, \ldots, n\}|$$

exists. Then P^* defines a randomly chosen integer X about which we can say $P(X \text{ is divisible by } 7) = 1/7$ or $P(X \text{ is a square number}) = 0$. But we are also assuming that each integer i has equal probability $p_i = \alpha$. If we extend P in the way we proposed, we would end up with $P(\Omega)$ equalling 0 or ∞, whereas the probability that the outcome is in Ω can only be 1. Similarly, it is possible to partition the unit interval into a countable number of uncountably denumerable sets $\mathcal{E}_1$ which are each modulo translations of a one member. Therefore, if we attempt to impose a uniform probability on the unit interval, we would require that $P(E)$ for each $E \in \mathcal{E}$ has the same probability, and we would similarly be forced to conclude that $P(\Omega)$ equals 0 or ∞. Both of these examples are the same in the sense that some principle of uniformity forces us to assign a common probability to an infinite number of disjoint outcome.

As we will next show, the solution to these problems differs somewhat for countable and uncountable Ω. For countable Ω, the object will be to extend P fully to $\mathcal{P}(\Omega)$, and the method for doing so will explicitly rule out examples such as the randomly chosen integer, by insisting at the start that $\sum_{i \in \Omega} p_i = 1$. It could be, and has been (Dubins and Savage (1976)), argued that this type of restriction (formally known as *countable additivity*, see below) is not really needed. It essentially forces P to be continuous in some sense, which might not be an essential requirement for a given application. We could have a perfectly satisfactory definition of a randomly chosen positive integer by restricting our definition to a subset of $\mathcal{P}(\Omega)$, as we have done. In fact, this is precisely how we deal with uncountable Ω, by first devising a rule for calculating $P(E)$ for intervals E, then extending P to sets which may be constructed from a countable number of set operations on the intervals, better known as the *Borel sets* (see below for formal definition). The final step adds all subsets of all Borel sets of probability zero. This class of sets is considerably smaller that $\mathcal{P}(\Omega)$ for uncountable Ω, and means that a probability set function is really no more complex an object than a function on Ω.

3.2.1 Formal construction of measures

Our discussion suggests that probabilities and integrals are similar objects, and both usually are based on the construction of a *measure space*. We do so here, although it is possible to construct an axiomatic theory of probability based on the integral operator, without the mediation of the measure space, the probability itself constructed as the expectation of an indicator function (Whittle (2000)).

We have already outlined the steps in the creation of a probability measure, or of measures in general (we need not insist that the measure of Ω is 1, or even finite). The first step is to define the sets on which the measure will be constructed.

Definition 3.6 *Let $\mathcal{F}$ be a collection of subsets of a set Ω. Then $\mathcal{F}$ is a* field *(or* algebra*) if the following conditions hold:*

(i) $\Omega \in \mathcal{F}$,
(ii) *if* $E \in \Omega$ *then* $E^c \in \Omega$,
(iii) *if* $E_1, E_2 \in \Omega$ *then* $E_1 \cup E_2 \in \Omega$.

If (iii) is replaced with

(iv) *if* $E_1, E_2, \ldots \in \Omega$ *then* $\cup_i E_i \in \Omega$.

Then $\mathcal{F}$ is a σ-field *(or* σ-algebra*).*

Condition (*iii*) extends to all finite unions, so we say that a field is closed under complementation and finite union, and a σ-field is closed under complementation and countable union. Both contain the empty set $\emptyset = \Omega^c$. By De Morgans' Law $(A \cup B)^c = A^c \cap B^c$, so that a field (or σ-field) is closed under all finite (or countable) set operations. For a field (or σ-field) $\mathcal{F}$, a *measurable partition* of $E \in \mathcal{F}$ is any partition of E consisting of elements of $\mathcal{F}$.

Then for any class of subsets $\mathcal{E}$ of Ω we define $\sigma(\mathcal{E})$ as the smallest σ-field containing $\mathcal{E}$, or equivalently the intersection of all σ-fields containing $\mathcal{E}$ (which must also be a σ-field). It is usually referred to as the *σ-field generated by $\mathcal{E}$*. This always exists, since $\mathcal{P}(\Omega)$ is a σ-field, but the intention is usually that $\sigma(\mathcal{E})$ will be considerably smaller.

If $E \subset \Omega$ and $\mathcal{F} = \sigma(\mathcal{E})$ then $\sigma(\mathcal{E} \cap E) = \mathcal{F} \cap E$. If $\mathcal{F}, \mathcal{F}'$ are two σ-fields on Ω and $\mathcal{F}' \subset \mathcal{F}$, we say that $\mathcal{F}'$ is a *sub σ-field* of $\mathcal{F}$.

Example 3.1 *Let $\mathcal{F}_0$ be a class of sets consisting of $\Omega = (\infty, \infty)$, and all finite unions of intervals $(a, b]$, including $(\infty, b]$ and $\emptyset = (b, b]$. This class of sets is closed under finite union and complementation, and so is a field on Ω. Then $\sigma(\mathcal{F}_0)$ is the σ-field consisting of all intervals, and all sets obtainable from countably many set operations on intervals. Note that $\sigma(\mathcal{F}_0)$ could be equivalently defined as the smallest σ-field containing all intervals in Ω, or all closed bounded intervals, all open sets, all sets $(\infty, b]$, and so on.*

We next define a *measure*:

Definition 3.7 *A set function $\mu : \mathcal{F} \to \bar{\mathbb{R}}_+$, where $\mathcal{F}$ is a σ-field on Ω, is a* measure *if $\mu(\emptyset) = 0$ and if it is* countably additive, *that is for any countable collection of disjoint sets $E_1, E_2, \ldots$ we have $\sum_i \mu(E_i) = \mu(\cup_i E_i)$. If $\mathcal{F}$ is a field, then μ is called a measure if countable additivity holds whenever $\cup_i E_i \in \mathcal{F}$.*

If Definition 3.7 did not require that $\mu(\emptyset) = 0$, then it would hold for $\mu \equiv \infty$. However, that $\mu(\emptyset) = 0$ for any other measure would follow from countable additivity, since we would have $\mu(E') < \infty$ for some E', and $\mu(E') = \mu(E') + \mu(\emptyset)$.

A measure μ is a *finite measure* if $\mu(\Omega) < \infty$, and is a *stochastic measure*, or *probability measure*, if $\mu(\Omega) = 1$. We sometimes need to consider a *substochastic measure*, for which $\mu(\Omega) \leq 1$. We say μ is a *σ-finite measure* if there exists a countable collection of subsets $E_i \in \mathcal{F}$ such that $\cup_i E_i = \Omega$ with $\mu(E_i) < \infty$. We refer to $(\Omega, \mathcal{F})$ as a *measurable space* if $\mathcal{F}$ is a σ-field on Ω, then $(\Omega, \mathcal{F}, \mu)$ is a *measure space* if μ is a measure on $(\Omega, \mathcal{F})$. We may also refer specifically to a finite, probability or σ-finite measure space as appropriate.

We have assumed that $\mu(E)$ is always nonnegative. Under some conventions the term *positive measure* is used instead. We will encounter *signed measures*, that is, set functions which share the properties of a measure, but are allowed to take negative values.

We have already referred to the countable additivity property as a type of continuity condition. Formally, we may define sequences of sets in terms of increasing unions. If $E_1 \subset E_2 \subset \ldots$ we write $E_i \uparrow E = \cup_j E_j$. For any sequence $A_1, A_2, \ldots$ we have $E_i = \cup_{j=1}^i A_j \uparrow \cup_{j=1}^\infty A_j = \cup_{j=1}^\infty E_j$, so that increasing sequences appear quite naturally. By taking complements, we equivalently have for any decreasing sequence $F_1 \supset F_2 \supset \ldots$ the limit $F_i \downarrow F = \cap_j F_j$, and any sequence $A_1, A_2, \ldots$ generates a decreasing sequence by setting $F_i = \cap_{j=1}^i A_j \downarrow \cap_{j=1}^\infty A_j = \cap_{j=1}^\infty F_j$.

This leads to a definition of continuity for measure spaces. It is important to note that continuity holds axiomatically for any countably additive measure (as all measures in this book will be), so this need not be verified independently. We summarize a number of such properties:

Theorem 3.4 *Suppose we are given a measure space $(\Omega, \mathcal{F}, \mu)$. The following statements hold.*

(i) $A \subset B$ *implies* $\mu(A) \leq \mu(B)$,
(ii) $\mu(A) + \mu(B) = \mu(A \cup B) + \mu(A \cap B)$,
(iii) $E_i \uparrow E$ *implies* $\lim_i \mu(E_i) = \mu(E)$,
(iv) $F_i \downarrow F$ *implies* $\lim_i \mu(F_i) = \mu(F)$,
(v) *For any sequence* $A_1, A_2, \ldots$ *in* $\mathcal{F}$ *we have* $\lim_i \mu\left(\cup_{j=1}^i A_j\right) = \mu\left(\cup_{j=1}^\infty A_j\right)$
(vi) *For any sequence* $A_1, A_2, \ldots$ *in* $\mathcal{F}$ *we have* $\lim_i \mu\left(\cap_{j=1}^i A_j\right) = \mu\left(\cap_{j=1}^\infty A_j\right)$.

Proof *(i)* Write the disjoint union $B = A \cup (B - A)$, then $\mu(B) = \mu(A) + \mu(B - A)$. *(ii)* Write the disjoint unions $A = (A - B) \cup AB$, $B = (B - A) \cup AB$, $A \cup B = (A - B) \cup (B - A) \cup AB$, then apply additivity. *(iii)* We write $D_1 = E_1$, $D_i = E_i - E_{i-1}$ for $i \geq 2$. The sequence $D_1, D_2, \ldots$ is disjoint, with $E_i = \cup_{i=1}^j D_j$ and $E = \cup_i D_i$. So, by countable additivity we have $\mu(E) = \mu(\cup_i D_i) = \sum_i \mu(D_i) = \lim_i \sum_{j=1}^i \mu(D_j) = \lim_i \mu(E_i)$. Then *(v)* follows after setting $E_i = \cup_{j=1}^i A_j$ and applying *(iii)*. Finally, *(iv)* and *(vi)* follow by expressing a decreasing sequence as an increasing sequence of the complements, then applying *(iii)* and *(iv)*. ///

3.2.2 Completion of measures

Any measure satisfies $\mu(\emptyset) = 0$ but $\emptyset$ need not be the only set of measure zero. We can refer to any set of measure zero as a *null set*. It seems reasonable to assign a measure of zero to any subset of a null set, since, if it was assigned a measure, it could only be 0 under the axioms of a measure. However, the definition of a measure space $(\Omega, \mathcal{F}, \mu)$ does not force $\mathcal{F}$ to contain all subsets of null sets, and counterexamples can be readily constructed. Accordingly, we offer the following definition:

Definition 3.8 *A measure space* $(\Omega, \mathcal{F}, \mu)$ *is* complete *if* $A \in \mathcal{F}$ *whenever* $A \subset B$ *and* $\mu(B) = 0$.

Any measure space may be *completed* by considering the class of subsets $\mathcal{M} = \{A \mid A \subset B, \ \mu(B) = 0\}$, and setting $\mu^*(A) = 0$ for all $A \in \mathcal{M}$ and $\mu^*(B) = \mu(B)$ for all $B \in \mathcal{F}$. It can be shown that $\mathcal{F}^* = \mathcal{F} \cup \mathcal{M}$ is a σ-field and μ^* is a measure, so that $(\Omega, \mathcal{F}^*, \mu^*)$ is a complete measure space.

3.2.3 Outer measure

Definition 3.9 *A set function* λ *on all subsets of* Ω *is an* outer measure *if the following properties are satisfied:*

(i) $\lambda(\emptyset) = 0$,
(ii) $A \subset B \Rightarrow \lambda(A) \leq \lambda(B)$,
(iii) $A \subset \cup_{i=1}^{\infty} A_i \Rightarrow \lambda(A) \leq \sum_{i=1}^{\infty} \lambda(A_i)$.

Property (ii) is referred to as monotonicity *and property (iii) is referred to as* countable subadditivity.

The outer measure differs from the measure in that it is defined on all subsets, so does not require a definition of measurability. However, it does induce a concept of measurability, and in fact directly induces a measure space.

Definition 3.10 *Given an outer measure* λ *on* Ω *a set E is* λ-measurable *if* $\lambda(A) = \lambda(A \cap E) + \lambda(A \cap E^c)$ *for all* $A \subset \Omega$. *By countable subadditivity, this condition can be replaced by* $\lambda(A) \geq \lambda(A \cap E) + \lambda(A \cap E^c)$ *for all* $A \subset \Omega$.

Theorem 3.5 *Given an outer measure* λ *on* Ω, *any set E for which* $\lambda(E) = 0$ *is* λ*-measurable.*

Proof Suppose $A \subset \Omega$ and $\lambda(E) = 0$. By monotonicity $0 \leq \lambda(AE) \leq \lambda(E) = 0$ and $\lambda(A) \geq \lambda(AE^c)$, so that Definition 3.10 holds. ///

We can always restrict λ to a class of subsets $\mathcal{E}$ of Ω. In fact, inducing a measure space from λ is quite straightforward:

Theorem 3.6 *Given an outer measure* λ *on* Ω, *the class* $\mathcal{B}$ *of* λ*-measurable sets is a* σ*-field in which* λ *is a complete measure.*

Proof See, for example, Theorem 12.1 of Royden (1968). ///

Many authors reserve a distinct symbol for a set function restricted to a class of subsets. Theorem 3.6 then describes a measure space $(\Omega, \mathcal{B}, \lambda_{\mathcal{B}})$ where $\lambda_{\mathcal{B}}$ is λ restricted to $\mathcal{B}$.

3.2.4 Extension of measures

Theorem 3.6 permits the construction of a measure space by restricting an outer measure λ to the class of λ-measurable sets, which can be shown to be a σ-field. The complementary procedure is to define a measure μ_0 on a simpler class of subsets $\mathcal{E}$, then to *extend* μ_0 to a measure μ on a σ-field which includes $\mathcal{E}$. The objective is to do this so that μ and μ_0 agree on the original sets $\mathcal{E}$ while μ satisfies Definition 3.7. In addition, we wish to know if any measure μ which achieves this is unique.

We have already introduced the field, in addition to the σ-field (Definition 3.6). In addition, the idea of extending any class of sets $\mathcal{E}$ to the σ-field $\sigma(\mathcal{E})$ is well defined. A class of sets, simpler than a field, from which an extension may be constructed is given by the following definition:

Definition 3.11 *A class of subsets $\mathcal{A}$ of Ω is a* semifield *(*semialgebra*) if the following conditions hold:*

(i) $A, B \in \mathcal{A} \Rightarrow A \cap B \in \mathcal{A}$.
(ii) $A \in \mathcal{A} \Rightarrow A^c$ *is a finite disjoint union of sets in* $\mathcal{A}$.

A σ-field is a field, which is a semifield. The latter is a quite intuitive object. The set of right closed intervals in $\mathbb{R}$, including $(-\infty, a]$ and (a, ∞) and $\emptyset$ is a semifield, which is easily extended into $\mathbb{R}^n$.

If $\mathcal{A}$ is a semifield, then the class of subsets $\mathcal{F}_0$ consisting of $\emptyset$ and all finite disjoint unions of sets in $\mathcal{A}$ can be shown to be a field, in particular, the *field generated by semifield* $\mathcal{A}$.

Theorem 3.7 *Suppose $\mathcal{A}$ is a semifield on Ω and $\mathcal{F}_0$ is the field generated by $\mathcal{A}$. Let μ be a nonnegative set function on $\mathcal{A}$ satisfying the following conditions:*

(i) *If $\emptyset \in \mathcal{A}$ then $\mu(\emptyset) = 0$,*
(ii) *if $A \in \mathcal{A}$ is a finite disjoint union of sets $A_1, \ldots, A_n$ in $\mathcal{A}$ then $\mu(A) = \sum_{i=1}^n \mu(A_i)$.*
(iii) *if $A \in \mathcal{A}$ is a countable disjoint union of sets $A_1, A_2, \ldots$ in $\mathcal{A}$ then $\mu(A) \leq \sum_{i=1}^n \mu(A_i)$.*

Then there exists a unique extension of μ to a measure on $\mathcal{F}_0$.

We have used the term *outer measure* to refer to a set of axioms applicable to a set function defined on all subsets of a space Ω. The term is also used to describe a specific constructed set function associated with Lebesgue measure (see Section 3.2.6).

Definition 3.12 *Suppose μ is a nonnegative set function defined on a class of subsets $\mathcal{A}$ of Ω which contains $\emptyset$ and covers Ω. Suppose $\mu(\emptyset) = 0$. The* outer measure *μ^* induced by μ is defined as*

$$\mu^*(E) = \inf_{E \subset \cup_i A_i} \sum_{i=1} \mu(A_i), \tag{3.1}$$

where the infimum is taken over all countable covers of $E \subset \Omega$ from $\mathcal{A}$.

It may be shown that under Definition 3.12 the set function (3.1) is an outer measure in the sense of Definition 3.9. Our main extension theorem follows (see, for example, Section 12.2, Royden (1968)).

Theorem 3.8 (Carathéodory Extension Theorem) *Suppose μ is a measure on a field $\mathcal{F}_0$ of subsets of Ω. Let μ^* be the outer measure induced by μ. Let $\mathcal{F}^*$ be the set of all μ^*-measurable sets, and let μ' be μ^* restricted to $\mathcal{F}^*$. Then $\mathcal{F}^*$ is a σ-field containing $\mathcal{F}_0$ on which μ' is a measure.*

If μ is finite, or σ-finite then so is μ'. Let μ'' be the restriction of μ^* to $\sigma(\mathcal{F}_0)$. If μ is σ-finite then μ'' is the unique extension of μ to $\sigma(\mathcal{F}_0)$.

The progression from a semifield to a field, and finally to a σ-field is a natural one. In $\mathbb{R}^n$ the semifield is adequate to describe assignment of measures to n-rectangles and their finite compositions. If this can be done in a coherent manner, extension to a measure space follows as described in Theorem 3.8.

However, it must be noted that several extensions are described in Theorem 3.8. The extension to the μ-measurable sets $\mathcal{F}^*$ is complete (see Theorem 3.6). Formally, this is not the same extension as that to $\sigma(\mathcal{F}_0)$. In other words the measure spaces $(\Omega, \mathcal{F}^*, \mu')$ and $(\Omega, \sigma(\mathcal{F}_0), \mu'')$ are not generally the same. In fact, this distinction plays a notable role in the theory of stochastic optimization.

3.2.5 Counting measure

We will usually encounter one of two types of measures. If we are given a countable set S, then $\mu(E) = |E \cap S|$ is called *counting measure*, and satisies the definition of a measure. It is important to note that E is not necessarily a subset of S, so that a counting measure can be defined on any space containing S. Many games of chance are good examples of counting measures.

3.2.6 Lebesgue measure

The second commonly encountered measure is the *Lebesgue measure*. On $\Omega = (-\infty, \infty)$ the set of intervals $(a, b]$, taken to include $(-\infty, b]$ and (a, ∞) is a semifield (we may also simply take the class of all intervals). We assign measure $m\left((a, b]\right) = b - a$ to all bounded intervals, and ∞ to all unbounded intervals. An application of Theorems 3.7 and 3.8 yields *Lebesgue measure*, which a completion of a measure which consistently measures the length of intervals. The same procedure may be used to construct Lebesgue measure in $\mathbb{R}^n$ by assigning the usual geometric volume to rectangles. Whether or not a set is Lebesgue measurable can be resolved by Definition 3.10 and the outer measure referenced in Theorem 3.8. A subset of $[0, 1)$ which is not Lebesgue measurable exists, given the axiom of choice, known as the *Vitali set* (Section 3.4, Royden (1968)).

3.2.7 Borel sets

For any topological space $(\Omega, \mathcal{O})$ the *Borel sets* are taken to be $\mathcal{B} = \sigma(\mathcal{O})$, that is, the smallest σ-field containing all open sets (or equivalently all closed sets). Suppose we are given a measure space $(\Omega, \mathcal{F}, \mu)$. If Ω is also a topological space $(\Omega, \mathcal{O})$, this will generally mean we expect all open sets to be measurable, so that $\mathcal{O} \subset \mathcal{F}$ and therefore $\mathcal{B} \subset \mathcal{F}$. Thus, when all open sets are measurable, μ may be restricted to the Borel sets. Any measure defined on $\mathcal{B}$ is a *Borel measure*.

There can be an important advantage to characterizing measurability in terms of a topology, provided the topology has sufficient structure. For example, if Ω is metrizable, then the Borel sets form the smallest class of subsets containing all open sets which is closed under a countable number of union and intersection operations (Proposition 7.11, Bertsekas and Shreve (1978)).

At this point, we have defined the properties sufficient to define a type of space possessing a useful balance of generality and structure, namely the *Polish space*.

Definition 3.13 *A* Polish space *is a separable completely metrizable topological space. A* Borel space *is a Borel subset of a Polish space.*

3.2.8 Dynkin system theorem

The *Dynkin system theorem* is a quite straightforward statement describing the relationship between various classes of subsets. It permits a number of quite elegant proofs, and turns out to play a specific role in the theory of dynamic programming.

Definition 3.14 *Given a set Ω and classes of subsets $\mathcal{E}$ and $\mathcal{L}$*

(i) $\mathcal{E}$ *is a* π-system *if* $A, B \in \mathcal{E}$ *implies* $AB \in \mathcal{E}$.
(ii) $\mathcal{L}$ *is a* λ-system *if*
  (a) $\Omega \in \mathcal{L}$,
  (b) $A, B \in \mathcal{L}$ *and* $B \subset A$ *implies* $A - B \in \mathcal{L}$,
  (c) $A_n \in \mathcal{L}$ *for* $n \geq 1$, *with* $A_n \uparrow A$ *implies* $A \in \mathcal{L}$.

Here, we refer to π- and λ-systems, while other conventions refer to a λ-system as a *Dynkin* system, or *D*-system.

A σ-field is both a π-system and a λ-system. A λ-system is closed under complementation. A λ-system that is also a π-system (or is closed under finite union) is also a σ-field. The main theorem follows (see, for example, Billingsley (1995) for a proof):

Theorem 3.9 (Dynkin System Theorem) *Given set Ω, if $\mathcal{E}$ is a π-system and $\mathcal{L}$ is a λ-system, then $\mathcal{E} \subset \mathcal{L}$ implies $\sigma(\mathcal{E}) \subset \mathcal{L}$.*

An important consequence is the following:

Theorem 3.10 *Given set Ω, if $\mathcal{E}$ is a π-system, and μ_1, μ_2 are two finite measures on $\sigma(\mathcal{E})$ for which $\mu_1(E) = \mu_2(E)$ for all $E \in \mathcal{E}$, then $\mu_1(E') = \mu_2(E')$ for all $E' \in \sigma(\mathcal{E})$.*

Proof Let $\mathcal{E}'$ be the collection of sets E' for which $\mu_1(E') = \mu_2(E')$. It is easily verified that $\mathcal{E}'$ is a λ-system. If $\mathcal{E} \subset \mathcal{E}'$ then by Theorem 3.9 $\sigma(\mathcal{E}) \subset \mathcal{E}'$. ///

A topology is also a π-system, so that any measures which agree on the open sets must agree on the Borel sets by Theorem 3.10. The intervals $(a, b]$, with $\emptyset$, also form a π-system.

3.2.9 Signed measures

Under Definition 3.7 the value of a measure is always nonnegative, which certainly conforms to the intuitive notion of measure. However, even when this is the intention we may have the need to perform algebraic operations on them, and it will prove quite useful to consider vector spaces of measures. In this case, an operation involving two measures such as $\mu_1 + \mu_2$ would result in a new measure, say $\nu = \mu_1 + \mu_2$. To be sure, ν could be evaluated by addition $\nu(E) = \mu_1(E) + \mu_2(E)$ in $\mathbb{R}$ for any measurable set E, but it is an entirely new measure. Subtraction seems just as reasonable, and we can define a set function by the evaluation $\nu(E) = \mu_1(E) - \mu_2(E)$, represented algebraically as $\nu = \mu_1 - \mu_2$. Of course, $\nu(E)$ might be negative, but we would expect it to share the essential properties of a measure.

Accordingly, Definition 3.7 can be extended to set functions admitting negative values.

Definition 3.15 *A set function $\mu : \mathcal{F} \to \bar{\mathbb{R}}$, where $\mathcal{F}$ is a σ-field on Ω, is a* signed measure *if $\mu(\emptyset) = 0$ and if it is* countably additive, *that is for any countable collection of disjoint sets $E_1, E_2, \ldots$ we have $\sum_i \mu(E_i) = \mu\left(\cup_j E_j\right)$, where the summation is either absolutely convergent or properly divergent.*

This definition does not appear to differ significantly from Definition 3.7, but the possibility of negative values introduces some new issues. For example, suppose we wish to modify Lebesgue measure m on $\mathbb{R}$ by assigning negative measure below 0, that is:

$$m^s(E) = -m(E \cap (-\infty, 0)) + m(E \cap [0, \infty)).$$

We must be able to assign a measure $m((\infty, -\infty))$, which by symmetry should be 0. However, countable additivity fails for the subsets $(i-1, i]$, $i \in \mathbb{I}$, since the implied summation is not absolutely convergent.

When signed measures are admitted, the notion of a positive measure must be clarified. It is possible, for example, to have $\mu(A) \geq 0$, with $\mu(B) < 0$ for some $B \subset A$. Accordingly, we say a measurable set A is *positive* if $\mu(B) \geq 0$ for all measurable $B \subset A$. A set is *negative* if it is positive for $-\mu$. A measure on $(\Omega, \mathcal{F})$ is positive (negative) if Ω is a positive (negative) set. A set is a *null* set if it is both positive and negative.

The monotonicity property does not hold for signed-measures. If A is positive and B is (strictly) negative, then we have $\mu(A \cup B) < \mu(A)$. If μ is a positive measure on $(\Omega, \mathcal{F})$ then $\mu(\Omega) < \infty$ forces all measurable sets to be of finite measure. Similarly, a signed measure is *finite* if all measurable sets are of finite measure. In fact, to define a signed measure as finite, it suffices to assume $\mu(\Omega)$ is finite. Otherwise, suppose for some $E \in \mathcal{F}$ we have $\mu(E) = \infty$. Definition 3.15 precludes assignment of a measure to $E^c \in \mathcal{F}$. The definition of the σ-finite property is the same for signed measures as for positive measures.

3.2.10 Decomposition of measures

We will encounter the situation in which we are given a single measurable space $\mathcal{M} = (\Omega, \mathcal{F})$ on which a class of measures is to be defined. Suppose for example, we have a probability measure P on $\Omega = [0, 1]$ on which P assigns a probability of 1 to outcome 1/2. For this case we need only a σ-field generated by $\{1/2\}$. However, we may wish to consider a family of probability measures defined on a single σ-field, as well as a method of calculating expected values. For greater generality, we might like $\mathcal{F}$ to be the Borel sets on $[0, 1]$ when continuous random variable arise, but we would also like to include our singular example. This poses no particular problem, since this probability measure is easily described by $P(E) = I\{1/2 \in E\}$ for all Borel sets E.

To clarify this issue, we introduce a few definitions.

Definition 3.16 *Let ν and μ be two measures on $\mathcal{M} = (\Omega, \mathcal{F})$. If $\mu(E) = 0 \Rightarrow \nu(E) = 0$ for all $E \in \mathcal{F}$, then ν is* absolutely continuous *with respect to μ. This is written $\nu \ll \mu$, and we also say ν is* dominated by *μ. If $\nu \ll \mu$ and $\mu \ll \nu$ then ν and μ are* equivalent. *Conversely ν and μ are* singular *if there exists $E \in \mathcal{F}$ for which $\nu(E) = \mu(E^c) = 0$, also written $\nu \perp \mu$.*

If ν is absolutely continous with respect to a counting measure on $S \subset \Omega = \mathbb{R}$, and μ is absolutely continuous with respect to Lebesgue measure, then $\nu \perp \mu$ since $\nu(S^c) = \mu(S) = 0$. Note that a pair of measures need not be either singular or equivalent (consider a measure describing a random waiting time W which is continuous above 0, but for which $P(W = 0) > 0$, and Lebesgue measure on the positive real numbers).

The *Lebesgue Decomposition Theorem* will prove useful:

Theorem 3.11 (Lebesgue Decomposition Theorem) *Suppose ν, μ are two σ-finite signed measures defined on a common measurable space $(\Omega, \mathcal{F})$. Then there exists a unique decomposition $\nu = \nu_0 + \nu_1$ for which $\nu_1 \ll \mu$ and $\nu_0 \perp \mu$.*

We have noted that signed measures arise naturally as differences of positive measures. It turns out that any signed measure can be uniquely represented this way. This is a consequence of the *Jordan-Hahn Decomposition Theorem.*

Theorem 3.12 (Jordan-Hahn Decomposition Theorem) *Suppose μ is a signed measure on $\mathcal{F}$.*

(i) *[Hahn Decomposition] There exists $E \in \mathcal{F}$ such that E is positive and E^c is negative. This decomposition is unique up to null sets.*
(ii) *[Jordan Decomposition] There exists a unique pair of mutually singular (positive) measures μ^+, μ^- for which $\mu = \mu^+ - \mu^-$.*

The uniqueness of the Jordan decomposition $\mu = \mu^+ - \mu^-$ permits the definition of the *total variation* measure $|\mu|(E) = \mu^+(E) + \mu^-(E)$.

3.2.11 Measurable functions

A commonly encountered assumption is that a function is a (usually Borel or Lebesgue) 'measurable mapping'. It is worth discussing what is meant by this, understanding that this definition does impose some restrictions on the functions we may consider. We have the definition:

Definition 3.17 *Given two measurable spaces $(\Omega_i, \mathcal{F}_i)$, $i = 1, 2$ a mapping $f : \Omega_1 \to \Omega_2$ is* measurable *if $f^{-1}(A) \in \mathcal{F}_1$ for all $A \in \mathcal{F}_2$.*

Rather like the definition of a continuous mapping between topologies, a measurable functions remains measurable if $\mathcal{F}_1$ is replaced by a strictly larger σ-field, or if $\mathcal{F}_2$ is replaced by a strictly smaller σ-field.

The following theorem is easily proven by noting that $f^{-1}(A \cup B) = f^{-1}(A) \cup f^{-1}(B)$ and $f^{-1}(A^c) = [f^{-1}(A)]^c$.

Theorem 3.13 *If f maps a measurable space $(\Omega, \mathcal{F})$ to range X then the collection $\mathcal{F}_X$ of sets $E \subset X$ for which $f^{-1}(E) \in \mathcal{F}$ is a σ-field.*

By Definition 3.17 the idea of a measurable function depends on separate definitions of measurability on the domain and the range. In many applications, when using real-valued functions it suffices to take the Borel sets as measurable on the range. We therefore adopt the convention that a function $f : X \to \mathbb{R}$ is $\mathcal{F}$-measurable, or a measurable mapping on $(X, \mathcal{F})$, if it is a measurable mapping from $(X, \mathcal{F})$ to $(\mathbb{R}, \mathcal{B})$. This simplifies somewhat the characterization of a real-valued measurable function. Suppose f is a real-valued function on $(X, \mathcal{F})$. Suppose further that

$$\{x \in X \mid f(x) \leq \alpha\} \in \mathcal{F} \quad \text{for all } \alpha \in \mathbb{R}. \tag{3.2}$$

By Theorem 3.13 the class of subsets $E \subset \mathbb{R}$ for which $f^{-1}(E) \in \mathcal{F}$ is a σ-field, and by assumption it contains all intervals $(\infty, \alpha]$, and so also contains the Borel sets (since this is the smallest σ-field containing these intervals). Of course, $<$, $>$ or $\geq$ could replace $\leq$ in the inequalities of (3.2). We therefore say a real-valued funtction f is *Borel measurable*, or *Lebesgue measurable*, if $\mathcal{F}$ are the Borel sets, or the Lebesgue sets. Similarly, measurablility of a mapping from a measurable space $(\Omega, \mathcal{F})$ to $\mathbb{R}^n$ will be defined *wrt* the Borel sets on $\mathbb{R}^n$.

Nonmeasurable mappings usually exist, and are easily constructed using indicator functions of nonmeasurable sets.

We note that composition preserves measurability.

Theorem 3.14 *If f, g are measurable mappings from $(\Omega_1, \mathcal{F}_1)$ to $(\Omega_2, \mathcal{F}_2)$, and from $(\Omega_2, \mathcal{F}_2)$ to $(\Omega_3, \mathcal{F}_3)$ respectively, then the composition $g \circ f$ is a measurable mapping from $(\Omega_1, \mathcal{F}_1)$ to $(\Omega_3, \mathcal{F}_3)$.*

Note that Theorem 3.14 does not state that compound mappings of Lebesgue measurable mappings are Lebesgue measurable, since only preimages of Borel sets (and not Lebesgue sets, a strictly larger class) need be Lebesgue measurable.

If $\mathcal{X}$ is a topological space (usually a metric space), then $\mathcal{F}(\mathcal{X})$ will be the set of mappings $f : \mathcal{X} \to \mathbb{R}$ which are measurable with respect to the Borel sets on $\mathcal{X}$ and $\mathbb{R}$.

Theorem 3.15 *If $f, g \in \mathcal{F}(\mathcal{X})$, then so are $f + g$, fg, $f \vee g$ and $f \wedge g$. If f_n is a countable sequence of measurable mappings, then* $\limsup_n f_n$, $\liminf_n f_n$, $\sup_n f_n$ *and* $\inf_n f_n$ *are measurable mappings.*

Closure of the class of measurable functions is given only for countable operations. Suppose $E \subset [0, 1]$ is not Lebesgue measurable. For each $z \in E$ define function $f_z(x) = I\{x = z\}$. Then $\sup_{z \in E} f_z = I\{x \in E\}$, which is not Lebesgue measurable, even though each f_z is.

In the context of a measure space, the notion of the equality of two functions is often usefully reduced to equivalence classes of functions. In particular, given a measure space $(\Omega, \mathcal{F}, \mu)$, we might find it useful to consider f and g equal if $f(x) = g(x)$ except for $x \in E$ for which $\mu(E) = 0$. Such properties are said to hold *almost everywhere* (that is, except on a set of measure zero) with respect to μ, usually shortened to $ae[\mu]$.

We have *Dini's Theorem* (Section 9.2, Royden (1968)):

Theorem 3.16 (Dini's Theorem) *If f_n is a sequence of usc functions on a countably compact set X which decreases monotonically to 0 then it also converges uniformly to 0.*

3.3 INTEGRATION

Suppose we have measure space $(\Omega, \mathcal{F}, \mu)$ and a measurable function $f : \Omega \to \mathbb{R}$. We say f is a *simple function* if there is a finite measurable partition $A_1, \ldots, A_m$ of Ω and finite constants $a_1, \ldots a_m$ such that $f(x) = \sum_{i=1}^m a_i I\{x \in A_i\}$. We should have little disagreement in defining the integral of a simple function as

$$\int_\Omega f d\mu = \sum_{i=1}^m a_i \mu(A_i).$$

Then, if f is the limit of a sequence of simple functions, the integral of f should be the limit of the integrals of the sequence. Accordingly, for nonnegative f we may define

$$\int_\Omega f d\mu = \sup\left\{ \int_\Omega s d\mu \mid \text{all simple functions } s\text{: } 0 \leq s \leq f \right\}. \tag{3.3}$$

If $E \in \mathcal{F}$ then we let $\int_E f d\mu = \int_\Omega f I_E d\mu$. This defines the *Lebesgue integral*, which is the standard method of constructing integrals on measure spaces. This contrasts with the *Riemann integral*, which uses a similar method, except that step functions replace simple functions as the approximators. A function is Riemann integrable with respect to Lebesque measure on a bounded interval if and only if it is continuous $ae[\mu]$, and equals the Lebesgue measure when this is the case (for example, Ash (1972) Theorem 1.7.1).

The definition is extended to general functions by the decomposition $f = f^+ - f^-$, where $f^+ = fI\{f > 0\}$ and $f^- = -fI\{f < 0\}$ are nonnegative, by evaluating

$$\int_\Omega f d\mu = \int_\Omega f^+ d\mu - \int_\Omega f^- d\mu.$$

Some care is needed when asserting that an integral exists. If f is positive it may be the case that $\int_\Omega f d\mu = \infty$. On the other hand, if $\int_\Omega f^+ d\mu = \int_\Omega f^- d\mu = \infty$, then we must regard $\int_\Omega f d\mu$ as undefined in the sense of $0/0$. However, the convention is usually to say that $\int_\Omega f d\mu$ *exists*, or that f is *μ-integrable* if $\int_\Omega f^+ d\mu$ and $\int_\Omega f^- d\mu$ are both finite. In this case, we would say that $\int_\Omega f d\mu$ is *defined* if at least one of the components is finite. Thus, the integral of $f(x) = x$ over Lebesgue measure on $\mathbb{R}$ is not defined, so we cannot assign a value to $\int x dx$. In contrast we say that $f(x) = x^2$ is not integrable because $\int x^2 dx = \infty$, but the integral is defined.

3.3.1 Convergence of integrals

If we are given a sequence $f_n \in \mathcal{F}(\mathcal{X})$ and a measure μ on $\mathcal{X}$, there will often be a need to relate the convergence properties of f_n to those of $\int_{\mathcal{X}} f_n d\mu$. The following theorems are therefore of considerable importance.

Theorem 3.17 (Fatou's Lemma) *For any sequence $f_n \geq 0$ ae$[\mu]$*

$$\liminf_{n\to\infty} \int_{\mathcal{X}} f_n d\mu \geq \int_{\mathcal{X}} \liminf_{n\to\infty} f_n d\mu.$$

In many statements of Fatou's lemma the sequence f_n is assumed to possess a limit, but this does not strengthen the conclusion.

Theorem 3.18 (Monotone Convergence Theorem) *For any sequence $f_n \geq 0$ for which $f_n \uparrow f$ ae$[\mu]$*

$$\lim_{n\to\infty} \int_{\mathcal{X}} f_n d\mu = \int_{\mathcal{X}} f d\mu.$$

Theorem 3.19 (Lebesgue Dominated Convergence Theorem) *Let f_n be a sequence with limit f ae$[\mu]$. If there exists function $h \geq |f_n|$ ae$[\mu]$ for which $\int_{\mathcal{X}} h d\mu < \infty$ then*

$$\lim_{n\to\infty} \int_{\mathcal{X}} f_n d\mu = \int_{\mathcal{X}} f d\mu.$$

An important special case of the dominated convergence theorem is often stated independently:

Theorem 3.20 (Bounded Convergence Theorem) *Let f_n be a sequence with limit f ae$[\mu]$. Suppose μ is a finite measure, and there exists constant $M \geq |f_n|$ ae$[\mu]$. Then*

$$\lim_{n\to\infty} \int_{\mathcal{X}} f_n d\mu = \int_{\mathcal{X}} f d\mu.$$

A counter-example is not difficult to construct. If μ is Lebesgue measure, $f(x) \equiv 0$ and $f_n(x) = nI\{x \in (0, 1/n)\}$ then $f_n \to_n f$, $\int_{\mathcal{X}} f_n d\mu = 1$, but $\int_{\mathcal{X}} f d\mu = 0$. The hypotheses of Theorems 3.18, 3.19 and 3.20 do not hold, while the hypothesis of Theorem 3.17 does, asserting that $\liminf_{n\to\infty} \int_{\mathcal{X}} f_n d\mu \geq \int_{\mathcal{X}} f d\mu$ as verified.

It is possible to weaken the assumption that $f_n \geq 0$ in Theorems 3.17 and 3.18, if there exists a function $f \leq f_n$ with a finite integral, by considering the sequence $f_n - f$ in place of f_n.

It will sometimes be useful to approximate integrals using simple functions. Accordingly, we state the following theorem (Ash (1972), Theorem 1.5.5):

Theorem 3.21 *(a) A nonnegative Borel measurable function is the limit of an increasing sequence of nonnegative, finite-valued simple functions. (b) Any Borel measurable function f is the limit of a sequence of finite-valued simple functions f_n, for which $|f_n| \leq |f|$ for all n.*

3.3.2 L^p spaces

Given a measure space $(\Omega, \mathcal{F}, \mu)$, we may define the norm

$$\|f\|_p = \left(\int_\Omega |f|^p d\mu \right)^{1/p}, \quad 0 < p < \infty,$$

for all measurable real valued functions f (see Section 6.4 for a formal definition of a norm). We may also define the *essential supremum*

$$\text{ess sup} f = \inf\{b \in \bar{\mathbb{R}} \mid \mu(\{f > b\}) = 0\},$$

and, we take the definition for $p = \infty$ to be

$$\|f\|_p = \text{ess sup}|f|.$$

This gives the *L^p space* as the collection of all measurable real valued functions for which $\|f\|_p < \infty$. It will sometimes be preferable to specify the measure associated with an L^p space, in which case we write $L^p[\mu]$. Two theorems associated with L^p spaces are of particular importance.

Theorem 3.22 (Hölder Inequality) *Let p, q be two positive numbers satisfying $p^{-1} + q^{-1} = 1$, including $p = 1$, $q = \infty$. If $f \in L^p$ and $g \in L^q$ then $fg \in L^1$ and $\|fg\|_1 \leq \|f\|_p \|g\|_q$.*

The pair p, q are referred to as *Hölder conjugates*. The case $p = q = 2$ is commonly referred to as the *Cauchy-Schwarz inequality*.

Theorem 3.23 (Minkowski Inequality) *Let $f, g \in L^p$, $p \geq 1$, then $f + g \in L^p$ and $\|f + g\|_p \leq \|f\|_p + \|g\|_p$.*

One important feature of L^p spaces is that the set of simple functions forms a dense subset, in the sense that for $p > 0$, including $p = \infty$, for any $f \in L^p$ and constant $\epsilon > 0$ there exists a simple function h for which $\|f - h\|_p < \epsilon$, and h may always be chosen to satisfy $|h| \leq |f|$ (see, for example, Ash (1972), Section 2.4).

If w is a positive measurable function we may also define the *weighted L^p* norm,

$$\|f\|_{p:w} = \|wf\|_p, \quad 0 < p \leq \infty.$$

The resulting space of finite normed functions is denoted L^p_w.

3.3.3 Radon-Nikodym derivative

In practice, random quantities are often expressed in terms of densities, which are real-valued functions rather than measures. Of course, calculations involving densities require integration, which depends on measure, usually Lebesgue or counting measure. One of the most powerful theorems in analysis allows a unified study of densities which does not depend on a specific measure.

Theorem 3.24 (Radon-Nikodym Theorem) *Suppose μ is a σ-finite measure and λ is a signed measure on a common measurable space $(\Omega, \mathcal{F})$. If $\lambda \ll \mu$ then there exists a $\mathcal{F}$-measurable function $f : \Omega \to \mathbb{R}$ for which*

$$\lambda(E) = \int_E f d\mu, \text{ for all } E \in \mathcal{F}.$$

If g is any other such function then $g = f$ $ae[\mu]$.

Then, if λ is a probability measure then f is its density.

3.4 PRODUCT SPACES

The construction of product spaces, and the projection from product spaces to their dimensional components, is a natural operation in mathemetical modeling. It is therefore necessary to examine the implications with respect to topological and measurabilty properties of extending and reducing dimension. As will be seen, these implications can be of some consequence.

In one sense, topological and measurability properties may be considered independently of dimension, since the dimension structure of the underlying space plays no role in their definition. Of course, analysis will usually involve various forms of projections onto lower dimensional subspaces. It will then be necessary to associate a topology and σ-algebra with each dimension. In this case, we would like there to be an unambiguous corresponce between all the systems of sets we will need to consider. The standard method of achieving this is through the construction of *product spaces*, although alternatives exists for models in which the resulting definition of measurability may be too coarse to capture some important model behavior (see, for example, the introduction to Chapter 5).

Let $\mathcal{A} = \{\Omega_t;\ t \in \mathcal{T}\}$ be an indexed family of sets. The *direct product* (*Cartesian product*) of $\mathcal{A}$ is written

$$\times_{t\in\mathcal{T}} \Omega_t,$$

and is the set of all indexed sets $\tilde{a} = \{a_t;\ t \in \mathcal{T}\}$ for which $a_t \in \Omega_t$ for all $t \in \mathcal{T}$. If $\Omega_t \equiv \Omega$ for some set Ω, then $\times_{t\in\mathcal{T}} \Omega_t = \times_{t\in\mathcal{T}} \Omega = \Omega^{\mathcal{T}}$ is interpretable as a set of vectors.

If we are given subsets $E_t \subset \Omega_t$, then the *rectangle* $\times_{t\in\mathcal{T}} E_t$, is the subset of $\times_{t\in\mathcal{T}} \Omega_t$ consisting of all elements $\tilde{a}$ for which $a_t \in E_t$.

When $\mathcal{T}$ is infinite, it will be useful to consider the *cylinder*, which is a rectangle $\times_{t\in\mathcal{T}} E_t \subset \times_{t\in\mathcal{T}} \Omega_t$ for which $E_t = \Omega_t$ for all but a finite set of indices.

With each index $t \in \mathcal{T}$ we define a *projection*, which is the mapping $proj_t : \times_{t\in\mathcal{T}} \Omega_t \to \Omega_t$, which is the coordinate of $\tilde{a} \in \times_{t\in\mathcal{T}} \Omega_t$ associated with t, that is, $proj_t(\tilde{a}) = a_t$ when $\tilde{a} = \{a_t;\ t \in \mathcal{T}\}$. As a notational convenience, we can accept as a subscript for the projection function any symbol identifying the component. For example, if $(x, y) \in X \times Y$ then $proj_X((x, y)) = x \in X$.

3.4.1 Product topologies

If we are given an indexed set of topologies $(\Omega_t, \mathcal{O}_t)$, $t \in \mathcal{T}$ we can construct a new topology on $\times_{t\in\mathcal{T}} \Omega_t$ by using as a base the class of all cylinders in $\times_{t\in\mathcal{T}} \Omega_t$ which are products of open sets.

Theorem 3.25 *Given an indexed set of topologies $(\Omega_t, \mathcal{O}_t)$, $t \in \mathcal{T}$ let $\mathcal{G}$ be the class of all cylinders in $\times_{t\in\mathcal{T}}\Omega_t$ which are products of open sets. Then $\mathcal{G}$ is a base for a topology on $\times_{t\in\mathcal{T}}\Omega_t$.*

Proof We make use of Theorem 3.1. Since $\mathcal{O}_t$ covers Ω_t condition (*i*) is satisfied. Suppose $\tilde{\omega} = \{\omega_t\} \in \times_{t\in\mathcal{T}}\Omega_t$. Let $G = \times_t O_t$ and $G' = \times_t O'_t$ be two elements of $\mathcal{G}$ for which $\tilde{\omega} = \in G \cap G' = \times_t O_t \cap O'_t$. We have $O_t \cap O'_t \in \mathcal{O}_t$ for all t, and $O_t \cap O'_t = \Omega_t$ for all but a finite number of t, so that condition (*ii*) holds. ///

The topology for which $\mathcal{G}$ defined in Theorem 3.25 is a base is known as the *product topology*.

3.4.2 Product measures

Two measurable spaces may form a natural product space:

Definition 3.18 *Given measurable spaces $(\Omega_i, \mathcal{F}_i)$, $i = 1, \ldots, n$ the* product σ-*field $\mathcal{F}_1 \times \cdots \times \mathcal{F}_n$ on $\Omega_1 \times \cdots \times \Omega_n$ is the smallest σ-field containing all* rectangles $E_1 \times \cdots \times E_n$, $E_i \in \mathcal{F}_i$.

Consider two measurable spaces $(X, \mathcal{X})$ and $(Y, \mathcal{Y})$ and the product σ-algebra $(X \times Y, \mathcal{X} \times \mathcal{Y})$. Given a subset $E \subset X \times Y$, there are various ways of reducing E to subsets of either X or Y which commonly arise in analysis, and we wish to know if the measurability of E implies the measurability of such reductions. Recall $proj_X\left((x,y)\right) = x$ and $proj_Y\left((x,y)\right) = y$. The image of the projection map of a subset $E \subset X \times Y$ is of considerable interest:

$$proj_X(E) = \{x \in X \mid \exists y \in Y \ni (x,y) \in E\} = \cup_{(x,y)\in E}\, proj_X\left((x,y)\right)$$
$$proj_Y(E) = \{y \in Y \mid \exists x \in X \ni (x,y) \in E\} = \cup_{(x,y)\in E}\, proj_Y\left((x,y)\right)$$

and is refered to as the *projection* of a subset of a product space. Intuitively, $proj_X(E)$ is the set of all $x \in X$ represented at least once as a coordinate in E. We may also define a *section*

$$sec_X(E \mid x) = \{y \in Y \mid (x,y) \in E\}$$
$$sec_Y\left(E \mid y\right) = \{x \in X \mid (x,y) \in E\}.$$

Intuitively, $sec_X(E \mid x) \subset Y$ is a cross-section, or slice, of E at fixed $x \in X$.

A number of statements may be asserted:

Theorem 3.26 *Given two measurable spaces $(X, \mathcal{X})$ and $(Y, \mathcal{Y})$, if $E \in \mathcal{X} \times \mathcal{Y}$ then $sec_X(E \mid x) \in \mathcal{Y}$ and $sec_Y\left(E \mid y\right) \in \mathcal{X}$ for each $x \in X$ and $y \in Y$. If $f: X \times Y \to \bar{\mathbb{R}}$ is measurable wrt $\mathcal{X} \times \mathcal{Y}$ then the functions $f_x(y) = f(x,y)$ or $f_y(x) = f(x,y)$ are measurable functions on Y and X.*

Proof Theorem 18.1 of Billingsley (1995). ///

Suppose $(X, \mathcal{X}, \mu_X)$ and $(Y, \mathcal{Y}, \mu_Y)$ are two σ-finite measure spaces. We may assign to any measurable rectangle $A \times B$ the measure

$$\nu(A \times B) = \mu_X(A)\mu_Y(B).$$

The collection of measurable rectangles $\mathcal{R}$ is a semifield. Following Royden (1968) (Section 12.4) it can be shown that ν is countably additive on $\mathcal{R}$, so that Theorem 3.7 applies, then the Carathéodory extension theorem (Theorem 3.8), and there exists a complete extension of ν from $\mathcal{R}$ to a σ-field containing $\mathcal{R}$. The σ-finite property of μ_X and μ_Y extends to the product measure, so that the extension is unique on the smallest σ-field containing $\mathcal{R}$, which is $\mathcal{X} \times \mathcal{Y}$.

Integration by parts

Suppose given $(X, \mathcal{X}, \mu_X)$ and $(Y, \mathcal{Y}, \mu_Y)$, we have a product measure ν on a σ-field $\mathcal{F}$ containing $\mathcal{X} \times \mathcal{Y}$. Suppose f is $\mathcal{F}$-measurable and integrable *wrt* ν. By *integration by parts* is meant the iterated evaluation of an integral

$$\int_{X\times Y} f(x,y)d\nu(x,y) = \int_X \left[\int_Y f(x,y)d\mu_Y(y)\right] d\mu_X(x)$$
$$= \int_Y \left[\int_X f(x,y)d\mu_X(x)\right] d\mu_Y(y). \tag{3.4}$$

The structure is clarified by defining, if possible, the functions

$$g_X(x) = \int_Y f(x,y)d\mu_Y(y),$$
$$g_Y(y) = \int_X f(x,y)d\mu_X(x) \tag{3.5}$$

so that (3.4) is more compactly written

$$\int_{X\times Y} f(x,y)d\nu(x,y) = \int_X g_X(x)d\mu_X(x) = \int_Y g_Y(y)d\mu_Y(y). \tag{3.6}$$

A few crucial assumptions are implicit in this ubiquitous method. First, it is assumed that if $f(x,y)$ is $\mathcal{F}$-measurable then $f_x(y) = f(x,y)$ and $f_y(x) = f(x,y)$ are measurable functions on Y and X. By Theorem 3.26 this is true if $\mathcal{F} = \mathcal{X} \times \mathcal{Y}$, but this is not the only case we need to consider. Of course, this assumption only verifies that g_X and g_Y are well defined. That they are intergrable, and that (3.6) holds is the subject of *Fubini's Theorem.*

Theorem 3.27 (Fubini's Theorem) *Suppose we are given measure spaces* $(X, \mathcal{X}, \mu_X)$ *and* $(Y, \mathcal{Y}, \mu_Y)$.

(i) *Suppose μ_X and μ_Y are σ-finite and ν is the product measure on $\mathcal{X} \times \mathcal{Y}$, and f is a ν-integrable function. Then f_x, f_y are integrable functions for almost all $x \in X$ and $y \in Y$, g_X and g_Y are integrable functions, and (3.6) holds.*

(ii) *Suppose μ_X and μ_Y are complete measures, and ν is the product measure extension from the measurable rectangles given in Theorem 3.8. Then the conclusion of (i) also holds.*

Proof Version (i) follows Billingsley (1995) (Section 18) and version (ii) follows Royden (1968) (Section 12.4). ///

Borel spaces

One important property of a Borel space is that a countable product of Borel spaces is also a Borel space under the product topology (Bertsekas and Shreve (1978), Proposition 7.12). If $\mathcal{X}$ and $\mathcal{Y}$ are two Borel spaces, this means that the Borel sets of the product topology are equivalent to $\mathcal{X} \times \mathcal{Y}$. Thus, defining products of Borel spaces poses no problem with respect to topology and measurablility.

This does not happen with Lebesgue measurable sets. To see this, suppose now that $\mathcal{X}$ and $\mathcal{Y}$ are the Lebesgue sets on $[0,1]$. Suppose E is a nonmeasurable subset of $[0,1]$ (the Vitali set, for example). The rectangle $E \times \emptyset$ is not in the product σ-field $\mathcal{X} \times \mathcal{Y}$, but, as a subset of a null set $[0,1] \times \emptyset$, it is Lebesgue measurable in $X \times Y$.

Projections and optimization

Fubini's theorem gives one example of a projection method that works as we would like under very general conditions. The situation can change with optimization. Suppose we are given a σ-field $\mathcal{F}$ containing $\mathcal{R}$, and a $\mathcal{F}$-measurable function f. We will be interested in subsequent chapters in functions of the form

$$h_X(x) = \inf_{y \in Y} f(x, y). \tag{3.7}$$

We have seen already that under various constructions the function $f_y(x) = f(x, y)$ is $\mathcal{X}$-measurable if f is $\mathcal{F}$-measurable. The same cannot be said for h_X. Recall that a real-valued function g is $\mathcal{G}$-measurable if $\{x \mid g(x) < \alpha\} \in \mathcal{G}$ for all real α. Then

$$\{x \mid h_X(x) < \alpha\} = \{x \mid \exists y \in Y \ni f(x, y) < \alpha\} = proj_X\left(\{f(x, y) < \alpha\}\right),$$

so that the measurability of h_X is resolved by establishing the measurability of projections. We have already seen that Lebesgue measurability is not preserved by projection, for example, by setting $f(x, y) = I\{E \times \emptyset\}$ for any nonmeasurable E.

The same is true for projections of Borel sets. In *descriptive set theory* (Kechris (1995)) projections of Borel sets (on Polish spaces) are termed *analytic sets* and are a strictly larger class of sets than the Borel sets. Construction of a non-Borel analytic subset is not straightforward, and requires theoretical ideas beyond the scope of this chapter (the reader can consult Appendix B of Bertsekas and Shreve (1978)). For our purposes, it is important to note that the Borel measurability of f in (3.7) does not imply the Borel measurability of h_X. The issue is an important one. We can characterize the dynamic programming algorithm introduced in Chapter 12 as the iterative construction of a function on a Borel product space followed by a projection onto X of the form (3.7), which is used in the construction of the subsequent iteration. It is therefore important to verify that measurability is preserved by such iterations.

It is worth discussing this issue in some more detail. Suppose the solution to our problem is of the form (3.7). If the infimum is always attained, the solution is expressible as a mapping $\phi : X \to Y$ for which $h_X(x) = f(x, \phi(x))$. Otherwise, we may be content with an ϵ-optimal solution ϕ, that is, $f(x, \phi(x)) \leq h_X(x) + \epsilon$. Even if f is measurable, or more generally all model elements are measurable, conditions must still be developed under which

(A1) the solution (that is, the pair (h_X, ϕ)) is measurable,
(A2) all operations required to determine the solution are well defined from the point of view of measurability.

We will see that there are several advantages to confining model definitions to Borel spaces. However, since the solution described in (A1) is obtained by pointwise optimization, and because projections need not preserve measurablility, the space of admissible solutions may reasonably include non-Borel measurable functions. It is important to note that this issue is not concerned with the distinction between optimal and ϵ-optimal solutions ('place this rock as close to that wall as possible without touching it'). The problem described here is the possible inability to construct an ϵ-optimal solution Borel measurable solution, that is, one uniformly close to a non-Borel measurable solution.

A number of approaches have been proposed. As our simple counter-example suggests, it seems reasonable to suppose that measurability can be preserved by imposing suitable topological regularity conditions related to continuity and compactness on the model components. It may then be possible to guarantee that a measurable solution exists, so that the algorithm can confine its search to measurable objects, using only measurable operations, so that (A1) and (A2) hold. This is the approach most commonly taken.

A second approach is to consider the measurability issues directly, and is exemplified in the seminal work of Bertsekas and Shreve (1978). Models are defined on Borel spaces, but the possibility of projections induced by (3.7) which are not Borel sets is not ruled out, eliminating the need for strict topological regularity conditions. In this approach, the solution space becomes the larger class of *universally measurable* functions. Definitions of this of varying generality exist in the literature. For our purposes, we can confine attention to Borel spaces. Then a set E is universally measurable if it is measurable *wrt* the completion of all Borel probability measures. This approach permits conditions (A1) and (A2) to hold, but would require significantly more study of measure theory than that presented here and some knowledge of descriptive set theory. The interested reader can consult Bertsekas and Shreve (1978), especially Sections 1.2 and B.5 (the measurability issues are made quite concrete by the 'two-stage problem' of Section 1.2).

For our purposes, we can accept the limits of Borel measurability, acknowledging the constraints this imposes. The measure theory discussed here suffices to define measure and integeration as well as a rigorous probability calculus. We are also able to define vector spaces of probability measures, define satisfactory concepts of continuity as well as proper norms. Certainly, it is easy to suspect that these constraints are stronger than should be required. However, weakening them significantly would require new ideas which would enlarge the class of algorithms we consider, but would otherwise not have much bearing on the main topic of this book.

This issue will be considered in more detail in the context of dynamic programming in Section 12.3.5.

3.4.3 The Kolmogorov extension theorem

That a product measure can be constructed from two measure spaces based on an inuitive construction for rectangles, that is, $\mu(A \times B) = \mu_1(A)\mu_2(B)$, was shown in Section 3.4.2. Of course this is only special case of a measure on a product space, albeit an important one. We also wish to construct a product space measure which is consistent with a rich enough class of measures defined on projections of the space. We confine attention to probability measures.

Suppose we are given an indexed set of measurable spaces $(\Omega_t, \mathcal{F}_t)$, $t \in \mathcal{T}$. Let $v \subset \mathcal{T}$ be a finite subset of indices $\{t_1, \ldots, t_n\}$. It is helpful to assume that if $\mathcal{T}$ is not a set of real numbers, then there exists some other canonical ordering that allows us to assume $t_1 < t_2 < \ldots < t_n$. For each such v we may denote the product σ-field $\mathcal{F}_v = \times_{i=1}^n \mathcal{F}_{t_i}$ on $\Omega_v = \times_{i=1}^n \Omega_{t_i}$. If $\tilde{a} \in \times_{t \in \mathcal{T}} \Omega_t$, we can let $a_{v|T} = (a_{t_1}, \ldots, a_{t_n}) \in \Omega_v$ be the vector of components of $\tilde{a}$ associated with the indices in v, in the canonical order. We then consider the index subset $u = \{t_{i_1}, \ldots, t_{i_m}\} \subset v$. If $a_v = (a_1, \ldots, a_n) \in \Omega_v$, then let $a_{u|v} = (a_{i_1}, \ldots, a_{i_m}) \in \Omega_u$.

Next, suppose P_v is a probability measure on $\mathcal{F}_v$, and P is a probability measure defined on $\times_{t \in \mathcal{T}} \mathcal{F}_t$. We can define the projections

$$P_{v|T}(A) = P\{\tilde{a} \in \times_{t \in \mathcal{T}} \Omega_t \mid a_{v|T} \in A\}, \quad A \in \mathcal{F}_v,$$
$$P_{u|v}(B) = P_v\{a_v \in \Omega_v \mid a_{u|v} \in B\}, \quad B \in \mathcal{F}_u.$$

We next assume that we may define for each finite index set a probability measure P_v on $\mathcal{F}_v$, and that these measures are *consistent* in the sense that for any two nested finite index sets $u \subset v$ we have $P_u = P_{u|v}$. We further assume that each measurable space $(\Omega_t, \mathcal{F}_t)$ is a Polish space (a separable completely metrizable space) so that $\mathcal{F}_t$ are the Borel sets. By the *Kolomogorov Extension Theorem* this suffices to construct a probability measure on the product space $\times_{t \in \mathcal{T}} \mathcal{F}_t$ consistent with the given system of measures.

Theorem 3.28 (Kolmogorov Extension Theorem) *We are given an indexed set of Borel spaces $(\Omega_t, \mathcal{F}_t)$, $t \in \mathcal{T}$, where $\mathcal{T}$ possesses an ordering. For each finite ordered subset of indices v there exists a measure P_v on $\mathcal{F}_v$. Suppose the system of measures is consistent in the sense that for any pair of finite indices for which $u \subset$ we have $P_u = P_{u|v}$. Then there exists a unique measure P on $\times_{t \in \mathcal{T}} \mathcal{F}_t$ for which $P_v = P_{v|T}$ for all finite subsets v.*

Proof This version of Kolmogorov's extension theorem is proven in Ash (1972), Section 4.4. ///

Chapter 4

Background – probability theory

The axiomatic foundation of modern probability theory was originally formalized in Andrey Kolmogorov's *Foundations of the Theory of Probability* (Kolmogorov (1933)) based on the concept of a measure, giving formal rules for the consistent assignment of probabilities. The set of all possible outcomes is well defined, denoted Ω, and the probability that a random outcome $\omega \in \Omega$ is in $E \subset \Omega$ is assigned a number $P(E)$ under the following three axioms:

(*i*) $P(E) \geq 0$,
(*ii*) $P(\Omega) = 1$,
(*iii*) If $E_1, E_2, \ldots$ is a countable collection of disjoint subsets of Ω then $P(\cup_i E_i) = \sum_{i \geq 1} P(E_i)$.

Thus, a normalized positive finite measure P on a σ-field $\mathcal{F}$ on Ω satisfies Komogorov's axioms, provided the evaluation of probabilites is restricted to sets in $\mathcal{F}$. Therefore, the resulting *probability measure space* $(\Omega, \mathcal{F}, P)$ is the universe on which any stochastic model is defined. A random outcome ω is observable up to a resolution defined by $\mathcal{F}$, that is, we observe whether or not $\omega \in E$ for all $E \in \mathcal{F}$. This doesn't necessarily mean we observe ω itself. For example, we may have $\Omega = [0, 1]$ and $\mathcal{F} = \{\emptyset, [0, 1/2), [1/2, 1], \Omega\}$, in which case we can only observe whether or not $\omega \in [0, 1/2)$. The obvious rejoinder is that we can avoid trivialities by redefining Ω to have only two outcomes, say $\Omega = \{0, 1\}$, with $\mathcal{F} = \{\emptyset, \{0\}, \{1\}, \Omega\}$, and then matching the probabilities of outcomes 0 and 1 with $[0, 1/2)$, $[1/2, 1]$. This is correct, but the larger point to be made is that what gives modern probability theory its rigor is the ability to construct a single probability measure $(\Omega, \mathcal{F}, P)$ on which a potentially quite complex system of partial observations of ω is defined, each of which must have a resolution defined by a specific sub σ-field of $(\Omega, \mathcal{F})$, which will often be considerbly more coarse than $\mathcal{F}$. It might be useful, therefore, to keep this trivial example in mind from time to time.

Note also that it will sometimes be convenient to consider *improper probability measures*, those for which $P(\Omega) < 1$ (the term *defective* is also used). When the distinction arises, P is a *proper probability measure* when $P(\Omega) = 1$.

A *random vector* $X \in \mathbb{R}^d$ is a measurable mapping from $(\Omega, \mathcal{F})$ to $(\mathbb{R}^d, \mathcal{B})$. If $d = 1$ then X is a *random variable* (RV). This construction is sometimes emphasized by writing $X = X(\omega)$, $\omega \in \Omega$, which is interpretable as a transformation of a random outcome ω. Possibly, X is a bijective mapping, but in most models of interest X often represents some partial observation of ω. Under these conditions $X^{-1}(B) \in \mathcal{F}$ when $B \in \mathcal{B}$. The collection of sets $\sigma(X) = \{X^{-1}(B) \mid B \in \mathcal{B}\}$ can be shown to be the smallest σ-field on Ω which makes X measurable, and is commonly referred to as the σ-field generated by X. This definition can play an important role in the study of stochastic processes. For example, if $\sigma(X_1) \subset \sigma(X_2)$, then the specific outcome $X_1(\omega)$ is completely known if $X_2(\omega)$ is.

The *marginal distribution* of a random vector is defined by the induced probability measure $(\mathbb{R}^d, \mathcal{B}, P_X)$, where $P_X(B) = P(X^{-1}(B))$, so that all marginal distributions are implicit in $(\Omega, \mathcal{F}, P)$. This distribution can be characterized by the *cumulative distribution function* (CDF)

$$F_X(x_1, \ldots, x_n) = P\left(\cup_{i=1}^n \{X_i \le x_i\}\right), \quad (x_1, \ldots, x_n) \in \bar{\mathbb{R}}^n.$$

We make use of the shorthand $\bar{F} = 1 - F$. We have seen in Chapter 3 (for example, Theorem 3.10) that two measures on $\mathbb{R}^n$ with equal CDFs must be equal on the Borel sets. The issue of uniqueness is not a trivial one. For example, two distinct random variables may be constructed which possess identical moments (see, for example, the discussion of the 'moment problem' in Feller (1971), Section VII.4).

The problem of constructing measures for infinite collections of random variables will be discussed in the next chapter.

4.1 PROBABILITY MEASURES – BASIC PROPERTIES

Other intuitive notions of probability are modeled by measure theoretic constructions. The *expected value* of a random variable X defined on $(\Omega, \mathcal{F}, P)$ is taken to be

$$E[X] = \int_{\omega \in \Omega} X(\omega) dP(\omega) = \int_{x \in \bar{\mathbb{R}}} x dP_X(x),$$

where $(\mathbb{R}, \mathcal{B}, P_X)$ is the marginal probability measure of X. The usual rules of integration apply, so that for a measurable function g

$$E[g(X)] = \int_{x \in \bar{\mathbb{R}}} g(x) dP_X(x) = \int_{y \in \bar{\mathbb{R}}} y dP_{g(X)}(y),$$

the logic being that $Y = g(X)$ is a random variable with its own marginal distribution $P_{g(X)} = P_Y$. This is a consequence of the theory of integration, and not a definition. We will sometimes need to clarify which measure is being used to define the integral. For example, $E_P[g(X)] = \int_{\omega \in \Omega} g(X(\omega)) dP(\omega)$, and $E_X[g(X)] = \int_{x \in \bar{\mathbb{R}}} g(x) dP_X(x)$.

The identification of $E[X]$ with the integral matches the intuititive notion of an average outcome weighted by probabilities, for which certain properties must hold, as described by the theory of integration. Accordingly, limit theorems for integrals apply to expected values directly:

Theorem 4.1 *Let X_n be a sequence of random variables on a probability measure space $(\Omega, \mathcal{F}, P)$.*

(i) ***Fatou's Lemma:*** *If $X_n \ge Y$, where $E[Y]$ is finite, then*

$$\liminf_{n\to\infty} E[X_n] \ge E[\liminf_{n\to\infty} X_n].$$

(ii) ***Monotone Convergence Theorem:*** *If $X_n \geq Y$, where $E[Y]$ is finite, and $X_n \uparrow X$ then*

$$\lim_{n\to\infty} E[X_n] = E[X].$$

(iii) ***Lebesgue Dominated Convergence Theorem:*** *Suppose X_n possesses limit X. If there exists random variable $Y \geq |X_n|$ for which $E[Y] < \infty$ then*

$$\lim_{n\to\infty} E[X_n] = E[X].$$

(iv) ***Bounded Convergence Theorem:*** *Suppose X_n possesses limit X. If there exists finite constant $M \geq |X_n|$ then*

$$\lim_{n\to\infty} E[X_n] = E[X].$$

We say an event E on $(\Omega, \mathcal{F}, P)$ holds *almost everywhere* if $P(E) = 1$, which could also be stated 'except on a set of measure 0' (the term *almost surely* is also used). This statement is made relative to a specific probability measure P, so we may choose to express this statement compactly as $ae[P]$. If we are given two random variables X, Y, it might be the case that $X(\omega) \leq Y(\omega)$ for all $\omega \in \Omega$, in which case we may simply write $X \leq Y$. If it happens that $P(X > Y) = 0$, this does not imply that $X \leq Y$, but that the set of ω for which $X(\omega) > Y(\omega)$ has probability measure 0, so we write $X \leq Y$ $ae[P]$. This convention is important when the properties of P itself are being investigated. When there is no ambiguity, the more intuitive shorthand $wp1$ (*with probability one*) will be used. All inequalities specified in Theorem 4.1 need only hold $ae[P]$.

Uniform integrability

The *uniform integrability* condition is sometimes imposed on a collection of random variables:

Definition 4.1 *A collection of random variables $\{X_t\}$, $t \in \mathcal{T}$, is* uniformly integrable *if* $\lim_{K\to\infty} \sup_{t\in\mathcal{T}} E[|X_t| I\{|X_t| > K\}] = 0$.

This condition implies that $\sup_t E[|X_t|] < \infty$, and it holds if $|X_t| \leq Y$ for all t for some integrable Y.

L^p norms

The L^p norm is given by $\|X\|_p = E[|X|^p]^{1/p}$, with the essential supremum defining L^∞ as before.

The Hölder inequality states that

$$E[|XY|] \leq E[|X|^p]^{1/p} E[|Y|^q]^{1/q} \quad \text{for conjugate pairs } p^{-1} + q^{-1} = 1,$$

while the Minkowski inequality states that

$$E[|X+Y|^p]^{1/p} \leq E[|X|^p]^{1/p} + E[|Y|^p]^{1/p} \quad \text{for } p \geq 1.$$

Densities

By the Radon-Nikodym theorem (Theorem 3.24), given a probability measure $(\Omega, \mathcal{F}, P)$ and a σ-finite measure μ on $(\Omega, \mathcal{F})$ for which $P \ll \mu$, there exists a measurable *density function* $f : \Omega \rightarrow \mathbb{R}$ for which

$$P(E) = \int_E f(\omega) d\mu(\omega),$$

and this density is unique $ae[\mu]$. Usually, a density represents a marginal distribution $(\mathbb{R}^d, \mathcal{B}, P_X)$ of a random vector X. It is important to note that this definition makes no explicit distinction between 'continuous' and 'discrete' distributions. The difference emerges in the choice of μ. To say that a random vector X is 'continuous' means that $P_X \ll \mu$ where μ is Lebesgue measure, so that $P_X(E) = 0$ for any set E of Lebesgue measure zero. To say that a random vector X is 'discrete' means that there exists a countable set $\mathcal{S}$ for which $P_X \ll \mu_{\mathcal{S}}$ where $\mu_{\mathcal{S}}$ is counting measure on $\mathcal{S}$, so that $P_X(E) = 0$ whenever $E \cap \mathcal{S} = \emptyset$. Of course, a random vector need not be entirely continuous or discrete. A relevant example would be the waiting time W of a customer entering a queueing system, since $P(W = 0)$ is usually greater than zero. In this case, μ would have to be chosen accordingly, and the appropriate integration method used. However, once this is done, we may still define a density function for W, evaluated by, for example

$$P_W(E) = f_W(0) I\{0 \in E\} + \int_{E \cap (0,\infty)} f_W(w) dw.$$

In fact, Lebesque measure μ on $\mathbb{R}^d$ and counting measure $\mu_{\mathcal{S}}$, $\mathcal{S} \subset \mathbb{R}^d$, are examples of singular measures (see Section 3.2.10), since $\mu(\mathcal{S}) = \mu_{\mathcal{S}}(\mathcal{S}^c) = 0$. We may have $P_X \ll \mu + \mu_{\mathcal{S}}$, in which case there exists a density f_X with respect to $\mu + \mu_{\mathcal{S}}$ for which

$$P_X(E) = \sum_{x \in \mathcal{S}} f_X(x) I\{x \in E\} + \int_E f_X(x) dx. \tag{4.1}$$

It is sometimes the convention to explicitly decompose $f_X(x) = f_X(x \mid \mu_{\mathcal{S}}) + f_X(x \mid \mu)$, using $f_X(x \mid \mu_{\mathcal{S}})$ only in the summation of (4.1) and $f_X(x \mid \mu)$ only in the integral. This might be preferable to aid clarity, but is not formally necessary, since the values of $f_X(x)$ at $x \in \mathcal{S}$ may be changed arbitrarily without changing the value of the integral portion of (4.1). As with the CDF, we will identify the density of X as f_X.

Independence

Independence is an intuitive probabilistic notion, defining unrelatedness of random outcomes. From a mathematical point of view it is sometimes counterintuitive.

Two sets A, B on $(\Omega, \mathcal{F}, P)$ are *independent* if $P(AB) = P(A)P(B)$. This is commonly written $A \perp B$ (independence resembles geometric perpendicularity in some important ways). We then say two collections of sets $\mathcal{F}$ and $\mathcal{G}$ are independent if $F \perp G$ for all $F \in \mathcal{F}$ and $G \in \mathcal{G}$. Finally, two random variables X, Y are independent if $P(\{X \in E\} \cap \{Y \in F\}) = P(\{X \in E\} \cap \{Y \in F\})$ for all measurable sets E, F. It can be seen that the definition of independence for random variables is implicit in the definition of independence for collections of sets by setting $\mathcal{F}=\sigma(X)$ and $\mathcal{G}=\sigma(Y)$. Independence of two random variables or collections of sets is similarly denoted $X \perp Y$ or $\mathcal{F} \perp \mathcal{G}$.

Suppose E_0, E_1 are any sets for which $P(E_0)=0$ and $P(E_1)=1$. Then for any set A we must have $P(AE_0)=0$, since $P(AE_0) \leq P(E_0)=0$. This means $A \perp E_0$. Similarly, $P(AE_1)=P(A)-P(AE_1^c)=P(A)$, since $P(AE_1^c) \leq P(E_1^c)=0$, and so $A \perp E_1$. In fact, $E \perp E$ implies $P(E)=P(E)^2$, that is $P(E)$ is 0 or 1.

A finite sequence of subsets $E_1,\ldots,E_n$ on $(\Omega,\mathcal{F},P)$ is *independent* if

$$P(\cap_{i\in I} E_i)=\prod_{i\in I} P(E_i), \tag{4.2}$$

for all nonempty index subsets $I \subset \{1,\ldots,n\}$. In fact, any sequence of subsets E_i for which $P(E_i) \in \{0,1\}$ is independent.

It is necessary to insist on the product rule for all nonempty selections from the sequence. To see this, consider three sets A,B,C. We may construct all joint probabilities by specifying suitable probability values for all eight regions of the Venn diagram. Suppose we set $P(ABC)=1/64$, $P(AB^cC^c)=P(A^cBC^c)=P(A^cB^cC)=15/64$ and $P(A^cB^cC^c)=18/64$. This gives $P(A)=P(B)=P(C)=1/4$, and so $P(ABC)=P(A)P(B)P(C)$. On the other hand, $P(AB)=1/64 \neq P(A)P(B)=1/16$, and so A and B are not independent.

Conversly, pairwise independence does not imply independence. For example, if $A \perp B$, $P(A)=P(B)=1/2$ and $C=AB \cup A^cB^c$, then it is easily verified that we also have $A \perp C$ and $B \perp C$, but that $P(ABC) \neq P(A)P(B)P(C)$.

Pairwise independence carries some implications. If $A \perp B$ then $A^c \perp B$, since $P(A^cB)=P(B)-P(AB)=P(B)-P(A)P(B)=P(B)(1-P(A))=P(B)P(A^c)$. This in turn implies $A^c \perp B^c$ and $A \perp B^c$. It is also true that $\Omega \perp A$ and $\emptyset \perp A$ for any set A. Therefore, $A \perp B$ implies $\sigma(\{A\}) \perp \sigma(\{B\})$.

A finite sequence of collections of subsets $\mathcal{E}_1,\ldots,\mathcal{E}_n$ is independent if all selections of sets $E_i \in \mathcal{E}_i$, $i=1,\ldots,n$ are independent as defined in (4.2). The independence of random variables $X_1,\ldots,X_n$ is equivalent to the independence of the σ-fields $\sigma(X_i)$, $i=1,\ldots,n$. We use the shorthand *iid* for independent, identically distributed RVs. If instead we define the collections of sets $\mathcal{E}_i=\{\{X_i \leq x\} \mid x \in \mathbb{R}\}$, it can be shown that independence of $\mathcal{E}_1,\ldots,\mathcal{E}_n$ implies the independence of $\sigma(X_1),\ldots,\sigma(X_n)$. This can be deduced from the Dynkin system theorem of Section 3.2.8 (Durrett (2010), Section 1.4). This means that random variables $\tilde{X}=(X_1,\ldots,X_n)$ are independent if and only if

$$F_{\tilde{X}}(x_1,\ldots,x_n)=\prod_{i=1}^{n} F_{X_i}(x_i) \textit{ or equivalently } f_{\tilde{X}}(x_1,\ldots,x_n)=\prod_{i=1}^{n} f_{X_i}(x_i). \tag{4.3}$$

It can be shown that if $E[|X_i|]<\infty$ for each i, independence implies:

$$E[\prod_{i=1}^{n} X_i]=\prod_{i=1}^{n} E[X_i]. \tag{4.4}$$

Moments

Let X be any random variable. The kth order moment and central moment are defined as

$$\mu_k[X] = \mu_k = E[X^k]$$
$$\bar{\mu}_k[X] = \bar{\mu}_k = E[(X - E[X])^k],$$

where the reference to X in the notation is omitted when there is no ambiguity.

The binomial theorem allows a comparison of the moments and central moments, setting $\mu = \mu_1[X]$,

$$\bar{\mu}_n[X] = E[(X - \mu)^n] = E\left[\sum_{i=0}^{n} \binom{n}{i} X^i \mu^{n-i}\right] = \sum_{i=0}^{n} \binom{n}{i} \mu_i[X] \mu^{n-i}.$$

Many such formulae are described in the literature, and the algebraic manipulation of moment expressions sometimes represents a significant technical challenge (a system of *tensors* for multivariate moments is proposed in McCullagh (1984)).

Some commonly used distributions

We refer to a set of densities indexed by a parameter $\theta \in \Theta \subset \mathbb{R}^d$ as a *parametric family of densities*. A specific member will be referenced by $f_X(x \mid \theta)$ or $f_\theta(x)$ as convenient. Then Θ is referred to as the *parameter space*.

Table 4.1 lists a number of commonly used parametric density families, with the notational conventions to be used in this book. The moment generating function $m(t)$ is given (Section 4.2 below). Also given are the most natural formulae for the higher order moments.

Density parameters are often given broad classifications, such as *location*, *scale*, *rate* or *shape* parameters. Suppose $f(x)$ is a density function on $\mathbb{R}$. For any $\mu \in (-\infty, \infty)$ and $\sigma \in (0, \infty)$ it is easily shown that $f(x \mid \mu, \sigma) = \sigma^{-1} f((x - \mu)/\sigma)$ is also a density, and we have generated a *location-scale* family of densities with *location parameter* μ and *scale parameter* σ (or a *location* family or *scale* family if only one of the parameters is involved). We expect the location and scale parameter to be in the same unit as the RV. The reciprocal of a scale parameter is a rate parameter, and so they are interchangeable. If a RV represents a random arrival time, a scale parameter might be the expected arrival time (or be related to it), while the rate parameter would represent the arrival rate (or be related to it). The gamma distribution in Table 4.1 is parametrized with a rate parameter λ, but many conventions replace λ with $1/\mu$, where μ would be a scale parameter.

A *shape parameter* changes the shape of a density, beyond the translation induced by a location parameter or the change in scale induced by a scale or rate parameter. The parameter α of the gamma density is an example.

We adopt the usual shorthand, writing $X \sim N(\mu, \sigma^2)$ to mean that X is a normally distributed random variable, and so on. In particular, $Z \sim N(0, 1)$ is a *unit normal* random variable.

Table 4.1 Notional conventions for commonly used distributions. The density f_X and the moment generating function $m(t)$ are given, as well as the most convenient formula for higher order moments.

Normal, $X \sim N(\mu, \sigma^2)$	
$\sigma > 0, \mu \in \mathbb{R}$	$f_X(x) = \dfrac{\exp(-(x-\mu)^2/(2\sigma^2))}{(2\pi\sigma^2)^{1/2}}, x \in \mathbb{R}$
Mean $= \mu$, Variance $= \sigma^2$	$m(t) = \exp(\mu t + \sigma^2 t^2/2), t \in \mathbb{R}$
$E[(X-\mu)^k] = \sigma^k \dfrac{k!}{2^{k/2}(k/2)!}, k = 2, 4, \ldots, E[(X-\mu)^k] = 0, k = 1, 3, \ldots$	
Gamma, $X \sim gamma(\alpha, \lambda)$	
$\alpha > 0, \lambda > 0$	$f_X(x) = \dfrac{\lambda(\lambda x)^{\alpha-1}\exp(-\lambda x)}{\Gamma(\alpha)}, x \geq 0$
Mean $= \alpha/\lambda$, Variance $= \alpha/\lambda^2$	$m(t) = (1 - t/\lambda)^{-\alpha}, t < \lambda$
$E[X^s] = \lambda^{-s}\dfrac{\Gamma(\alpha+s)}{\Gamma(\alpha)}, s \geq 0$	
Exponential, $X \sim exp(\lambda)$ Equivalent to $X \sim gamma(1, \lambda)$	
χ^2 (*chi-squared*), $X \sim \chi^2(d)$ Equivalent to $X \sim gamma(d/2, 1/2)$	
Double Exponential, $X \sim DE(\lambda)$	
$\mu \in \mathbb{R}, \lambda > 0$	$f_X(x) = \dfrac{\lambda}{2}\exp(-\lambda\vert x-\mu\vert), x \in \mathbb{R}$
Mean = μ, Variance = $2/\lambda^2$	$m(t) = \exp(\mu t)(1 - (t/\lambda)^2)^{-1}, \vert t\vert < \lambda$
$E[(X-\mu)^k] = \lambda^{-k}k!, k = 2, 4, \ldots, E[(X-\mu)^k] = 0, k = 1, 3, \ldots$	
Poisson, $X \sim pois(\lambda)$	
$\lambda > 0$	$f_X(x) = \dfrac{\lambda^x}{x!}\exp(-\lambda), x \in \mathbb{N}_0$
Mean $= \lambda$, Variance $= \lambda$	$m(t) = \exp(\lambda(\exp(t) - 1)), t \in \mathbb{R}$
$E[X(X-1)\cdots(X-k+1)] = \lambda^k, k \in \mathbb{N}$	
Binomial, $X \sim bin(n,p)$	
$p \in [0,1], n \in \mathbb{N}$	$f_X(x) = \dbinom{n}{x} p^x(1-p)^{n-x}, x \in \{0, 1, \ldots, n\}$
Mean $= np$, Variance $= np(1-p)$	$m(t) = (p\exp(t) + (1-p))^n, t \in \mathbb{R}$
$E[X(X-1)\cdots(X-k+1)] = p^k\dfrac{n!}{(n-k)!}, k \in \{1, 2, \ldots, n\}$	
Bernoulli, $X \sim bern(p)$ Equivalent to $X \sim bin(1,p)$	

Note that a number of commonly used distributions are special cases, in particular the χ^2 and exponential distributions (of the gamma distribution). In addition, a χ^2 RV of d *degrees of freedom* (*df*) equals in distribution the sum of squares of d *iid* unit normal RVs.

Multivariate distributions

Multivariate distributions describe distributions of random vectors $\tilde{X} = (X_1, \ldots, X_n)$. The *marginal distribution* of a component of $\tilde{X}$ is, for example,

$$f_{X_1}(x_1) = \int_{\mathbb{R}^{n-1}} f(x_1, x_2, \ldots, x_n) dx_2 \cdots dx_n,$$

or more generally the distribution of any strict subset of components.

Suppose we are given a probability distribution $P = (p_1, \ldots, p_m)$ on $\mathcal{S} = \{1, \ldots, m\}$. If we are given an *iid* sample of size n from P, and we let $\tilde{N} = (N_1, \ldots, N_m)$ be the vector of sample frequencies for each outcome, then $\tilde{N}$ has a *multinomial distribution* with density given by

$$\begin{aligned} f_{\tilde{N}}(n_1, \ldots, n_m) &= P(N_1 = n_1, \ldots, N_m = n_m) \\ &= \binom{n}{n_1, \ldots, n_m} \prod_{i=1}^{m} p_i^{n_i}, \quad \min_i n_i \geq 0, \quad n_1 + \cdots + n_m = n. \end{aligned}$$

Recall the multinomial coefficient of Section 2.1.12. The marginal distributions are $N_i \sim bin(n, p_i)$.

Suppose $\tilde{X} = (X_1, \ldots, X_m)$ is a random vector. The *mean vector* can be written $\mu_{\tilde{X}} = E[\tilde{X}] = (E[X_1], \ldots, E[X_m])$ in the appropriate context. In matrix algebra $\mu_{\tilde{X}}$ is usually interpreted as a column vector.

The $m \times m$ *variance matrix* (also referred to as the *covariance matrix*) of $\tilde{X}$ is defined elementwise as

$$\left[\Sigma_{\tilde{X}}\right]_{i,j} = cov[X_i, X_j],$$

where we denote covariance $cov[X, Y] = E[(X - E[X])(Y - E[Y])]$ and consequently $var[X] = cov[X, X]$. Two RVs may be refered to as *linearly independent* if their covariance is zero, although this does not by itself imply independence under the formal definition.

When the context permits we may write $var[\tilde{X}] = \Sigma_{\tilde{X}}$. Since $cov[X, Y] = cov[Y, X]$, $\Sigma_{\tilde{X}}$ is always symmetric. For any linear combination $Y = a_1 X_1 + \cdots + a_m X_m$ based on constant coefficients a_i it may be shown that

$$var[Y] = \tilde{a}^T \Sigma_{\tilde{X}} \tilde{a}, \tag{4.5}$$

where $\tilde{a} = (a_1, \ldots, a_m)$ is taken to be a column vector. Since a variance is always non-negative this must mean $\Sigma_{\tilde{X}}$ is positive semidefinite, and is positive definite unless a subset of the elements of $\tilde{X}$ are linearly dependent *wp*1.

Next, suppose b is a $k \times 1$ constant column vector, A is a $k \times m$ constant matrix, and $\tilde{X}$ is a $m \times 1$ random vector. Then

$$\tilde{Y} = b + A\tilde{X}$$

is a linear tranformation yielding a $k \times 1$ random vector, consisting of k linear combinations of $\tilde{X}$. The mean and variance matrices of $\tilde{X}$ and $\tilde{Y}$ are always related by

$$E[\tilde{Y}] = b + AE[\tilde{X}] \quad \text{and} \quad var[\tilde{Y}] = A\left(var[\tilde{X}]\right)A^T.$$

Suppose $var[\tilde{X}]$ is positive definite. Then there exists an invertible symmetric square root matrix $var[\tilde{X}]^{1/2}$ (Section 2.3.3). If $\tilde{Y} = var[\tilde{X}]^{-1/2}\tilde{X}$ then

$$\begin{aligned} var[\tilde{Y}] &= var[\tilde{X}]^{-1/2} var[\tilde{X}] var[\tilde{X}]^{-1/2} \\ &= var[\tilde{X}]^{-1/2} var[\tilde{X}]^{1/2} var[\tilde{X}]^{1/2} var[\tilde{X}]^{-1/2} \\ &= I. \end{aligned}$$

Thus, any random vector with a positive definite variance matrix $var[\tilde{X}]$ possesses a linear transformation yielding linearly independent coordinates of unit variance.

Suppose $\tilde{\mu}$ is a $m \times 1$ column vector and Σ is a positive defnite $m \times m$ matrix. The *multivariate normal density* function is defined as

$$\begin{aligned} f(x \mid \tilde{\mu}, \Sigma) &= (2\pi)^{-m/2} \det(\Sigma)^{-1/2} \exp(-Q/2), \quad x \in \mathbb{R}^m, \text{ where} \\ Q &= (x - \tilde{\mu})^T \Sigma^{-1} (x - \tilde{\mu}). \end{aligned} \tag{4.6}$$

Then $\tilde{X} = (X_1, \ldots, X_m)$ is a *multivariate normal random vector* if it possesses this density, in which case it may be shown that $E[\tilde{X}] = \tilde{\mu}$, $var[\tilde{X}] = \Sigma$. In addition, the marginal distributions are $X_i \sim N(\tilde{\mu}_i, \Sigma_{i,i})$. The $m = 2$ case is often referred to as the *bivariate normal* distribution.

It is important to note that a random vector with marginal normal densities is not necessarily multivariate normal. For example, if $X \sim N(0,1)$ and $Y = SX$ where S is an independent random sign, then $Y \sim N(0,1)$, $cov[X, Y] = 0$, but (X, Y) does not possess a multivariate normal density.

The definition of a multivariate normal random vector can be generalized to include any random vector of the form $\tilde{X} = \tilde{\mu} + A\tilde{Z}$, where $\tilde{\mu}$ is an $k \times 1$ column vector, A is an $k \times m$ matrix, and $\tilde{Z}$ is a $m \times 1$ column vector of independent unit normal random variables. In this case $var[\tilde{X}]$ need not be positive definite, so (4.6) cannot be used directly.

4.2 MOMENT GENERATING FUNCTIONS (MGF) AND CUMULANT GENERATING FUNCTIONS (CGF)

For a random variable X the *moment generating function* MGF and *cumulant generating function* CGF are defined, where possible, as

$$\begin{aligned} m_X(t) &= E\left[\exp(tX)\right] \\ c_X(t) &= \log(m_X(t)), \end{aligned}$$

where the reference to X in the notation is omitted when there is no ambiguity. The MGF is related to the Laplace transform (formally equal to $m(-t)$), however, the convention in probability theory is to say that the MGF exists only when $m(t)$ is finite on a set $\mathcal{T}$ containing $t = 0$ in the interior. It is easily verified that if $m(t) < \infty$, then $m(s) < \infty$ for $0 < s < t$ or $0 > s > t$. Therefore, existence of the MGF can be checked by verifying that $m(t) < \infty$ for one positive and one negative t.

The MGF and CGF are of value because of their tractability with respect to standard calculus. This often means freely exchanging the order of differentiation, summation or integration. This can be justified, as discussed in Section 2.4 of Casella and Berger (2002).

Convexity of the CGF

The CGF is convex. Its second derivative is given by

$$\left.\frac{d^2 c_X(t)}{dt^2}\right|_{t=0} = \frac{E\left[X^2\exp(tX)\right]E\left[\exp(tX)\right] - E\left[X\exp(tX)\right]^2}{E\left[\exp(tX)\right]^2}. \tag{4.7}$$

Suppose all the expected values in (4.7) exist. Let $Y = X\exp(tX)$. By Hölder's inequality we may write

$$|E[Y]| \leq E[|Y|] = E[\sqrt{YX}\sqrt{Y/X}] \leq E[YX]^{1/2}[Y/X]^{1/2},$$

since $YX \geq X^2\exp(Xt) \geq 0$ and $Y/X = \exp(Xt) > 0$. This means the second derivative (4.7) is nonnegative, so that the CGF is convex.

Expansion of the MGF

Expanding the MGF within the expectation operator gives

$$m_X(t) = E\left[\exp(tX)\right] = E\left[\sum_{i=0}^{\infty}\frac{(tX)^i}{i!}\right] = \sum_{i=0}^{\infty}\mu_i[X]\frac{t^i}{i!}. \tag{4.8}$$

If X is a positive RV then exchange of integration and summation is justified by the bounded convergence theorem (Theorem 4.1), provided $m_X(t) < \infty$, in which case all moments $E\left[X^i\right]$ must exist. Otherwise we have,

$$\left|\sum_{i=0}^{n}\frac{(tX)^i}{i!}\right| \leq \sum_{i=0}^{n}\frac{(t|X|)^i}{i!} \leq \exp(t|X|)$$

so that for general RVs (4.8) holds by the dominated convergence theorem (Theorem 4.1) if $m_{|X|}(t) < \infty$.

Existence of the MGF

The existence of all moments does not guarantee $m_X(t) < \infty$, as is the case for the log-normal distribution, constructed by setting $\exp(X)$ where $X \sim N(\mu, \sigma^2)$.

Suppose $X \geq 0$, and let $\mu = E[X]$. Then the MGF of X exists if there is some $t > 0$ for which

$$\sum_{k=0}^{\infty}\frac{\mu_k[X]}{k!}t^k < \infty.$$

By Jensen's inequality we must expect $\mu_k[X] \geq \mu^k$, and in fact $\mu_k[X]^{1/k} \to_k \|X\|_\infty$, so that if X is not bounded above $\mu_k[X]$ must supergeometric in k, since $\|X\|_\infty = \infty$.

For example, the moments of an exponential RV with $\lambda = 1$ are exactly $E[X^k] = k!$, and so it possesses a MGF (set $t \in (0,1)$). On the other hand $Y = X^2$ does not possess a MGF, since $E[Y^k] = E[X^{2k}] = (2k)!$, and the MGF series is not summable for any $t > 0$. A simple sufficient condition for the existence of a MGF can be given by the sequence of moment bounds

$$\mu_k[|X|] \leq k!\theta^k, \quad k = 1, 2, \ldots$$

for any finite $\theta > 0$.

4.2.1 Moments and cumulants

From (4.8) the following important property is apparent:

$$\left.\frac{d^k m_X(t)}{dt^k}\right|_{t=0} = \mu_k[X].$$

The CGF is a direct transformation of the MGF, but this straightforward device can lead to considerable simplification of an analysis. The *cumulants* of X may be defined in terms of the expansion

$$c_X(t) = \sum_{i=1}^{\infty} \kappa_i[X] \frac{t^i}{i!}. \tag{4.9}$$

Similar to the MGF, we have

$$\left.\frac{d^k c_X(t)}{dt^k}\right|_{t=0} = \kappa_k[X].$$

Note that the order t^0 term in (4.9) is necessarily 0, since $m(0) = 1$. The cumulants may be deduced by comparing the order t^k term in (4.9) to that obtained by substituting (4.8) into an expansion of $\log(x)$ about $x = 1$. This straightforward (if cumbersome) process yields

$$\begin{aligned}
\kappa_1[X] &= \mu_1[X] \\
\kappa_2[X] &= \bar{\mu}_2[X] \\
\kappa_3[X] &= \bar{\mu}_3[X] \\
\kappa_4[X] &= \bar{\mu}_4[X] - 3\bar{\mu}_2[X]^2 \\
&\vdots\ .
\end{aligned}$$

A number of properties of the cumulants may be noted. First, for any constant b

$$\begin{aligned}\kappa_k[bX] &= b^k \kappa_k[X], \\ \kappa_1[X+b] &= \kappa_1[X]+b, \\ \kappa_k[X+b] &= \kappa_k[X], \quad k \geq 2,\end{aligned}$$

so that the CGF is location invariant after the first order term, unlike the MGF.

4.2.2 MGF and CGF of independent sums

If $X_1, \ldots, X_n$ are independent random variables then by (4.4) we have

$$m_{X_1+\cdots+X_n}(t) = \prod_{i=1}^{n} m_{X_i}(t), \tag{4.10}$$

and for *iid* RVs this reduces to

$$m_{X_1+\cdots+X_n}(t) = \left[m_{X_1}(t)\right]^n. \tag{4.11}$$

In addition, following (4.10) we have

$$c_{X_1+\cdots+X_n}(t) = \sum_{i=1}^{n} c_{X_i}(t),$$

and for the *iid* case

$$c_{X_1+\cdots+X_n}(t) = n\, c_{X_1}(t).$$

4.2.3 Relationship of the CGF to the normal distribution

One property of the CGF we will find useful is the fact that $c_X(t) = \mu t + \sigma^2 t^2/2$ for $X \sim N(\mu, \sigma^2)$, and the simplicity of this form will permit some greater precision in subsequent analysis. Furthermore, when the CGF of X can be bounded by $\mu + \sigma^2 t^2/2$ in some sense then a useful approximation method becomes available. This need not mean that X is 'approximately normal'. In fact, by Taylor's theorem (Theorem 2.3):

$$c_X(t) = \kappa_1[X]t + \left.\frac{d^2 c_X(t)}{dt^2}\right|_{t=\eta} \frac{t^2}{2},$$

where $0 \leq \eta \leq t$ or $t \leq \eta \leq 0$. Noting that $\lim_{t\to 0} d^2 c_X(t)/dt^2 = var[X]$, we can see that the CGF of *any* RV X resembles that of the normal in a small enough neighborhood of $t = 0$. Viewed in this way, the extension of properties of the normal distribution will depend on how far from $t = 0$ the second order term of $c_X(t)$ remains dominant. This in turn depends on the size of $\kappa_2[X]$ relative to the higher order cumulants $\kappa_k[X]$, $k > 2$.

For example, suppose $\bar{X}_n$ is the average of an *iid* sample from a distribution of mean zero, which also possesses a MGF. We may normalize $Z_n = n^{1/2}\bar{X}_n$, which has cumulants $\kappa_1[Z_n] = 0$ and $\kappa_k[Z_n] = n^{1-k/2}\kappa_k[X_1]$. Thus $\kappa_2[Z_n]$ does not depend on n, while all higher order cumulants approach 0 as $n \to 0$, on which basis we may conclude that Z_n is asymptotically normal.

4.2.4 Probability generating functions

When a RV X is distributed on $\mathbb{N}_0$ it is sometimes convenient to work with a *probability generating function* (PGF):

$$h_q(s) = \sum_{i=0}^{\infty} s^i q_i = E[s^X] \tag{4.12}$$

for distribution $q = (q_0, q_1, \ldots)$, $P(X = i) = q_i$. We recover the MGF through $m(t) = h_q(e^t)$. The PGF may be also be taken as a function of the complex numbers. Otherwise, we may recover the distribution q and the moments by the following evaluations:

$$\left.\frac{d^i h_q(s)}{ds^i}\right|_{s=0} = \frac{q_i}{i!}, \quad \text{and}$$
$$\left.\frac{d^i h_q(s)}{ds^i}\right|_{s=1} = E\left[\frac{X!}{(X-i)!}\right]. \tag{4.13}$$

Note that $h_q(s)$ and its derivatives exist for $|s| < 1$, but at $s = 1$ only the left derivative may exist, which may be used in (4.13).

4.3 CONDITIONAL DISTRIBUTIONS

The conventional definition of a conditional probablity is

$$P(A \mid B) = P(AB)/P(B), \qquad \text{when } P(B) > 0,$$

understood to refer to the probability that A occurs given that (conditional on the event that) B has occured. Extending this idea to a *condition distribution* poses no problems for discrete random variables. For example, if $f_{XY}(x, y)$ is a distribution function on $\mathbb{I}^2$ for the random vector (X, Y) we may write

$$f_{X|Y}(x \mid y) = \frac{P(X = x \text{ and } Y = y)}{P(Y = y)} = \frac{f_{XY}(x, y)}{\sum_x f_{XY}(x, y)} = \frac{f_{XY}(x, y)}{f_Y(y)}.$$

This formula is usually extended to continuous densities even though the conditional event $\{Y = y\}$ has probability 0. In fact $f_{XY}(x, y)$ is not a probability, so the correspondence to the original definition $P(A \mid B) = P(AB)/P(B)$ needs to be made. For $\mathbb{R}^k$, this is not particularly difficult if $f_{X|Y}(x \mid y)$ is understood as a limit of $P(X \in B_\epsilon(x) \mid Y \in B_\epsilon(y))$ as $\epsilon \downarrow 0$. Essentially, we are stating that $f_{X|Y}(x \mid y)dx = f_{XY}(x, y)dxdy/f_Y(y)dy$. However, this informal calculus depends on the geometry of the coordinate system and can lead to contradictions in other coordinates. A well known example is *Borel's Paradox*, which illustrates the difficulty of using this formula to characterize densities on the surface of a sphere (the interested reader can be referred to Section 33 of Billingsley (1995)).

Under such a formula, conditional expectations become integrals calculated in the usual way by using conditional densities, that is,

$$E[X \mid Y = y] = \int x f_{X|Y}(x \mid y) d\mu(x),$$

where μ is the appropriate measure.

If X is defined on $(\Omega, \mathcal{F}, P)$, for $A \in \mathcal{F}$, $P(A) > 0$, the conditional distribution is essentially constructed by replacing Ω with A, by confining the definition of the original density $f_X(x)$ to $x \in A$ then renormalizing, giving conditional density:

$$f_X(x \mid A) = \frac{f_X(x)}{P(A)} I\{x \in A\},$$

and conditional expectation

$$E[x \mid A] = P(A)^{-1} \int_{x \in A} x f_X(x) d\mu = P(A)^{-1} E\left[XI\{X \in A\}\right].$$

This justifies an expression such as $E\left[XI\{X \in A\}\right] = E[x \mid A]P(A)$, and the liberal use of indicator functions in this manner will often be a useful way to apply the concepts of conditioning.

Independence may also be characterized by the statement

$$X \perp Y \iff f_{X|Y}(x \mid y) = f_X(x),$$

as a consequence of (4.3).

There exists a definition of conditional probability which is independent of any specific properties of a measurable space, which is stated in the following theorem.

Theorem 4.2 *Let Y be a random variable on $(\Omega, \mathcal{F}, P)$ for which E[Y] exists, and let $\mathcal{G}$ be a sub σ-field of $\mathcal{F}$. Then there exists a unique ae[P] random variable $E[Y \mid \mathcal{G}]$ which is measurable wrt $(\Omega, \mathcal{G})$ and which satisfies*

$$\int_E Y dP = \int_E E[Y \mid \mathcal{G}] dP \tag{4.14}$$

for all $E \in \mathcal{G}$.

Proof The result is a consequence of the Radon-Nikoym theorem (Theorem 3.24). See, for example, Ash (1972), Chapter 6. ///

Recall that a measurable mapping from $\mathcal{F}_1$ to $\mathcal{F}_2$ remains measurable if $\mathcal{F}_1$ is replaced by a larger, but not smaller, σ-field. Therefore, Y itself will generally not satisfy the definition of $E[Y \mid \mathcal{G}]$. This construction is referred to as *conditioning on a σ-field*, and it is important to stress that the sub σ-field has replaced the conditional event. To see this, suppose we calculate $E[Y \mid A]$. The analogous conditional expectation would be defined by specifying $\mathcal{G} = \{A, A^c, \emptyset, \Omega\}$ and defining $Y_A = E[Y \mid \mathcal{G}]$ as in Theorem 4.2. Since Y_A is measurable *wrt* $\mathcal{G}$, it must be constant on A and also on A^c $ae[P]$. From (4.14) these values must be $Y_A = E[Y \mid A]$ and $Y_A = E[Y \mid A^c]$, that is,

$$Y_A = E[Y \mid A]I_A + E[Y \mid A^c]I_{A^c}$$

satisfies the definition of $E[Y \mid \mathcal{G}]$.

Conditional probabilities are naturally defined as $P(E \mid \mathcal{G}) = E[Y \mid \mathcal{G}]$ with $Y = I_E$, so there is no important difference between conditional expectation and conditional probability. Theorem 4.2 suffices for both. Furthermore, the object $E[X \mid \mathcal{G}]$ will usually correspond to the appropriate explicit construction. For example $E[X \mid Y = y]$ is formally a function of y, say $h(y)$. On the other hand $E[X \mid \sigma(Y)]$ is a RV which is measurable with respect to $\sigma(Y)$, and so is a function of Y. In fact, $E[X \mid \sigma(Y)] = h(Y) = E[X \mid Y]$ satisfies (4.14).

Some important properties of conditional expectations are given in the following theorem.

Theorem 4.3 *Suppose we are given a probability measure* $(\Omega, \mathcal{F}, P)$.

(i) *If* $X = a$ $ae[P]$ *for some constant* a, *then* $E[X \mid \mathcal{G}] = a$ $ae[P]$.
(ii) *Conditional expectations are linear:*

$$E[aX + bY \mid \mathcal{G}] = aE[X \mid \mathcal{G}] + bE[Y \mid \mathcal{G}], \quad ae[P].$$

(iii) *If* $X \geq 0$ $ae[P]$ *then* $E[X \mid \mathcal{G}] \geq 0$, $ae[P]$.
(iv) *If* $X_n \to X$ $ae[P]$, $|X_n| < Y$, *and* $E[Y] < \infty$, *then*

$$E[X_n \mid \mathcal{G}] \to E[X \mid \mathcal{G}], ae[P].$$

(v) *If* X *is measurable wrt* $\mathcal{G}$, *and* $E[Y]$ *and* $E[XY]$ *exist, then*

$$E[XY \mid \mathcal{G}] = XE[Y \mid \mathcal{G}], ae[P].$$

(vi) *If* $E[Y]$ *exists, and* $\mathcal{G}_1 \subset \mathcal{G}_2$, *then*

$$E\left[E[Y \mid \mathcal{G}_2] \mid \mathcal{G}_1\right] = E[Y \mid \mathcal{G}_1], ae[P].$$

Proof See, for example, Billingsley (1995), Section 34. ///

Note that *(i)* and *(v)* of Theorem 4.3 imply that $E[X \mid \mathcal{G}] = X$ $ae[P]$ under the hypothesis of *(v)*. This apparently straightforward fact, along with the other statements of Theorem 4.3, prove to be crucial to many arguments involving conditional expectations.

If $\mathcal{G} = \{\Omega, \emptyset\}$, then $E[Y \mid \mathcal{G}] = E[Y]$, the unconditional expectation. Thus, a consequence of *(vi)* is that $E\left[E[Y \mid \mathcal{G}']\right] = E[Y]$ for any sub σ-field $\mathcal{G}'$.

Higher order conditional moments are usually calculated simply by replacing $P(\cdot)$ with $P(\cdot \mid \mathcal{G})$ so we may write $var(X \mid \mathcal{G}) = E[X^2 \mid \mathcal{G}] - E[X \mid \mathcal{G}]^2$.

The advantage of conditioning on σ-fields is that a consistent construction method is available which does not depend on the details of a particular measure. Separate definitions do not need to be made for continuous and discrete random variables, and the problem of conditioning on events of measure zero does not arise. Certainly, we will make use of explicit constructions as needed. But until they are, the approach of Theorem 4.2 will usually permit arguments that are at the same time more general and more concise.

4.4 MARTINGALES

Suppose we observe some sequence of random occurrences $X_1, X_2, \ldots$. It is sometimes useful to consider a *history process* $H_k = (X_1, \ldots, X_k)$, $k \geq 1$, representing all accumulated information up to and including the kth outcome. Obviously, knowledge of H_k implies knowledge of H_j if $j \leq k$.

It is sometimes worth considering a formal generalization of this process, expressible in terms of σ-fields, which, as discussed in the previous section, have a close connection with the notion of information. Consider the following definition:

Definition 4.2 *Suppose we are given a probability measure $(\Omega, \mathcal{F}, P)$. A sequence $\mathcal{F}_1, \mathcal{F}_2, \ldots$, of sub σ-fields of $\mathcal{F}$ is a* filtration *if $\mathcal{F}_k \subset \mathcal{F}_{k+1}$, for all $k \geq 1$. A sequence of random variables X_k is* adapted *to $\mathcal{F}_k$ if $\sigma(X_k) \subset \mathcal{F}_k$ (that is X_k is measurable wrt $\mathcal{F}_k$). We refer to X_i, $i = 1, 2, \ldots$ as an* adapted process. *It may be assumed that $\mathcal{F}_0 = \{\Omega, \emptyset\}$.*

The last statement of Definition 4.2 permits any reference to $E[X_k \mid \mathcal{F}_{k-1}]$ to include $k = 1$, in which case $E[X_1 \mid \mathcal{F}_0] = E[X_1]$.

Here, the filtration $\mathcal{F}_k$ assumes the role of a history process, except that it need not be constructed explicitly from a specific process. Rather, we instead consider (and compare) the entire class of processes adapted to a common filtration. This often proves to yield some considerable simplification of an analysis. The theory of *martingales* is a particularly elegant example of this approach. The formal definition follows.

Definition 4.3 *Suppose we are given a probability measure $(\Omega, \mathcal{F}, P)$, on which exists filtration $\mathcal{F}_n$, $n \geq 1$. Suppose for a sequence of random variables S_n the following conditions hold:*

(i) $E[|S_n|] < \infty$,
(ii) S_n *is* adapted *to filtration* $\mathcal{F}_n$,
(iii) $E[S_{n+1} \mid \mathcal{F}_n] = S_n$ *ae for all* $n \geq 1$,

then S_n is a martingale *(on filtration $\mathcal{F}_n$). If equality in (iii) is replaced with $\leq$ or $\geq$ then S_n is a* supermartingale *or* submartingale *respectively.*

A martingale is both a supermartingale and a submartingale.

It is always possible to define the filtration $\mathcal{F}_n = \sigma(S_1, \ldots, S_n)$, but a richer theory is available by allowing the more general definition implied by the adapted process. Monotonicity of the sequence S_n implies Definition 4.3:

Theorem 4.4 *If S_n is adapted to filtration $\mathcal{F}_n$, and $S_{n+1} \geq S_n$ ae for all $n \geq 1$, then S_n is a submartingale. If $\geq$ is replaced with $\leq$ then S_n is a supermartingale.*

Proof From Theorem 4.3 we conclude that $0 \leq E[S_{n+1} - S_n \mid \mathcal{F}_n] = E[S_{n+1} \mid \mathcal{F}_n] - E[S_n \mid \mathcal{F}_n] = E[S_{n+1} \mid \mathcal{F}_n] - S_n$ *ae*. A similar argument holds for supermartingales. ///

Given sequence S_n the *differences* are denoted $\epsilon_n = S_n - S_{n-1}$ (set $S_0 = 0$ if needed). They become *martingale differences* when S_n is a martingale, so the terminology is somewhat dependent on context. Whatever the case, these differences assume definite properties under Definition 4.3.

A (sub-, super-) martingale is an L^p process if $E[|S_n|^p] < \infty$ for all $n \geq 1$. By (*i*) of Definition 4.3 all martingales are L^1 processes. An L^2 martingale is often referred to as *square integrable martingale*. By Minkowski's inequality, the L^p criterion can be equivalently applied to the sequence S_n, or to the differences ϵ_n.

In general, the assumption of stronger integrability conditions confers more structure. The assumption of uniform integrability will be considered below in Section 4.10, and some important convergence properties follow from this assumption.

Alternatively, the assumption of square integrability confers on martingales the character of a random noise process, without requiring the sometimes unrealistic assumption of independence. By Hölder's inequality, if each squared difference ϵ_n^2 is integrable then so is each product $\epsilon_m \epsilon_n$. In fact, a square integrable martingale is an *orthogonal process*, that is, the differences ϵ_n are uncorrelated and of zero mean:

Theorem 4.5 *If S_n is a martingale on filtration $\mathcal{F}_n$, and ϵ_n are the martingale differences, then $E[S_n \mid \mathcal{F}_m] = S_m$, and consequently $E[\epsilon_n \mid \mathcal{F}_m] = 0$ ae for all $n > m$. If in addition S_n is a square intergrable martingale then S_n is an orthogonal process, in the sense that $E[\epsilon_n] = 0$ and $E[\epsilon_n \epsilon_m] = 0$ for all $m \neq n$.*

If S_n is a submartingale $E[S_n \mid \mathcal{F}_m] \geq S_m$, and consequently $E[\epsilon_n \mid \mathcal{F}_m] \geq 0$ ae for all $n > m$.

If S_n is a supermartingale $E[S_n \mid \mathcal{F}_m] \leq S_m$, and consequently $E[\epsilon_n \mid \mathcal{F}_m] \leq 0$ ae for all $n > m$.

Proof (Assume equalities hold *ae* where appropriate). Suppose $n > m$. By (*vi*) of Theorem 4.3 we have $E[S_n \mid \mathcal{F}_m] = E[E[S_n \mid \mathcal{F}_{n-1}] \mid \mathcal{F}_m] = E[S_{n-1} \mid \mathcal{F}_m]$. The argument is completed by repeating until $n - 1 = m$, so that $E[S_n \mid \mathcal{F}_m] = E[S_m \mid \mathcal{F}_m] = S_m$. Then for $n > m$ we have $E[\epsilon_n \mid \mathcal{F}_m] = E[S_n - S_{n-1} \mid \mathcal{F}_m] = S_m - S_m = 0$. We may write $E[\epsilon_n] = E[E[\epsilon_n \mid \mathcal{F}_{n-1}]] = E[0] = 0$. Then, for $n > m$ $E[\epsilon_n \epsilon_m \mid \mathcal{F}_m] = \epsilon_m E[\epsilon_n \mid \mathcal{F}_m] = \epsilon_m \cdot 0$, so that $E[\epsilon_n \epsilon_m] = 0$, assuming square intergrability.

The remainder of the proof holds after a similar argument for submartingales and supermartingales. ///

Convex transformations of martingales tend to preserve some martingale properties (see, for example, Section 4.2 of Durrett (2010)):

Theorem 4.6 (*i*) *If S_n is a martingale, h is convex, and $h(S_n)$ are each integrable, then $h(S_n)$ is a submartingale on the same filtration.* (*ii*) *If S_n is a submartingale, h is a convex nondecreasing function, and $h(S_n)$ are each integrable, then $h(S_n)$ is a submartingale on the same filtration.*

Thus, if S_n is a martingale or a positive submartingale then $|S_n|^p$ is a submartingale for any $p \geq 1$, where integrability holds.

4.4.1 Stopping times

One more simple but consequential idea associated with the theory of martingales is the stopping time. Suppose we are given a filtration $\mathcal{F}_n$. A random variable τ is a *stopping time* if $\{\tau = n\} \in \mathcal{F}_n$ for all $n \geq 1$. The decision to stop a process must be made on currently available information, which is what this definition forces. A set E is *prior* to stopping time τ if $E \cap \{\tau = n\} \in \mathcal{F}_n$ for all $n \geq 1$. If S_n is adapted to $\mathcal{F}_n$, then a set depending on $(S_1, \ldots, S_\tau)$, such as $E = \{\max_{i \leq \tau} S_i > 0\}$ is prior to τ. If we let $\mathcal{F}^\tau$ be the class of all sets prior to τ, it may be verified that $\mathcal{F}^\tau$ is a σ-field.

Next, suppose τ_n is a sequence of finite stopping times which is nondecreasing *ae*. We can then define a new process $Y_n = S_{\tau_n}$, where $\tau_n < \infty$. One of the central results in martingale theory is the *Optional Sampling Theorem* due to Doob (See, for example, Section 7.7 of Ash (1972)).

Theorem 4.7 (Optional Sampling Theorem) *Let S_n be a submartingale on filtration $\mathcal{F}_n$, and let τ_n be a nondecreasing sequence of finite stopping times. Then $\mathcal{F}^{\tau_n}$, $n = 1, 2, \ldots$, is a filtration. Set $Y_n = S_{\tau_n}$. If (i) $E[|Y_n|] < \infty$ for all n, and (ii) $\liminf_{i \to \infty} E[|X_i| I\{\tau_n > i\}] = 0$ for all n, then Y_n is a submartingale on filtration $\{\mathcal{F}^{\tau_n}\}$. If S_n is a martingale, so is Y_n.*

4.5 SOME IMPORTANT THEOREMS

A large part of probability theory relies on a relatively small number of classical theorems, some of which we introduce here.

Boole's Inequality is a direct consequence of countable additivity:

Theorem 4.8 (Boole's Inequality) *For any countable collection of sets $\{E_i\}$ on a probability measure $(\Omega, \mathcal{F}, P)$ it always holds that*

$$P(\cup_i E_i) \leq \sum_i P(E_i).$$

Very few theorems have quite the same combination of utility and simplicity as does *Markov's Inequality*.

Theorem 4.9 (Markov's Inequality) *If $X \geq 0$, $t > 0$, then*

$$P(X \geq t) \leq \frac{E[X]}{t}.$$

Proof This follows from the inequality $tI\{x \geq t\} \leq x$ for $x \in [0, \infty)$, and the monotonicity of the expectation operator. ///

Chebyshev's Inequality is a special case of Markov's inequality which appeared earlier.

Theorem 4.10 (Chebyshev's Inequality) *For any random variable with mean μ and variance σ^2 the following inequality holds for any $t > 0$:*

$$P(|X - \mu| \geq t\sigma) \leq \frac{1}{t^2}.$$

Proof Apply Theorem 4.9 to the event $\{(X-\mu)^2/\sigma^2 \geq t^2\}$. ///

If we are given a countable sequence of events $E_1, E_2, \ldots$ on $(\Omega, \mathcal{F}, P)$ the following limits may be defined

$$\limsup_{n\to\infty} E_n = \cap_{n=1}^{\infty} \cup_{m\geq n} E_m = \lim_{n\to\infty} \cup_{m\geq n} E_m = \{E_n \text{ i.o.}\},$$

where i.o. means 'infinitely often', that is, the event $\{E_n \text{ i.o.}\}$ occurs if an infinite number of the events E_n occur. A definition often accompanying this is

$$\liminf_{n\to\infty} E_n = \cup_{n=1}^{\infty} \cap_{m\geq n} E_m = \lim_{n\to\infty} \cap_{m\geq n} E_m = \{E_n \text{ a.f.}\},$$

where a.f. means 'all but finitely often', that is, the event $\{E_n \text{ a.f.}\}$ occurs if for some finite n all E_m, $m \geq n$ occur. By De Morgan's law $\{E_n^c \text{ i.o.}\} = \{E_n \text{ a.f.}\}^c$ and $\{E_n^c \text{ a.f.}\} = \{E_n \text{ i.o.}\}^c$.

The *Borel-Cantelli Lemmas* apply to events of this type:

Theorem 4.11 (Borel-Cantelli Lemma I) *If* $\sum_{n\geq 1} P(E_n) < \infty$ *then* $P(E_n \text{ i.o.}) = 0$.

Proof By Boole's inequality $P\left(\cup_{m\geq n} E_m\right) \leq \sum_{m\geq n} P(E_m)$. By hypothesis this upper bound approaches 0 as $n \to \infty$ so the result holds by the continuity of P. ///

Theorem 4.12 (Borel-Cantelli Lemma II) *If the events* E_n *are independent and* $\sum_{n\geq 1} P(E_n) = \infty$ *then* $P(E_n \text{ i.o.}) = 1$.

Proof By independence we may write, for any $N \geq n$, $P\left(\cap_{m\geq n} E_m^c\right) \leq P\left(\cap_{m=n}^{N} E_m^c\right) = \prod_{m=n}^{N} (1 - P(E_m)) \leq \exp\left(-\sum_{m=n}^{N} P(E_m)\right)$. By hypothesis, the upper bound approaches 0 as $N \to \infty$. Thus, $P\left(E_n^c \text{ a.f.}\right) = 0$, which concludes the proof. ///

Example 4.1 *On day* n, $n = 1, 2, \ldots$, *the numbers* $\{1, \ldots, n\}$ *are randomly ordered, and each ordering is independent of the previous orderings. The probability that* 1 *occurs first i.o. is 1, but the probability that* 1 *and* 2 *occur first and second i.o. is zero. The number of days on which any four consecutive integers appear consecutively anywhere in the random ordering is finite* $wp1$.

Jensens's Inequality is commonly used, which states:

Theorem 4.13 (Jensen's Inequality) *If* X *is a random variable with support* $\mathcal{T}$, *and* $\phi(t)$ *is a convex function on* $\mathcal{T}$ *then* $E[\phi(X)] \geq \phi(E[X])$. *If* ϕ *is strictly convex, then* $\phi(E[X]) = E[\phi(X)]$ *implies* $X = E[x]$ $ae[P]$, *when the expectations exist.*

The hypothesis that ϕ is a convex function includes the assumption that $\mathcal{T}$ is a convex set.

Example 4.2 *Since* $\phi(x) = x^q$, $q \geq 1$, *is a convex function on* $\mathbb{R}_+$, *Jensen's inequality implies that the* L^p *norm for random variables is increasing in* p, *that is, for* $r < s$, *set* $\phi(x) = x^{s/r}$. *Then* $E[|X|^s] = E[\phi(|X|^r)] \geq \phi(E[|X|^r]) = E[|X|^r]^{s/r}$, *which is equivalent to* $E[|X|^s]^{1/s} \geq E[|X|^r]^{1/r}$.

4.6 INEQUALITIES FOR TAIL PROBABILITIES

One of the most important technical problems arising in probability theory is the bounding of tail probabilities of random variables, that is, the construction of statements such as

$$P(X \geq t) \leq g(t), \quad t \geq t_0$$

for some function $g(t)$. We have already seen two examples in the Markov and Chebyshev inequalities. The work of bounding tail probabilities is usually meant to result in functions $g(t)$ which converge to 0 as quickly as possible as $t \to \infty$. A direct application of Markov's inequality to a positive RV X yields $g(t) \propto 1/t$, which is a quite slow rate of convergence in comparison to most commonly used distributions. However, there is some freedom in considering transformations of X. Suppose $\mathcal{X}$ is the support of X, and $h : \mathcal{X} \to \mathbb{R}_+$ is a strictly increasing function. Markov's inequality gives

$$P(X \geq t) = P(h(X) \geq h(t)) \leq \frac{E[h(X)]}{h(t)}, \quad t \in \mathcal{X}. \tag{4.15}$$

Note that we need not assume $X \geq 0$, since it suffices that $h \geq 0$.

To take a common application of this method, suppose $X \geq 0$ has all kth order moments. Using $h(x) = x^k$, we have $P(X \geq t) \leq t^{-k}\mu_k[X]$. Thus, the rate of convergence of the bound to 0 as $t \to \infty$ can always be improved by increasing k, although the fact that $\mu_k[X]$ may increase quickly in k will often hamper the utility of this approach. We have already seen in Section 4.2 that the rate of increase of $\mu_k[X]$ is of great importance, and for unbounded RVs that rate must be supergeometric.

Example 4.3 *Returning to the example of the exponential RV with $\lambda = 1$, we have moments $\mu_k[X] = k!$, and so using Markov's inequality with power transformation $h(x) = x^k$ gives inequality*

$$P(X \geq t) \leq \frac{k!}{t^k}, \quad t > 0, k = 1, 2, \ldots. \tag{4.16}$$

The bound holds for all k, and it can be seen that the value of k yielding the sharpest bound varies by t. Therefore, rather than accepting (4.16) *for some fixed k, we can do better by using*

$$P(X \geq t) \leq \inf_k \frac{k!}{t^k}, \quad t > 0. \tag{4.17}$$

In fact, the strategy of Example 4.3 is commonly used to obtain bounds similar to (4.17), the *Chernoff bound*, discussed in the next section, being a well known example.

4.6.1 Chernoff bounds

Here we consider the class of RVs possessing an MGF. Using (4.15), for any $t > 0$:

$$\begin{aligned} P(X \geq x) &= P\left(\exp(tX) \geq \exp(tx)\right) \\ &\leq \exp(-tx) m_X(t) = \exp(c_X(t) - tx)\,, \end{aligned}$$

and similarly

$$\begin{aligned} P(X \leq x) &= P\left(\exp(-tX) \geq \exp(-tx)\right) \\ &\leq \exp(tx) m_X(-t) = \exp(c_X(-t) + tx)\,. \end{aligned}$$

This holds for all $t > 0$, so we may substitute that value yielding the smallest uppper bound:

$$P(X \geq x) \leq \inf_{t>0} \exp(c_X(t) - tx) = \exp\left\{\inf_{t>0}(c_X(t) - tx)\right\}, \tag{4.18}$$

and similarly

$$P(X \leq x) \leq \inf_{t>0} \exp(c_X(-t) + tx) = \exp\left\{\inf_{t>0}(c_X(-t) + tx)\right\}. \tag{4.19}$$

This often turns into a simple exercise in calculus. We note that the CGF is defined on a set $\mathcal{T}$ containing an open neighborhood of $t = 0$, on which it is infinitely divisible and convex. This must also hold for $h(t) = c_X(t) - tx$ and $h(-t)$. From the point of view of optimization, we may take $c_X(t) = \infty$ for $t \notin \mathcal{T}$. Thus, if a stationary point of $h(t)$ exists in $(0, \infty) \cap \mathcal{T}$ then it must be the value minimizing (4.18). Similarly, a stationary point of $h(-t)$ in $(0, \infty) \cap -\mathcal{T}$ will minimize (4.19).

On the other hand, a RV X defined by $P(X = 1) = P(X = -1) = 1/2$ gives an example for which this procedure cannot be used, although of course, (4.18) and (4.19) still hold.

4.6.2 Chernoff bound for the normal distribution

Suppose $X \sim N(\mu, \sigma^2)$, and assume $x > \mu$, then

$$\inf_{t>0} c_X(t) - tx = \inf_{t>0} \mu t + \sigma^2 \frac{t^2}{2} - tx = -\frac{(x-\mu)^2}{2\sigma^2},$$

since $h(t)$ possesses stationary point $t_0 = (x - \mu)/\sigma^2 > 0$. The same holds for (4.19), so we have

$$\begin{aligned} &P(X \geq x) \leq \exp\left(-2^{-1}(x-\mu)^2/\sigma^2\right), \quad x > \mu, \quad \text{and} \\ &P(X \leq x) \leq \exp\left(-2^{-1}(x-\mu)^2/\sigma^2\right), \quad x < \mu. \end{aligned} \tag{4.20}$$

It turns out that we can obtain somewhat sharper bounds by exploiting some elementary concepts of stochastic ordering, to be introduced in Section 4.7, but this method will prove very useful when approximations are used.

4.6.3 Chernoff bound for the gamma distribution

Consider $X \sim gamma(\alpha, \lambda)$. The CGF for the gamma distribution is $c_X(t) = -\alpha \log(1 - t/\lambda)$, $t < \lambda$, so the minimum is taken over $t \in (0, \lambda)$. Then $h(t) = -\alpha \log(1 - t/\lambda) - tx$, which possesses stationary point $t_0 = \lambda - \alpha/x$. Then $t_0 \in (0, \lambda)$ if and only if $x > \mu = \alpha/\lambda$, giving bound:

$$P(X \geq x) \leq \exp\left(-\lambda x + \alpha + \alpha \log(x/\mu)\right), \quad x > \mu. \tag{4.21}$$

As in the case of the normal distribution, stochastic ordering may be used to refine the bound, as will be discussed below.

4.6.4 Sample means

Suppose we are given independent zero mean RVs $X_1, \ldots, X_n$, with sum $S_n = \sum_{i=1}^n X_i$. Furthermore, suppose there exist constants σ_i^2, and $t^* > 0$ for which

$$c_{X_i}(t) \leq \frac{\sigma_i^2 t^2}{2}, \quad i = 1, \ldots n, \quad t \in (0, t^*). \tag{4.22}$$

Then, letting $\bar{\sigma^2} = n^{-1} \sum_i \sigma_i^2$, we may write

$$P(S_n \geq n^{1/2} x) \leq \inf_{t \in (0, t^*)} \exp\left(n \bar{\sigma^2} \frac{t^2}{2} - t n^{1/2} x\right).$$

Note that the bound (4.18) holds if the infimum operation is taken over a strict subset of $(0, \infty)$. As shown above, the stationary point of the preceding upper bound is $t' = x/(n^{1/2} \bar{\sigma^2})$, and so if $x > 0$ and $t' < t^*$ we may conclude

$$P(n^{-1/2} S_n \geq x) \leq \exp\left(-x^2/(2\bar{\sigma^2})\right). \tag{4.23}$$

The requirement that $t' < t^*$ is quite reasonable. Suppose we are given a sequence of RVs for increasing n. We need not insist that the marginal distributions are equal, but we instead impose a more informal type of uniformity. Suppose (4.22) holds for all X_i for a common t^*, and that $\bar{\sigma^2} \geq \sigma_{min}^2$ as $n \to \infty$. The stationary point t' satisfies

$$t' = x/(n^{1/2} \bar{\sigma^2}) \leq x/(n^{1/2} \sigma_{min}^2),$$

so that for all large enough n we will have $t' < t^*$, so that (4.23) eventually holds.

This can be made more precise. Define for any CGF $c(t)$:

$$\kappa_2^*(t) = \begin{cases} \sup_{0 \le t' \le t} \left.\frac{d^2 c(s)}{ds^2}\right|_{s=t'} & ; \quad t \ge 0, |c(t)| < \infty \\ \infty & ; \quad t \ge 0, |c(t)| = \infty \end{cases}.$$

Note that since $d^2 c_X(s)/ds^2\big|_{s=0} = var[X]$ we must have $\kappa_2^*(t) \ge var[X] \ge 0$.

Theorem 4.14 *Suppose for a RV X we have* $E[X] = 0$, *and* $c_X(t)$ *exists. Then for* $x > 0$ *we have*

$$P(X \ge x) \le \inf\{\exp(-x^2/(2\kappa_2^*(t^*))) \mid t^* > 0, \quad t^*\kappa_2^*(t^*) \ge x\}. \tag{4.24}$$

Proof We proceed by verifying

$$P(X \ge x) \le \exp(-x^2/(2\kappa_2^*(t^*))) \tag{4.25}$$

for any $t^* > 0$ satisfying $t^*\kappa_2^*(t^*) \ge x$, then (4.24) follows. First, if $\kappa_2^*(t^*) = \infty$ then (4.25) holds trivially. Otherwise, by Taylor's theorem we must have

$$c_X(t) \le \frac{\kappa_2^*(t^*)t^2}{2}, \quad t \in [0, t^*].$$

A stationary point of the resulting Chernoff bound exists in $(0, t^*]$ if $x \le t^*\kappa_2^*(t^*)$, under which (4.25) holds, which completes the proof. ///

4.6.5 Some inequalities for bounded random variables

Suppose X is a bounded RV, that is, there exists finite a, b for which $P(a \le X \le b) = 1$. Since $E[|X|^k] \le \max(|a|, |b|)^k$ the MGF always exists. This permits some specialized results for bounded RVs, although if the bounds a, b are truly outliers, there may be no special advantage over more generally applicable methods.

Hoeffding's Lemma provides a quadratic bound for the MGF of the bounded RV.

Lemma 4.1 (Hoeffding's Lemma) *Suppose for a RV X,* $P(a \le X \le b) = 1$, *for finite constants* a, b *and* $E[X] = 0$. *Then*

$$m_X(t) \le \exp(t^2(b-a)^2/8). \tag{4.26}$$

Proof The lemma may be proven by first noting that from the convexity of $\exp(tx)$ we have $\exp(t[pa + (1-p)b]) \le p\exp(ta) + (1-p)\exp(tb)$ for any $p \in [0,1]$. Since $X \in [a,b]$ $wp1$, and $E[X] = 0 \in [a,b]$ we may then write $E[\exp(tX)] \le p\exp(ta) + (1-p)\exp(tb) = \exp(g(t))$ with $p = b/(b-a)$. The proof follows by verifying that $g'(0) = 0$, $g''(t) \le 1/4$ for all $t \in \mathbb{R}$, then applying Taylor's theorem. ///

Thus, the CGF of a bounded RV is always dominated by the CGF of some normal RV.

The following bound for independent bounded independent random variables is given in Hoeffding (1963):

Theorem 4.15 *If $X_1, \ldots, X_n$ are independent random variables with finite bounds $a_i \leq X_i \leq b_i$ then*

$$P\left(\sum_{i=1}^{n}(X_i - E[X_i]) \geq t\right) \leq \exp\left(-2t^2/\sum_{i=1}^{n}(b_i - a_i)^2\right), \text{ and}$$
$$P\left(\left|\sum_{i=1}^{n}(X_i - E[X_i])\right| \geq t\right) \leq 2\exp\left(-2t^2/\sum_{i=1}^{n}(b_i - a_i)^2\right)$$

for all $t > 0$.

In particular, if $X \in bin(n, p)$ then

$$P\left((X/n) - p \geq \epsilon\right) \leq \exp(-2n\epsilon^2), \text{ and}$$
$$P\left(|(X/n) - p \geq \epsilon| \geq \exp\right) \leq 2\exp(-2n\epsilon^2)$$

The following extension of Hoeffding's inequality due to McDiarmid (1989) will prove useful:

Theorem 4.16 (McDiarmid's Inequality) *If $\tilde{X} = (X_1, \ldots, X_n)$ are independent random variables with $X_i \in \mathcal{X}$, and $f : \mathcal{X}^n \to \Re$ is a function satisfying, for some fixed constants c_i,*

$$\sup_{x_j = x'_j, j \neq i} |f(\tilde{x}) - f(\tilde{x}')| \leq c_i, \quad i = 1, \ldots, n,$$

(that is, vectors $\tilde{x}, \tilde{x}' \in \mathcal{X}^n$ may differ only for the ith coordinate). Then

$$P\left(f(\tilde{X}) - E[f(\tilde{X})] \geq t\right) \leq \exp\left(-2t^2/\sum_{i=1}^{n}c_i^2\right) \tag{4.27}$$

for all $t > 0$.

The following elegant bound on the variance of the random variable studied in (4.27) is due to Devroye (1991).

Theorem 4.17 *Under the hypothesis of Theorem 4.16*

$$var\left[f(\tilde{X})\right] \leq 4^{-1}\sum_{i=1}^{n}c_i^2. \tag{4.28}$$

4.7 STOCHASTIC ORDERING

Let X, Y be two RVs. There will be some interest in characterizing the tendency of one random variable to be larger or of greater variability than the other. A *stochastic ordering* may be regarded as a formal comparison of two probability distributions, as opposed to a concern with an event such as $\{X \geq Y\}$. Informally, we have already made use of such an ordering, since if we may claim $m_X(t) \leq m_Y(t)$ for all $t > 0$, then a Chernoff bound for Y also holds for X.

The conventions used in Ross (1996) will be used here. We say X is *stochastically larger* than Y, written $X \geq_{st} Y$, if $\bar{F}_X(t) \leq \bar{F}_Y(t)$ for all $t \in \mathbb{R}$. It may be shown (Proposition 9.1.2, Ross (1996)) that $X \geq_{st} Y$ if and only if $E[h(X)] \geq E[h(Y)]$ for all increasing functions h, and this characterization forms a convenient point of comparison for other forms of stochastic order. Here, and in what follows, we say that the ordering is *strict* if at least one of the defining inequalities is strict.

We say X is *stochastically more variable* than Y, written $X \geq_v Y$, if $E[h(X)] \geq E[h(Y)]$ for all increasing convex functions h. If X, Y are nonnegative with $E[X] = E[Y]$ then $X \geq_v Y$ if and only if $E[h(X)] \geq E[h(Y)]$ for all convex functions h (Corollary 9.5.2, Ross (1996)).

A number of orderings apply specifically to nonnegative X, Y with densities f_X, f_Y. In this case the *hazard rate* (or *failure rate*) of X is $\lambda_X(t) = f_X(t)/\bar{F}_X(t)$, which is interpretable as the rate at time t which a random lifetime X ends (or a component fails) given survival up to t. We may define *hazard rate ordering*, denoted $X \geq_{HR} Y$ if $\lambda_X(t) \leq \lambda_Y(t)$ for all $t \geq 0$. We may also define *likelihood ratio ordering*, denoted $X \geq_{LR} Y$, as the condition $f_X(t)/f_Y(t)$ is nondecreasing in t over $[0, \infty)$.

We will make use in subsequent chapters of *MGF ordering*, that is, $X \geq_{MGF} Y$ if $m_X(t) \geq m_Y(t)$ for all $t > 0$ (it may also be specified that $m_X(t)$ is finite in a neighborhood of zero). Suppose $E[X] = E[Y] = \mu$, and $X \geq_{MGF} Y$. Then $m_X(t) = 1 + \mu t + var[X] t^2/2 + o(t^2)$ and $m_Y(t) = 1 + \mu t + var[Y] t^2/2 + o(t^2)$, where the $o(t^2)$ hold uniformly in a neighborhood of $t = 0$. By allowing $t \to 0^+$ we can see that $var[X] \geq var[Y]$, which we can use to define *variance ordering* $X \geq_{var} Y$ between two random variables of equal mean.

We have the following implications (since $\exp(tx)$ is an increasing convex function on $x \in \mathbb{R}$ when $t > 0$), where applicable:

$$\begin{aligned} X \geq_{LR} Y &\Rightarrow X \geq_{HR} Y \Rightarrow X \geq_{st} Y \Rightarrow X \geq_v Y \\ &\Rightarrow X \geq_{MGF} Y \Rightarrow X \geq_{var} Y \end{aligned} \tag{4.29}$$

Finally, we may define *moment ordering*, that is, $X \geq_{moment} Y$ if $E[X^k] \geq E[Y^k]$ for all $k \in \mathbb{N}$. Clearly, $X \geq_{moment} Y \Rightarrow X \geq_{MGF} Y$. In addition, for nonnegative X, Y, $X \geq_v Y \Rightarrow X \geq_{moment} Y$. Since we will make considerable use of the ordering $X \geq_{MGF} Y$, the implications (4.29) may be quite useful in establishing regularity conditions, since the stronger forms of ordering may be simpler to verify. The reader interested in further detail on MGF ordering may be referred to, for example, Li (2004) or Wang and Ma (2009).

4.7.1 MGF ordering of the gamma and exponential distribution

The MGF of a $gamma(\alpha, \beta)$ distribution is $m(t) = (1 - t/\beta)^{-\alpha}$ for $t < \beta$. There will be some interesting in establishing for a RV Y the ordering $X \geq_{MGF} Y$ where $X \sim gamma(\alpha, \beta)$, in which case we say Y is *gamma dominated*. If $\alpha = 1$ we say Y is *exponential dominated*. If in addition $\beta^{-1} = E[Y]$ we say that the exponential domination is *tight*. This terminology is motivated by the fact that this tightness may be useful and achievable in many important cases, which can be related to well known concepts in stochastic lifetime modeling.

We say X is IFR (DFR) if it has an increasing (decreasing) failure rate. If X is the time until repair of a machine, we expect X to be IFR, since the probability of requiring repair after usage time t should increase with t. On the other hand, the hazard rate of infant survival of many species tends to be decreasing near birth, then increasing in maturity, so both IFR and DFR properties appear naturally. The exponential density (uniquely) has a constant hazard rate, so if, for example Y is IFR, and $\lambda_Y(0) > 0$, then there exists an exponentially distributed random variable X for which $X \geq_{HR} Y$. We may have Y both IFR and DFR, in which case λ_Y is constant, and Y is exponentially distributed. If only one of the properties holds, it holds strictly, which is usually what is intended.

The IFR/DFR properties may be weakened somewhat. A positive random variable X is *new better than used in expectation* (NBUE) if $E[X - t \mid X \geq t] \leq E[X]$ for all $t \geq 0$, and is *new worse than used in expectation* (NBWE) is the inequality is reversed. If Y is NBUE, it may be shown that $X \geq_v Y$ where $X \sim exp(E[Y]^{-1})$, so that Y is exponential dominated (Proposition 9.6.1, Ross (1996)). It may also be shown that if Y is IFR it is also NBUE, so we may find $X_i \sim exp(\mu_i^{-1})$, $i = 1, 2$ for which $X_1 \geq_{HR} Y$ and $X_2 \geq_v Y$. The former is the stronger ordering (as stated in (4.29)), however, it is important to note that we can set $\mu_2 = E[Y]$, but would have to set $\mu_1 > E[Y]$ if the IFR property is strict. Thus, even if Y is IFR it is still preferrable to rely on the weaker NBUE property (which follows from IFR) from the point of view of verifying exponential dominance.

4.7.2 Improved bounds based on hazard functions

The monotoncity properties of the hazard rate $h_X(x) = f_X(x)/\bar{F}_X(x)$ are well known for certain distributions, and we can exploit the fact that for IFR distributions

$$\frac{f_X(x_0)}{\bar{F}_X(x_0)} \leq \frac{f_X(x)}{\bar{F}_X(x)}, \quad \text{when} \quad x_0 \leq x,$$

and therefore

$$\bar{F}_X(x) \leq \frac{\bar{F}_X(x_0)}{f_X(x_0)} f_X(x), \quad \text{for } x \geq x_0. \tag{4.30}$$

If X is a DFR distribution, suppose we have positive limit $\lim_{x \to \infty} h_X(x) = h_{min}$, or h_X otherwise has a positive lower bound h_{min}, then

$$\bar{F}_X(x) \leq h_{min}^{-1} f_X(x), \quad \text{for all } x. \tag{4.31}$$

For the normal case, the Chernoff bound is conservative. For $X \sim N(\mu, \sigma^2)$, it may be shown that $h_X(x)$ is increasing for $x \geq \mu$, so setting $x_0 = \mu$ in (4.30) yields

$$\begin{aligned}
\bar{F}_X(x) &\leq \frac{\bar{F}_X(\mu)}{f_X(\mu)} f_X(x) \\
&= \frac{1/2}{(2\pi\sigma^2)^{-1/2}} (2\pi\sigma^2)^{-1/2} \exp\left(-(x-\mu)^2/2\sigma^2\right) \\
&= (1/2)\exp\left(-(x-\mu)^2/2\sigma^2\right), \quad x \geq \mu,
\end{aligned} \tag{4.32}$$

which, after normalization, improves (4.20) by a factor of 1/2.

In addition, for $Z \sim N(0,1)$ it is well known that for $\bar{F}_Z(z)/f_Z(z) \approx 1/z$ for large z. The Chernoff bound is proportional to $f_Z(z)$ and so will be considerably larger than $\bar{F}_Z(z) \approx z^{-1} f_Z(z)$ as $z \to \infty$. However, when given statistical estimators which are asymptotically normal, such as sample means, we can generally not assume that tail probabilities decrease at the same rate as the normal distribution, but we will be able to construct bounds of the form (4.20).

The gamma distribution $X \sim gamma(\alpha, \lambda)$ is IFR for $\alpha \geq 1$ and is DFR for $\alpha \leq 1$. For $\alpha \geq 1$ set $x_0 = \alpha/\lambda = \mu$ in (4.30), which gives

$$\begin{aligned}\bar{F}_X(x) &\leq \bar{F}_X(\mu) \exp\big(-\lambda(x-\mu) + (\alpha - 1)\log(x/\mu)\big) \\ &= \bar{F}_X(\mu) \exp\big(-\lambda x + \alpha + (\alpha - 1)\log(x/\mu)\big), \quad x \geq \mu, \end{aligned} \tag{4.33}$$

which can be seen to be a strict improvement over (4.21), since $\bar{F}_X(\mu) < 1$, and $(\alpha - 1)\log(x/\mu) < \alpha \log(x/\mu)$ when $\alpha > 1$ and $x/\mu > 1$. When $\alpha \leq 1$ it may be shown that $h_{min} = \lambda$, so that for all $x > 0$ (4.31) implies

$$\bar{F}_X(x) \leq \Gamma(\alpha)^{-1} \exp\big(-\lambda x + (\alpha - 1)\log(\lambda x)\big), x \geq 0, \tag{4.34}$$

which improves (4.21) for (at least) $x > \lambda^{-1}$.

4.8 THEORY OF STOCHASTIC LIMITS

There are many forms of limiting processes associated with stochastic models, but these are generally of two types, those concerning sequences of probability measures, and those concerning sequences of random variables $X_1, X_2, \ldots$ considered jointly. Most of these definitions have a hierarchical relationship, but there are exceptions.

All forms of convergence described here will play a role in models developed in this book.

4.8.1 Covergence of random variables

Notions of the convergence of random variables can be defined in terms of the convergence of measurable functions (Section 3.2.11). Suppose $X_1, X_2, \ldots$ is a sequence of random variables defined on a probability measure space $(\Omega, \mathcal{F}, P)$. The following definitions of convergence are standard:

Definition 4.4 *Let X and $X_1, X_2, \ldots$ be random variables on a probability measure space $(\Omega, \mathcal{F}, P)$.*

(i) *If $P(|X_n - X| > \epsilon) \to 0$ for all $\epsilon > 0$, then X_n* converges in probability *to X. This is denoted*

$$X_n \xrightarrow{i.p.} X.$$

(ii) *If $E[|X_n - X|^p] \to 0$ then X_n* converges in L^p *to X, denoted*

$$X_n \xrightarrow{L^p} X.$$

(iii) *If* $P\left(\lim_{n\to\infty} X_n = X\right) = 1$ *then* X_n *converges to* X with probability one *(wp1), or* almost surely *(a.s.), denoted*

$$X_n \stackrel{wp1}{\to} X.$$

The implications

$$X_n \stackrel{wp1}{\to} X \Rightarrow X_n \stackrel{i.p.}{\to} X,$$

$$X_n \stackrel{L^p}{\to} X \Rightarrow X_n \stackrel{i.p.}{\to} X, \quad p > 0,$$

$$X_n \stackrel{L^r}{\to} X \Rightarrow X_n \stackrel{L^s}{\to} X, \quad 0 < s < r, \tag{4.35}$$

are verified in most probability textbooks. Counterexamples involving L^p or $wp1$ convergence are readily constructed using random variables of the form $P(X_n = x_n) = p_n = 1 - P(X_n = 0)$, with limit $X = 0$ $wp1$, using suitably chosen sequences of constants x_n, p_n.

4.8.2 Convergence of measures

We next consider convergence defined exclusively in terms of measures. These measures may exist on a common measure space, but this need not be the case. If they are, joint distributional properties play no role.

The notion of *weak convergence* relies on topologicial properties, in particular, those associated with Borel spaces. We will make use of the following defintion:

Definition 4.5 *Given a topological space* Ω*, the* boundary ∂A *of* $A \subset \Omega$ *is the set of all limits of sequences in* A *which are also limits of sequences in* A^c.

Theorem 4.18 *Let* Ω *be a metric space, and let* μ $\mu_1, \mu_2, \ldots$ *be finite measures on the Borel sets of* Ω*. The following conditions are equivalent:*

(i) $\lim_{n\to\infty} \int_\Omega f d\mu_n = \int_\Omega f d\mu$ *for all bounded continuous* $f : \Omega \to \mathbb{R}$.
(ii) $\liminf_{n\to\infty} \int_\Omega f d\mu_n \geq \int_\Omega f d\mu$ *for all bounded lower semicontinuous* $f : \Omega \to \mathbb{R}$.
(iii) $\limsup_{n\to\infty} \int_\Omega f d\mu_n \leq \int_\Omega f d\mu$ *for all bounded upper semicontinuous* $f : \Omega \to \mathbb{R}$.
(iv) $\liminf_{n\to\infty} \mu_n(A) \geq \mu(A)$ *for every open set* $A \subset \Omega$*, and* $\lim_{n\to\infty} \mu_n(\Omega) = \mu(\Omega)$.
(v) $\limsup_{n\to\infty} \mu_n(A) \leq \mu(A)$ *for every closed set* $A \subset \Omega$*, and* $\lim_{n\to\infty} \mu_n(\Omega) = \mu(\Omega)$.
(vi) $\lim_{n\to\infty} \mu_n(A) = \mu(A)$ *for any Borel set* $A \subset \Omega$ *for which* $\mu(\partial A) = 0$.

Any sequence of measures μ_n satisfying any of the conditions of Theorem 4.18 is said to *converge weakly* to μ. The notation $\mu_n \Rightarrow \mu$ is conventionally used for this assertion. We also write $X_n \Rightarrow X$ if $P_{X_n} \Rightarrow P_X$. Note that this can be expressed in terms of the CDFs, that is $X_n \Rightarrow X$ if and only if $F_{X_n}(x) \to_n F_X(x)$ for all continuity points of F_X. Note also that Theorem (4.18) holds for improper probability measures $P(\Omega) < 1$.

It may be shown that

$$X_n \stackrel{i.p.}{\to} X \quad \text{implies} \quad X_n \Rightarrow X,$$

and so from (4.35) we know that L^p and $wp1$ convergence also imply weak convergence.

A stronger definition of measure convergence is given by the following:

Definition 4.6 *Given a measurable space* $(\Omega, \mathcal{F})$, *a sequence of measures* μ_n *converges* setwise *to* μ *if and only if* $\lim_{n\to\infty} \mu_n(E) = \mu(E)$ *for all* $E \in \mathcal{F}$.

Essentially, setwise convergence is obtained from weak convergence by strengthening condition (*i*) of Theorem 4.18, in that the class of continuous functions is replaced by the larger class of bounded measurable functions.

Theorem 4.19 *Given a measurable space* $(\Omega, \mathcal{F})$ *let* μ_n *be a sequence of measures which converge setwise to* μ. *Let* f_n, g_n *be two sequences of measurable functions with pointwise limits* f, g. *If* $|f_n| \leq g_n$ *and* $\lim_{n\to\infty} \int_\Omega g_n d\mu_n = \int_\Omega g d\mu < \infty$ *then* $\lim_{n\to\infty} \int_\Omega f_n d\mu_n = \int_\Omega f d\mu < \infty$.

Proof See Proposition 18, in Section 11.4 of Royden (1968). The characterization of setwise convergence in given by the following theorem.

Theorem 4.20 *Suppose we are given a measurable space* $(\Omega, \mathcal{F})$. *A sequence of measures* μ_n *converges setwise to a finite measure* μ *if and only if*

$$\lim_{n\to\infty} \int_\Omega f d\mu_n = \int_\Omega f d\mu \tag{4.36}$$

for all bounded measurable functions.

Proof If (4.36) holds for all bounded measurable functions, then setwise convergence follows by setting $f = I_E$ for all $E \in \mathcal{F}$. To prove the converse, suppose f is a measurable function with bound $|f| \leq M < \infty$. Then Theorem 4.19 implies (4.36) by setting $g_n = g \equiv M$ and $f_n = f$ for all n. ///

4.8.3 Total variation norm

An alternative form of convergence is based on the total variation norm. For any finite signed measure on measurable space $(\Omega, \mathcal{F})$ this is defined as

$$\|\mu\|_{TV} = |\mu|(\Omega) = \mu^+(\Omega) + \mu^-(\Omega), \tag{4.37}$$

where $|\mu|$ is the total variation measure and μ^+, μ^- are the positive measures defining the Jordan decomposition (Section 3.2.10). An equivalent definition is given by

$$\|\mu\|_{TV} = \sup_{E\in\mathcal{F}} \mu(E) - \inf_{E\in\mathcal{F}} \mu(E),$$

where equality is obtainable from the Hahn decomposition.

If μ_1, μ_2 are two proper probability measures it can be shown that

$$\|\mu_1 - \mu_2\|_{TV} = 2 \sup_{E\in\mathcal{F}} |\mu_1(E) - \mu_2(E)|,$$

again, a consequence of the Hahn decomposition.

The total variation norm is related to the L^1 norm (Section 3.3.2), which provides an evaluation method. Suppose for measure μ we have $\mu \ll \nu$, where ν is a positive measure on Ω, and f is a density of μ *wrt* ν. Then (4.37) becomes

$$\|\mu\|_{TV} = \int_\Omega |f| d\nu. \tag{4.38}$$

Note that the definition (4.37) is independent of the measure of integration used in (4.38), and does not rely on the notion of a density function at all, so that if $\|\mu\|_{TV}$ is calculated using density functions, the dependence on ν should be noted (which is usually Lebesgue measure or counting measure). It is also possible to define a weighted total variation norm, where w is a positive measurable function

$$\|\mu\|_{TV(w)} = \int_\Omega w d|\mu| = \int_\Omega w|f| d\nu,$$

and we note that $\|\mu\|_{TV(w)} = \|\mu\|_{TV}$ for weight function $w \equiv 1$. The distance between two measures may be given as

$$\|\mu_1 - \mu_2\|_{TV(w)} = \int_\Omega w|f_1 - f_2| d\nu, \tag{4.39}$$

where f_i are the respective densities of μ_i *wrt* ν.

The following theorem will be important to note:

Theorem 4.21 *For any signed measures μ and positive weight function w*

$$\|\mu\|_{TV(w)} = \sup_{|v| \leq w} |\int_\Omega v d\mu|. \tag{4.40}$$

Proof Using the Jordan decomposition $\mu = \mu^+ - \mu^-$ we may write

$$\int_\Omega v d\mu = \int_\Omega v d\mu^+ - \int_\Omega v d\mu^- \leq \int_\Omega |v| d\mu^+ + \int_\Omega |v| d\mu^- = \int_\Omega |v| d|\mu|.$$

If we take the supremum, noting that $|\mu|$ is a positive measure we have

$$\sup_{|v| \leq w} \int_\Omega v d\mu \leq \sup_{|v| \leq w} \int_\Omega |v| d|\mu| = \int_\Omega w d|\mu| = \|\mu\|_{TV(w)} \, .$$

Applying essentially the same argument to bound $\int_\Omega v d\mu$ from below over $|v| \leq w$ gives

$$\sup_{|v| \leq w} \left| \int_\Omega v d\mu \right| \leq \|\mu\|_{TV(w)} \, . \tag{4.41}$$

Finally, by the Hahn decomposition there exists $E \in \mathcal{F}$ which is positive for μ and for which E^c is negative. Equality in (4.41) can be attained by setting $v' = wI_E - wI_{E^c}$, which completes the proof. ///

Convergence in the total variation norm of a sequence of measures μ_n to μ is defined by the limit $\lim_{n\to\infty} \|\mu_n - \mu\|_{TV(w)} = 0$. The following implications hold:

convergence in $\|\cdot\|_{TV} \Rightarrow$ setwise convergence $\Rightarrow$ weak convergence.

It is important to note that convergence in $\|\cdot\|_{TV(w)}$ must imply that the measures are nonsingular, in which case convergence of measures is equivalent to L^1_w convergence of densities. Suppose μ_1, μ_2 are two measures such that μ_1 possesses a density with respect to a positive measure ν, and $\mu_2 = \mu_2^+ + \mu_2^-$ for which $\mu_2^+ \ll \nu$ and $\mu_2^- \perp \nu$, which may always be done according to the Lebesgue decomposition theorem (Theorem 3.11). Suppose f_1 and f_2^+ are the respective densities of μ_1 and μ_2^+ *wrt* ν. Then (4.39) becomes

$$\|\mu_1 - \mu_2\|_{TV(w)} = \int w|f_1 - f_2^+| d\nu + \int w d\mu_2^-, \tag{4.42}$$

and so μ_1 and μ_2 can only be 'arbitrarily close' if they are equivalent according to Definition 3.16.

Total variation and the span seminorm

The *span seminorm* of a function is defined by

$$\|f\|_{SP} = \sup_x f(x) - \inf_x f(x)$$

(which will be discussed in more detail in Section 6.8.4). We may also denote the midpoint of the range of f as

$$mid(f) = (\sup_x f(x) + \inf_x f(x))/2.$$

There is an important relationship between the total variation norm and the span seminorm, which is summarized in the following theorem:

Theorem 4.22 *Suppose μ is a measure on $(\Omega, \mathcal{F})$. Let w be any positive function on $\mathcal{X}$. Then*

$$\left|\int g d\mu\right| \leq \frac{1}{2} \|\mu\|_{TV(w)} \|w^{-1}g\|_{SP} + |mid(w^{-1}g)| \left|\int w d\mu\right|. \tag{4.43}$$

In addition, if P_1 and P_2 are two proper probability measures, then

$$\left|\int g dP_1 - \int g dP_2\right| \leq \frac{1}{2} \|P_1 - P_2\|_{TV} \|g\|_{SP}. \tag{4.44}$$

Proof Let μ be a measure with Jordan decomposition $\mu = \mu^+ - \mu^-$. Let $g^w_{sup} = \sup_x w(x)^{-1}g(x)$, $g^w_{inf} = \inf_x w^{-1}g(x)$, then write

$$\begin{aligned}\int g d\mu &= \int w^{-1} g w d\mu^+ - \int w^{-1} g w d\mu^- \\ &\leq g^w_{sup} \int w d\mu^+ - g^w_{inf} \int w d\mu^- \\ &= \|w^{-1}g\|_{SP} \int w d\mu^+ + g^w_{inf} \int w d\mu. \end{aligned} \tag{4.45}$$

A similar argument gives

$$\int g d\mu \leq \|w^{-1}g\|_{SP} \int w d\mu^- + g^w_{sup} \int w d\mu. \tag{4.46}$$

We may take the average of the upper bounds (4.45)–(4.46), then apply the argument to $-g$ to yield (4.43) after which (4.44) follows by noting that $\int d(P_1 - P_2) = 0$. ///

4.9 STOCHASTIC KERNELS

Suppose we have Borel space $\mathcal{X}$ (recall Definition 3.13), with Borel sets $\mathcal{B}(\mathcal{X})$. The set of all measurable real valued functions on $\mathcal{X}$ is denoted $\mathcal{F}(\mathcal{X})$. Let $\mathcal{M}_\mathcal{X}$ the set of finite signed measures. Suppose we are given another Borel space $\mathcal{Y}$. A type of object of considerable importance is a mapping of the form $Q : \mathcal{Y} \to \mathcal{M}_\mathcal{X}$, an indexed set of finite signed measures, or *measure kernel*. The range of Q may be restricted, as appropriate, to $\mathcal{M}^1_\mathcal{X}$, $\mathcal{M}^{1-}_\mathcal{X}$ or $\mathcal{M}^+_\mathcal{X}$, the set of stochastic, substochastic or positive finite measures, respectively. We then refer to a *stochastic kernel*, *substochastic kernel*, or *positive measure kernel* (or more simply, *positive kernel* when the context is clear). The possibility that Q is not a positive kernel will sometimes be emphasized by reference to a *signed measure kernel*, or simply *signed kernel*.

Notational conventions

It is important to establish a consistent notation for measure kernels and related operations. Given $Q : \mathcal{Y} \to \mathcal{M}_\mathcal{X}$, the notation $Q(\cdot \mid y)$ denotes the specific measure indexed by $y \in \mathcal{Y}$. This is slightly more cumbersome than the notation $Q(y)$ often used, but we opt for the greater clarity. For any measurable E, $Q(E \mid \cdot) : \mathcal{Y} \to \mathbb{R}$ is a real-valued function on $\mathcal{Y}$ evaluated pointwise by $Q(E \mid y)$.

In general, if $V \in \mathcal{F}(\mathcal{X})$ and μ is a measure on $\mathcal{X}$ then we may use the compact notation

$$\mu V = \int_{x \in \mathcal{X}} V(x) d\mu \quad \in \bar{\mathbb{R}}$$

to represent integration (always over the entire space). Both forms will be used, according to which seems most intuitive in the context. The operation $W = QV$ yields a mapping $W : \mathcal{Y} \to \bar{\mathbb{R}}$ evaluated pointwise by

$$W(y) = Q(\cdot \mid y) V = \int_{x \in \mathcal{X}} V(x) dQ(\cdot \mid y) \in \bar{\mathbb{R}}, \quad y \in \mathcal{Y}.$$

If μ is a measure on $\mathcal{Y}$, the operation μQ yields a measure $\mu' = (\mu Q)$ on $\mathcal{X}$ evaluated setwise by

$$\mu'(E) = \mu Q(E \mid \cdot) = \int_{y \in \mathcal{Y}} Q(E \mid y) d\mu, \quad E \subset \mathcal{X}.$$

When $\mathcal{X} = \mathcal{Y}$ a kernel may transform another kernel by combining the preceding operations. Suppose for $i = 1, 2$ we have measure kernels $Q_i : \mathcal{X} \to \mathcal{M}_{\mathcal{X}}$. Then $Q_1 Q_2$ yields kernel $Q' = (Q_1 Q_2) : \mathcal{X} \to \mathcal{M}_{\mathcal{X}}$, evaluated by

$$\begin{aligned} Q'(E \mid x) &= Q_1(\cdot \mid x) Q_2(E \mid \cdot) \\ &= \int_{x' \in \mathcal{X}} Q_2(E \mid x') dQ_1(\cdot \mid x), \quad x \in \mathcal{X}, E \subset \mathcal{X}. \end{aligned} \tag{4.47}$$

Of course, there are assumptions underlying these operations, which we consider next.

4.9.1 Measurability of measure kernels

For any $V \in \mathcal{F}(\mathcal{X})$ and measure kernel Q the mapping $W = QV : \mathcal{Y} \to \mathbb{R}$ is well defined. We will of course be interested in determining which properties of W follow from V, and would at the very least like to conclude that $W \in \mathcal{F}(\mathcal{Y})$ (V must be measurable in order that the operation QV be defined), so we will reserve the term *measure kernel* for mappings Q guaranteeing this minimal requirement.

Construction of the Lebesgue integral follows from limits of approximating simple functions (Section 3.3). Suppose we are given simple function $V_s = \sum_{i=1}^n a_i I_{E_i} \in \mathcal{F}(\mathcal{X})$. Then $W_s = QV_s = \sum_{i=1}^n a_i Q(E_i \mid y)$. Thus, W_s will be measurable if $h_E(y) = Q(E \mid y)$ is a measurable function for each $E \in \mathcal{B}(\mathcal{X})$. Any measurable function is a bounded limit of simple functions (Theorem 3.21), so that Q preserves measurability under this assumption. In the remainder of this section we will make this idea formal.

The essential condition is that $Q(E \mid y)$ is a measurable function on $\mathcal{Y}$ for all measurable $E \subset \mathcal{X}$. The Dynkin system theorem (Section 3.2.8) may be used to simplify the verification of this assumption. Recall that a π-system $\mathcal{E}$ is a class of subsets that is closed under finite intersection. Under given conditions we may conclude that a property which holds for all sets in $\mathcal{E}$ must hold for all measurable sets. For example, the set of intervals in $\mathbb{R}$, or the set of measurable rectangles in a product space is a π system.

Lemma 4.2 *Suppose we are given Borel spaces $\mathcal{X}, \mathcal{Y}$ and mapping $Q : \mathcal{Y} \to \mathcal{M}_{\mathcal{X}}$ for which $Q(\mathcal{X} \mid y)$ is a measurable function on $\mathcal{Y}$. Suppose $\mathcal{E}$ is a π-system for which $\mathcal{B}(\mathcal{X}) \subset \sigma(\mathcal{E})$. Then if $h_E(y) = Q(E \mid y)$ is a measurable function on $\mathcal{Y}$ for all $E \in \mathcal{E}$, it follows that $h_{E'}(y)$ is a measurable function for all $E' \in \mathcal{B}(\mathcal{X})$.*

Proof The proof follows Theorem 3.10. The class $\mathcal{E}'$ of subsets E' for which $h_{E'}$ is measurable is clearly a λ-system (Definition 3.14). By Theorem 3.9, if $\mathcal{E} \subset \mathcal{E}'$ then $\mathcal{B}(\mathcal{X}) \subset \sigma(\mathcal{E}) \subset \mathcal{E}'$, which completes the proof. ///

A measure kernel preserves pointwise convergence of measurable functions:

Lemma 4.3 *If Q is a measure kernel, and $V_n \to V$, with $|V_n| \leq V$ then $W_n = QV_n$ converges to $W = QV$.*

Proof The result is a direct application of the dominated convergence theorem (Theorem 3.19). ///

Thus we can give essentially the same notion of kernel measurability proposed in Bertsekas and Shreve (1978) (Definition 7.12 and Proposition 7.26):

Definition 4.7 *Suppose we are given Borel spaces $\mathcal{X}, \mathcal{Y}$, and suppose $\mathcal{E}$ is a π-system for which $\mathcal{B}(\mathcal{X}) \subset \sigma(\mathcal{E})$. Then the mapping $Q : \mathcal{Y} \to \mathcal{M}_{\mathcal{X}}$ is a* measure kernel *if $Q(E \mid \cdot)$ is a measurable mapping from $\mathcal{Y}$ to $\mathbb{R}$ for all $E \in \mathcal{E}$. The definition extends naturally to stochastic, substochastic and positive measure kernels.*

That a measure kernel preserves measurability, that is $QV \in \mathcal{F}(\mathcal{Y})$ for any $V \in \mathcal{F}(\mathcal{X})$, is given in the following theorem:

Theorem 4.23 *A measurable stochastic kernel (Definition 4.7) preserves measurability.*

Proof First suppose $h_E(y) = Q(E \mid y)$ is a measurable mapping on $\mathcal{Y}$ for all $E \in \mathcal{B}(\mathcal{X})$. Then for any simple function $V_s \in \mathcal{F}(\mathcal{X})$ we may conclude that $W_s = QV_s$ is also measurable. Then, by Theorem 3.21 any $V \in \mathcal{F}(\mathcal{X})$ is the limit of simple functions V_n for which $|V_n| \leq |V|$. Using Lemma 4.3 we have $W = QV = \lim_{n\to\infty} QV_n$, that is, W is the limit of measurable functions, and is therefore measurable. To complete the proof, by Lemma 4.2 it suffices to claim only that h_E is measurable for all $E \in \mathcal{E}$. ///

This also resolves the measurablility of the composition of two measure kernels. Consider the composition defined in (4.47). If Q_1, Q_2 satisfy Definition 4.7 then $Q_2(E \mid \cdot) \in \mathcal{F}(\mathcal{X})$, Q_1 is measure preserving, which implies $Q'(E \mid \cdot) \in \mathcal{F}(\mathcal{X})$, so that Q' also satisfies Definition 4.7.

4.9.2 Continuity of measure kernels

The measure preserving property of the measure kernel of Definition 4.7 does not rely on metric space properties, although this is included in the definition of a Borel space. However, metric structure becomes important in defining continuity properties, and so is an important part of the definition.

We then give the following definitions of continuity for a positive measure kernel.

Definition 4.8 *Suppose we are given Borel spaces $\mathcal{X}, \mathcal{Y}$, and assume $\mathcal{X}$ is a metric space. The mapping $Q : \mathcal{Y} \to \mathcal{M}^+_{\mathcal{X}}$ is a* weakly continuous measure kernel *if the sequence of measures $Q(\cdot \mid y_n)$ converges weakly to $Q(\cdot \mid y)$ whenever $y_n \to y$. Then Q is a* strongly continuous measure kernel *if weak convergence is replace by setwise convergence.*

We then have

Theorem 4.24 *(i) If $Q : \mathcal{Y} \to \mathcal{M}^+_{\mathcal{X}}$ is a weakly continuous measure kernel, then*

(a) V is bounded and continuous $\Rightarrow W = QV$ is continuous,
(b) V is bounded below and lsc $\Rightarrow W = QV$ is lsc,
(c) V is bounded above and usc $\Rightarrow W = QV$ is usc.

(ii) *If Q is a strongly continuous measure kernel, then*
V is bounded and measurable $\Rightarrow W = QV$ is continuous.

Proof The theorem follows from Theorems 4.18 and 4.20. ///

4.10 CONVERGENCE OF SUMS

Given a sequence of RVs $\epsilon_1, \epsilon_2, \ldots$, it is sometimes necessary resolve the convergence properties of the partial sums $S_n = \sum_{i=1}^{n} \epsilon_i$. One straightforward problem, and one which will suffice for some applications, is to determine if S_n has a finite limit $wp1$. Suppose we may assume that $\sup_n E[S_n] = K < \infty$ and that S_n is bounded below by an integrable random variable Y. If we first assume that S_n is nondecreasing (for example, if $\epsilon_n \geq 0$), then by the monotone convergence theorem we may conclude directly that S_n converges $wp1$ to some S_∞ for which $E[S_\infty] = K$.

We give two results with which to test convergence of S_n. For independent ϵ_n *a.s.* convergence is exactly resolved by the *Kolmogorov Three Series Theorem* (Durrett (2010), Chapters 1-2).

Theorem 4.25 (Kolmogorov Three Series Theorem) *Let ϵ_n be an independent sequence, and let $Y_n = \epsilon_n I\{|\epsilon_n| \leq 1\}$. Then S_n converges a.s. if and only if (i) $\sum_{n\geq 1} P(|\epsilon_n| > 1) < \infty$; (ii) the series $\sum_{n\geq 1} E[Y_n]$ is convergent; (iii) $\sum_{n\geq 1} var[Y_n] < \infty$.*

The assumption that the ϵ_n are independent will often be too restrictive. In such cases, it may help to regard ϵ_n as an adapted process on a filtration $\mathcal{F}_n$ representing process history. If it can be verified that ϵ_n possesses any martingale properties, as given in Definition 4.3, then the convergence properties of S_n may be easily resolved. For example, a standard result for submartingales is given by the *Doob Martingale Convergence Theorem* (See, for example, Durrett (2010), Section 4.2).

Theorem 4.26 (Doob Martingale Convergence Theorem) *If S_n is a submartingale, and $\sup_n E[S_n I\{S_n \geq 0\}] < \infty$, then S_n converges a.s. to some random variable S_∞, with $E[|S_\infty|] < \infty$.*

Note that neither Theorems 4.25 or 4.26 imply

$$E\left[\lim_{n\to\infty} S_n\right] = \lim_{n\to\infty} E[S_n],$$

(an interesting counter-example occurs in branching processes, see Billingsley (1995), Section 35). So, if interest is in establishing convergence in the L^p norm, additional argument is needed. It turns out that the uniform integrability property (Definition 4.1) establishes equivalence between L^1 and other forms of convergence.

The next two theorems are proven in Durrett (2010) (Section 4.5, Theorems 5.2, 5.3):

Theorem 4.27 *Suppose a sequence $X_1, X_2, \ldots$ possesses limit $X_n \overset{i.p.}{\to} X$. Then the following are equivalent:*

(i) *the sequence X_n is uniformly integrable,*

(ii) $X_n \overset{L^1}{\to} X$,

(iii) $E[|X_n|] \to E[|X|] < \infty$.

Since *a.s.* convergence implies convergence in probability Theorem 4.27 is directly applicable to submartingales satisfying the hypothesis of Theorem 4.26.

Theorem 4.28 *For a submartingale S_n the following are equivalent:*

(i) S_n *is uniformly integrable,*
(ii) S_n *converges a.s. and in* L^1,
(iii) S_n *converges in* L^1.

Convergence properties may also follow from a uniform bound on higher order moments. For example, it can be shown (Lemma 7.6.9 of Ash (1972)) that if $\sup_n E\left[|S_n|^p\right] < \infty$ for any $p > 1$ then S_n is uniformly integrable. Recall also, by Theorem 4.6, that if S_n is a martingale or nonnegative submartingale, then for $p \geq 1$, given integrability, $|S_n|^p$ is a submartingale. As a consequence, conditions for L^p convergence can also be given. We then have the following theorem (Theorem 7.6.10, Ash (1972)):

Theorem 4.29 *If S_n is a martingale or a nonnegative submartingale for which $\sup_n E[|S_n|^p] < \infty$, $p > 1$ then S_n converges a.s. and in L^p to a limit S_∞.*

4.11 THE LAW OF LARGE NUMBERS

Given partial sums $S_n = \sum_{i=1}^n \epsilon_i$, define the sample means $\bar{S}_n = n^{-1} S_n$. Suppose $E[\epsilon_n] = \mu$ for all n. A *Strong Law of Large Numbers* (SLLN) asserts that $\bar{S}_n \overset{a.s.}{\to} \mu$, and may also specify convergence rates. In contrast a *Weak Law of Large Numbers* (WLLN) asserts that $\bar{S}_n \overset{i.p.}{\to} \mu$, and we may also define L^p laws when $\bar{S}_n \overset{L^p}{\to} \mu$. We know that a SLLN or an L^p law for any $p > 0$ implies a WLLN (see Section 4.8.1).

Generally, μ can be taken as 0 by replacing ϵ_n with $\epsilon_n - \mu$. Many forms of the SLLN exist, involving both the properties of ϵ_n and the aggregation method of $\bar{S}_n$.

Suppose the ϵ_n are independent. If $\sup_n E[\epsilon_n^4] < \infty$, the SLLN follows by applying Markov's inequality and the Borel-Cantelli lemma I. Assume $\mu = 0$, then $\sum_{n\geq 1} P(\bar{S}_n^4 > \epsilon^4) \leq \sum_{n \geq 1} E[\bar{S}_n^4]/\epsilon^4 \propto \sum_{n\geq 1} n^{-2} < \infty$, therefore $P(|\bar{S}_n| > \epsilon \text{ i.o.}) = 0$, which implies $\bar{S}_n \overset{a.s.}{\to} 0$.

If ϵ_n is an orthogonal process, that is, $E[\epsilon_n] = 0$ and the sequence ϵ_n is uncorrelated, and $\sup_n E[\epsilon_n^2] < \infty$, then $E[\bar{S}_n^2] \propto n^{-1}$, and so $\bar{S}_n \overset{L^2}{\to} 0$.

For identically distributed ϵ_n a somewhat more detailed argument (but still based on elementary probability theory) permits a weakening of the moment bound to $E\left[|\epsilon_1|\right] < \infty$, while requiring only pairwise independence (this result is quite recent, introduced in Etemadi (1981)):

Theorem 4.30 *Let $\epsilon_1, \epsilon_2, \ldots$ be an identically distributed pairwise independent sequence. If $E\left[|\epsilon_1|\right] < \infty$ then $\bar{S}_n \overset{a.s.}{\to} E\left[\epsilon_1\right]$.*

Proof See, for example, Section 1.7 of Durrett (2010) for a proof. ///

This result may be extended to the infinite mean case.

Theorem 4.31 *Let $\epsilon_1, \epsilon_2, \ldots$ be an iid sequence with $\epsilon_1 \geq 0$ and $E[\epsilon_1] = \infty$ then $\bar{S}_n \overset{a.s.}{\to} \infty$.*

Proof Suppose $P(\epsilon_1 = \infty) > 0$. Then by the Borel-Cantelli lemma II, *wp*1 we must have $\epsilon_n = \infty$ i.o., so the result holds. If $P(\epsilon_1 = \infty) = 0$ define, for any finite $K > 0$, the sequence $\epsilon_1^K = \min(\epsilon_1, K)$, and let $\bar{S}_n^K$ be the sample means for this sequence. Since $E[\epsilon_1^K] \leq K < \infty$, by Theorem 4.30 we must have *wp*1

$$\liminf_{n\to\infty} \bar{S}_n \geq \lim_{n\to\infty} \bar{S}_n^K = E[\epsilon_1^K]. \tag{4.48}$$

However, by the monotone convergence theorem, $E[\epsilon_1^K] \to \infty$ as $K \to \infty$, which means that the lower bound of (4.48) may be made arbitrarilly large, completing the proof. ///

We will make use of the following SLLN for martingales (Feller (1971), Section VII.8):

Theorem 4.32 *Let $S_n = \epsilon_1 + \cdots + \epsilon_n$ be a martingale. If b_n is an unbounded increasing sequence of constants , and $\sum_{n\geq 1} b_n^{-2} E[\epsilon_n^2] < \infty$, then $b_n^{-1} S_n \overset{a.s.}{\to} 0$, and the sequence $Y_n = \sum_{k=1}^n b_k^{-1} \epsilon_k$ is convergent a.s..*

If the terms defining S_n are independent with zero mean, then S_n is a martingale, so Theorem 4.32 applies. If we have bound $\sup_n E[\epsilon_n^2] < \infty$ then we generally set $b_n = n$, which gives $\bar{S}_n \to 0$.

More generally, suppose ϵ_n is adapted to filtration $\mathcal{F}_n$. If ϵ_n represent a noise process we may be able to claim that that $E[\epsilon_n \mid \mathcal{F}_{n-1}] = 0$ and that $E[\epsilon_n^2 \mid \mathcal{F}_{n-1}] \leq \sigma_n^2$ for some sequence σ_n^2. Then by Theorem 4.3 $E[\epsilon_n^2] \leq \sigma_n^2$, so that Theorem 4.32 is applicable. If the sequence σ_n^2 is bounded we can similarly conclude that $\bar{S}_n \overset{a.s.}{\to} 0$. Stronger results are generally possible for the *iid* case, but for our purposes we lose little by modeling noise processes as martingales, while gaining considerably more flexibility, especially in adaptive control models.

We have seen that the behavior of a random sum S_n can be determined by the behavior of its expected value. The same is not exactly true of the sample mean process $\bar{S}_n$. Under the hypothesis of Theorem 4.32 (and many other SLLNs) ϵ_n will be an orthogonal process. Suppose we have $E[\epsilon_n^2] \leq \sigma^2 < \infty$. Then $E[\bar{S}_n^2] \leq n^{-1}\sigma^2$, and we would conclude $E[|\bar{S}_n|] = O(n^{-1/2})$. It turns out that the convergence rate of $\bar{S}_n$ itself is slightly slower. The best possible rate is known, and any convergence statement giving this rate is known as a *Law of the Iterated Logarithm* (LIL). A proof of the following version may be found in Durrett (2010), (Section 7.9).

Theorem 4.33 (Law of the Iterated Logarithm) *If $X_1, X_2, \ldots$ is an iid sequence with $E[X_i] = 0$, $E[X_i^2] = 1$ then*

$$\limsup_{n\to\infty} \frac{\sum_{i=1}^n X_i}{(2n \log\log n)^{1/2}} = 1, \quad wp1.$$

In other words, for L^2 convergence we have rate $E[|\bar{S}_n|] = O(n^{-1/2})$, but for almost sure convergence we have the slightly larger LIL rate $\bar{S}_n = \Omega(2n^{-1/2}(\log\log n)^{1/2})$. There are a variety of conditions beyond those given in Theorem 4.33 under which the LIL holds, and the reader can be referred to a summary of the quite interesting history of the LIL in Durrett (2010) and to Hall and Heyde (1980) for a comprehensive treatment of martingale limit theory.

In the meantime, we can offer the following compromise. The LIL implies that $\bar{S}_n = o(n^{-1/2+\delta})$ for any $\delta > 0$, and this can be tested by Theorem 4.32.

Theorem 4.34 *If S_n is a martingale for which $\sup_n E[\epsilon_n^2] = \sigma^2 < \infty$, then $\bar{S}_n = o(n^{-1/2+\delta})$ wp1 for any $\delta > 0$.*

Proof In Theorem 4.32 set $b_n = n^{1/2+\delta}$ where $\delta > 0$. Under the stated conditions the hypothesis is satisfied, therefore $S_n/n^{1/2+\delta} \overset{a.s.}{\to} 0$, which completes the proof. ///

4.12 EXTREME VALUE THEORY

Suppose we are given an indefinite *iid* sequence $X_1, X_2, \ldots$, where X_i has CDF F_X, and we are interested in the process $M_n = \max_{1 \leq i \leq n} X_i$. Suppose Q_p is the p-quantile, that is, $p = F_X(Q_p)$ (with a suitable adjustment when F_X is not continuous), and we set $Q_1 = \sup\{Q \mid F_X(Q) < 1\}$. The CDF of M_n is

$$F_{M_n}(x) = P(M_n \leq x) = P(\cap_i \{X_i \leq x\}) = \prod_{i=1}^{n} P(X_i \leq x) = F_X(x)^n.$$

If for x we have $F_X(x) < 1$, so that $M_n > x$ has nonzero probability, then $\lim_n F_{M_n}(x) = \lim_n F_X(x)^n = 0$. In fact, by Borel-Cantelli lemma I *wp*1 there exists $m < \infty$ for which $M_n > x$ for all $n > m$ (in other words, $M_n \overset{a.s.}{\to} Q_1$). A possible normalization for M_n is easily deduced by noting

$$F_{M_n}(Q_{1-1/n}) = (1 - 1/n)^n \to_n e^{-1},$$

so that we can at least expect M_n to be located with high probability near $Q_{1-1/n}$. In principle, asymptotic tail probabilities for M_n, suitably normalized, can be constructed in this way.

It turns out that the study of limiting distributions of this type for the *iid* case possesses a remarkably refined theory, based on the *Fisher-Tippett-Gnedenko Theorem*. A statement about the limiting distribution of M_n, using a more refined normalization method, can take the form:

$$\lim_{n \to \infty} F_X(a_n x + b_n)^n = G(x) \text{ at all continuity point of } G, \tag{4.49}$$

for some sequence $(a_n, b_n), n \geq 1$ and CDF G. Then a_n, b_n are the normalizing constants for M_n, and G is the limiting distribution, since

$$P\left(\frac{M_n - b_n}{a_n} \leq x\right) = P\left(M_n \leq a_n x + b_n\right) = F_X(a_n x + b_n)^n \approx G(x).$$

We then have, assuming that expectations exist,

$$E\left[\frac{M_n - b_n}{a_n}\right] \approx E_G[X], \quad \text{or} \quad E[M_n] \approx b_n + a_n E_G[X],$$

the important fact being that $E_G[X] = O(1)$.

Much of the theory underlying statements of the form (4.49) follows from the fact that there exist only 3 possible candidates for G:

$$G_1(x) = \exp\left(-e^{-x}\right), \quad -\infty < x < \infty,$$

$$G_2(x \mid \alpha) = \begin{cases} 0 & ; \quad x \le 0 \\ \exp\left(-x^{-\alpha}\right); & x > 0 \end{cases},$$

$$G_3(x \mid \alpha) = \begin{cases} \exp(-(-x)^{\alpha}); & x < 0 \\ 1 & ; \quad x \ge 0 \end{cases},$$

for some parameter $\alpha > 0$. Location and scale parameters may be introduced into these distributions, but no generality is lost by normalizing the original sequence X_i instead, since for $a > 0$, $b \in \mathbb{R}$, we have $\max_{1 \le i \le n} a(X_i - b) = a(M_n - b)$. The dependence on a single parameter can then be emphasized, and so it is therefore possible to regard G_1, G_2, G_3 as special cases of a single parametric family, refered to as the *generalized extreme value distribution*. Otherwise, G_1, G_2 and G_3 are referred to as the *Gumbel*, *Fréchet* and *Weibull*, or Type I, II and III extreme value distributions.

Theorem 4.35 (Fisher-Tippett-Gnedenko) *Suppose for F_X there exists some normalizing sequence (a_n, b_n), $n \ge 1$, with $a_n > 0$, for which* (4.49) *holds for some nondegenerate CDF G. Then G must be one of the extreme value distributions G_1, G_2 or G_3.*

This means the limiting distribution G follows once normalizing constants can be deduced. For our purposes, the most important special case will involve the Gumbel distribution G_1. Suppose $f_X(x) > 0$ and is differentiable for all $x \in (x_0, Q_1)$ for some large enough x_0. Let $h_X(x)$ be the hazard rate for F_X. If

$$\lim_{x \to Q_1} dh_X^{-1}(x)/dx = 0,$$

then the limit (4.49) exists for $G = G_1$, and we may take $b_n = Q_{1-1/n}$ and $a_n = [nf_X(b_n)]^{-1}$. This condition is satisfied by the normal distribution. This summarizes part of Theorem 10.5.2 of David and Nagaraja (2003), and readers can consult this reference for more details.

4.13 MAXIMUM LIKELIHOOD ESTIMATION

Suppose $f(x \mid \theta)$ defines a parametric family of densities on sample space $\mathcal{S}$, where $\theta \in \Theta \subset \mathbb{R}^d$. Assume also that $f(x \mid \theta)$ possesses all first order partial derivatives. The $d \times d$ *information matrix* $I(\theta)$ is defined element-wise by

$$I_{ij}(\theta) = E_\theta\left[\frac{\partial}{\partial \theta_i} \log f(X \mid \theta) \times \frac{\partial}{\partial \theta_j} \log f(X \mid \theta)\right]$$

where X has density function $f(x \mid \theta)$. It may be shown that for all θ

$$E_\theta\left[\frac{\partial}{\partial \theta_i} \log f(X \mid \theta)\right] = 0,$$

so that $I_{ij}(\theta)$ is interpretable as the covariance,

$$I_{ij}(\theta) = cov_\theta\left(\frac{\partial}{\partial\theta_i}\log f(X\mid\theta), \frac{\partial}{\partial\theta_j}\log f(X\mid\theta)\right).$$

This means $I(\theta)$ is a covariance matrix, and is therefore positive semidefinite, and positive definite unless the partial derivatives of $\log f(X\mid\theta)$ are linearly dependent. If $\log f(X\mid\theta)$ possesses second order partial derivatives, we have

$$I_{ij}(\theta) = -E_\theta\left[\frac{\partial^2}{\partial\theta_i\theta_j}\log f(X\mid\theta)\right],$$

which will often be easier to evaluate.

The *log-likelihood function* is defined as

$$L(\theta\mid X) = \log f(X\mid\theta),$$

and is regarded as a function of θ, rather than X. If we are given a random outcome X from density $f(x\mid\theta)$, the *maximum likelihood estimate* (MLE) of θ is

$$\hat{\theta}_{MLE} = \text{argmin}_\theta L(\theta\mid X),$$

assuming a unique minimum exists.

Suppose we are given an *iid* random sample $\tilde{X} = (X_1, \ldots, X_n)$ from $f(x\mid\theta)$. By independence the density of $\tilde{X}$ is $f_{\tilde{X}}(x_1, \ldots, x_n\mid\theta) = \prod_i f(x_i\mid\theta)$, and the likelihood function of θ given the sample $\tilde{X}$ is

$$L(\theta\mid\tilde{X}) = \sum_i \log f(X_i\mid\theta),$$

with MLE $\hat{\theta}_{MLE} = \text{argmin}_\theta L(\theta\mid\tilde{X})$. Under quite general conditions (see, for example, Lehmann and Casella (1998) or Casella and Berger (2002)) we have for the MLE based on $\tilde{X}$

$$n^{1/2}(\hat{\theta}_{MLE} - \theta) \to_n N(0, I^{-1}(\theta)),$$

or equivalently,

$$n(\hat{\theta}_{MLE} - \theta)^T I(\theta)(\hat{\theta}_{MLE} - \theta) \to_n \chi^2_d.$$

Note that $I(\theta)$ remains the information matrix for $f(x\mid\theta)$, whereas the information for the density of the sample would be $nI(\theta)$, directly from the definition.

This leads to stochastic bounds of the form

$$\left\|\hat{\theta}_{MLE} - \theta\right\| \leq n^{-1}\lambda_{max}\left[I^{-1}(\theta)\right]\chi^2_d,$$

where $\lambda_{max}\left[I(\theta)^{-1}\right]$ is the maximum eigenvalue of $I(\theta)^{-1}$.

4.14 NONPARAMETRIC ESTIMATES OF DISTRIBUTIONS

Given a sample $X_1, \ldots, X_n$ from distribution F on $\mathbb{R}^d$, it is generally possible to construct a new distribution $\tilde{F}$ which is close to F in some sense, and may therefore form the basis for an approximation model. The empirical distribution is defined as

$$\hat{F}(x_1, \ldots, x_d) = n^{-1} \sum_{i=1}^{n} I\{X_{i,1} \leq x_i, \ldots, X_{i,d} \leq x_d\}. \tag{4.50}$$

When the support $\mathcal{S}$ of F is countable $\hat{F}$ is equivalent to the collection of empirical frequencies

$$\hat{p}_j = n^{-1} Z_j, \quad Z_j = \sum_{i=1}^{n} I\{X_i = j\}, \quad j \in \mathcal{S}.$$

For continuous distributions a wider range of alternatives are generally considered. Kernel density estimates are based on a density kernel $K(x)$, which is formally a density function on $\mathbb{R}^d$, and may be written

$$\hat{f}(x; K, h) = (nh^d)^{-1} \sum_{i=1}^{n} K((x - X_i)/h),$$

where h is a positive constant. The kernel K is typically chosen to be a standardized uniform or Gaussian density, while h approaches 0 for increasing n. When the support of F is a bounded set $\mathcal{S} \subset \mathbb{R}^d$ a histogram density may be used. Suppose $S = (S_1, \ldots, S_m)$ a partition of $\mathcal{S}$. This generates a histogram approximation

$$\hat{f}(x; S) = \sum_{i=1}^{m} |S_i|^{-1} \frac{|\{X_i \in S_i\}|}{n} I\{x \in S_i\},$$

so that $\hat{f}(x; S)$ is the uniform density conditional on each partition element S_i. Typically, the partition is allowed to become more refined as n increases. Often, S_i is defined by a grid discretization of $\mathcal{S}$ of sides h, in which case this parameter plays a role comparable to h in the definition of $\hat{f}(x; K, h)$, although the two types of approximations remain distinct even when K is the uniform density.

Approximation theory requires the definition of a metric for distributions. The Glivenko-Cantelli (or Kolmogorov) distance between two CDFs F_1, F_2 on $\mathbb{R}^d$ is

$$d_{GC}(F_1, F_2) = \sup_x |F_1(x) - F_2(x)|,$$

while the total variation distance is

$$d_{TV}(F_1, F_2) = \sup_A \left| \int_A dF_1 - \int_A dF_2 \right|, \tag{4.51}$$

the supremum being taken over all measurable sets. This is equivalent to the total variation norm introduced in Section 4.8.3 in the sense that

$$\|F_1 - F_2\|_{TV} = 2d_{TV}(F_1, F_2).$$

Some authors introduce a factor of 2 into (4.51), thus equating the total variation norm and distance. Without this factor the maximum achievable value of d_{TV} is 1. When F_1, F_2 are both absolutely continuous with respect to a common measure μ we have

$$d_{TV}(F_1, F_2) = \frac{1}{2}\int |f_1 - f_2| d\mu, \tag{4.52}$$

where f_1, f_2 are the respective Radon-Nikodym derivatives of F_1, F_2.

The quantity $d_{GC}(\hat{F}, F)$ plays an important role in mathematical statistics. It is well known that it converges to 0 as $n \to \infty$ for *iid* samples from F. This forms the basis for the Kolmogorov-Smirnov statistic $\sqrt{n} d_{GC}(\hat{F}, F_0)$ for testing against the null hypothesis $H: F = F_0$, the distribution of which is well understood.

We further note that since d_{GC} is calculable as the supremum over a subset of measurable sets, we necessarily have $d_{GC}(F_1, F_2) \leq d_{TV}(F_1, F_2)$ for any two distributions, so that d_{TV} is the more stringent distance, and this difference is important. The quantity $d_{GC}(\hat{F}, F)$ can be used to compare an empirical distribution $\hat{F}$ to a continuous distribution F, even though they are singular measures (see Definition 3.16). In contrast, in this case we would have $d_{TV}(\hat{F}, F) = 1$. Thus, when approximation theory necessitates the use of d_{TV}, any estimate F' of a distribution F must itself be a measure equivalent to F, that is, it is the densities *wrt* a common measure which must be estimated. We will consider this type of estimate further in Section 15.2.

4.15 TOTAL VARIATION DISTANCE FOR DISCRETE DISTRIBUTIONS

Suppose F_X has support $\mathcal{S} = \{1, 2, \ldots\}$, and is defined by probability vector $P = (p_1, p_2, \ldots)$ where $p_j = P(X = j)$. The natural choice of approximate distribution is $\hat{F}$, defined by corresponding empirical frequences $\hat{P} = (\hat{p}_1, \hat{p}_2, \ldots)$. Then

$$\begin{aligned} 2d_{TV}(F, \hat{F}) &= \|P - \hat{P}\|_1 \\ &= \sum_{i \geq 1} |p_i - \hat{p}_i|, \end{aligned}$$

where $\|\cdot\|_1$ is the L^1 norm *wrt* counting measure. We also wish to incorporate weighted distances, and can do so in a manner for which the theory differs little from the unweighted case. Suppose $w = (w_1, w_2, \ldots)$ is a system of positive weights with finite maximum w^*. The weighted L^1 norm is then

$$\left\|P - \hat{P}\right\|_{1:w} = \sum_{i \geq 1} w_i |p_i - \hat{p}_i|.$$

Denote the standardized weighted L^1 distance based on n samples by

$$W_n = (w^*)^{-1} \left\| P - \hat{P} \right\|_{1:w}.$$

First note that we must have $W_n \leq 2$ for any n, even for unbounded support. Then, recall that if $X \sim bin(n,p)$, the variance of X is $np(1-p)$, so by Jensen's inequality we have the convenient bound

$$E[W_n] \leq n^{-1/2} A(P,w), \quad \text{where}$$
$$A(P,w) = (w^*)^{-1} \sum_{i \geq 1} w_i \sqrt{p_i(1-p_i)}. \tag{4.53}$$

The bound is sharp up to a finite constant in the following sense. In Blyth (1980) it was shown that

$$E\left[|X/n - p|\right] = n^{-1/2}\sqrt{2/\pi}\sqrt{p(1-p)} + O(n^{-3/2})$$

so the bound (4.53) is asymptotically sharp up to a fixed multiplicative constant $\sqrt{2/\pi}$. The quantity $A(P,w)$ is unbounded over the space of all distributions on unbounded $\mathcal{S}$. If the support is of cardinality m, then a Lagrange multiplier argument gives the bound:

$$A(P,w) \leq m^{1/2}(1 - m^{-1})^{1/2},$$

which is shown to be sharp by considering the uniform distribution under uniform weighting, giving the slightly weaker, but more intuitive form

$$E[W_n] \leq (m/n)^{1/2}. \tag{4.54}$$

It is therefore possible for this bound to be larger that the upper bound of W_n, so that these approximations become useful only when n is large compared to the support size (in particular, the tolerance is a function of m/n). It should also be noted that $A(P,w)$ may be infinite when the support is unbounded, although $E[W_n]$ must be finite even in this case. The issue is that the bound does not hold uniformly over p, which proves crucial for the unbounded support case. An exact expression for $E[W_n]$ was actually derived by Abraham De Moivre in 1730 (Diaconis and Zabell (1991)). However, this derivation uses mathematical techniques quite distinct from the more familiar ones used in this section, as this apparently straightforward quantity exhibits some quite interesting and counterintuitive behavior. As it happens we may conclude that the bounds derived here are suitable for our purposes, as long as $A(P,w) < \infty$. Of course, it is entirely possible that this will not hold, in which case a new argument would be required. Whatever the case, Blyth (1980) and Diaconis and Zabell (1991) are can be highly recommended to the reader who is curious about this interesting problem.

Regarding the distribution of W_n, by Theorem 4.15 we have the bound $P(|p_i - \hat{p}_i| > \epsilon) \leq 2\exp(-2n\epsilon^2)$. Applying Booles' inequality to the events $\{|p_i - \hat{p}_i| > \epsilon/m\}$ indexed by i extends the bound to $P(\|P - \hat{P}\|_1 > \epsilon) \leq 2m\exp(-2n\epsilon^2/m^2)$. However, the dependence on m can be all but eliminated by applying McDiarmid's inequality (Theorem 4.16):

Theorem 4.36 *For an empirical distribution $\hat{P}$ based on an iid sample from probability vector $P=(p_1,p_2,\dots)$, we have*

$$P(W_n - E[W_n] \geq \epsilon) \leq \exp(-n\epsilon^2/2), \tag{4.55}$$
$$P(W_n - E[W_n] \leq -\epsilon) \leq \exp(-n\epsilon^2/2), \tag{4.56}$$
$$P(|W_n - E[W_n]| \geq \epsilon) \leq 2\exp(-n\epsilon^2/2). \tag{4.57}$$

In addition,

$$P(W_n \geq \epsilon) \leq \exp\left(-n\left(\epsilon - n^{-1/2}A(P,w)\right)^2/2\right), \quad \textit{for } \epsilon \geq n^{-1/2}A(P,w)$$

which implies, for example, that for $P=(p_1,\dots,p_m)$

$$P(W_n \geq \epsilon) \leq \exp\left(-n\epsilon^2/8\right), \quad \textit{for } \epsilon \geq 2(m/n)^{-1/2}.$$

Proof In Theorem 4.16, set function f of the hypothesis equal to $\|P-\hat{P}\|_1$ interpreted as a function of the random sequence $X_1,\dots,X_n$. Clearly, altering a single element X_i in any way forces a change in exactly two empirical frequencies $\hat{p}_i$, $\hat{p}_j$, and this change is no larger than $1/n$. This implies that the hypothesis of Theorem 4.16 is satisfied by setting each $c_j=2/n$, which gives (4.55) directly, with (4.56) and (4.57) following after applying the theorem to $-f$. The remainder of the lemma follows after noting that $E[W_n] \leq \sqrt{m/n}$ for any distribution on m support points. ///

Thus, for example, if $m<\infty$, then tail probabilities for W_n can be obtained which do not depend on P, as long as n is a large enough multiple of m.

One useful feature of Theorem 4.36 is that upper bounds for higher order absolute central moments are easily obtained, leading to a bound on the moment generating function.

Lemma 4.4 *For any distribution P on $\mathcal{S}=\{1,2,\dots\}$ the following inequality holds*

$$E\left[|W_n - E[W_n]|^k\right] \leq \begin{cases} n^{-k/2} & ; \quad k=1,2 \\ 2(2/n)^{k/2}\Gamma(k/2+1); & k=3,4,\dots \end{cases}. \tag{4.58}$$

In addition, if $\chi_n = n(W_n - E[W_n])^2$ then

$$m_{\chi_n}(t) \leq 2(1-2t)^{-1}, \quad t<1/2. \tag{4.59}$$

Proof Since $|W_n - E[W_n]|^k$ is a positive random variable, its expectation may be evaluated by

$$\begin{aligned}
E\left[|W_n - E[W_n]|^k\right] &= \int_{t>0} P\left(|W_n - E[W_n]|^k \geq t\right) dt \\
&= \int_{t>0} P\left(|W_n - E[W_n]| \geq t^{1/k}\right) dt \\
&\leq \int_{t>0} 2\exp\left(-nt^{2/k}/2\right) dt \\
&= 2(2/n)^{k/2}\Gamma(k/2+1)
\end{aligned}$$

using the inequality (4.57) from Theorem 4.36. We then note that the upper bound can be improved for $k = 1, 2$ by noting that Theorem 4.17 (using a Doob martingale construction) gives a sharper bound for the variance $var[W_n] \leq n^{-1}$ than that obtained by the preceding inequality ($var[W_n] \leq 4n^{-1}$) which also improves the bound for $k = 1$ using Jensen's inequality, giving (4.58).

Then (4.59) follows after noting that $E[\chi_n^k] \leq 2 \times 2^k k!$. ///

A bound on $E\left[W_n^k\right]$ can be obtained using an argument such as:

$$\begin{aligned}
E\left[W_n^k\right] &= E\left[(W_n - E[W_n] + E[W_n])^k\right] \\
&\leq E[W_n]^k + \sum_{j=2}^{k} E\left[|W_n - E[W_n]|^j\right] E[W_n]^{k-j}
\end{aligned}$$

involving quantities for which bounds have been obtained in this section. It can be concluded, for example that $E\left[W_n^k\right] = O(n^{-k/2})$, and that the contribution from P is $O(A(P,w)^k)$ for fixed n.

Chapter 5

Background – stochastic processes

A *stochastic process* may be defined as a (possibly uncountable) indexed collection of random variables $\{X_t\}$, $t \in \mathcal{T}$. The index set $\mathcal{T}$ usually represents time or space, and may be discrete or continuous. Whatever the case, we will at least assume $\mathcal{T}$ can be ordered (as can any subset of $\mathbb{R}^k$). The central problem usually concerns the interaction of the dependence structure among the X_t with the index variable t.

Suppose we may define for each finite set of indices $v = (t_1, \dots, t_n)$, $t_1 < \cdots < t_n$, a Borel measurable marginal distribution P_v for $\tilde{X}_v = (X_{t_1}, \dots, X_{t_n})$. We say this system of probability measures is *consistent* if whenever $u \subset v$, the distribution P_u is exactly the marginal distribution of $\tilde{X}_u$ obtainable from P_v. Then by the Kolmogorov extension theorem (Theorem 3.28) there exists a unique probability measure P on the product space $\mathcal{B}^T$ with marginal distributions P_v, where $\mathcal{B}$ are the Borel sets on $\mathbb{R}$.

This method is quite general, in particular $\mathcal{T}$ need not be countable. However, the limits of this extension must be understood. Theorem 3.28 limits the extension of finite projections to the smallest σ-field containing these projections, which is relatively coarse. The problem is that 'sample path' properties which depend on observing $x(t)$ everywhere on $\mathcal{T}$ cannot be reduced to a countable number of projections. This is because such properties, including continuity, monotonicity or anything involving extrema, can be altered by changing a single value of $x(t)$. This is not a problem if $\mathcal{T}$ is countable. Otherwise, more specialized methods may be used to construct probablity measures for widely used stochastic process on $\mathcal{T} = [0, \infty)$ such as Poisson or Gaussian processes (see, for example, Billingsley (1995); Durrett (2010)).

5.1 COUNTING PROCESSES

Counting processes form an important class of stochastic processes, which may be defined as follows:

Definition 5.1 A counting process *is a stochastic process* $N(t)$ *defined on* $t \in [0, \infty)$ *satisfying (i)* $N(0) = 0$*;* $N(t) \in \mathbb{I}$*;* $N(t)$ *is nondecreasing in* t *wp*1.

Usually, $N(t)$ represents the number of events in a sequence which have occurred by time t. It is helpful to think of $N(t)$ as an *arrival process*, as though we were marking the arrival of customers at a queue starting at $N(0) = 0$ at time $t = 0$. Then for $t > s$ $N(t) - N(s)$ is the number of arrivals in time interval $(s, t]$. Note that $N(t)$ contains perfect information regarding the times at which events occur. Definition 5.1 says little about the distributional properties of $N(t)$, so we need to discuss probabilistic models which permit the description of intuitive properties of $N(t)$, such as an arrival rate $\lim_{t\to\infty} t^{-1}N(t)$. Many counting processes conform to one of two models, which we briefly consider next (discussions of these processes in Parzen (1962), Karlin and Taylor (1975) or Ross (1996) can be recommended to the interested reader).

5.1.1 Renewal processes

Let $X_1, X_2, \ldots$ be an *iid* sequence of positive random variables. Let $S_n = \sum_{i=1}^n X_i$. The counting process

$$N(t) = \sup\{n : S_n \leq t\}, \quad \text{with } N(0) = 0 \tag{5.1}$$

is referred to as a *renewal process*. Conventions may differ. Some definitions require that $P(X_1 = 0) = 0$ others only that $P(X_1 = 0) < 1$. In addition, some definitions require that $E[X_1] < \infty$, since most of the important results rely on this assumption.

The terminology suggests a process which indefinitely restarts itself stochastically, with the nth renewal occurring at time S_n. Often, a renewal process is embedded in a more complex stochastic process with this character. Alternatively, a renewal process may be interpreted as an arrival process with *iid* interarrival (renewal) times. Note that if $P(X_1 = 0) > 0$ the formulation of (5.1) permits instantaneous increments of $N(t)$ of value greater than 1, as though instantaneous multiple arrivals are permitted. We give the following lemma:

Theorem 5.1 *If $N(t)$ is a renewal process, then*

(i) $P(X_1 = \infty) = 0$ *implies* $P\left(\lim_{t\to\infty} N(t) = \infty\right) = 1$,

(ii) $P(X_1 = \infty) > 0$ *implies* $P\left(\lim_{t\to\infty} N(t) < \infty\right) = 1$,

(iii) $P\left(\lim_{t\to\infty} t^{-1}N(t) = E[X_1]^{-1}\right) = 1$.

Proof (*i*)–(*ii*) From (5.1) is follows that $\lim_{t\to\infty} N(t) < \infty$ if and only if $X_m = \infty$ for some finite m. By Boole's inequality $P(\cup_m X_m = \infty) = 0$ if $P(X_1 = \infty) = 0$, and by the Borel-Cantelli lemma II $P(\cup_m X_m = \infty) = 1$ if $P(X_1 = \infty) > 0$, which completes the proof.

(*iii*) If $P(X_1 = \infty) > 0$ then by (*ii*) we have $\lim_t N(t) < \infty$ *wp*1, and $E[X_1]^{-1} = 0$ so (*iii*) holds.

Otherwise, $\lim_t N(t) = \infty$ *wp*1, and by Theorem 4.30 or 4.31 we may conclude $n^{-1}S_n \overset{a.s.}{\to} E[X_1]$. We may then write $\lim_{t\to\infty} N(t)^{-1}S_{N(t)} = E[X_1]$ *wp*1. Next, by definition we must have $S_{N(t)} \leq t < S_{N(t)+1}$, so that

$$\frac{N(t)}{S_{N(t)}} \geq \frac{N(t)}{t} > \frac{N(t)+1}{S_{N(t)+1}} \frac{N(t)}{N(t)+1}.$$

The result holds by allowing $t \to \infty$ and noting the limits just described. ///

There may be interest in the *renewal function*

$$m(t) = E[N(t)],$$

about which the following statements hold.

Theorem 5.2 *For any renewal process with* $E[X_1]=\mu>0$ *we have* $m(t)<\infty$ *for all* $t\in[0,\infty)$, *with limit*

$$\lim_{t\to\infty}\frac{m(t)}{t}=\mu^{-1}. \tag{5.2}$$

Proof See Proposition 3.2.2 and Theorem 3.3.4 of Ross (1996). ///

It is important to note the generality of Theorem 5.2, requiring only that $P(X_1=0)<1$. If $\mu=\infty$ the limit in (5.2) is $\mu^{-1}=0$.

5.1.2 Poisson process

One of the most important stochastic process models is the *Poisson process*, which relies on the following definitions. A counting process $N(t)$ has *independent increments* if the quantities $N(t_1)-N(s_1)$ and $N(t_2)-N(s_2)$ are independent whenever $s_1<t_1<s_2<t_2$. In addition, $N(t)$ is *stationary* (or has *stationary increments*) if the distribution of $N(t)-N(s)$ depends only on $t-s$ for any $s<t$.

Definition 5.2 *A counting process* $N(t)$ *is a (homogenous) Poisson process with rate* λ *if the following conditions hold:*

(*i*) $N(t)$ *has independent and stationary increments,*
(*ii*) $P(N(s)=1)=\lambda s+o(s)$,
(*iii*) $P(N(s)>1)=o(s)$.

If for some function $\lambda:[0,\infty)\to\mathbb{R}_+$ (*i*)–(*iii*) *are replaced by*

(*i*)′ $N(t)$ *has independent increments,*
(*ii*)′ $P(N(t+s)-N(t)=1)=\lambda(t)s+o(s)$,
(*iii*)′ $P(N(t+s)-N(t)>1)=o(s)$.

then $N(t)$ *is a* nonhomogenous *Poisson process with intensity function* λ.

The usual convention is to assume that a Poisson process is homogenous unless explictly stated otherwise. The properties (*ii*)–(*iii*) essentially state that the expected number of arrivals between times t and $t+s$ is approximately λs, so that λ becomes the arrival rate. Definition 5.2 is a strong one, since it may shown that when it holds we must conclude that $N(t+s)-N(t)$ has a Poisson distribution with mean λs (Ross (1996), Chapter 2). If this is the case then, by the independent increment property, $P(X_2>s\mid X_1=t)=P(N(t+s)-N(t)=0)=\exp(-\lambda s)$, where X_i is the time between arrivals $i-1$ and i (the 0th arrival occurs at time $t=0$). That is, a Poisson process is a renewal process with exponentially distributed renewal times.

The Poisson process is of importance in stochastic modeling for much the same reason that the normal distribution is of importance in modeling noise processes. The latter is the limit of a superposition of an arbitrarily large number of independent (or weakly dependent) noise processes. Similarly, the Poisson process is a limit of a superposition of an arbitrily large number of renewal processes. This means that an arrival process which is really an aggregation of a large number of essentially separate arrival processes of general type will resemble a Poisson process. Derivations of this fact can be found in Section 5.9 of Karlin and Taylor (1975) or Section XI.3 of Feller (1971).

5.2 MARKOV PROCESSES

Given a Borel space $\mathcal{X}$ a stochastic process $X_n \in \mathcal{X}$ on a filtration $\mathcal{F}_n$, $n \geq 0$, is a discrete time *Markov process* if for some sequence of stochastic kernels $Q_n : \mathcal{X} \to \mathcal{M}_{\mathcal{X}}$, $n \geq 1$ we have $P(X_n \in E \mid \mathcal{F}_{n-1}) = Q_n(E \mid X_{n-1})$, with initial state X_0 distributed as Q_0. A Markov process is *time homogenous* if $Q_n \equiv Q$, $n \geq 1$, for some stochastic kernel Q. The elements of $\mathcal{X}$ are usually referred to as *states*. By convention a Markov process is assumed to be time homogeneous unless otherwise stated. Markov processes may also be defined on continuous time. The term *Markov chain* usually refers to a discrete time Markov process for which $\mathcal{X}$ is countable, although reference is often made to Markov chains with general (uncountable) state spaces (see Durrett (2010), Section 5.6). The theory of continuous-time Markov processes assumes a different character, and we will be concerned primarily with Markov chains.

Given two probability measures $P_0, P_1 \in \mathcal{M}^1_{\mathcal{X}}$, the equation $P_1 = P_0 Q$ means that if X_0 has distribution P_0, and X_1 is distributed conditionally as $Q(\cdot \mid X_0)$ then X_1 has distribution P_1. We have defined composition of stochastic kernels $Q : \mathcal{X} \to \mathcal{M}_{\mathcal{X}}$ in Section 4.9, and they play an important role in the theory of Markov chains. This leads to the J-step transition probabilities, defined recursively using

$$Q^J(E \mid x) = \int_{x' \in \mathcal{X}} Q^{J-1}(E \mid x') dQ(\cdot \mid x) = \int_{x' \in \mathcal{X}} Q(E \mid x') dQ^{J-1}(\cdot \mid x),$$

which represents the conditional probability $P(X_{n+J} \in E \mid X_n = x)$ for any n. The usual convention is to associate $J = 0$ with the identity transformation $Q^0(E \mid x) = I\{x \in E\}$.

We say that μ is an *invariant measure* of Q if

$$\mu = \mu Q.$$

Note that μ need not be a finite measure. If it is, it may be normalized to $\pi = \mu / \mu(\mathcal{X})$, in which case π is a *stationary distribution* for Q. The existence of such a distribution has important implications. Since $\pi = \pi Q$, if X_0 has distribution π, then so does X_1, and hence all X_n. In this case, we say that the Markov chain is in a *steady state*, which generally implies stable and predictable behavior.

Of course, ensuring that a Markov chain possesses an initial distribution π will usually not be a practical goal. The important question is whether or not the Markov chain has a tendency to approach π from any initial distribution, in the sense that $Q^n(\cdot \mid x) \Rightarrow \pi$ as $n \to \infty$ for any initial state $x \in \mathcal{X}$. When this holds, we may be able to replace weak convergence with the stronger convergence in the total variation norm, and possibly establish rates of convergence.

Alternatively, a weaker version of this property may hold. Suppose a Markov chain X_n proceeds indefinitely, and the empirical distribution of the observed states converges to a well defined distribution π. This may occur without $Q^n(\cdot \mid x)$ possessing a limit, and in fact will be the case for many models of practical interest.

The property under which a Markov chain possesses stable limiting behavior is referred to as *ergodicity*, and is also associated with the tendency of a Markov process towards recurrent behavior. Ideally, a Markov chain is a renewal process, with renewals defined by transition to some fixed state. If renewal times have finite expectation, it becomes possible at least to define long run frequencies.

For convenience we say a Markov chain is *discrete* if the state space $\mathcal{X}$ is countable. The important ideas can be developed for discrete Markov chains, following which we will consider the general state space model.

5.2.1 Discrete state spaces

We will rely primarily on the conventions used in Parzen (1962) and Ross (1996), two particularly elegant introductions to this subject. We first define a family of counting processes

$$N_j(t) = \sum_{n=0}^{t} I\{X_n = j\} \tag{5.3}$$

for each state $j \in \mathcal{X}$. The total number of visits $N_j(\infty) = \lim_{t\to\infty} N_j(t)$ is well defined. If we condition on $X_0 = j$, then by the Markov property $N_j(t)$ is a renewal process (Section 5.1.1), with renewal times given by transition into j. Note that for $N_j(t)$ the time domain may remain continuous ($t \in [0, \infty)$), with renewal times as integer valued random variables. Define

$$m_{ij}(t) = E[N_j(t) \mid X_0 = i].$$

Directly from (5.3) we have

$$m_{ij}(t) = \sum_{n=0}^{\lfloor t \rfloor} E[X_n = j \mid X_0 = i] = \sum_{n=0}^{\lfloor t \rfloor} Q^n(j \mid i).$$

Then $m_{jj}(t)$ is a renewal function for $N_j(t)$ given $X_0 = j$, and permits the definition of a rate of occurrence, or long run frequency, of a state j as $\lim_{t\to\infty} t^{-1} m_{jj}(t) = \pi_j$, where the limit exists. In fact, this type of limit is precisely characterized for a renewal process.

Define the random variable $T_j = \inf\{n \geq 1 \mid X_n = j\}$ as the minimum number of transitions required to reach state j from inititial state X_0. It will be convenient to set

$$f(i,j) = P(T_j < \infty \mid X_0 = i),$$

which, by the Markov property, is the probability that $X_{n+m} = j$ for some $m \geq 1$ given that $X_n = i$. We may then set

$$\mu_{ij} = E[T_j \mid X_0 = i].$$

Given $X_0 = j$, T_j models a renewal time for $N_j(t)$, with expected value μ_{jj}. By Theorem 5.2 we always have $m_{jj}(t) < \infty$. Otherwise, we have three possible cases.

Transient $f(j,j) < 0$. Then $\mu_{jj} = \infty$ and $wp1$ there is eventually an infinite renewal time, so directly from (5.1) we have $P(N_j(\infty) < \infty \mid X_0 = j) = 1$, and $\pi_j = 0$.

Null Recurrent $f(j,j)=1$ but $\mu_{jj}=\infty$. By Theorem 5.1 we have $P(N_j(\infty)=\infty \mid X_0=j)=1$, and by Theorem 5.2 we have $\pi_j=0$.

Positive Recurrent $\mu_{jj}<\infty$ (which implies $f(j,j)=1$). By Theorem 5.1 we have $P(N_j(\infty)=\infty \mid X_0=j)=1$, and by Theorem 5.2 we have $\pi_j=\mu_{jj}^{-1}>0$.

The three cases are labeled according to the standard terminology (Ross (1996)). A state j is *transient* or *recurrent* if $f(j,j)<1$ or $f(j,j)=1$, respectively. Recurrence is then classified as *null recurrence* or *positive recurrence* if $\mu_{jj}=\infty$ or $\mu_{jj}<\infty$, respectively.

Most attention is focused on Markov chains with positive recurrent states, since they possess well defined long run frequencies π_j. This case can be further subdivided. We may have limit $Q^n(j \mid j)\to_n \pi_j$, as well as $Q^n(j \mid i)\to_n \pi_j$ for certain states i (that this limit might not depend on the initial state is an important fact). In other cases $Q^n(j \mid j)$ does not possess a limit, so long run frequency is interpreted as a limit of average probabilities,

$$\pi_j=\lim_{n\to\infty} n^{-1}\sum_{k=1}^{n} Q^k(j \mid i).$$

5.2.2 Global properties of Markov chains

Although some of the important properties of a Markov chain can be resolved by renewal theory applied to specific states, a more powerful theory is available by defining global properties. These depend on the network structure implied by the kernel Q (the material in Sections 2.1.11, 2.3.4 and 2.3.6 are very much related).

We first note the *Chapman-Kolmogorov equations*:

$$Q^{m+n}(j \mid i)=\sum_{k\in\mathcal{X}} Q^m(j \mid k)Q^n(k \mid i) \tag{5.4}$$

for any $n,m\geq 0$, which follow from a standard conditioning argument.

Then we may define a directed graph $G(Q)$ with nodes labeled by the states in $\mathcal{X}$, which possesses a directed edge from i to j if and only if $Q(j \mid i)>0$. We can have $i=j$, permitting self edges, or directed edges from i to i. A Markov chain is then a random walk on $G(Q)$ along directed paths, with transitions from X_{n-1} to X_n represented by single edges, and governed conditionally by distribution $Q(\cdot \mid X_{n-1})$. Many global properties of a Markov chain follow from the connectivity properties of this graph (see, for example, Section 2.4 of Brémaud (1999)).

We say state j is *accessible* from i if $Q^n(j \mid i)>0$ for some $n\geq 0$. Following Theorem 2.2 and (5.4), if $i\neq j$ this holds if and only if there is at least one directed path in $G(Q)$ from i to j. If $i=j$ then $Q^0(i \mid i)=1$, that is i is always accessible from i. This convention holds even if $Q(i \mid i)=0$ (there is no self edge associated with node i in $G(Q)$).

If i,j are accessible to each other then they *communicate*, written $i\leftrightarrow j$. As previously discussed, we always have $i\leftrightarrow i$ for all $i\in\mathcal{X}$. This property is also symmetric and transitive, that is $i\leftrightarrow j$ if and only if $j\leftrightarrow i$; in addition $i\leftrightarrow j$ and $j\leftrightarrow k$ implies $i\leftrightarrow k$ (this follows from (5.4)). Thus, communication is an equivalence relation (Definition 2.1). This means that $\mathcal{X}$ can be partitioned into disjoint (equivalence) *classes*, such that i,j are in the same class if and only if $i\leftrightarrow j$. A Markov chain is *irreducible* if $\mathcal{X}$ is a single class.

A property that may be associated with a state is a *class property* if whenever it holds for one member of a class it must also hold for all other members. Transience, null recurrence and positive recurrence are all class properties.

The *period* of a state is defined as $d = \gcd\{n \geq 1 \mid Q^n(i \mid i) > 0\}$ (where gcd is the greatest common divisor). If there exists at least one finite n' for which $Q^{n'}(i \mid i) > 0$ then $1 \leq d \leq n'$, otherwise, $d = \infty$. If $d = 1$ then Q is *aperiodic*. Note that if a Markov chain has period d, it does not follow that $Q^d(i \mid i) > 0$. However, there must be some $n' < \infty$ for which $Q^{nd}(i \mid i) > 0$ for all $n \geq n'$, in particular, for an aperiodic state $Q^n(i \mid i) > 0$ for all large enough n (see, for example, complement 8D, Section 6.8, Parzen (1962)). It may be shown that perodicity is a class property in the sense that all states in a class have the same period.

Theorem 5.3 *For a discrete state Markov chain Q, state i is recurrent if and only if $E[N_i(\infty) \mid X_0 = i] = \sum_{n=1}^{\infty} Q^n(i \mid i) = \infty$.*

Proof If i is nonrecurrent then $P(N_i(\infty) \geq k \mid X_0 = i) = (1 - f(i,i))^k$ for all $k \geq 0$, which implies $E[N_i(\infty) \mid X_0 = i] < \infty$. If i is recurrent, then we must have $N_i(\infty) = \infty$ $wp1$. Therefore, $E[N_i(\infty) \mid X_0 = i] = \infty$ if and only if i is recurrent. Then

$$E[N_i(\infty) \mid X_0 = i] = \sum_{n=0}^{\infty} E[X_n = i \mid X_0 = i] = \sum_{n=0}^{\infty} Q^n(i \mid i),$$

which completes the proof. ///

We may have an *absorbing state* i, definied by the property $Q(i \mid i) = 1$. The Markov chain will not leave an absorbing state, once entered. A recurrence class E must also be absorbing in the sense that the Markov chain can never leave it once it is entered. To see this, suppose the Markov chain leaves E, then returns to E from some state $j \notin E$. This means that j is accessible for some state $i \in E$, and some state $i' \in E$ is accessible from j. But $i \leftrightarrow i'$, which implies $j \leftrightarrow i$, so j must be in E. This makes the important point that recurrence classes may in some sense be studied independently, and if more than one exists the Markov chain will select one of them, and remain in it indefinitely. For this reason, many (if not all) of the important properties of Markov chains can be studied by assuming irreducibility.

We have already seen that the advantage of positive recurrence lies in the natural definition of a steady state. If state i is visited infinitely often, with a finite recurrence time μ_{ii}, then in the long run we can expect to find the process in state i with a frequency of μ_{ii}^{-1}, giving us a steady state distribution. We introduced the steady state distribution as a solution to the equation $\pi = \pi Q$. For discrete Markov chains this leads to the following definition:

Definition 5.3 *A probability distribution π on $\mathcal{X}$ is* stationary *for a Markov chain Q if*

$$\pi_j = \sum_{i \in \mathcal{X}} Q(j \mid i) \pi_i \tag{5.5}$$

for all $j \in \mathcal{X}$.

This means that if a random variable X_1 has distribution $\pi = \{\pi_i\}$, and $P(X_2 = j \mid X_1 = i) = Q(j \mid i)$, then X_2 also has distribution π. A stationary distribution need not exist, but when it does it of considerable importance, representing the long run occupancy of the states.

Recall the definition of counting process $N_j(t)$ as the number of transitions into j by time t. We give the following theorem (Theorem 4.3.1 of Ross (1996)):

Theorem 5.4 *For a given Markov chain, if $i \leftrightarrow j$ then*

$$
\begin{aligned}
&(i) \quad P\left\{\lim_{t\to\infty} N_j(t)/t = \mu_{jj}^{-1} \mid X_0 = i\right\} = 1,\\
&(ii) \quad \lim_{n\to\infty} n^{-1} \textstyle\sum_{k=1}^{n} Q^k(j \mid i) = \mu_{jj}^{-1},\\
&(iii) \quad \textit{If } j \textit{ is aperiodic then } \lim_{n\to\infty} Q^n(j \mid i) = \mu_{jj}^{-1},\\
&(iv) \quad \textit{If } j \textit{ has period } d \textit{ then } \lim_{n\to\infty} Q^{nd}(j \mid j) = d\mu_{jj}^{-1}.
\end{aligned}
$$

It is important to note that although statements *(i)*–*(iii)* of Theorem 5.4 refer to some initial state i, all of the related limits are independent of i. In this sense, if a Markov chain is irreducible the dependence on the initial state vanishes in the limit.

The application of Theorem 5.4 is most clear if we assume that a Markov chain is *irreducible and aperiodic* (the definition makes sense, because irreducibility implies one class, and aperiodicity is a class property). In this case we conclude (Theorem 4.3.3 of Ross (1996)):

Theorem 5.5 *Assume that Markov chain is irreducible and aperiodic. Exactly one of the two statements holds:*

(i) *All states are transient, or all states are null recurrent, in which case* $\lim_{n\to\infty} Q^n(j \mid i) = 0$ *for all* i, j, *and no stationary distribution exists.*
(ii) *All states are positive recurrent, in which case* $\lim_{n\to\infty} Q^n(j \mid i) = \pi_j > 0$ *for all* i, j, *and* π_j *is the unique stationary distribution for* Q.

Following Theorem 5.5 an irreducible aperodic Markov chain on a countable state space $\mathcal{X}$ satisfying *(ii)* of Theorem 5.5 is called *ergodic*, and it is ergodic if and only if it possess a stationary distribution, which is unique when it exists.

If Q is irreducible, positive recurrent but periodic, then (5.5) will hold for $\pi_j = \mu_{jj}^{-1}$ (see Theorem 5.4 *(iv)*), but π is a set of long run frequencies, rather than a formal distribution.

Example 5.1 *Suppose we are given state space* $\mathbb{N}_0$, *and permit* $Q(j \mid i) > 0$ *only when* $|i - j| = 1$, *with the exception that* $Q(0 \mid 0)$ *may be nonzero. In particular, we set* $a_i = Q(i+1 \mid i)$, *and the constraint forces* $Q(i-1 \mid i) = 1 - a_i$ *for all* $i \geq 1$ *and* $Q(0 \mid 0) = 1 - a_0$ *The steady state equation (5.5) becomes*

$$
\begin{aligned}
\pi_0 &= (1 - a_0)\pi_0 + (1 - a_1)\pi_1,\\
\pi_i &= a_{i-1}\pi_{i-1} + (1 - a_{i+1})\pi_{i+1}, \quad i \geq 1,
\end{aligned} \tag{5.6}
$$

the successive addition of which yields solution

$$\pi_i = \frac{a_{i-1}}{1-a_i}\pi_{i-1}, \ i \geq 1, \text{ or}$$

$$\pi_i = \frac{\prod_{k=0}^{i-1} a_k}{\prod_{k=1}^{i}(1-a_k)}\pi_0, \ i \geq 1. \tag{5.7}$$

Clearly, the Markov chain is irreducible on $\mathbb{N}_0$ *if and only if* $a_i \in (0,1)$ *for* $i \geq 1$ *and* $a_0 \in (0,1]$. *If* $a_0 < 1$ *it will also be aperiodic, but will have period* $d = 2$ *if* $a_0 = 1$, *since transitions could only take place between even and odd numbers. In the aperiodic case a normalized solution to (5.6), if it exists, is interpretable as a steady state distribution, but in the periodic case it would be a set of long run frequencies, and not a distribution (steady state or otherwise). The question of periodicity often seems to interfere with an otherwise elegant theory. Many important theorems assume aperiodicity, but many commonly studied Markov chains are periodic. That periodicity might not greatly affect some of the important properties of a Markov chain is suggested by the ease with which it can be removed, in this case merely by ensuring that* $a_0 < 1$.

To fix ideas, suppose $a_i \equiv \alpha \in (0,1)$, $i \geq 0$. *The Markov chain is aperiodic and irreducible, and the solution to (5.6) becomes*

$$\pi_i = \left(\frac{\alpha}{1-\alpha}\right)^i \pi_0, \ i \geq 1.$$

Applying the normalization constraint $\sum_{i \geq 0} \pi_i = 1$ *yields*

$$\pi_0 = \left(1 + \sum_{i \geq 1}\left(\frac{\alpha}{1-\alpha}\right)^i\right)^{-1},$$

but as can be seen this is possible only if $\alpha < 1/2$, *in which case a solution to (5.6) exists, so that* π_i *defines a steady state distribution. If* $\alpha > 1/2$, *there will be a tendency for the Markov chain to drift upwards indefinitely, and will in this case be transient. If* $\alpha = 1/2$ *the Markov chain can be recognized as the absolute value of a symmetric random walk, which will be null recurrent (it will return to state* $i = 0$ *infinitely often* $wp1$*) but not ergodic (see further discussion below).*

We next define a finite state Markov chain on $\mathcal{X} = \{0, \ldots, M\}$ *by setting* $a_0 = (0,1]$, $a_i \in (0,1)$ *for* $i = 1, \ldots, M-1$, *and* $a_i = 0$ *for* $i \geq M$. *This Markov chain will be irreducible, and a solution to (5.6) can be obtained from (5.7) and a finite normalization constant.*

Thus, for finite state Markov chains, the ergodicity property will follow from irreducibility and aperodiocity, and a suitable modification of the definition of ergodicity can be made for periodic Markov chains (see below). For infinite state Markov chains, ergodicity often becomes an important point of inquiry, and does not follow from irreducibility and aperiodicity alone. In addition, as we have seen, even when these conditions do hold it does not suffice to verify that all states are visited infinitely often $wp1$.

5.2.3 General state spaces

The concept of ergodicity described in Theorem 5.5 involves positive recurrence, under which each state is visited infinitely often according to a renewal process, the renewal times possessing finite expected values. When the state space $\mathcal{X}$ is a more general metric space we cannot expect recurrence in this sense. There do, however, exist analogous definitions of irreducibility, recurrence and ergodicity which permit a unified theory.

First, suppose $\mathcal{X}$ is a Borel space, and Q is a stochastic kernel $Q:\mathcal{X}\to\mathcal{M}^1_{\mathcal{X}}$ satisfying Definition 4.7. If we are given $P_0\in\mathcal{M}^1_{\mathcal{X}}$ then $P_1=P_0Q$ is evaluated by

$$P_1(A)=\int_{\mathcal{X}} Q(A\mid y)dP_0(y), \quad \text{for all measurable } A\subset\mathcal{X}. \tag{5.8}$$

That P_1 is well defined follows from the fact that $Q(A\mid\cdot)$ is a measurable real-valued function on $\mathcal{X}$. That P_1 satisfies the definition of a probability measure is then easily verified. Any measure ν satisfying the fixed point equation $\nu=\nu Q$ is an *invariant measure*. Clearly, all scalar multiples of an invariant measure are also invariant measures. Thus, if a finite invariant measure exists in $\mathcal{M}^+_{\mathcal{X}}$, then there exists an *invariant probability measure* $P=PQ$.

We have seen that for discrete state spaces, ergodicity may be characterized by, in addition to recurrence, the convergence of the n-step transition probabilities to a proper distribution π, which is an invariant (i.e. stationary) distribution. Under suitable regularity conditions this concept extends to general state spaces.

Definition 5.4 *Let Q be a stochastic kernel on Borel space $\mathcal{X}$ (as in Definition 4.7). A measure P on $\mathcal{X}$ is* strictly positive *if $P(E)>0$ for all open sets E. Then Q is* strictly positive *if $Q(\cdot\mid x)$ is strictly positive for all $x\in\mathcal{X}$.*

In addition, Q is regular *if the sequence of real-valued functions on $\mathcal{X}$ evaluated by $u_n(x)=\int u_{n-1}dQ(\cdot\mid x)$ is equicontinuous whenever u_0 is uniformly continous (see Section 3.1.2).*

Finally, Q is ergodic *if there exists a strictly positive probability measure P such that for any probability measure P_0 the sequence P_0Q^n converges weakly to P (see Section 4.8.2).*

The following is a summary of Theorems 1-2 in Feller (1971), Section VIII.7:

Theorem 5.6 *Let Q be a strictly positive regular kernel on Borel space $\mathcal{X}\subset\mathbb{R}^p$. Then:*

(*i*) *If $\mathcal{X}$ is closed and bounded then Q is ergodic.*
(*ii*) *Q is ergodic if and only if it possesses a strictly positive stationary distribution P.*

The assumption in Theorem 5.6 that a stationary distribution is strictly positive is a natural one. In contrast, the assumption that Q is strictly positive is quite restrictive. It would be similar to the requirement that $Q(j\mid i)>0$ for all $i,j\in\mathcal{X}$ for a discrete state Markov chain, which would rule out most of the examples used in this book. The assumption can be relaxed somewhat by considering a finite iteration Q^k, which may be strictly positive even if Q isn't. The implication of this is that there exists a single k for which the k-step transition probability $Q^k(E\mid x)$ is positive for all initial states x and open sets E.

A much less restrictive characterization of ergodicity is available through the idea of *Harris recurrence* (Harris (1956)). It will clarify matters to first summarize how the recurrence properties of a discrete state space Markov chain can be established by considering a single state. The notion of recurrence implies an embedded renewal process, and we may select any state i, and define any transition into that state as a renewal time. Then i is recurrent if $f(i,i)=1$, or, by Theorem 5.3, if $E[T_i \mid X_0=i]=\sum_{n=1}^{\infty} Q^n(i \mid i)=\infty$. If another state j is accessible from i we must have $i \leftrightarrow j$, since otherwise we would have $1-f(i,i) \geq f(i,j) > 0$. Therefore, all states accessible from i form a recurrence class. In addition, since positive recurrence and periodicity are class properties whatever holds for i holds for the entire class (we will make use of this device again in Section 12.7).

This approach cannot be used directly for general state spaces in which singletons have measure 0. However, it might be possible to define conditions under which an *auxiliary state* α may be embedded into a Markov chain such that (*i*) $Q^n(\{\alpha\} \mid \alpha) > 0$ for large enough n; and (*ii*) the resulting process retains the Markovian property. If this is possible, then recurrence may be defined in terms of the auxiliary state α.

For any $A \subset \mathcal{X}$, let $f(x,A)$ be the probability that a Markov chain visits any state $y \in A$, given initial state x, after a finite number of transitions. This is comparable to the quantity $f(i,j)$ introduced earlier. Similarly, $N_A(t)$ is the counting process of visits to A, with limit $N_A(\infty)$.

We then define the *Harris chain*:

Definition 5.5 *A stochastic kernel Q on Borel space $\mathcal{X}$ defines a* Harris chain *if there exist measurable sets $A, B \subset \mathcal{X}$, a probability measure ν for which $\nu(B)=1$, and a constant $\epsilon > 0$ for which the following statements hold:*

(*i*) $f(x,A) > 0$ *for all* $x \in \mathcal{X}$,
(*ii*) *If* $x \in A$ *and* $C \subset B$ *then* $Q(C \mid x) \geq \epsilon\nu(C)$.

The definition of a Harris chain may seem unintuitive, at least statement (*ii*). It is best understood in relation to its purpose, which is the construction of the type of auxiliary state α just described. Clearly, the intention is to associate renewal times, and therefore α, with visits to an aggregation of states A. Unfortunately, we cannot simply collapse A into a single state. Certainly, there is nothing preventing us from replacing a Markov chain X_n with $X_n^\delta = X_n I\{X_n \notin A\} + \delta I\{X_n \in A\}$, making use of dummy state label $\delta \notin \mathcal{X}$. We would merely replace any $X_n \in A$ with an indicator that X_n is in A. The problem with this approach is that X_n^δ need no longer be a Markov chain, as is generally the case when states are aggregated in this way.

It is statement (*ii*) of Definition 5.5 which permits the construction of an auxiliary state $\alpha \notin \mathcal{X}$ with which to define recurrence. Based on Q, we construct stochastic kernel $\bar{Q}$ on Borel space $\mathcal{X} \cup \{\alpha\}$ as follows:

$$\begin{aligned}
\bar{Q}(C \mid x) &= Q(C \mid x) \quad \text{for } x \notin A, \ C \subset \mathcal{X}, \\
\bar{Q}(\{\alpha\} \mid x) &= \epsilon \quad \text{for } x \in A, \\
\bar{Q}(C \mid x) &= Q(C \mid x) - \epsilon\nu(C) \quad \text{for } x \in A, \ C \subset \mathcal{X}, \\
\bar{Q}(C \mid \alpha) &= \int_B \bar{Q}(C \mid x) d\nu \quad \text{for } C \subset \mathcal{X} \cup \{\alpha\}.
\end{aligned}$$

Note that $\bar{Q}$ is not a stochastic kernel unless statement (*ii*) of Definition 5.5 holds.

Essentially, Q and $\bar{Q}$ define the same Markov chain. The distinction may be seen by defining a third Markov chain X'_n. If $X'_n = x \notin A$ suppose X'_{n+1} is selected by $Q(C \mid x)$. On the other hand, if $X'_n = x \in A$ suppose the subsequent state is determined by the mixture

$$Q_A(\cdot \mid x) = (1 - \epsilon)Q_1 + \epsilon Q_2$$

where

$$Q_1 = \frac{Q(\cdot \mid x) - \epsilon\nu}{1 - \epsilon} \quad \text{and} \quad Q_2 = \nu.$$

This is equivalent to generating an independent random indicator variable Z_n equalling 1 with probability ϵ, then selecting X'_{n+1} from Q_2 if $Z_n = 1$ and from Q_1 otherwise. Of course, as is easily verified, $Q_A = Q$, so X'_n is identical to the Markov chain defined by Q, except that we now have a sequence of independent indicator variables Z_n.

Suppose we then derive the process $\bar{X}_n$ from X'_n by setting $\bar{X}_n = X'_n$, unless we had $X'_{n-1} \in A$ with the associated indicator variable $Z_{n-1} = 1$, in which case we set $\bar{X}_n = \alpha$ (and X'_n is distributed as ν). In this case $\bar{X}_n$ is a Markov chain with kernel $\bar{Q}$, and we have a single state α, with nonzero probability, which may be used to define recurrence. That $\bar{X}_n$ remains a true Markov chain while possessing a single state which might be recurrent is attributable to the ingenuity of Definition 5.5.

A Harris chain is recurrent if $f(\alpha, \{\alpha\}) = 1$. Theorem 5.3 is directly applicable, that is, a Harris chain is recurrent if and only if

$$\sum_{n=0}^{\infty} \bar{Q}^n(\{\alpha\} \mid \alpha) = \infty. \tag{5.9}$$

Similarly, the period of α is $d = \gcd\{n \geq 1 \mid \bar{Q}^n(\{\alpha\} \mid \alpha) > 0\}$ and a Harris chain is aperiodic if $d = 1$.

Recurrence may also be given in terms of set A. Suppose for all $x_0 \in A$ we have $f(x_0, A) = 1$. Then for any initial state $X_0 = x_0 \in A$ we must have $N_A(\infty) = \infty$ *wp*1. Clearly, $E[N_\alpha(\infty) \mid X_0 = x_0] \geq \epsilon E[N_A(\infty) \mid X_0 = x_0] = \infty$, so α must be recurrent.

We summarize Theorems 6.5 and 6.8 of Durrett (2010):

Theorem 5.7 *A recurrent Harris chain possesses an invariant measure. In addition, if a recurrent aperiodic Harris chain Q possesses a stationary distribution $P = PQ$ then*

$$\lim_{n\to\infty} \left\| Q^n(\cdot \mid x) - P \right\|_{TV} = 0$$

whenever $f(x, \{\alpha\}) = 1$.

If the invariant measure described in Theorem 5.7 is finite, the Harris chain is *positive recurrent*. The invariant measure can then be normalized to define a stationary distribution (see Hernández-Lerma and Lasserre (2001)). Otherwise the Harris chain is *null recurrent*.

Associate with $\mathcal{X}$ a σ-finite measure μ, permitting stochastic kernel Q^n, for any $n \geq 1$ to be represented by a *density kernel* $q^n(x' \mid x)$, which is assumed to be a Borel measurable mapping from $\mathcal{X}^2$ to $\bar{\mathbb{R}}_+$. This gives

$$Q^n(E \mid x) = \int_{x' \in E} q^n(x' \mid x) d\mu. \tag{5.10}$$

The intention is usually that μ is Lebesgue measure on $\mathcal{X} \subset \mathbb{R}^p$ or counting measure on any countable $\mathcal{X}$.

If μ is Lebesgue measure on $\mathcal{X} \subset \mathbb{R}^p$, we can generally find x_0, y_0 for which $q(y_0 \mid x_0) > 0$, with $q(y \mid x)$ continuous on $\mathcal{X}^2$ at (x_0, y_0). In this case, we can find $\epsilon > 0$ and small enough neighborhoods A and B of x_0 and y_0 so that statement (*ii*) of Definition 5.5 holds when ν is Lebesgue measure concentrated on B and normalized so that $\nu(B) = 1$. Thus, verifying that Q is a Harris chain involves selecting a state x_0 possessing a neighborhood to which the Markov chain will indefinitely return, the remainder of the definition following from sufficient continuity conditions for q.

A similar construction for countable $\mathcal{X}$ is simpler, setting $A = \{x_0\}$ where x_0 is accessible from all other states, and $B = \{y_0\}$ for any state satisfying $Q(\{y_0\} \mid x_0) > 0$. It may be shown that if any sets A, B satisfy Definition 5.5, there exists a pair of singletons A', B' which also satisfy the definition (see Example 6.1, Chapter 5 of Durrett (2010)).

5.2.4 Geometric ergodicity

Definition 5.4 relies on weak convergence $Q^n(\cdot \mid x) \Rightarrow_n \pi$ to define ergodicity. In contrast Theorem 5.7 gives conditions for convergence in the total variation norm (recall from Section 4.8.3 that convergence of measures in $\|\cdot\|_{TV}$ implies weak convergence). Norm convergence permits the definition of convergence rates, and a rich theory exists which yields $O(\rho^n)$ convergence. The constant $\rho < 1$ is itself of considerable interest.

Definition 5.6 *An irreducible aperiodic Markov chain is* geometrically ergodic *if it possess a stationary distribution π, and there is a constant $\rho < 1$ and π-integrable function M such that*

$$\left\| Q^n(\cdot \mid x) - \pi \right\|_{TV} \leq M(x)\rho^n$$

for all $n \geq 1, x \in \mathcal{X}$. The total variation norm may be weighted by w, in which case we refer to w-geometric ergodicity. For countable state spaces the condition may be stated

$$\left| Q^n(j \mid i) - \pi_j \right| \leq C_{ij}\rho^n$$

where $C_{ij} < \infty$ for all i, j.

Recurrence of a Harris chain is a consequence of the assumption that $f(x, A) = 1$ for all $x \in \mathcal{X}$. A stronger ergodicity property follows from the assuption that for some $A \subset \mathcal{X}$ we have the uniform lower bound $Q(A \mid x) \geq \delta > 0$ for all $x \in \mathcal{X}$. This can be seen to be a stronger assumption than that require for Harris recurrence. The following theorem is derived from Case (b), Section V.5 of Doob (1953).

Theorem 5.8 *Suppose we are given stochastic kernel Q, with Borel measurable density kernels q^n as defined in (5.10). Suppose there exists $C \subset \mathcal{X}$, $\delta > 0$ and integer m for which $0 < \mu(C) < \infty$, and*

$$q^m(y \mid x) \geq \delta \quad \textit{for all } y \in C \textit{ and } x \in \mathcal{X}. \tag{5.11}$$

Then Q possesses a stationary distribution π, for which $\pi(C') \geq \delta\mu(C')$ when $C' \subset C$, and for which

$$\begin{aligned} 2^{-1} \left\| Q^n(\cdot \mid x) - \pi \right\|_{TV} &= \sup_{E \subset \mathcal{X}} \left| Q^n(E \mid x) - \pi(E) \right| \\ &\leq (1 - \delta\mu(C))^{(n/m)-1} \end{aligned} \tag{5.12}$$

for all $n \geq 1$, $x \in \mathcal{X}$.

The condition (5.11) is known as *Doeblin's condition* (see, for example, Isaac (1963) for further discussion of this condition). For discrete Markov chains this is equivalent to the existence of $j \in \mathcal{X}$, $\delta > 0$ and integer $m \geq 1$ for which $Q^m(\{j\} \mid i) \geq \delta$ for all $i \in \mathcal{X}$.

An alternative approach is given by necessary and sufficient conditions for geometric ergodicity for countable state spaces, due to Popov (1977):

Theorem 5.9 *If $\mathcal{X}$ is countable and Q is an ergodic Markov chain, then Q is also geometrically ergodic if and only if there is a finite real valued function f on $\mathcal{X}$, a number $\rho < 1$, and a finite subset $B \subset \mathcal{X}$ which satisfy*

$$E[f(X_1) \mid X_0 = i] \leq \rho f(i) \quad \textit{for all } i \notin B, \textit{ and}$$

$$\max_{i \in B} E[f(X_1) \mid X_0 = i] < \infty.$$

Conditions analagous to those given in Theorem 5.9 for recurrent Harris chains with general state space are given in Nummelin and Tuominen (1982).

The interested reader can be referred to Section 7.3 of Hernández-Lerma and Lasserre (1999), and to a concise set of conditions given in Chan (1989). The discussion of ergodicity for Markov chains on general state spaces in Section V.5 of Doob (1953) extends beyond the conclusion of Theorem 5.8.

Geometric ergodicity of finite state Markov chains

Suppose $\mathcal{X}$ is finite, and Q is irreducible and aperidodic. If case (*i*) of Theorem 5.5 holds then $Q^n(j \mid i) \to_n 0$ for all pairs i, j. However, this cannot be the case, since $Q^n(\cdot \mid i)$ is a proper probability distribution for all $n \geq 1$. By aperiodicity we may then identify N for which $Q^n(\{i\} \mid i) > 0$ for all $n \geq N$ and $i \in \mathcal{X}$. By irreducibility, for any pair of states i, j there exists n for which $Q^n(\{j\} \mid i) > 0$. If $Q^{n'}(\{j\} \mid j) > 0$, by the Chapman-Kolmogorov equations (5.4) we have $Q^{n+n'}(\{j\} \mid i) > 0$. This in turn implies the existence of N_{ij} for which $Q^n(\{j\} \mid i) > 0$ for all $n \geq N_{ij}$. Since the number of pairs i, j is finite, Doeblin's condition (5.11) holds, so that (5.12) of Theorem 5.8 holds. We have just proven:

Theorem 5.10 *Any irreducible aperiodic Markov chain on a finite state space is geometrically ergodic.*

It is important to emphasize that the essential condition here is irreducibility. With some additional technical argument, periodic Markov chains can be analyzed as aperiodic Markov chains after *cyclic decomposition* (see, for example, Durrett (2010), Section 5.5).

We expect stable control systems to be ergodic, but many other important Markov chains will not be.

Example 5.2 *A random walk on* $\mathbb{R}$ *can be defined as a cumulative sum* $S_n = \sum_{i=1}^{\infty} X_i$ *where* X_i *is an iid sequence. Suppose* $P(X_1 = 1) = P(X_1 = -1) = 1/2$. *Then* S_n *is irreducible and periodic (*$d = 2$ *since* S_n *must have the same parity as* n*). The kernel can be given by* $Q(\{j\} \mid i) = 1/2$ *if* $|i - j| = 1$ *and* $Q(\{j\} \mid i) = 0$ *otherwise. It is easily seen that counting measure* μ *on* $\mathcal{I}$ *is an invariant measure for* Q. *It solves (5.5) but is not a probability distribution. Despite this, it does describe intuitively the limiting behavior of* S_n *as increasingly diffuse. The long run frequencies are easy to deduce from the marginal distribution of* $S_n = 2Z_n - n$, *where* $Z_n \sim bin(n, 1/2)$, *forcing us to conclude that* $\lim_{n\to\infty} Q^n(j \mid i) = 0$ *and so* S_n *is not positive recurrent. However, it can be shown that* $\sum_{n=1}^{\infty} Q^n(i \mid i) = \infty$ *(use the binomial distribution and Stirling's approximation) so that all states are null recurrent. Even without ergodicity, much can be said about the long run properties of* Q.

If we would rather S_n *be aperiodic we may alter the support of* X_1, *for example,* $P(X_1 = 1) = P(X_1 = -1) = \alpha < 1/2$ *and* $P(X_1 = 0) = 1 - 2\alpha$. *Then counting measure remains an invariant measure. Interestingly, the symmetric random walk remains null recurrent when extended to two dimensions, but becomes transient in any higher dimension. This is known as Pólya's Theorem (Example 8.6, Billingsley (1995)).*

5.2.5 Spectral properties of Markov chains

Suppose $\mathcal{X} = \{1, \ldots, N\}$. In this case a stochastic kernel Q can be represented as a *stochastic matrix* (Section 2.3.5). Taking liberties with notation, we have $Q \in M_N(\mathbb{R})$, setting $Q_{i,j} = Q(\{j\} \mid i)$. Each row of Q consists of a probability distribution on $\mathcal{X}$, which is the defining property of a stochastic matrix. If π is a steady state distribution for Q, then it is a left eigenvector, since $\pi = \pi Q$, with associated eigenvalue 1. In addition $\vec{1}$ is a right eigenvector, since for any stochastic matrix $\vec{1} = Q\vec{1}$, also with associated eigenvalue 1.

The definition of an irreducible nonnegative matrix (Definition 2.2) corresponds exactly to the definition of an irreducible Markov chain. The discussion of Section 2.3.6 makes clear that this is precisely what is implied by the mutual communication between all pairs of states. The stronger definition of the primitive matrix (Definition 2.3) is equivalent in this sense to aperiodicity. Therefore, the Perron-Frobenius theorem (Theorem 2.13) is directly relevant to the theory of Markov chains.

Finally, the convergence rate given by (2.19) is applicable under general conditions (see, for example, Section 6.1 of Brémaud (1999)).

5.3 CONTINUOUS-TIME MARKOV CHAINS

The definition of a Markov chain extends naturally from discrete to continuous time for countable state spaces. A stochastic process $\{X(t) : t \in [0, \infty)\}$ is a *continuous-time*

Markov chain on $\mathcal{X}$ if:

$$P(X(t+s)=j \mid X(t)=i \wedge X(t')=x(t'), t' \in [0,t)) \\ = P(X(t+s)=j \mid X(t)=i) \quad (5.13)$$

for any $i,j \in \mathcal{X}$, $t,s \geq 0$, and mapping $x:[0,t) \to \mathcal{X}$. It may be assumed that $X(t)$ is right-continuous $wp1$. The Markovian property (5.13) actually has two independent aspects. We may identify a sequence of transition times $t_0, t_1, t_2, \ldots$ in which $t_0 = 0$ and t_i is the ith point of discontinuity of $X(t)$. Then $\{X(t_i) : i \geq 0\}$ is referred to as the *embedded* Markov chain. Of course, the embedded Markov chain will be a true Markov chain under the assumption (5.13). In addition, the intertransition times must also satisfy the appropriate Markovian property. In particular, at any point in time the distribution of the time remaining until a transition cannot depend on the time since the last transition. Any waiting time with this property is referred to as *memoryless*. For this reason, the transition law of a continuous-time Markov chain may be constructed using elementary probability theory, based on the remarkable properties of the exponential distribution, which defines a random variable X on support $[0,\infty)$ using density function $f(x \mid \lambda) = \lambda \exp(-\lambda x)$ for any $\lambda > 0$. The parameter λ is appropriately referred to as the *rate*, since a Poisson process with rate λ has interarrival times of density $f(x \mid \lambda)$. Note that, as we would expect, $E[X] = \lambda^{-1}$.

Theorem 5.11 *The following properties hold for the exponential distribution:*

(*i*) *The exponentially distributed random variable is the unique memoryless waiting time with support* $[0,\infty)$.
(*ii*) *If* $X_1, \ldots, X_n$ *are independent exponentially distributed random variables with rates* $\lambda_1, \ldots, \lambda_n$ *then* $Y = \min_i X_i$ *is exponentially distributed with rate* $\sum_i \lambda_i$.
(*iii*) *In addition,* $P(Y = X_j) = \lambda_j / \sum_i \lambda_i$.

Proof (*i*) A waiting time X on $[0,\infty)$ is memoryless if and only if $P(X > t+s \mid X > t) = P(X > s)$, or $P(X > t+s) = P(X > s)P(X > t)$, for all $s,t \geq 0$. That this property is satisfied by the exponential distribution is easily verified by substituting $\bar{F}(x) = \exp(-\lambda x)$. The prove the converse, suppose a memoryless waiting time on support $[0,\infty)$ has distribution function F. Letting $S(u) = \log(\bar{F}(u))$, the memoryless property implies $S(t+s) = S(t) + S(s)$. Since S is monotone, a solution to this equation must be of the form $S(t) = ct$, which completes the proof.

(*ii*) $P(Y > t) = P(\cap_i \{X_i > t\}) = \prod_i P(X_i > t) = \prod_i \exp(-\lambda_i t) = \exp\left(-t \sum_i \lambda_i\right)$.

(*iii*) For $n=2$, we may evaluate $P(Y = X_1) = P(X_1 < X_2) = \int_{x_1 < x_2} f(x_1 \mid \lambda_1) f(x_2 \mid \lambda_2) dx_1 dx_2 = \lambda_1/(\lambda_1 + \lambda_2)$. The extension to $n \geq 2$ follows from (*ii*). ///

The proof of (*i*) of Theorem 5.11 involves a solution to *Cauchy's equation* ($S(t+s) = S(t) + S(s)$). This clearly includes $S(t) = ct$ for any constant c, but includes additional nonlinear solutions. Fortunately, the regularity conditions for S required to preclude nonlinear solutions are quite minimal, namely that S is continuous at one point (Kuczma (2009)).

A construction of (5.13) may be based on Theorem 5.11. For countable $\mathcal{X}$ we define a finite transition rate $\lambda_{ij} \geq 0$ for any pair of states i,j. To avoid complications we may assume that for each $i \in \mathcal{X}$ there is at least one $j \neq i$ for which $\lambda_{ij} > 0$, otherwise i would be an absorbing state. If this holds, it can be shown that whenever $\lambda_{ii} > 0$ it will be possible to reset λ_{ii} to zero and adjust the remaining rates to yield an equivalent process. We may therefore assume $\lambda_{ii} \equiv 0$ without loss of generality. Let E_i be all states j for which $\lambda_{ij} > 0$ and define $\lambda_i^* = \sum_{j \in E_i} \lambda_{ij}$. When the process is in state i we associate with each $j \in E_i$ an exponentially distributed waiting time of rate λ_{ij}. All waiting times are independent of each other and of the process history. The process remains in state i until completion of any waiting time, at which time the process moves to the state associated with that waiting time. By Theorem 5.11, this process is a continuous-time Markov chain, for which the transition time from state i is exponentially distributed with rate λ_i^*, and for which the embedded Markov chain has transition kernel $Q(\{j\} \mid i) = \lambda_{ij}/\lambda_i^*$.

A distinction has to be made between the steady state of the embedded Markov chain and the continuous-time Markov chain $X(t)$, and the two can be quite different. To see this, suppose for some state i all transitions rates λ_{ij} in E_i are divided by a common factor M. This does not change the transition probabilities $Q(\{j\} \mid i)$ of the embedded Markov chain, but the expected time that $X(t)$ remains in state i pending a transition has expected value M/λ_i^*, and so can be made arbitrarily large while leaving the distributional properties of the embedded Markov chain unchanged.

The steady state occupation distribution is given by $P_j = \lim_{t\to\infty} P(X(t) = j \mid X(0) = i)$, assuming the limit exists and does not depend on i, and is interpretable as the proportion of time the system spends in state j. Again, this is distinct from the steady state distribution π of the embedded Markov chain.

A formal derivation would be based on the Chapman-Kolmogorov equations for continuous-time Markov chains, which are essentially the same set of equations as (5.4) for Markov chains:

$$P_{ij}(t+s) = \sum_{k \in \mathcal{X}} P_{ik}(t) P_{kj}(s) \tag{5.14}$$

for $s, t \geq 0$ and all $i, j \in \mathcal{X}$, where $P_{ij}(t) = P(X(t) = j \mid X(0) = i)$ represents the transition probabilities for time increments of length t. In addition, $P_{ij}(t)$ defines a proper distribution over $j \in \mathcal{X}$ for all $i \in \mathcal{X}$ and $t \geq 0$, with $\lim_{t\to 0} P_{ij}(t) = 1$ if $i = j$, and is zero otherwise (the last assumption is comparable to *(ii)*–*(iii)* of Definition 5.2 of a Poisson process).

This elementary construction suffices for our purposes, but does not resolve all issues related to the type of model defined by (5.13). The interested reader may be referred to Chapter 14 of Karlin and Taylor (1981) or to Davis (1993) for a more general treatment.

Accepting the elementary model, necessary conditions for a steady state may be obtained using the *balance equations*. These are based on the observation that the number of times $X(t)$ enters any state i (eventually) equals the number of times it leaves that state. We may consider the counting process of exits from state i. This would have rate λ_i^* while $X(t)$ is at state i, so that if the occupancy frequency of state i was P_i, the exit rate would be $\lambda_i^* P_i$. Of these exits, a proportion $Q(\{j\} \mid i)$ are to state j, where Q is the transition kernel of the embedded Markov chain, so transitions from

i to j occur at rate $Q(\{j\} \mid i)\lambda_i^* P_i = (\lambda_{ij}/\lambda_i^*)\lambda_i^* P_i = \lambda_{ij} P_i$. Then the entrance rate for state i is the sum of all transition rates into i, leading to balance equations

$$\lambda_i^* P_i = \sum_{k \neq i} \lambda_{ki} P_k, \quad i \in \mathcal{X}. \tag{5.15}$$

With additional regularity conditions the equations (5.15) can be derived as a steady state solution to a system of differential equations obtained by allowing s to approach 0 in the Chapman-Kolmogorov equations (5.14) (Karlin and Taylor (1981)).

5.3.1 Birth and death processes

One important model for which the balance equations (5.15) do characterize the steady state is the *birth and death process*. Here, $\mathcal{X} = \mathbb{N}_0$, and transitions occur only between states i, j for which $|i - j| = 1$. Using the elementary model, with each state i we associate a *birth rate* $\lambda_i = \lambda_{i,i+1}$ and *death rate* $\mu_i = \lambda_{i,i-1}$, setting $\mu_0 = 0$, $\lambda_0 > 0$ and, for the moment, assuming $\lambda_i, \mu_i > 0$ for $i \geq 1$. The balance equations become

$$\begin{aligned} \lambda_0 P_0 &= \mu_1 P_1, \\ (\lambda_i + \mu_i) P_i &= \lambda_{i-1} P_{i-1} + \mu_{i+1} P_{i+1}, \quad i \geq 1, \end{aligned} \tag{5.16}$$

for which a solution must take the form

$$P_i = \frac{\lambda_{i-1}}{\mu_i} P_{i-1} = \frac{\prod_{k=0}^{i-1} \lambda_k}{\prod_{k=1}^{i} \mu_k} P_0, \quad i \geq 1. \tag{5.17}$$

The steady state distribution can be expressed exactly after calculating a normalizing constant from $\sum_i P_i = 1$ (the same procedure as in Example 5.1). We may also restrict the process to a finite state space $\mathcal{X} = \{0, 1, \ldots, M\}$, in which case we would need to define positive birth and death rates λ_{i-1} and μ_i for $i = 1, \ldots, M$, and note that (5.17) would hold for this range.

The steady state distribution of the embedded Markov chain may be obtained from (5.7) after setting $a_i = \lambda_i / (\mu_i + \lambda_i)$ for $i \geq 1$ and $a_0 = 1$.

5.4 QUEUEING SYSTEMS

Queueing systems form a class of stochastic process of considerable importance in operations research, and present a rich set of applications in control and optimization. They also possess an elegant and intuitive parametric modeling theory, and will therefore serve well to illustrate some of the techniques described in this volume. A queueing system is easy to describe. It contains a queue into which an arrival process of customers enter. It also contains m servers, who service customers. The time of service has a specified distribution. A customer in the queueing system is either being served, or is in the queue waiting to be served. A server is either busy serving a customer, or is free. Upon service completion the customer exits the system and the server immediately begins service of some customer from the queue if it is not empty. Accordingly, a customer entering the system begins service immediately if there is a free server, or enters the queue otherwise to wait for service. A variety of *queueing disciplines* exist to determine which customer in the queue enters service at the next service completion, the most common being the normally observed FIFO/FCFS discipline (*first in first out* or *first come first served*). The queue discipline affects some system properties (waiting time of an arriving customer) but not others (busy period of server), at least absent additional structure, and so is specified only when relevant.

5.4.1 Queueing systems as birth and death processes

Considerable insight can be gained by modeling queueing systems as birth and death processes on $\mathbb{N}_0$. The state variable will be the number of customers in the system (in service and in queue), and either increases by one when a customer arrives, or decreases by one when a service ends. Service times and interarrival times are exponentially distributed (that is, memoryless), but the service time rates μ_i and the arrival rates λ_i (using the notation of Section 5.3.1) are allowed to depend on the state variable i. The steady state distribution, if it exists, is obtained directly from (5.17) as

$$\begin{aligned} P_i &= \frac{\prod_{k=0}^{i-1} \lambda_k}{\prod_{k=1}^{i} \mu_k} P_0, \quad i \geq 1, \\ P_0 &= \frac{1}{1 + \sum_{i=1}^{\infty} \frac{\prod_{k=0}^{i-1} \lambda_k}{\prod_{k=1}^{i} \mu_k}} \end{aligned} \tag{5.18}$$

We have directly that $P_0 \leq 1$, so the essential remaining condition is that $P_0 > 0$, equivalently, that the sum $\sum_{i=1}^{\infty} (\prod_{k=0}^{i-1} \lambda_k)/(\prod_{k=1}^{i} \mu_k)$ is finite. A steady state distribution exists if and only if this condition holds. Similarly, the steady state distribution π of the embedded Markov chain follows from (5.7) as

$$\begin{aligned} \pi_i &= \frac{\prod_{k=0}^{i-1} \lambda_k/(\mu_k + \lambda_k)}{\prod_{k=1}^{i} \mu_k/(\mu_k + \lambda_k)} \pi_0 \\ &= \left(\frac{\mu_i + \lambda_i}{\lambda_0}\right)\left(\frac{\prod_{k=0}^{i-1} \lambda_k}{\prod_{k=1}^{i} \mu_k}\right) \pi_0, \quad i \geq 1, \\ \pi_0 &= \frac{1}{1 + \sum_{i=1}^{\infty} \left(\frac{\mu_i + \lambda_i}{\lambda_0}\right)\left(\frac{\prod_{k=0}^{i-1} \lambda_k}{\prod_{k=1}^{i} \mu_k}\right)}. \end{aligned} \tag{5.19}$$

A few features of (5.18) and (5.19) are worth noting. First, the term λ_0 cancels in (5.19), so that the calculation of π_i does not involve λ_0, unlike P_i. However, its inclusion in (5.19) makes the point that if π_i is replaced with $\pi_i = \lambda_i^* P_i$ (noting that $\lambda_0^* = \lambda_0$) we obtain (5.18), that is, the respective solutions are related by $\pi_i \propto \lambda_i^* P_i$. This is expected, since $1/\lambda_i^*$ is the expected time the process remains in state i during a single visit.

The simplest queueing model asumes constant arrival and service rates $\lambda_i \equiv \lambda$ and $\mu_i = \mu$ with $m = 1$ server. By substitution into (5.18) or (5.19) we can see that the process is ergodic if and only if $\lambda < \mu$.

The birth and death model easily accomodates modifications. Suppose we have $m > 1$ servers, and each is capable of serving at a rate μ. In this case μ_i models not a single server rate, but the system service rate. If there are i customers in service, then by Theorem 5.11 the system service rate is $i\mu$. If there are i customers in the system, there are $\min(i, m)$ customers in service, so the birth and death process is defined by $\lambda_i \equiv \lambda$ and $\mu_i = \min(i, m)\mu$, and these values may be substituted into (5.18) or (5.19). Similarly, if the queue has finite capacity, that is, it can hold no more than $K < \infty$ customers, this can be modeled by setting μ_i, λ_i to zero for all large enough i. In addition, the number of customers may be a finite number M. If each potential customer enters the queueing system at rate λ, then the system arrival rate is $\lambda_i = (M - i)\lambda$ when there are i customers in the system.

5.4.2 Utilization factor

The *utilization factor* of a queueing system may be defined as

$$\rho = \lambda / \mu$$

where λ is the average arrival rate of customers, and μ is the service rate. The precise definition depends on the system, since the arrival and service characteristics need not be time homogenous. In such cases, μ may be taken as the maximum service rate. This quantity is fundamental to queueing systems, since we expect the service rate to be smaller than the arrival rate, otherwise the queue size will increase indefinitely (this generally holds even when $\rho = 1$). In the single server model of Section 5.4.1, this idea is made precise by the observation that the queueing system is ergodic if and only $\rho < 1$. See also equation (5.20) below. The utilization factor will be of some importance in subsequent examples, since even within the constraint $\rho < 1$ distribution properties affecting approximation methods can vary greatly, and computational challenges may arise when ρ is close to 1.

5.4.3 General queueing systems and embedded Markov chains

Birth and death queueing system models offer considerable flexibility and insight, but will clearly not be adequate for all systems. As discussed in Section 5.1.2 the Poisson process will approximate an aggregation of many independent arrival processes, which seems a reasonable assumption for many actual queueing systems. The assumption that the service time distribution is exponential is more tentative. The exponential density is defined by only one parameter, so that the mean μ and standard deviation σ obey a fixed relationship, namely that the *coefficient of variation* is always $\sigma/\mu = 1$. There is certainly no reason to think that this value is inherent to service times in any given queueing system. The coefficient of variation will surely differ significantly between for example, the time required to process a fixed payment, and the time required for general repair services.

Queueing models are commonly classified using *Kendall's notation* which originally took form $A/B/m$, as originally proposed in Kendall (1953). It is assumed that the customer arrival process is a renewal process with a renewal distribution described by A. Then B refers in the same way to the service time (which are assumed to be independent), and m is the number of servers. The convention has since been extended, most commonly to $A/B/m/K/M$, in which K is the system capacity, and M is the number of customers in a finite population. The last two parameters are often omitted when they equal ∞. The symbols for A, B are standard, with M denoting the exponential distribution (M for 'memoryless'), D a deterministic, or constant distribution, and G denoting a general distribution (the assumption of independence is sometimes indicated by the symbol GI).

As discussed above the queue $M/M/m/K/M$ may be modeled as a birth and death process, and so the distributional properties can generally be obtained explicitly, with sufficient algebra. If we require a greater variety of distributional properties, this might be done within the context of continuous-time Markov chains by using the 'method of stages', due to A.K. Erlang. For example, we may replace a single exponential service time of rate λ with r exponential service times in series, each with rates $\lambda_1, \ldots, \lambda_r$. Thus, completion of service occurs after the sequential completion of r stages. The resulting distribution of the total service time is refered to as the *Erlangian distribution*, E_r in Kendall's notation. Note that this system may be modeled as a continuous-time Markov chain, as long as the state space is extended to include the current stage of a customer. There is a considerable variety of density shapes within E_r, including the gamma distribution. The coefficient of variation can be made arbitrarily small, but can never exceed 1. If E_r is generated by constructing stages in series, it is also possible to construct stages in parallel. Here, service consists of selecting one of r stages according to a fixed probability distribution, and then completing service after an exponential waiting time with the rate associated with that stage. This may also be modeled with continuous-time Markov chains, and the resulting service time is refered to as the *hyper-exponential* distribution, denoted H_r in Kendall's notation, and is formally a mixture of exponential densities. These distributions may of course apply also to arrival times. In general, a queue $G/G/m/K/M$ can be modeled as a continuous-time Markov chain if the arrival time and service times are either M, E_r or H_r.

Next, consider a $M/G/1$ queue. In the absence of any Markovian structure a continuous-time Markov chains model cannot be used. A commonly used approach is the *embedded Markov chain* approach (Kendall (1953)), in which a *semi-Markov process* (a discrete time Markov process with random inter-transition times) is defined by taking transition epochs to be service completions. The state space of the embedded Markov chain X_n is then interpretable as the number of customers left behind in the system by a departing customer. If G is the service time distribution, then

$$P\{\text{number of arrivals during service} = k\} = \alpha_k = \int_0^\infty \frac{(\lambda t)^k}{k!} e^{-\lambda t} dG(t).$$

The quantities α_k suffice to determine the transition kernel for X_n. To see this, suppose $X_n = 0$, that is, the system is empty after the nth service period. At that point the next event must be an arrival, at which point a service period begins immediately. The state X_{n+1} is determined when this service period ends, and must equal the number of arrivals during this period (since the current customer leaves the system). Thus, the transition distribution is $Q(\{j\} \mid 0) = \alpha_j$, and similar reasoning yields $Q(\{j\} \mid i) = \alpha_{j+i-1}$ for $i \geq 1$ and $j \geq i - 1$. Adjustments may be made for variable arrival rates (for example, with finite capacity $K < \infty$).

A general form for the steady state distribution π_i of X_n is known in the form a probability generating function (Section 4.2.4):

$$h_\pi(s) = \frac{(1-\rho)(s-1)h_\alpha(s)}{s - h_\alpha(s)}, \tag{5.20}$$

where $h_\pi(s) = \sum_{i=0}^{\infty} \pi_i s^i$, $h_\alpha(s) = \sum_{i=0}^{\infty} \alpha_i s^i$ and the utilization factor $\rho = \lambda m_G$, where λ is the arrival rate, and m_G is the mean service time. Clearly, we must have $\rho < 1$.

Equation (5.20) is known as the *Pollaczek-Khinchin transform equation.* Discussion of this equation, and of the $M/G/1$ more generally, may be found in, for example, Kleinrock (1975).

5.5 ADAPTED COUNTING PROCESSES

We have defined the counting process in Section 5.1. A discrete time counting process N_n can be represented by $N_n = \sum_{i=1}^{n} Z_i$ for a binary process $Z_1, Z_2, \ldots$. Suppose Z_n is a nonhomogeneous Markov chain with transition matrix

$$Q_n = \begin{bmatrix} 1-\alpha_n & \alpha_n \\ 1-\gamma & \gamma \end{bmatrix}, \tag{5.21}$$

governing the transition from Z_{n-1} to Z_n. We can think of this process as a sequence of blocks of consecutive occupancy in state $Z_k = 1$, into which the process enters at a rate α_n when the process is outside a block. The blocks have a length with a common geometric distribution with parameter γ, so it would be natural to think of the number of blocks as an embedded counting process, particularly when α_n is small, possibly approaching 0. Such a process can be useful in control policies requiring intermittent exploration periods of bounded length, the frequency of which is to be decreased at some predetermined rate. We will show that this can be achieved by (5.21). However, we may wish to allow Z_n to depend on process history, both to exploit available information, and to avoid the type of completely randomized exploration that might not be feasible in a working control system.

For any binary process Z_n, taking $Z_0 = 0$, define

$$\begin{aligned} B_n &= I\{Z_n = 1\} I\{Z_{n-1} = 0\}, \quad n \geq 1, \\ S_n &= \sum_{i=1}^{n} B_i, \quad n \geq 1, \\ J_k &= \inf\{j : S_j = k\}, \quad k \geq 1, \\ I_k &= \inf\{m \geq 1 : Z_{J_k + m} = 0\} I\{J_k < \infty\}, \quad k \geq 1, \\ \hat{I}_k &= k^{-1} \sum_{i=1}^{k} I_i, \quad k \geq 1. \end{aligned} \tag{5.22}$$

If S_n remains bounded we will have $J_k = \infty$ for all large enough k so we set $I_k = 0$ if $J_k = \infty$. Note that a block (a maximal sequence of consecutive 1's) begins at stage n if

$B_n = 1$, S_n is the number of blocks begun by stage n, J_k is the stage at which the kth block begins, I_k is the duration of the kth block, and $\hat{I}_k$ is the cumulative average block duration. The process is defined on probability measure $(\Omega, \mathcal{F}, P)$, and we assume Z_n is adapted to filtration $\mathcal{F}_n$. By analogy with (5.21) we can define the adapted process $\alpha_n(\mathcal{F}_{n-1})$ by

$$P(B_n = 1 \mid \mathcal{F}_{n-1}) = \alpha_n(\mathcal{F}_{n-1}) I\{Z_{n-1} = 0\}, \quad n \geq 1, \tag{5.23}$$

so that under (5.21) $\alpha_n(\mathcal{F}_{n-1}) \equiv \alpha_n$.

We make use of the following theorem (Dubins and Freedman (1965)), which is essentially a statement of Borel-Cantelli lemmas I and II for adapted sequences of events:

Theorem 5.12 *Suppose a sequence of events $E_1, E_2, \ldots$ is adapted to a filtration $\mathcal{F}_i$, $i \geq 0$, defined on probability measure space $(\Omega, \mathcal{F}, P)$. Then*

$$L_n = \frac{\sum_{i=1}^n I\{E_i\}}{\sum_{i=1}^n P(E_i \mid \mathcal{F}_{i-1})}$$

converges to a finite limit L $wp1$ as $n \to \infty$, and in addition

$$\sum_{i \geq 1} P(E_i \mid \mathcal{F}_{i-1}) = \infty \textit{ implies } L = 1, \quad wp1.$$

5.5.1 Asymptotic behavior

We first consider the asymptotic behavior of S_n. Theorem 5.12 gives necessary and sufficient conditions for which $S_n \to \infty$, and we may give a number of special cases.

Theorem 5.13 *Suppose $I_k < \infty$ for all $k \geq 1$. Then $wp1$*

$$\begin{aligned} \lim_{n\to\infty} S_n = \infty &\iff \sum_{i=1}^{\infty} \alpha_i(\mathcal{F}_{i-1}) = \infty \\ &\iff \sum_{i=1}^{\infty} \alpha_i(\mathcal{F}_{i-1}) I\{Z_{i-1} = 0\} = \infty. \end{aligned} \tag{5.24}$$

In addition the following statements hold:

(*i*) *If the following bounds hold almost surely*

$$\limsup_{k\to\infty} \hat{I}_k = \mu_I < \infty \tag{5.25}$$

$$\liminf_{n\to\infty} \frac{\sum_{i=1}^n \alpha_i(\mathcal{F}_{i-1}) I\{Z_{i-1} = 0\}}{\sum_{i=1}^n I\{Z_{i-1} = 0\}} = \zeta > 0 \tag{5.26}$$

then

$$\liminf_{n\to\infty} n^{-1}S_n = (\mu_I + \zeta^{-1})^{-1} > 0. \tag{5.27}$$

If the sequences in (5.25) *and* (5.26) *possess limits* μ_I *and* ζ *then*

$$\lim_{n\to\infty} n^{-1}S_n = (\mu_I + \zeta^{-1})^{-1} > 0. \tag{5.28}$$

(*ii*) *If* $\sum_{i\geq 1}\alpha_i(\mathcal{F}_{i-1}) < \infty$ *almost surely, then* S_n *is bounded.*

(*iii*) *If in addition to (5.25) the following conditions hold almost surely:*

$$\lim_{n\to\infty}\sum_{i=1}^{n}\alpha_i(\mathcal{F}_{i-1}) = \infty, \tag{5.29}$$

$$\lim_{n\to\infty}\alpha_n(\mathcal{F}_{n-1}) = 0, \tag{5.30}$$

then

$$\lim_{n\to\infty}\frac{S_n}{\sum_{i=1}^{n}\alpha_i(\mathcal{F}_{i-1})} = 1 \tag{5.31}$$

almost surely.

Proof The equivalence

$$\lim_{n\to\infty} S_n = \infty \iff \sum_{i=1}^{\infty}\alpha_i(\mathcal{F}_{i-1})I\{Z_{i-1}=0\} = \infty \tag{5.32}$$

follows from Theorem 5.12 applied to the events $\{B_n = 1\}$ and the formulation of equation (5.23), since the limits of the numerator and denominator of L_n must be both finite or both infinite.

Next, suppose S_n is bounded. Then either $Z_n = 1$ or $Z_n = 0$ for all large enough n. The first case is ruled out by the hypothesis $I_k < \infty$. If the second case holds then $Z_n = 0$ for all $n \geq N$ for some finite N. In this case we have, for $n \geq N$

$$\sum_{i=1}^{n}\alpha_i(\mathcal{F}_{i-1})I\{Z_{i-1}=0\} \geq \sum_{i=N}^{n}\alpha_i(\mathcal{F}_{i-1}).$$

By (5.32) S_n cannot be bounded unless $\sum_{i\geq 1}\alpha_i(\mathcal{F}_{i-1}) < \infty$, which completes the proof of (5.24).

(*i*) If $\sum_{i=1}^{n} I\{Z_{i-1}=0\} < \infty$, then $I_k = \infty$ for some k, which contradicts (5.25). We again apply Theorem 5.12 applied to the events $\{B_n = 1\}$. By (5.26) the denominator of L_n approach ∞, so $S_n \to \infty$ and L_n approaches 1, giving

$$\liminf_{n\to\infty}\frac{S_n}{\zeta\sum_{i=1}^{n} I\{Z_{i-1}=0\}} = 1.$$

We must have on $\{S_n > 1\}$ the inequalities

$$\hat{I}_{S_n-1}[S_n - 1] + \sum_{i=1}^{n} I\{Z_{i-1} = 0\} \leq n \leq \hat{I}_{S_n} S_n + \sum_{i=1}^{n} I\{Z_{i-1} = 0\},$$

from which we have

$$\frac{1}{\hat{I}_{S_n-1}[1 - S_n^{-1}] + S_n^{-1} \sum_{i=1}^{n} I\{Z_{i-1} = 0\}} \geq \frac{S_n}{n}$$

$$\geq \frac{1}{\hat{I}_{S_n} + S_n^{-1} \sum_{i=1}^{n} I\{Z_{i-1} = 0\}},$$

which, when combined with (5.33), and the fact that $S_n \to \infty$ gives (5.27), and (5.28) follows from a similar argument.

(*ii*) The result holds from (5.24), and the fact that

$$\sum_{i=1}^{\infty} \alpha_i(\mathcal{F}_{i-1}) I\{Z_{i-1} = 0\} \leq \sum_{i=1}^{\infty} \alpha_i(\mathcal{F}_{i-1}).$$

(*iii*) We may write

$$\sum_{i=1}^{n} \alpha_i(\mathcal{F}_{i-1}) I\{Z_{i-1} = 0\} = \sum_{i=1}^{n} \alpha_i(\mathcal{F}_{i-1}) - \sum_{i=1}^{n} \alpha_i(\mathcal{F}_{i-1}) I\{Z_{i-1} = 1\}. \qquad (5.33)$$

Fix $\epsilon > 0$. From (5.25) and (5.30) there is N_ϵ such that $S_n^{-1} \sum_{i=1}^{n} I\{Z_i = 1\} < \mu_I + \epsilon$ and $\alpha_n(\mathcal{F}_{n-1}) < \epsilon$ for $n > N_\epsilon$. The final summation in (5.33) satisfies

$$\sum_{i=1}^{n} \alpha_i(\mathcal{F}_{i-1}) I\{Z_{i-1} = 1\} < \epsilon(\mu_I + \epsilon) S_n + K_\epsilon$$

for $n > N_\epsilon$ and some finite K_ϵ. Dividing (5.33) by $S_n > 0$ gives

$$0 \leq \frac{\sum_{i=1}^{n} \alpha_i(\mathcal{F}_{i-1})}{S_n} - \frac{\sum_{i=1}^{n} \alpha_i(\mathcal{F}_{i-1}) I\{Z_{i-1} = 0\}}{S_n} < \epsilon(\mu_I + \epsilon) + \frac{K_\epsilon}{S_n}. \qquad (5.34)$$

By (5.29) and (5.24) we may conclude that $S_n \overset{a.s.}{\to} \infty$. Letting $n \to \infty$ in (5.34), using Theorem 5.12 then letting $\epsilon \to 0$ completes the proof. ///

Somewhat more precision can be given for Case (*iii*) of Theorem 5.13 by imposing bounds α_n^u on $\alpha_n(\mathcal{F}_n)$.

Theorem 5.14 *Suppose we are given Case (iii) of Theorem 5.13, and the following conditions hold:*

(*i*) *There exists finite positive* N, m *and constant* $\gamma \in [0, 1)$ *such that*

$$P(Z_{n+m} = \cdots = Z_{n+1} = 1 \mid Z_n = 1) \leq \gamma, \qquad (5.35)$$

for all $n \geq N$.

(*ii*) *There exists a sequence of constants* α_n^u, *with partial sums* $\xi_n = \sum_{i=1}^n \alpha_i^u$, *for which the following hold:*

$$\alpha_n(\mathcal{F}_{n-1}) \leq \alpha_n^u, \quad n \geq 1, \tag{5.36}$$

$$\lim_{n\to\infty} \alpha_n^u = 0, \tag{5.37}$$

$$\lim_{n\to\infty} \xi_n = \infty, \tag{5.38}$$

$$\liminf_{n\to\infty} \frac{\sum_{i=1}^n \alpha_i(\mathcal{F}_{i-1})}{\sum_{i=1}^n \alpha_i^u} = K_\alpha, \tag{5.39}$$

$$\lambda^l\{\alpha_n^u\} = 1 \tag{5.40}$$

almost surely for a finite constant $K_\alpha > 0$. *Then almost surely*

$$\liminf_{n\to\infty} \frac{S_n}{\xi_n} > 0, \tag{5.41}$$

and

$$\liminf_{n\to\infty} \frac{m\alpha_n^u}{P(Z_n = 1)} \geq (1-\gamma). \tag{5.42}$$

Proof By condition (*ii*), conditions (5.29) and (5.30) of Theorem 5.13, Case (*iii*) are satisfied. Additionally, condition (*i*) implies condition (5.25), so that (5.31) holds, which in turn implies (5.41).

Next, if $m > 1$ we may write

$$\begin{aligned}&\{Z_{n+m} = 1 \wedge Z_n = 1\} \subset \\ &\{Z_{n+m} = \cdots = Z_n = 1\} \cup \left[\cup_{i=1}^{m-1}\{Z_{n+i+1} = 1 \wedge Z_{n+i} = 0 \wedge Z_n = 1\}\right].\end{aligned}$$

Using Theorem 4.3, equation (5.23) and the assumption (5.36) we have, for $i \geq 1$,

$$\begin{aligned}P\left(Z_{n+i+1} = 1 \wedge Z_{n+i} = 0 \wedge Z_n = 1 \mid \mathcal{F}_{n+i}\right) &= P\left(B_{n+i+1} \mid \mathcal{F}_{n+i}\right) I\{Z_n = 1\} \\ &\leq \alpha_{n+i+1}^u I\{Z_n = 1\}.\end{aligned}$$

Taking the expectation gives

$$P(Z_{n+i+1} = 1 \wedge Z_{n+i} = 0 \wedge Z_n = 1) \leq \alpha_{n+i+1}^u P(Z_n = 1),$$

from which we may conclude

$$\begin{aligned}P(Z_{n+m} = 1 \mid Z_n = 1) &\leq P(Z_{n+m} = \cdots = Z_{n+1} = 1 \mid Z_n = 1) + \xi_{n+m} - \xi_{n+1}, \\ &\leq \gamma + \xi_{n+m} - \xi_{n+1},\end{aligned}$$

by condition (*i*). A similar argument may be used to show that

$$P(Z_{n+m} = 1 \mid Z_n = 0) \leq \xi_{n+m} - \xi_n.$$

We may therefore write

$$P(Z_{n+m}=1) = P(Z_{n+m}=1 \mid Z_n=0)P(Z_n=0) + P(Z_{n+m}=1 \mid Z_n=1)P(Z_n=1),$$
$$\leq \xi_{n+m} - \xi_n + (\gamma + \xi_{n+m} - \xi_{n+1})\, P(Z_n=1).$$

It may then be directly verified that the preceding inequality holds also for $m=1$.

Fix $\epsilon > 0$, then there is finite n_ϵ for which $\xi_{n+m} - \xi_{n+1} < \epsilon$ for $n \geq n_\epsilon$. An iterative argument gives

$$P(Z_{n+jm}=1) \leq \sum_{i=1}^{j} (\gamma + \epsilon)^{j-i} (\xi_{n+im} - \xi_{n+(i-1)m}), \quad n \geq n_\epsilon.$$

By (5.40) we have $\lambda^l\{\xi_{n+im} - \xi_{n+(i-1)m}\} = 1$ for any fixed n, then applying Lemma 9.8 (to be discussed in Section 9.4) gives

$$\liminf_{i\to\infty} \frac{\xi_{n+im} - \xi_{n+(i-1)m}}{P(Z_{n+im}=1)} \geq (1-\gamma), \quad n \geq 1,$$

from which (5.42) follows directly. ///

5.5.2 Relationship to adapted events

We next consider interactions between adapted counting processes and other processes $Y_n \in \mathcal{Y}$ adapted to $\mathcal{F}_n$. For $E \subset \mathcal{Y}$, we define the counting process

$$M_n(E) = \sum_{i=1}^{n} I\{Y_i \in E\},$$

representing the cumulative number of visits to E. Suppose the intention is to associate visits to E with blocks. Define the event

$$K_n(E) = \{B_n = 1\} \cap \left[\cup_{j=1}^{I_{S_n}} \{Y_{n+j-1} \in E\} \right]$$

which is the event that a block starts at stage n, and a visit to E occurs during this block. Our objective is to verify that if this occurs with a minimum probability, then $M_n(E)$ increases at the same rate as S_n.

Theorem 5.15 *Suppose the following conditions hold:*

(*i*) $S_n \overset{a.s.}{\to}_n \infty$.

(*ii*) *For some constant* $\delta > 0$,

$$P(K_n(E) \mid \mathcal{F}_n) \geq \delta I\{B_n = 1\}.$$

Then

$$\liminf_{n\to\infty} \frac{M_n(E)}{S_n} \geq \delta$$

almost surely.

Proof Noting that $J_1, J_2, \ldots$ forms a sequence of increasing finite stopping times, by Theorem 4.7 $\mathcal{F}^{J_k}$, $k \geq$ is a filtration. We then have, for $S_n > 1$,

$$M_n(E) \geq \sum_{k=1}^{S_n-1} I\{K_{J_k}(E)\}, \quad n \geq 1. \tag{5.43}$$

We then argue that $K_{J_k}(E) \in \mathcal{F}^{J_{k+1}}$, since the occurrence of $K_{J_k}(E)$ is resolved before stopping time J_{k+1}. On the other hand we have $P(K_{J_k}(E) \mid \mathcal{F}^{J_k}) \geq \delta$, $k \geq 1$. Applying Theorem 5.12 gives

$$1 = \lim_{n\to\infty} \frac{\sum_{k=1}^{S_n-1} I\{K_{J_k}(E)\}}{\sum_{k=1}^{S_n-1} P(K_{J_k}(E) \mid \mathcal{F}^{J_k})} \leq \liminf_{n\to\infty} \frac{M_n(E)}{\delta(S_n - 1)}, \quad w.p.1$$

which proves the theorem. ///

Finally, we consider conditions under which the limit condition (5.39) of Theorem 5.14 holds. It would be possible to set $\alpha_n(\mathcal{F}_{n-1}) = \alpha_n^u$, but this would mean that any stage could be in a block with positive probability, independently of the history, as would be the case with the model defined by (5.21). We may wish to impose restrictions on the initiation of a block, in the form of a sequence of events $A_n \in \mathcal{F}_n$, so that if stage n is not in a block, a block may be initiated in the next stage only if A_n occurs. This is summarized in the following theorem. Note that condition (5.44) holds for any $\alpha_n^u = n^{-p}$, $p > 0$.

Theorem 5.16 *Suppose, given the sequence α_n^u of Theorem 5.14, we set*

$$\alpha_n(\mathcal{F}_{n-1}) = \alpha_n^u I_{A_{n-1}}$$

where $A_n \in \mathcal{F}_n$. Then if the following conditions hold:

$$\liminf_n \frac{n \min_{i \leq n} \alpha_i^u}{\sum_{i=1}^n \alpha_i^u} > 0, \tag{5.44}$$

$$\liminf_n \frac{M_n^A}{n} > 0 \tag{5.45}$$

almost surely, where $M_n^A = \sum_{i=1}^n I_{A_n}$, then (5.39) of Theorem 5.14 holds

Proof We may write

$$\frac{\sum_{i=1}^n \alpha_i(\mathcal{F}_{i-1})}{\sum_{i=1}^n \alpha_i^u} \geq \frac{M_{n-1}^A \min_{i \leq n} \alpha_i^u}{\sum_{i=1}^n \alpha_i^u}$$

from which the theorem holds directly from (5.44)–(5.45). ///

Chapter 6

Functional analysis

Underlying functional analysis is the specification of a set of objects $\mathcal{V}$ on which is imposed some additional structure, typically representing generalizations of notions natural to multidimensional Euclidean vector geometry. The objects in $\mathcal{V}$ may be functions, infinite dimensional sequences or random variables, in addition to Euclidean vectors, but may all be conveniently referred to simply as *vectors*. We therefore need to define abstract notions of vector algebra, as well as quantitative abstractions of concepts such as the distance between vectors, the length of a vector, and the angle between vectors. There is a hierarchical progression to the definition of this structure, and applications will differ in the amount of structure required.

We start first with a *metric space* $(\mathcal{X}, d)$ which imposes a *metric* $d(x,y)$ between two objects x, y from the set $\mathcal{X}$, representing distance (Definition 3.3). The definition of a vector algebra is the next step in the progression (which is not strictly needed for the metric space), yielding a *vector space* $\mathcal{V}$. With a vector algebra comes the zero vector (the identity with respect to vector addition), which centers the space with a point of reference, permitting the notion of the length $\|x\|$ of a vector x, called a *norm*. Paired with $\mathcal{V}$ we have a *normed vector space* $(\mathcal{V}, \|\cdot\|)$. The norm in turn defines a metric $d(x,y) = \|x - y\|$, so that a normed vector space is also a metric space.

The final step needed for our analysis is the introduction of angle coupled with length, in the form of an inner product $\langle x, y\rangle$, which is a generalization of the familiar 'dot product' for Euclidean vectors. A vector space coupled with an inner product is known as an *inner product space* $(\mathcal{V}, \langle\cdot,\cdot\rangle)$. Conveniently for our hierarchy, an inner product generates a norm $\|x\| = \langle x, x\rangle^{1/2}$, which in turn generates a metric, as already described. Thus, an inner product space is also a normed vector space and a metric space.

Since each of these spaces is also a metric space, the notion of a convergent sequence is naturally defined. We may also consider *Cauchy sequences* (Definition 3.4), which are sequences in $\mathcal{X}$ with convergent behavior, but which need not possess a limit in $\mathcal{X}$. An example of a Cauchy sequence is $1/1, 1/2, 1/3, \ldots$, which exists in $(0, \infty)$ but does not possess a limit in $(0, \infty)$ (and its convergent behavior can be characterized without reference to the limit 0). A metric space is *complete* if all Cauchy sequences possess a limit in $\mathcal{X}$. A normed vector space or inner product space which is complete with respect to the induced metric is better known as a *Banach space* or *Hilbert space* respectively.

6.1 METRIC SPACES

If $(\mathcal{X}, d)$ is a metric space (Definition 3.3), then a sequence $x_n \in \mathcal{X}$ converges to $x \in \mathcal{X}$ if $\lim_{n\to\infty} d(x_n, x) = 0$. If all Cauchy sequences in $\mathcal{X}$ possess limits in $\mathcal{X}$, then $(\mathcal{X}, d)$ is a complete metric space (Definition 3.4).

It will sometimes be useful to be able to define a distance function d which does not strictly satisfy the identifiability axiom in the sense that we may have $d(x, y) = 0$ for some $x \neq y$, but satisfies all the other requirements of a metric. This is usually referred to as a *semimetric*.

Definition 6.1 *If in Definition 3.3 the identifiability axiom is replaced with the weaker axiom which holds that $d(x, x) = 0$ for all $x \in \mathcal{X}$, then d is called a* semimetric *(or* pseudometric*), and $(\mathcal{X}, d)$ is a* semimetric space *or (*pseudometric space*).*

For a semimetric we may have $d(x, y) = 0$ if $x \neq y$.

6.1.1 Contractive mappings

We next consider mappings on a metric space $T : \mathcal{X} \to \mathcal{X}$. The notion of the continuity of T is natural on a metric space. We say that T is continuous at $x_0 \in \mathcal{X}$ if for all $\epsilon > 0$ there exists $\delta > 0$ such that $d(x, x_0) < \delta$ implies $d(Tx, Tx_0) < \epsilon$. Then $\lim_{n\to\infty} Tx_n = Tx_0$ whenever $\lim_{n\to\infty} x_n = x_0$. We may recursively define the mapping $T^J : \mathcal{X} \to \mathcal{X}$ as the Jth *iteration* $T^J x = T^{J-1}(Tx)$. If T is continuous, so is T^J.

A mapping T is *Lipschitz continuous* (is *Lipschitz*) if there exists a finite constant L (referred to as a *Lipschitz constant*) for which $d(Tx, Ty) \leq Ld(x, y)$ for all $x, y \in \mathcal{X}$. If T is Lipschitz continuous, then the infimum of all Lipschitz constants must also be a Lipschitz constant. This value is sometimes referred to as the *smallest*, *best* or simply *the* Lipschitz constant (as opposed to *a* Lipschitz constant) when the context is clear. In many instances any Lipschitz constant suffices, so the distinction should be kept in mind. In addition, T is *locally Lipschitz continuous* (is *locally Lipschitz*) if every $x \in \mathcal{X}$ possesses a neighborhood B_x on which T is Lipschitz continuous. Clearly, the Lipschitz condition implies continuity.

The Lipschitz constant of mapping T on metric space $(\mathcal{X}, d)$ is *submultiplicative* in the sense that

$$d(T^2x, T^2y) = d(T(Tx), T(Ty)) \leq Ld(Tx, Ty) \leq L^2 d(x, y), \tag{6.1}$$

so that if L is a Lipschitz constant of T, L^2 is a Lipschitz constant of T^2. By applying this argument iteratively, we conclude that if T is Lipschitz continuous so is any iterate T^J. Clearly then, we can associate with any Lipschitz mapping a sequence of constants ρ_J, $J \geq 1$, defined as the smallest Lipschitz constant of T^J. The submultiplicative property can then be characterized by the following theorem:

Theorem 6.1 *If T is a* Lipschitz continuous *mapping on metric space $(\mathcal{X}, d)$ and ρ_J is the smallest Lipschitz constant of T^J, then $\rho_{m+n} \leq \rho_m \rho_n$.*

Proof The proof follows by applying (6.1). ///

It will be convenient to standardize the constants ρ_m by defining

$$\bar{\rho}_m = (\rho_m)^{1/m},$$

so that, because $\bar{\rho}_J^J$ and ρ_1^J are both Lipschitz constants for T^J, $\bar{\rho}_J$ can be directly compared to ρ_1. We state the following consequence of Theorem 6.1.

Theorem 6.2 *If T is a* Lipschitz continuous *mapping on metric space $(\mathcal{X}, d)$ and ρ_J is the smallest Lipschitz constant of T^J, then for any positive integers m, k*

$$\bar{\rho}_{km} \leq \bar{\rho}_m$$

and in particular

$$\bar{\rho}_m \leq \rho_1$$

for all $m \geq 1$.

Proof The proof follows directly from Theorem 6.1. ///

Note that it does not follow that $\bar{\rho}_m$ is nonincreasing in m.

The notion of a complete metric space, coupled with the notion of a contraction mapping, plays a crucial role in the theorem of iterative algorithms. Specifically, a mapping $T : \mathcal{X} \to \mathcal{X}$ on a metric space $(\mathcal{X}, d)$ is a *contraction mapping* (is *contractive*) if it has a Lipschitz constant $\rho < 1$. Then ρ is referred to as a *contraction constant*, and we refer to the smallest contraction constant as *the* contraction constant. If T has Lipschitz constant $\rho = 1$ then it is a *nonexpansive mapping*. By Theorem 6.1, if T is nonexpansive then ρ_J is nonincreasing in J, and if T is contractive then ρ_J is decreasing. Clearly, if T is contractive (or nonexpansive) then so is any iterate T^J.

Then write, by Theorem 6.2,

$$d(T^J x, T^J y) = \bar{\rho}_J{}^J d(x, y) \leq \rho_1^J d(x, y). \tag{6.2}$$

In the simplest applications of the contraction property, the upper bound in (6.2) based on a single stage contraction constant $\rho_1 < 1$ would be used. In effect, when this is done the bound on the standardized contraction constant $\bar{\rho}_J \leq \rho_1$ is accepted as being sharp, or at least approximately so. However, this will not always be the case, and we will encounter important examples in which $\bar{\rho}_J < \rho_1$ by a significant factor (whether or not $\rho_1 < 1$). Therefore, it will be important to characterize a 'best possible' contraction constant (or Lipschitz constant in general) for T, which is independent of the iteration order T^J. This is given in the following theorem:

Theorem 6.3 *If T is a Lipschitz continuous mapping on a metric space $(\mathcal{X}, d)$, and ρ_J is the smallest Lipschitz constant for T^J, then there exists a finite constant ρ such that*

$$\lim_{n \to \infty} \bar{\rho}_n = \rho. \tag{6.3}$$

Furthermore, ρ is the best possible contraction rate in the sense that

$$\bar{\rho}_n \geq \rho, \quad \textit{for all } n \geq 1, \tag{6.4}$$

and for any $\epsilon > 0$, there exists n_ϵ for which

$$\bar{\rho}_n \leq \rho + \epsilon, \quad \textit{for all } n \geq n_\epsilon. \tag{6.5}$$

Proof By Theorem 6.2 for any n we have $\bar{\rho}_n \leq \rho_1$, which means there is a finite constant $\rho = \liminf_{n\to\infty} \bar{\rho}_n \leq \rho_1$. Then for any $\epsilon > 0$ there is some m for which $\rho_m < (\rho + \epsilon)^m$. For any $n > m$ write $n = n_1 m + n_2$ where $n_1 = \lfloor n/m \rfloor$ and $0 \leq n_2 < m$. Then, by the submultiplicative property

$$\rho_n \leq \rho_{n_1 m} \rho_{n_2} \leq \rho_m^{n_1} \rho_1^{n_2} \leq (\rho + \epsilon)^{n_1 m} \rho_1^{n_2} = (\rho + \epsilon)^n (\rho_1/(\rho + \epsilon))^{n_2} \leq K(\rho + \epsilon)^n,$$

where $K = \max\left((\rho_1/(\rho + \epsilon))^{m-1}, 1\right)$, and so does not depend on n. Letting $n \to \infty$ gives

$$\limsup_{n\to\infty} \bar{\rho}_n \leq \limsup_{n\to\infty} \left[(\rho + \epsilon)K^{1/n}\right] = \rho + \epsilon.$$

Then (6.3) follows by letting ϵ approach 0, and noting the definition of ρ.

Suppose (6.4) does not hold, so that for some m we have $\bar{\rho}_m < \rho$. By Theorem 6.2 we have $\bar{\rho}_{km} \leq \bar{\rho}_m < \rho$ for all $k \geq 1$, which contradicts (6.3), so that (6.4) must hold. Then (6.5) follows directly from (6.3). ///

We may refer to the limit in (6.3) as the *asymptotic Lipschitz constant* for mapping T, or *asymptotic contraction rate* as appropriate. It clearly plays a role analogous to the *spectral radius* of a linear operator (see Section 6.8.2), but requires only a metric space structure for its definition.

6.2 THE BANACH FIXED POINT THEOREM

An element x of a metric space is a *fixed point* of T if $Tx = x$, which is referred to as a *fixed point equation*. Much of the theory of this book concerns a straightforward idea. The fixed point equation is ubiquitous in a wide range of applications in applied mathematics, so general methods for obtaining its solution (that is, the fixed point) are of considerable importance. When the mapping possesses certain properties, it can be proven that a fixed point exists, and that an iterative sequence $x_k = Tx_{k-1} = T^k x_0$, $k \geq 1$ will converge to it. More stringent conditions may guarantee the uniqueness of the fixed point. The seminal case is that of the contraction mapping defined on a complete metric space. In this case it can be verified that a unique fixed point exists, which is the limit of the iterative sequence $T^k x_0$ for any starting point.

Theorem 6.4 (Banach Fixed-Point Theorem) *If T is a contraction mapping on a complete metric space $(\mathcal{X}, d)$ then a unique fixed point exists, which is the limit of the iterative sequence $T^k x_0$ for any initial point x_0.*

The result holds also if T is continuous and T^J is contractive.

Proof Suppose ρ is the contraction constant for T, and set $x_n = T^n x_0$ for any $x_0 \in \mathcal{X}$. By the triangle inequality we must have

$$d(x_n, x_0) \leq \sum_{i=1}^{n} d(x_i, x_{i-1}).$$

Then note that for $i > 1$ we have

$$d(x_i, x_{i-1}) = d(Tx_{i-1}, Tx_{i-2}) \leq \rho d(x_{i-1}, x_{i-2}),$$

which applied iteratively implies $d(x_i, x_{i-1}) = \leq \rho^{i-1} d(x_1, x_0)$ for all $i \geq 1$. Substituting into the previous inequality gives $d(x_n, x_0) \leq (1-\rho)^{-1} d(x_1, x_0)$.

Next, suppose $n \geq m$. Then

$$\begin{aligned} d(x_n, x_m) &= d(T^m x_{n-m}, T^m x_0) \\ &\leq \rho^m d(x_{n-m}, x_0) \\ &\leq \rho^m (1-\rho)^{-1} d(x_1, x_0), \end{aligned}$$

from which it follows that x_n is a Cauchy sequence, and therefore possesses a limit x in $\mathcal{X}$ under the completeness assumption. Since T is Lipschitz, it is continuous. It then follows that

$$Tx = T \lim_n x_n = \lim_n Tx_n = \lim_n x_{n+1}, \tag{6.6}$$

so that x is a fixed point of T. Finally, suppose y is any fixed point of T. Then $d(x,y) = d(Tx, Ty) \leq \rho d(x,y)$. Since $\rho < 1$ this implies $d(x,y) = 0$.

Next, suppose T is continuous and T^J is contractive. The preceding conclusion then applies to the J subsequences x_{Ji+k}, $i \geq 1$, for $k = 0, 1, \ldots, J-1$, so that each possesses a common limit. Then equation (6.6) holds by the continuity of T, which completes the proof. ///

The contraction property by itself guarantees that there cannot be more than one fixed point in $\mathcal{X}$, and that if it exists it will be the limit of the iteration $T^k x_0$. The role of the completion assumption is to guarantee the existence of a fixed point. The theory is essentially the same when d is a semimetric, except that multiple fixed points may exists. In this case, fix starting point x_0 and let x be the limit of $T^k x_0$, which will be a fixed point. Then all fixed points must belong to a unique equivalence class $\{y \in \mathcal{X} \mid d(y,x) = 0\}$, which is obtainable in this way using any starting point x_0.

We will sometimes have the weaker property of *pseudocontraction*, that is, that there exists some element $x^* \in \mathcal{X}$ and constant $\rho < 1$ such that $d(Tx, x^*) \leq \rho d(x, x^*)$ for all $x \in \mathcal{X}$. In this case, the essential features of the Banach fixed point theorem will hold, and the proof is somewhat simpler than for Theorem 6.4, since the hypothesis has come somewhat closer to implying the existence of a fixed point.

Theorem 6.5 *Suppose T is a pseudocontractive mapping, so that for some $x^* \in \mathcal{X}$ and $\rho < 1$ we have $d(Tx, x^*) \leq \rho d(x, x^*)$ for all $x \in \mathcal{X}$. Then x^* is the unique fixed point of T, which is the limit of the iterative sequence $T^k x_0$ for any initial point x_0. The result holds also if T^J is pseudocontractive.*

Proof By hypothesis $d(Tx^*, x^*) \leq \rho d(x^*, x^*) = 0$, so that $x^* = Tx^*$. If x' is *any* fixed point, then $d(x', x^*) = d(Tx', x^*) \leq \rho d(x', x^*)$, which implies $d(x', x^*) = 0$ if $\rho < 1$. Then $d(T^n x_0, x^*) \leq \rho d(T^{n-1} x_0, x^*)$, and eventually $d(T^n x_0, x^*) \leq \rho^n d(x_0, x^*)$. The remainder of the theorem follows an argument similar to that used in Theorem 6.4. ///

It is important to note that the terminology can vary considerably by field. We take the meaning of 'pseudocontraction' used in, for example, Bertsekas and Tsitsiklis (1996) and other texts concerned with control theory and dynamic programming. In contrast, in the context of functional analysis the terms 'contraction' and 'pseudo-contraction' are part of a general classification system of operators, as exemplified in Berinde (2007). In effect, there is a family of 'contraction properties' which are similar but distinct. We use the simpler meaning of these terms, while introducing our own classification system in Section 10.4.

6.2.1 Stopping rules for fixed point algorithms

A contraction mapping is also a pseudocontraction, so in either case the convergence of the iteration algorithm $T^n x_0$ to fixed point x^* is easily characterized as $d(T^n x_0, x^*) \leq \rho^n d(x_0, x^*)$. This leaves a practical problem, in particular that the calculation of x^* requires an infinite number of iterations, so that in practice $x_N = T^N x_0$ will have to suffice as an approximation. So far, we can state the approximation error $d(x_N, x^*) \leq \rho^N d(x_0, x^*)$. This has the disadvantage of depending on x^*, which we are trying to estimate. On the other hand, we may be content to bound $d(x_N, x^*)/d(x_0, x^*) \leq \rho^N$. If $d(x_0, x^*)$ is interpretable as a magnitude of x^* (which we must remember is a concept not defined in a metric space), then ρ^N would be interpretable as a relative error. This idea will be very natural in normed vector spaces, which we will discuss below.

Fortunately, a much more refined and practical bound on the approximation error is possible.

Theorem 6.6 *If T is the pseudocontractive mapping of Theorem 6.5, and $x_n = T^n x_0$ then*

$$d(x_n, x^*) \leq (1 - \rho)^{-1} d(Tx_n, x_n). \tag{6.7}$$

In addition, if T is contractive then

$$d(x_n, x^*) \leq \rho(1 - \rho)^{-1} d(x_n, x_{n-1}) \leq \rho^n (1 - \rho)^{-1} d(x_1, x_0). \tag{6.8}$$

Proof By the triangle inequality, we may write

$$d(x_n, x^*) \leq d(Tx_n, x^*) + d(Tx_n, x_n) \leq \rho d(x_n, x^*) + d(Tx_n, x_n)$$

from which (6.7) follows directly. Under the stronger contraction assumption, (6.8) follows from (6.7) after noting that $d(Tx_n, x_n) = d(Tx_n, Tx_{n-1}) \leq \rho d(x_n, x_{n-1})$. ///

The advantage of (6.8) over (6.7) is the introduction of the factor $\rho < 1$, and the fact that the upper bound for $d(x_n, x^*)$ does not require the extra iteration Tx_n. In many cases ρ will be close to 1, but we will see important examples in which ρ is quite small. The importance of either bound is that they are both calculable exactly within a finite number of iterations.

Stopping rules for multistage contractions

In principal, stopping rules (6.7) or (6.8) may be applied to multistage contractive operators by applying them to the embedded iterations $x_{kJ} = T^J x_{(k-1)J}$ to achieve essentially the same result. However, given the result of Theorem 6.3, there may be some advantage to considering such schemes even when T is single stage contractive. To simplify the analysis suppose $\rho_J < 1$. There is a choice between regarding x_{kJ} as k iterations of T^J, or one iteration of T^{kJ}. In the former case, (6.8) becomes

$$d(x_{kJ}, x^*) \leq \rho_J (1 - \rho_J)^{-1} d(x_{kJ}, x_{(k-1)J}) \leq \rho_J^k (1 - \rho_J)^{-1} d(x_J, x_0). \tag{6.9}$$

As a single iteration of T^{kJ} we have

$$\begin{aligned} d(x_{kJ}, x^*) &\leq \rho_{kJ} (1 - \rho_{kJ})^{-1} d(x_{kJ}, x_0) \\ &\leq \rho_{kJ} (1 - \rho_{kJ})^{-1} \sum_{i=1}^{k} d(x_{iJ}, x_{(i-1)J}) \\ &\leq \rho_{kJ} (1 - \rho_{kJ})^{-1} \sum_{i=1}^{k} \rho_J^{i-1} d(x_J, x_0) \\ &\leq \rho_{kJ} (1 - \rho_{kJ})^{-1} (1 - \rho_J)^{-1} d(x_J, x_0). \end{aligned} \tag{6.10}$$

The upper bounds of (6.9) and (6.10) may be directly compared by the ratio

$$R = \frac{(\bar{\rho}_J)^{kJ}}{(\bar{\rho}_{kJ})^{kJ} \left[1 - (\bar{\rho}_{kJ})^{kJ}\right]^{-1}},$$

with $R < 1$ favoring bound (6.9). By Theorem 6.3, $\bar{\rho}_{kJ} \to_k \rho$, the asymptotic contraction rate (Theorem 6.3), so we must have $\liminf_k R \geq 1$, with strict inequality if $\bar{\rho}_J > \rho$. Thus bound (6.10) is asymptotically no worse than (6.9), and strictly better if $\bar{\rho}_J > \rho$. On the other hand, (6.9) may be the better choice when $\bar{\rho}_J \approx \rho$, and will usually be simpler to implement.

6.3 VECTOR SPACES

Suppose we wish to evaluate Tx on a metric space, but we only have available an approximate mapping $\bar{T}$, along with an analysis method which yields some bound $d(Tx, \bar{T}x) \leq \epsilon$. A satisfactory result may be obtained by analyzing $\bar{T}$ in place of T.

An alternative point of view is to express the approximation as something like $\bar{T}x = Tx + u$, which requires an addition operation. Unfortunately, addition is not defined on a metric space. Since our theory is based on the idea that, despite the obvious nature of the formulation, the expression $Tx + u$ yields certain advantages over $\bar{T}x$, we will need to use a space in which additive error may be defined, which is precisely what the vector space allows.

The formal definition of a vector space requires some basic definitions of abstract algebra, in particular, the *group*, *abelian* or *commutative group* and the *field* (Definitions 6.2, 6.3 and 6.4).

Definition 6.2 *A* group *is a pair* $(G, *)$ *where* G *is a set and* $*$ *is a binary operation on* G *satisfying the axioms:*

> ***Closure*** *If* $a, b \in G$ *then* $a * b \in G$,
> ***Associativity*** *For all* $a, b, c \in G$ *we have* $(a * b) * c = a * (b * c)$,
> ***Existence of identity*** *There exists* $e \in G$ *such that for all* $a \in G$ *we have* $a * e = e * a = a$,
> ***Existence of inverse*** *For each* $a \in G$ *there exists* $b \in G$ *such that* $a * b = b * a = e$.

Definition 6.3 *An* abelian group *(or* commutative group*) is a group* $(G, *)$ *which satisfies the additional axiom*

> ***Commutativity*** *For any* $a, b \in G$ *we have* $a * b = b * a$.

Definition 6.4 *A* field *is a triplet* $(\mathbb{K}, +, \times)$ *where* $\mathbb{K}$ *is a set and* $+$ *and* $\times$ *are binary operations on* $\mathbb{K}$ *(by analogy referred to as* addition *and* multiplication*) satisfying the following axioms:*

> ***Group structure of addition*** $(\mathbb{K}, +)$ *is an abelian group,*
> ***Group structure of multiplication*** $(\mathbb{K} - \{0\}, \times)$ *is an abelian group, where* 0 *is the additive identity,*
> ***Distributivity of multiplication*** *For all* $a, b, c \in \mathbb{K}$ *we have*
>
> $$a \times (b + c) = (a \times b) + (a \times c).$$

The definition of a vector space follows:

Definition 6.5 *Suppose we are given a field* $\mathbb{K}$ *of* scalars *and an abelian group* $(\mathcal{V}, +)$ *of* vectors *(by analogy,* $+$ *is referred to as vector addition). Suppose also that for each pair* $a \in \mathbb{K}$ *and* $x \in \mathcal{V}$ *there exists a unique composite product* $a \circ x \in \mathcal{V}$*. The collection* $(\mathbb{K}, \mathcal{V}, +, \circ)$ *is a* vector space *(or* linear space*) if the following additional axioms are satisfied:*

> ***Existence of identity for composite product*** *For any vector* $x \in \mathcal{V}$ *we have* $1 \circ x = x$ *where* 1 *is the multiplicative identity of* $\mathbb{K}$,
> ***Compatibility of scalar and composite product*** *For all* $a, b \in \mathbb{K}$ *and* $x \in \mathcal{V}$ *we have* $a \circ (b \circ x) = (a \times b) \circ x$,
> ***Distributivity over scalar addition*** *For all* $a, b \in \mathbb{K}$ *and* $x \in \mathcal{V}$ *we have* $(a + b) \circ x = a \circ x + b \circ x$,
> ***Distributivity over vector addition*** *For all* $a \in \mathbb{K}$ *and* $x, y \in \mathcal{V}$ *we have* $a \circ (x + y) = (a \circ x) + (a \circ y)$.

The vector space is intended to be a generalization of the standard algebra for Euclidean vectors, with the scalar field $\mathbb{K}$ representing the set of real numbers under the standard binary operations, while the vector group $(\mathcal{V}, +)$ represents the space of Euclidean vectors coupled with vector addition. In this book the scalar field will usually be the set of real numbers, with the usual operations. On the other hand, the set of vectors $\mathcal{V}$ will be a variety of objects in addition to Euclidean vectors. In general, it will be convenient to denote a vector space using $\mathcal{V}$ alone, assuming all the axioms of Definition 6.5.

A subset E of a vector space $\mathcal{V}$ is a *vector subspace* (or *linear manifold*, or simply *subspace*) if it is closed under linear composition, that is $x, y, \in \mathcal{V}$, $a, b \in \mathbb{K}$ implies $ax + by \in \mathcal{V}$. A subspace is also a vector space.

Vector spaces consisting of real valued functions will be of special interest. For any set $\mathcal{X}$ the space of all real valued functions $f : \mathcal{X} \to \mathbb{R}$ will be denoted $\mathcal{R}(\mathcal{X})$. This notation will sometimes be introduced without defining $\mathcal{X}$, when properties of $\mathcal{X}$ play no role.

6.3.1 Quotient spaces

Let $\mathcal{N}$ be a subspace of $\mathcal{V}$. For any $x \in \mathcal{V}$ the *coset* of x modulo $\mathcal{N}$ is the subset $[x] = \{x + n \mid n \in \mathcal{N}\}$. If $y \in [x]$, then $y = x + n_y$ for some $n_y \in \mathcal{N}$. Since $\mathcal{N}$ is a subspace, we have $-n_y \in \mathcal{N}$, so that $x \in [y]$. This means that $[x] = [y]$ for any $y \in [x]$, so that the definition of a coset does not depend on the representative vector. In addition, it may be verified that the space of cosets is a vector space under the linear composition $a[x] + b[y] = [ax + by]$, with additive identity $[\vec{0}] = \mathcal{N}$, where $\vec{0}$ is the additive identity of $\mathcal{V}$. Note that $0[x] = [0x] = \{0x + n \mid n \in \mathcal{N}\} = \mathcal{N}$, which is not the same as elementwise multiplication by 0 of the set $[x]$.

The condition $x - y \in \mathcal{N}$ defines an equivalence relationship $x \sim y$, under which the cosets $[x]$ are equivalence classes (see Section 2.2). The vector space of equivalence sets is referred to as a *quotient space* $\mathcal{V}/\mathcal{N}$.

6.3.2 Basis of a vector space

The definition of a basis follows that for vectors in $\mathbb{R}^m$ discussed in Section 2.3. Elements $V_1, \ldots, V_m$ of a vector space $\mathcal{V}$ are *linearly independent* if $\sum_{i=1}^{m} a_i V_i = 0$ implies $a_i = 0$ for all i. Equivalently, no V_i is a linear combination of the remaining vectors. The *span* of a set of vectors $\tilde{V} = (V_1, \ldots, V_m)$, denoted $span(\tilde{V})$, is the set of all linear combinations of these vectors. The span is a vector space, and $\tilde{V}$ is referred to as a *basis* whenever the defining vectors are linearly independent. The number of vectors in $\tilde{V}$ is the *dimension* of both the basis and its span. This definition is consistent, because while a basis of a vector space is not unique, any basis must be of the same dimension. Recognizing the dimension of a vector space is important, since it will be the case that any m-dimension vector space will be equivalent to $\mathbb{R}^m$ in many important respects.

6.3.3 Operators

An *operator* is a mapping $T : \mathcal{V} \to \mathcal{W}$ between vector spaces $\mathcal{V}, \mathcal{W}$. The *domain* and *range* of T are

$$\begin{aligned} domain\,(T) &= \{V \in \mathcal{V} \mid TV \text{ is defined}\}, \\ range\,(T) &= \{TV \in \mathcal{W} \mid V \in domain\,(T)\}. \end{aligned}$$

Unless otherwise stated, assume that *range* $(T) = \mathcal{W}$. The terms *injective*, *surjective* and *bijective* (Section 2.1.6) apply to operators, that is, $T : \mathcal{V} \to \mathcal{W}$ is injective (or one-to-one) if $V_1 \neq V_2 \Rightarrow TV_1 \neq TV_2$, is surjective if *range* $(T) = \mathcal{W}$, and is bijective if it is both injective and surjective. A bijective operator possesses an *inverse operator* $T^{-1} : \mathcal{W} \to \mathcal{V}$ for which $TV = W \iff T^{-1}W = V$. An injective mapping $T : \mathcal{V} \to \mathcal{W}$ induces a bijective mapping $T : \mathcal{V} \to$ *range* (T). Equality of two operators $T_1 = T_2$ with the same domains and ranges means that $T_1 V = T_2 V$ for all $V \in \mathcal{V}$.

Given three vector spaces $\mathcal{V}, \mathcal{W}, \mathcal{Y}$, and operators $T_1 : \mathcal{V} \to \mathcal{W}$ and $T_2 : \mathcal{W} \to \mathcal{Y}$ we may define compound operator $T_2 T_1 : \mathcal{V} \to \mathcal{Y}$ by the evaluation method $T_2 T_1 V = T_2(T_1 V)$.

If $E \subset$ *domain* (T), we take the *image* of E to be the set

$$TE = \{TV \in \mathcal{W} \mid V \in E\}.$$

It will be clear from the context when the image TE is intended.

For convenience we say T is an operator on $\mathcal{V}$ when $\mathcal{V} = \mathcal{W}$. In this case we always have the identity operator $I : \mathcal{V} \to \mathcal{V}$ for which $IV = V$. If T is a bijective operator, we have $I = T^{-1}T = TT^{-1}$. Operators T_1, T_2 on $\mathcal{V}$ *commute* if $T_1 T_2 = T_2 T_1$.

If $\mathcal{V}$ is a vector space over field $\mathbb{K}$, then $f : \mathcal{V} \to \mathbb{K}$ is a *functional*. Given that $\mathbb{K}$ is itself a vector field (with scalar field $\mathbb{K}$), a functional is a type of operator. If $\mathcal{V}, \mathcal{W}$ are vector spaces over the same scalar field $\mathbb{K}$, then $T : \mathcal{V} \to \mathcal{W}$ is a *linear operator* if $T(ax + by) = aTx + bTy$ for all $x, y \in \mathcal{V}$ and scalars $a, b \in \mathbb{K}$. The definition applies also to functionals. This assumption defines linear compositions of operators, so that the families of linear operators are themselves vectors spaces with scalar field $\mathbb{K}$.

Two vector spaces $\mathcal{V}$, $\mathcal{W}$ are *isomorphic* if there is a linear bijective operator $Q : \mathcal{V} \to \mathcal{W}$, and any such mapping is an *isomorphism*.

The *null set* of an operator $T : \mathcal{V} \to \mathcal{W}$ is defined as $\mathcal{N}(T) = \{V \in \mathcal{V} \mid TV = \vec{0}\}$.

When orderings are defined on vector spaces $\mathcal{V}$ and $\mathcal{W}$, then $T : \mathcal{V} \to \mathcal{W}$ is a *monotone operator* if $V_1 \leq V_2$ implies $TV_1 \leq TV_2$.

6.4 BANACH SPACES

The notion of the length of a vector is not implicit in a metric, unless we fix a reference vector e, which would naturally become the additive identity element in a vector space (typically $e = \vec{0}$), and take the length of a vector x to be $d(e, x)$. However, the usual approach in functional analysis is to define the notion of vector length explictly, using the *norm*, from which metric properties will follow. This will require vector addition, so norms must be defined on vector spaces.

Definition 6.6 *Suppose $\mathcal{V}$ is a vector space over field $\mathbb{K} \subset \mathbb{C}$, and suppose we have mapping $\|\cdot\| : \mathcal{V} \to [0, \infty)$. Then $\|\cdot\|$ is a* norm, *and $(\mathcal{V}, \|\cdot\|)$ is a* normed vector space *if the following axioms hold:*

> ***Identifiability*** *For any $x \in \mathcal{V}$ we have $\|x\| = 0$ if and only if $x = e$, the additive identity element of $\mathcal{V}$.*
>
> ***Scalar homogeneity*** *For any scalar a and vector $x \in \mathcal{V}$ we have $\|ax\| = |a| \|x\|$,*
>
> ***Triangle inequality*** *(or* ***subadditivity****) For any $x, y \in \mathcal{V}$ we have $\|x + y\| \leq \|x\| + \|y\|$.*

The field $\mathbb{K}$ in Definition 6.6 will usually be $\mathbb{K} = \mathbb{R} \subset \mathbb{C}$. We have already encountered examples of norms, such as the L^p norm or total variation norm, which both satisfy Definition 6.6. One further norm which we will encounter repeatedly is the *supremum norm* $\|\cdot\|_{sup}$. Suppose $\mathcal{V} \subset \mathcal{R}(\mathcal{X})$, containing vectors of the form $f : \mathcal{X} \to \mathbb{R}$. Then $\|f\|_{sup} \equiv \sup_{x \in \mathcal{X}} |f(x)|$, which is easily seen to satisfy Definition 6.6. This in turn generates a class of norms, known as the *weighted supremum norm* $\|\cdot\|_w$. If w is a positive function on $\mathcal{X}$, then $\|f\|_w \equiv \|f/w\|_{sup}$, which must also be a norm.

It must be remembered that the formal definition of a norm includes the vector space on which it is defined, and it must be assumed that $\|V\| < \infty$. Of course, a norm is often associated with a specific method of evaluation, for which we may have $\|V\| = \infty$ (the supremum norm of any unbounded function, for example). A certain flexibility in convention is therefore needed. For example, a normed vector space is sometimes defined as a subspace on which $\|V\| < \infty$, which implies the possibility that $\|V\| = \infty$. When we speak of a 'vector with finite norm', we are really referring to the evaluation method associated with the norm. This should cause no contradictions, provided that it is understood that the formal definition of a norm includes a vector space, and that $\|\cdot\|$ may be a norm on one vector space, but not another.

As in the case of the semimetric, it will sometimes be convenient to define a quantity which is a norm in every way except that there may not be a unique zero norm vector. Such an object will be called a *seminorm* (or *pseudonorm*).

Definition 6.7 *If in Definition 6.6 the identifiability axiom is replaced with the weaker axiom which holds that $\|e\| = 0$, then $\|\cdot\|$ is called a* seminorm *(or* pseudonorm*), and $(\mathcal{V}, \|\cdot\|)$ is a* seminormed vector space *(or* pseudonormed vector space*).*

Convergence $V_n \to V$ in a normed vector space $\mathcal{V}$ follows from the limit $\lim_{n\to\infty} \|V_n - V\| = 0$. Sometimes, it will be necessary to distinguish *convergence in the norm* $\|\cdot\|$ from other forms of convergence which may be defined for the objects in $\mathcal{V}$ (pointwise convergence, for example). When the context is clear, convergence in $\mathcal{V}$ will be assumed to be with respect to the given norm.

We then note that norm convergence of V_n to V is implied by $\|V_n - V\| \to_n 0$, that is $(V_n - V) \to_n \vec{0}$. So, all convergence statements can be reexpressed as convergence to $\vec{0}$.

If $T : \mathcal{V} \to \mathcal{W}$ is an operator between normed vector spaces $(\mathcal{V}, \|\cdot\|_{\mathcal{V}})$ and $(\mathcal{W}, \|\cdot\|_{\mathcal{W}})$, then T is *continuous* if $TV_n \to TV$ in $\mathcal{W}$ whenever $V_n \to V \in \mathcal{V}$, where convergence is in the respective norms. In addition, T is *bounded* if for all $\epsilon > 0$ there exists $L_\epsilon < \infty$ for which $\|V\| \le \epsilon \Rightarrow \|TV\| \le L_\epsilon$.

Some care must be taken in distinguishing between an operator on general vector spaces, and a operator on a normed vector space. To say that V is in normed vector space $(\mathcal{V}, \|\cdot\|)$ implies $\|V\| < \infty$, and so T cannot be an operator on a normed vector space unless $\|V\| < \infty \Rightarrow \|TV\| < \infty$. If $\|TV\| = \infty$ for some $V \in \mathcal{V}$, it may still be possible to say that T is an operator on the vector space $\mathcal{V}$, if not the normed vector space $\mathcal{V}$.

Of course, a test for closure of an operator T on a normed vector space is straightforward, reducing to $\|TV - V\| < \infty$ for all $V \in \mathcal{V}$, since by the triangle inequality we have $\|TV\| \le \|TV - V\| + \|V\|$.

6.4.1 Banach spaces and completeness

If we define the distance function $d(x,y) = \|x - y\|$, the axioms of the norm (or seminorm) imply that d is a true metric (or semimetric). It follows that d is *translation invariant* ($d(x+z, y+z) = d(x,y)$ for all $x,y,z \in \mathcal{V}$) and *scalar homogeneous* ($d(ax, ay) = |a|d(x,y)$ for all scalars a and $x,y \in \mathcal{V}$). Conversely, if d is translation invariant and scalar homogenous, then $\|x\| = d(\vec{0}, x)$ is a norm.

The notions of Lipschitz continuity and contractivity extend to a normed vector space $(\mathcal{V}, \|\cdot\|)$ through this metric, so that a mapping $T : \mathcal{V} \to \mathcal{V}$ is Lipschitz continuous if $\|Tx - Ty\| \leq L\|x - y\|$ for all $x,y \in \mathcal{V}$, locally Lipschitz continuous if for all $x \in \mathcal{V}$ it is Lipschitz continuous on some neighborhood of x, and is a contractive (or nonexpansive) mapping if $\|Tx - Ty\| \leq \rho\|x - y\|$ for all $x,y \in \mathcal{V}$, for some constant $\rho \in [0,1)$ (or $\rho \in [0,1]$).

Since a normed vector space $(\mathcal{V}, \|\cdot\|)$ is also a metric space, it may also be completed to include all limits of Cauchy sequences. Therefore, the kind of algorithm considered by the Banach fixed point theorem (Theorem 6.4) may be defined on $\mathcal{V}$, supplemented with vector algebra and the rich theory of linear operators.

We then have the definition of a *Banach space*:

Definition 6.8 A Banach space *is a normed vector space* $(\mathcal{V}, \|\cdot\|)$ *which is complete with respect to the metric* $d(x,y) = \|x - y\|$.

It is well known that L^p spaces are Banach spaces for $0 < p \leq \infty$ (this is often referred to as the *Riesz-Fischer Theorem* for historical reasons, although this reference is sometimes specific to L^2). Verifying the completeness of a normed linear space may proceed using the device of absolutely summable series:

Definition 6.9 *Suppose* $\mathcal{V}$ *is a normed vector space. A sequence* $\{V_n\}$ *is* summable *if for some* $V \in \mathcal{V}$ *we have* $\|V - \sum_{i=1}^{n} V_i\| \to_n 0$. *In addition,* $\{V_n\}$ *is* absolutely summable *if* $\sum_{i=1}^{\infty} \|V_i\| < \infty$.

Theorem 6.7 *A normed vector space* $\mathcal{V}$ *is complete if and only if every absolutely summable series is summable.*

Proof A proof may be found in Section 6.3 of Royden (1968), the point of which is that any Cauchy sequence may be represented by an absolutely summable series. ///

A Banach space must be confined to elements V for which $\|V\| < \infty$, since if we are given a sequence V_n with $\|V_n\| < \infty$ and V for which $\|V - V_n\| \to_n 0$, then the triangle inequality implies $\|V\| \leq \|V_n\| + \|V - V_n\| < \infty$.

Suppose $\mathcal{G}(\mathcal{X}) \subset \mathcal{R}(\mathcal{X})$ is some vector space of real valued functions on $\mathcal{X}$, and let $\mathcal{G}(\mathcal{X}, \|\cdot\|) = \{V \in \mathcal{G}(\mathcal{X}) \mid \|V\| < \infty\}$. Verification that $\mathcal{G}(\mathcal{X}, \|\cdot\|)$ is complete involves confirming that any Cauchy sequence converges to some limit in the norm, and that the limit is in $\mathcal{G}(\mathcal{X})$. We will be especially concerned with the supremum norm on a vector space of real valued functions.

Theorem 6.8 *If* $\mathcal{G}(\mathcal{X}) \subset \mathcal{R}(\mathcal{X})$ *is a vector space with additive identity* $\vec{0}$ *then* $\mathcal{G}(\mathcal{X}, \|\cdot\|_{sup})$ *is a Banach space.*

Proof It is easily verified that $\mathcal{G}(\mathcal{X}, \|\cdot\|_{sup})$ is a vector space with additive identity $\vec{0}$. Then suppose $\{V_n\}$ is absolutely summable, and let $S_n = \sum_{i=1}^{n} V_i$. Each sequence

$S_n(x)$ possesses a limit $V(x)$, which is bounded by $|V(x)| \leq M = \sum_{i=1}^{\infty} \|V_i\|_{sup} < \infty$, so that S_n possesses a pointwise limit V bounded in $\|\cdot\|_{sup}$. Furthermore, we have $|V(x) - S_n(x)| \leq \sum_{i>n}^{\infty} \|V_i\|_{sup}$, and the upper bound vanishes as $n \to \infty$, therefore $\|V - S_n\|_{sup} \to_n 0$, which completes the proof. ///

We will be especially interested in the vector space of measurable functions $\mathcal{F}(\mathcal{X})$ on a Borel space $\mathcal{X}$, which is closed under limits, so that $\mathcal{F}(\mathcal{X}, \|\cdot\|_{sup})$ is a Banach space.

6.4.2 Linear operators

Suppose we are given two normed vector spaces $(\mathcal{V}, \|\cdot\|_{\mathcal{V}})$, $(\mathcal{W}, \|\cdot\|_{\mathcal{W}})$. Let $L(\mathcal{V}, \mathcal{W})$ be the class of all linear operators $A : \mathcal{V} \to \mathcal{W}$. For any $A \in L(\mathcal{V}, \mathcal{W})$ we may define the operator norm

$$|||A|||_{\mathcal{W}|\mathcal{V}} = \sup\{\|Ax\|_{\mathcal{W}} \mid \|x\|_{\mathcal{V}} \leq 1\}.$$

If $|||A|||_{\mathcal{W}|\mathcal{V}} < \infty$ then A is a *bounded linear operator*. The class of all bounded linear operators is denoted $B(\mathcal{V}, \mathcal{W})$. The operator norm may be equivalently defined as the minimum constant c for which $\|Ax\|_{\mathcal{W}} \leq c\|x\|_{\mathcal{V}}$ for all $x \in \mathcal{V}$ (because of the scalar homogeneity of norms of Definition 6.6). Since A is linear, we have

$$\|Ax_1 - Ax_2\|_{\mathcal{W}} = \|A(x_1 - x_2)\|_{\mathcal{W}} \leq |||A|||_{\mathcal{W}|\mathcal{V}} \|x_1 - x_2\|_{\mathcal{V}},$$

so if A is bounded it is also Lipschitz continuous, and hence uniformly continuous.

If $(\mathcal{W}, \|\cdot\|_{\mathcal{W}}) = (\mathcal{V}, \|\cdot\|_{\mathcal{V}})$, so that A is an operator on $(\mathcal{V}, \|\cdot\|_{\mathcal{V}})$, the operator norm may be written $|||A|||_{\mathcal{V}}$.

Theorems 6.9, 6.10 and 6.11 state some of the important properties of linear operators (see, for example, Chapter 10 of Royden (1968)).

In a normed vector space, for a linear operator boundedness, continuity and Lipschitz continuity are equivalent.

Theorem 6.9 *For any* $A \in L(\mathcal{V}, \mathcal{W})$

(*i*) *If A is bounded then it is uniformly continuous,*
(*ii*) *If A is continuous anywhere then it is bounded.*

Linear operators themselves may be considered vectors under linear combination, that is, if $A, B \in L(\mathcal{V}, \mathcal{W})$ then given scalars a, b we have $aA + bB \in L(\mathcal{V}, \mathcal{W})$ under the usual evaluation rules $(aA + bB)V = aAV + bBV$. In fact, Banach space structure occurs naturally:

Theorem 6.10 *The vector space $B(\mathcal{V}, \mathcal{W})$, where $\mathcal{V}$ is a normed vector space and $\mathcal{W}$ is a Banach space, is a Banach space in the operator norm.*

Given normed linear spaces $\mathcal{V}, \mathcal{W}$ a bijective operator $Q \in B(\mathcal{V}, \mathcal{W})$ is an *isomorphism* if it possesses inverse $Q^{-1} \in B(\mathcal{W}, \mathcal{V})$. In this case $\mathcal{V}$ and $\mathcal{W}$ are isomorphic. If $\mathcal{V}$ is a Banach space then $\mathcal{W}$ must be as well. On the other hand, an important property of Banach spaces is that any inverse of a bounded linear operator must also be a bounded linear operator.

Theorem 6.11 (Bounded Inverse Theorem) *Given Banach spaces $\mathcal{V}, \mathcal{W}$, if $Q : \mathcal{V} \to \mathcal{W}$ is a bijective continuous linear operator, then Q^{-1} is a continuous linear operator.*

As a true norm, the operator norm is subadditive,

$$|||A + B|||_{\mathcal{W}|\mathcal{V}} \leq |||A|||_{\mathcal{W}|\mathcal{V}} + |||B|||_{\mathcal{W}|\mathcal{V}}.$$

The operator norm directly forces the Lipschitz property

$$\|Ax\|_{\mathcal{W}} \leq |||A|||_{\mathcal{W}|\mathcal{V}} \|x\|_{\mathcal{V}},$$

and is equal to the smallest Lipschitz constant. For linear operators on $\mathcal{V}$, the *submultiplicative* property holds:

$$\|ABx\|_{\mathcal{V}} \leq |||A|||_{\mathcal{V}} \|Bx\|_{\mathcal{V}} \leq |||A|||_{\mathcal{V}} |||B|||_{\mathcal{V}} \|x\|_{\mathcal{V}}, \quad \text{for all } x \in \mathcal{V},$$

that is, $|||AB|||_{\mathcal{V}} \leq |||A|||_{\mathcal{V}} |||B|||_{\mathcal{V}}$, which will prove to be an important property. However, the fact that the submultiplicative property is sometimes not sharp may necessitate some amount of additional analysis, in order to compare $|||AB|||_{\mathcal{V}}$ and $|||A|||_{\mathcal{V}} |||B|||_{\mathcal{V}}$.

If Q is a linear operator, then it is easily verified that the null set $\mathcal{N}(Q)$ is a vector subspace, so we refer to it as the *null space* of Q. If $\mathcal{N}(Q) = \{\vec{0}\}$ then Q is *nonsingular*, and is otherwise *singular*. The following theorem is easy to prove, but worth noting.

Theorem 6.12 *If $Q : \mathcal{V} \to \mathcal{W}$ is a nonsingular linear operator then it is injective. Therefore*

(*i*) *Any solution V to the equation $QV = w$ is unique.*
(*ii*) *If Q is surjective, then it possesses an inverse Q^{-1}.*

Proof Suppose $QV_1 = QV_2$. Then $Q(V_1 - V_2) = \vec{0}$, and therefore $V_1 - V_2 \in \mathcal{N}(Q)$, that is, $V_1 = V_2$. ///

Theorem 6.12 can be compared to Theorem 2.5.

Positive linear operators

Definition 6.10 *A linear operator Q on a vector space $\mathcal{V} \subset \mathcal{R}(\mathcal{X})$ is* positive *if $QV \geq 0$ whenever $V \geq 0$.*

A positive linear operator is monotone:

Lemma 6.1 *A positive linear operator Q on a vector space $\mathcal{V} \subset \mathcal{R}(\mathcal{X})$ is monotone in the sense that $V_2 \leq V_1$ implies $QV_2 \leq QV_1$. In addition*

$$|Q(V_2 - V_1)| \leq Q|V_2 - V_1|$$

for any two elements $V_1, V_2 \in \mathcal{V}$.

Proof If $V_2 - V_1 \geq 0$ then $0 \leq Q(V_2 - V_1) = QV_2 - QV_1$. For any two $V_1, V_2 \in \mathcal{V}$. $V_2 - V_1 \leq |V_2 - V_1|$, so $QV_2 - QV_1 = Q(V_2 - V_1) \leq Q|V_2 - V_1|$, with the proof completed by exchanging V_1 and V_2. ///

Diagonal operator

Suppose we are given vector space $\mathcal{V} \subset \mathcal{R}(\mathcal{X})$, and another mapping $h \in \mathcal{R}(\mathcal{X})$. We may define the *diagonal operator* I_h, evaluated for $V \in \mathcal{V}$ by $(I_h V)(x) = h(x)V(x)$. It is not expected that h exist either in the domain or range of I_h. If $h \equiv 1$ then I_h is the identity operator (in analogy to the identity matrix).

6.5 NORMS AND NORM EQUIVALENCE

When analyzing properties of norms, the notion of *norm equivalence* will be crucial for our applications. Since the defining property of a Banach space depends on norm convergence, if it can be claimed that on a common vector space $\mathcal{V}$ two distinct norms $\|\cdot\|_\alpha$ and $\|\cdot\|_\beta$ imply exactly the same convergence statements, then they are *equivalent norms*. This is clearly an equivalence relationship (Section 2.2), expressible $\|\cdot\|_\alpha \sim \|\cdot\|_\beta$, which defines equivalence classes of norms which possess identical convergence properties on a common vector space.

Certainly, two equivalent norms define essentially the same Banach space. This point of view has some quite practical advantages. Because equivalent norms may be based on very different evaluation methods, it can be quite advantageous to be able to select the norm most suitable for a given task.

A norm $\|\cdot\|$ may be thought of as a family of spheres $S_t = \{x \mid \|x\| \leq t\}$, $t \in [0, \infty)$, where $S_0 = \{\vec{0}\}$. We may take $\|x\| \leq 1$ to define a canonical sphere (or *unit sphere*), and since $\|x\| = 1$ implies $\|tx\| = |t|$, any other sphere S_t is a scale transformation of S_1. The spheres are clearly concentric, that is $S_t \subseteq S_s$ if $t \leq s$ (here, a subset of a sphere is meant to include the interior). Clearly, $V_n \to_n \vec{0}$ if and only if there is a sequence $t_n \downarrow 0$ for which $V_n \subset S_{t_n}$.

Suppose the spheres S_t^α and S_t^β induced by two norms $\|\cdot\|_\alpha$ and $\|\cdot\|_\beta$ are *mutually concentric*, in sense that for all S_t^α and $S_{t'}^\beta$ there exists $S_s^\beta \subset S_t^\alpha$ and $S_{s'}^\alpha \subset S_{t'}^\beta$. This suffices to establish norm equivalence.

Theorem 6.13 *Suppose the spheres induced by norms* $\|\cdot\|_\alpha$ *and* $\|\cdot\|_\beta$ *on a common vector space* $\mathcal{V}$ *are mutually concentric. Then* $\|\cdot\|_\alpha \sim \|\cdot\|_\beta$.

Proof Assume $V_n \to_n \vec{0}$ *wrt* $\|\cdot\|_\alpha$. Choose t_1 for which $V_1 \subset S_{t_1}^\alpha$. There exists s' for which $S_{s'}^\beta \subset S_{t_1}^\alpha$. Make $0 < s_1 < s'$ as small as we like. Select t_2 for which $S_{t_2}^\alpha \subset S_{s_1}^\beta$. By convergence *wrt* $\|\cdot\|_\alpha$ there exists n_1 for which $V_{n_1} \subset S_{t_2}^\alpha \subset S_{s_1}^\beta$. We may select $s_2, s_3, \ldots$ in the same way, ensuring that $s_n \to_n 0$. This means that $V_n \to_n \vec{0}$ *wrt* $\|\cdot\|_\beta$ as well. The proof concludes by exchanging $\|\cdot\|_\alpha$ and $\|\cdot\|_\beta$. ///

Norm monotonicity

Consider the following norm property:

Definition 6.11 *A norm* $\|\cdot\|$ *on a vector space* $\mathcal{V} \subset \mathcal{R}(\mathcal{X})$ *is* monotone *if* $|V_2(x)| \geq |V_1(x)|$ *for all* $x \in \mathcal{X}$ *implies* $\|V_2\| \geq \|V_1\|$.

Although this property seems quite intuitive, the construction of a counterexample is straightforward. The monotonicity property implies that if $|V_2(x)| \geq |V_1(x)|$ for all x then V_2 is not in the interior of the sphere containing V_1 on its surface. Let $\|\cdot\|_{sup}$ be the supremum norm on $\mathbb{R}^2$, which is monotone. The unit sphere is a square with sides parallel to an axis. To create a counterexample, we only need to rotate this square by an angle less than 45 degrees.

It is tempting to conclude from an example like this that the properties of a norm may be very sensitive to its specific geometric properties. However, it also suggests that when a norm does not possess a certain property, such as monotonicity, it is possible that a similar norm does. In this case, the modified norm is not monotone, but from Theorem 6.13 it is clearly equivalent to one that is.

6.5.1 Norm dominance

While Theorem 6.13 gives an intuitive characterization of norm equivalence, the notion of *norm dominance* offers greater precision.

Definition 6.12 *Suppose we are given two norms* $\|\cdot\|_\alpha$, $\|\cdot\|_\beta$ *on a vector space* $\mathcal{V}$. *We say* $\|\cdot\|_\alpha$ dominates $\|\cdot\|_\beta$ *if there is a finite constant b for which* $\|V\|_\beta \leq b\|V\|_\alpha$ *for all* $V \in \mathcal{V}$, *which is written* $\|\cdot\|_\beta \leq \|\cdot\|_\alpha$.

Dominance may be used to characterize norm equivalence.

Theorem 6.14 *Given two norms* $\|\cdot\|_\alpha$, $\|\cdot\|_\beta$ *on vector space* $\mathcal{V}$, $\|\cdot\|_\beta \leq \|\cdot\|_\alpha$ *if and only if all sequences converging wrt* $\|\cdot\|_\alpha$ *also converge wrt* $\|\cdot\|_\beta$.

As a consequence $\|\cdot\|_\beta \sim \|\cdot\|_\alpha$ *if and only if* $\|\cdot\|_\beta \leq \|\cdot\|_\alpha$ *and* $\|\cdot\|_\alpha \leq \|\cdot\|_\beta$.

Proof If $\|\cdot\|_\beta \leq \|\cdot\|_\alpha$ and $\|V_n\|_\alpha \to 0$ for some sequence V_n then $\limsup_n \|V_n\|_\beta \leq \limsup_n b\|V_n\|_\alpha = 0$ for some finite b.

Next suppose there exists a sequence V_n for which $\|V_n\|_\alpha \to 0$ but $\limsup_n \|V_n\|_\beta = r > 0$. Then we cannot have $\|\cdot\|_\beta \leq \|\cdot\|_\alpha$, since the sequence of ratios $\|V_n\|_\beta / \|V_n\|_\alpha$ is unbounded as $n \to \infty$. This completes the proof. ///

Norm equivalence can be equivalently defined as the existence of two positive scalars a, b such that

$$a\|V\|_\alpha \leq \|V\|_\beta \leq b\|V\|_\alpha$$

for all $V \in \mathcal{V}$. This set of inequalities are understood to imply that $\|V\|_\alpha = \infty$ if and only if $\|V\|_\beta = \infty$.

If $\mathcal{V}$ is complete with respect to two norms, one dominance relationship suffices to establish norm equivalence.

Theorem 6.15 *Given a vector space* $\mathcal{V}$ *which is complete in norms* $\|\cdot\|_\alpha$ *and* $\|\cdot\|_\beta$, *if* $\|\cdot\|_\alpha \leq \|\cdot\|_\beta$ *then* $\|\cdot\|_\alpha \sim \|\cdot\|_\beta$.

Proof Since $\|\cdot\|_\alpha \leq \|\cdot\|_\beta$, the identity map is a bijective continuous linear operator between Banach spaces $(\mathcal{V}, \|\cdot\|_\alpha)$ and $(\mathcal{V}, \|\cdot\|_\beta)$, therefore the theorem holds by applying the bounded inverse theorem (Theorem 6.11). ///

6.5.2 Equivalence properties of norm equivalence classes

We refer to a property as a *norm equivalence property* if it necessarily holds for all norms in an equivalence class. The definition of convergence is a norm equivalence property (the defining one). All norms in an equivalence class are dominated by any single member, so dominance by a single norm is a norm equivalence property. We have seen that monotonicity in $\mathcal{V} \subset \mathcal{R}(\mathcal{X})$ is not a norm equivalence property, but also that a norm that is not monotone may be equivalent to one that is.

The contraction properties of an operator T on $\mathcal{V}$ are defined relative to a norm. We will show below that the contraction property is not a norm equivalence property, but that multistage contraction is.

We define the quantity

$$\eta(\|\cdot\|_\beta \mid \|\cdot\|_\alpha) = \sup_{V : \|V\|_\alpha > 0} \frac{\|V\|_\beta}{\|V\|_\alpha}$$

which is identical to the operator norm of the identity map from $(\mathcal{V}, \|\cdot\|_\alpha)$ to $(\mathcal{V}, \|\cdot\|_\beta)$ used in the proof of Theorem 6.15. Then $\eta(\|\cdot\|_\beta \mid \|\cdot\|_\alpha) < \infty$ if and only if $\|\cdot\|_\beta \leq \|\cdot\|_\alpha$, and is in fact the smallest constant b for which $\|V\|_\beta \leq b\|V\|_\alpha$ uniformly.

We will also make use of the quantity

$$\eta(\|\cdot\|_\alpha, \|\cdot\|_\beta) = \eta(\|\cdot\|_\alpha \mid \|\cdot\|_\beta)\eta(\|\cdot\|_\beta \mid \|\cdot\|_\alpha).$$

Clearly, $\eta(\|\cdot\|_\alpha, \|\cdot\|_\beta) < \infty$ if and only if $\|\cdot\|_\alpha \sim \|\cdot\|_\beta$. We also note the following lower bound:

Lemma 6.2 *The quantity* $\eta(\|\cdot\|_\alpha, \|\cdot\|_\beta) \geq 1$.

Proof For any nonzero $V \in \mathcal{V}$ we may write

$$1 = \frac{\|V\|_\beta}{\|V\|_\alpha} \frac{\|V\|_\alpha}{\|V\|_\alpha} = \left\| \frac{V}{\|V\|_\alpha} \right\|_\beta \left\| \frac{V}{\|V\|_\beta} \right\|_\alpha \leq \eta(\|\cdot\|_\alpha, \|\cdot\|_\beta)$$

///

It will be useful to construct a distance function to define the following relationship between two extended valued scalars a, b. Suppose we are given a third, fixed finite scalar η, and that the following inequalities hold:

$$a \leq \eta b \quad \text{and} \quad b \leq \eta a. \tag{6.11}$$

Then a and b are either both finite or both infinite. In the former case this is equivalent to $\max(a/b, b/a) \leq \eta$. Finally, if only one of a, b is finite, then we must have $\eta = \infty$. This suggests a scalar distance function between positive extended scalars

$$d_s(a, b) = \begin{cases} \max(a/b, b/a); & a < \infty \text{ or } b < \infty \\ 1 \qquad ; & a = \infty \text{ and } b = \infty \end{cases}.$$

Then the inequalities in (6.11) are equivalent to $d_s(a, b) \leq \eta$. Note that $d_s(a, b) \geq 1$.

Then $\|\cdot\|_\alpha \sim \|\cdot\|_\beta$ if and only if $\sup_{V \in \mathcal{V}} d_s(\|V\|_\alpha, \|V\|_\beta) < \infty$.

Theorem 6.16 *Suppose we are given a vector space $\mathcal{V}$ and $\|\cdot\|_\alpha$ and $\|\cdot\|_\beta$ are alternative norms which complete $\mathcal{V}$. Then*

(*i*) *For any $V \in \mathcal{V}$ $\|V\|_\beta \leq \eta(\|\cdot\|_\beta \mid \|\cdot\|_\alpha)\|V\|_\alpha$.*
(*ii*) *Let T be an operator on $\mathcal{V}$. If T possesses Lipschitz constant L for $\|\cdot\|_\alpha$, then it possesses Lipschitz constant $L\eta(\|\cdot\|_\alpha, \|\cdot\|_\beta)$ for $\|\cdot\|_\beta$.*
(*iii*) *For a linear operator Q on $\mathcal{V}$ the respective operator norms for Banach spaces $(\mathcal{V}, \|\cdot\|_\alpha)$ to $(\mathcal{V}, \|\cdot\|_\beta)$ satisfy*

$$d_s\left(|||Q|||_\alpha, |||Q|||_\beta\right) \leq \eta(\|\cdot\|_\alpha, \|\cdot\|_\beta).$$

Proof (*i*) follows from the definition of $\eta(\|\cdot\|_\beta \mid \|\cdot\|_\alpha)$. To show (*ii*), suppose $V, V' \in \mathcal{V}$, then write

$$\begin{aligned}\|TV - TV'\|_\beta &\leq \eta(\|\cdot\|_\beta \mid \|\cdot\|_\alpha)\|TV - TV'\|_\alpha \\ &\leq \eta(\|\cdot\|_\beta \mid \|\cdot\|_\alpha)L\|V - V'\|_\alpha \\ &\leq \eta(\|\cdot\|_\beta, \|\cdot\|_\alpha)L\|V - V'\|_\beta.\end{aligned}$$

Then (*iii*) follows from (*ii*), given the interpretation of the operator norm as a Lipschitz constant. ///

Essentially, the properties of a Banach space important to our application will be norm equivalence properties. This includes the boundedness of a linear operator Q and the Lipschitz continuity of an operator T. Strict contractivity is not a norm equivalence property but multistage contractivity is, as well as a specific the asymptotic contraction rate (Theorem 6.3). This issue will be considered in detail in Section 7.1, but it is worth noting the following. Suppose we have $\|T^J V_1 - T^J V_2\|_\alpha \leq K\rho^J$ for finite K and $\rho < 1$. This suffices to establish multistage contractivity. Then by Theorem 6.16 we also have $\|T^J V_1 - T^J V_2\|_\beta \leq K\eta(\|\cdot\|_\alpha, \|\cdot\|_\beta)\rho^J$, giving multistage contractivity for any other norm $\|\cdot\|_\beta \sim \|\cdot\|_\alpha$. The existence of a contractive fixed point algorithm itself is therefore a norm equivalence property.

6.6 QUOTIENT SPACES AND SEMINORMS

A seminormed vector space is not formally a normed vector space, since the seminorm does not possesses the identifiability property $\|V\| = 0 \iff V = \vec{0}$. In particular, $\|V_n\| \to_n 0$ does not imply that V_n possesses limit $\vec{0}$.

However, a related normed vector space can be constructed. Since any seminorm $\|\cdot\|_\alpha$ satisfies the triangle inequality, if $\|V - U\|_\alpha = \|U - W\|_\alpha = 0$ we must have $\|V - W\|_\alpha = 0$. This means that the relationship $\|V - W\|_\alpha = 0$ is transitive, and satisfies the other properties of an equivalence relationship (Section 2.2). Therefore, $\|\cdot\|_\alpha$ partitions a vector space into *equivalence classes* $E_V = \{W \in \mathcal{V} \mid \|W - V\|_\alpha = 0\}$.

We may also represent this partition as a system of cosets, used to define a quotient space in Section 6.3.1. Set $\mathcal{N} = \{V \in \mathcal{V} \mid \|V\|_\alpha = 0\}$. Then $W \in E_V$ if and only if $\|V - W\|_\alpha = 0$, or equivalently $V - W = n \in \mathcal{N}$, so that $E_V = [V]$, where $[V]$ is the coset of V modulo $\mathcal{N}$.

The construction of a normed vector space follows.

Theorem 6.17 *Suppose* $(\mathcal{V}, \|\cdot\|_\alpha)$ *is a seminormed vector space. Set null space* $\mathcal{N} = \{V \in \mathcal{V} \mid \|V\|_\alpha = 0\}$. *Then* $\mathcal{N}$ *is a subspace of* $\mathcal{V}$, *and therefore* $\mathcal{V}/\mathcal{N}$ *is a quotient space.*

Then $\|V_1\|_\alpha = \|V_2\|_\alpha$ *for all* $V_1, V_2 \in [W]$ *and therefore the quantity* $\|[W]\|_\alpha = \|W\|_\alpha$ *is well defined (that is, it does not depend on the representative vector* W*). Furthermore* $(\mathcal{V}/\mathcal{N}, \|\cdot\|_\alpha)$ *is a normed vector space.*

Proof If $V, W \in \mathcal{N}$, then $\|aV + bW\|_\alpha \leq a\|V\|_\alpha + b\|W\|_\alpha = 0$, so that $\mathcal{N}$ is a subspace, and therefore $\mathcal{V}/\mathcal{N}$ is a quotient space.

Next, suppose $V_1, V_2 \in [W]$. Then $\|V_1\|_\alpha \leq \|V_1 - V_2\|_\alpha + \|V_2\|_\alpha = \|V_2\|_\alpha$, and reversing the indices gives $\|V_2\|_\alpha \leq \|V_1\|_\alpha$, so that $\|V_1\|_\alpha = \|V_2\|_\alpha$, therefore $\|[W]\|_\alpha = \|W\|_\alpha$ is well defined. For scalar a we have

$$\|a[W]\|_\alpha = \|[aW]\|_\alpha = \|aW\|_\alpha = |a|\|W\|_\alpha = |a|\|[W]\|_\alpha.$$

Also,

$$\|[V] + [W]\|_\alpha = \|[V + W]\|_\alpha = \|V + W\|_\alpha \leq \|W\|_\alpha + \|V\|_\alpha = \|[W]\|_\alpha + \|[V]\|_\alpha.$$

By definition $V \in \mathcal{N}$ if and only if $\|V\|_\alpha = 0$, so the additive identity $\mathcal{N}$ uniquely satisfies $\|\mathcal{N}\|_\alpha = 0$, which completes the proof. ///

By Theorem 6.17 a seminorm $\|\cdot\|_\alpha$ defines a normed quotient vector spaces $\mathcal{V}/\mathcal{N}$ for null subspace $\mathcal{N} = \{V \in \mathcal{V} \mid \|V\|_\alpha = 0\}$. No reference to a formal normed vector space is needed. However, construction of a seminormed quotient space will often be based on a formal normed vector space, so it will be useful to know which properties of the latter hold for the former.

If $\mathcal{N}$ is a vector subspace of a normed vector space $(\mathcal{V}, \|\cdot\|)$, define for the coset of V modulo $\mathcal{N}$:

$$\|[V]\|_\alpha = \inf\{\|W\| \mid W \in [V]\} = \inf_{n \in \mathcal{N}} \|V - n\|. \tag{6.12}$$

Then $\|[V]\|_\alpha$ can be interpreted as the minimum normed distance of V from $\mathcal{N}$. Clearly, $\|\mathcal{N}\|_\alpha = 0$, but not necessarily uniquely.

Theorem 6.18 *If* $\mathcal{N}$ *is a vector subspace of a normed vector space* $(\mathcal{V}, \|\cdot\|)$, *then the quantity* $\|\cdot\|_\alpha$ *defined in* (6.12) *is a seminorm on the quotient space* $\mathcal{V}/\mathcal{N}$ *with additive identity* $\mathcal{N}$.

If $\mathcal{N}$ *is closed (wrt* $\|\cdot\|$*) then* $\|\cdot\|_\alpha$ *is also a norm, and if in addition* $(\mathcal{V}, \|\cdot\|)$ *is a Banach space, then so is* $(\mathcal{V}/\mathcal{N}, \|\cdot\|_\alpha)$.

Proof Since $\mathcal{N}$ is a vector space we have for any scalar a

$$\begin{aligned}
\|a[W]\|_\alpha &= \|[aW]\|_\alpha \\
&= \inf_{n \in \mathcal{N}} \|aW - n\| \\
&= \inf_{n \in \mathcal{N}} \|aW - an\| \\
&= \inf_{n \in \mathcal{N}} |a|\|W - n\| \\
&= |a|\|[W]\|.
\end{aligned}$$

Similarly,

$$\begin{aligned}
\|[V]+[W]\|_\alpha &= \|[V+W]\|_\alpha \\
&= \inf_{n\in\mathcal{N}} \|V+W-n\| \\
&= \inf_{n,n'\in\mathcal{N}} \|V-n+W-n'\| \\
&\le \inf_{n,n'\in\mathcal{N}} \|V-n\|+\|W-n'\| \\
&= \inf_{n\in\mathcal{N}} \|V-n\| + \inf_{n'\in\mathcal{N}} \|W-n'\| \\
&= \|[V]\|_\alpha + \|[W]\|_\alpha
\end{aligned}$$

In addition, $\|\mathcal{N}\|_\alpha = 0$, which completes the first part of the proof.

If $\|[V]\|_\alpha = 0$ then V must be a limit *wrt* $\|\cdot\|$ of a sequence in $\mathcal{N}$. If $\mathcal{N}$ is closed then $V \in \mathcal{N}$, which implies $[V] = \mathcal{N}$ and therefore that $\|\cdot\|_\alpha$ is a norm.

To complete the proof we make use of Theorem 6.7. Suppose $[V_n]$ is an absolutely summable series, that is, $\sum_{n\ge 1} \|[V_n]\|_\alpha < \infty$. By construction, for each $n \ge 1$ we may select $n_n \in \mathcal{N}$ for which $\|V_n - n_n\| \le \|[V_n]\|_\alpha + (1/2)^n$. This means $V_n - n_n$ is an absolutely summable sequence in the Banach space $(\mathcal{V}, \|\cdot\|)$, and therefore there exists limit $V' \in \mathcal{V}$ for which $\|V' - \sum_{i=1}^n (V_i - n_i)\| \to_n 0$. Then

$$\begin{aligned}
\|[V'] - \sum_{i=1}^n [V_i]\|_\alpha &= \inf_{n\in\mathcal{N}} \|V' - \sum_{i=1}^n V_i - n\| \\
&\le \|V' - \sum_{i=1}^n (V_i - n_n)\|.
\end{aligned}$$

The proof is completed by allowing $n \to \infty$. ///

To summarize, suppose we are given a seminorm $\|\cdot\|_\alpha$ on a Banach space $(\mathcal{V}, \|\cdot\|)$. If its null space $\mathcal{N}$ is closed with respect to $\|\cdot\|$, and $\|\cdot\|_\alpha$ is equivalent to the seminorm (6.12), then $(\mathcal{V}/\mathcal{N}, \|\cdot\|_\alpha)$ is a Banach space.

6.7 HILBERT SPACES

When fixed points are calculated by repeated iterations, and each iteration contributes error, we expect the effect on the algorithm to be some aggregation of these errors. In some cases, these errors will be signed and roughly distributed about zero, and the aggregation will permit some 'error canceling' effect, especially when the errors are stochastic. This introduces a new notion of direction, or angle, in addition to distance and length. In fact, for random variables statistical independence is analogous to orthogonality, with the linear correlation coefficient interpetable as a *cosine*.

These concepts are not naturally expressed in Banach spaces. Suppose we are given the sum of two random variables $X_1 + X_2$. While this is well defined on a vector space, we know that the properties of this sum are influenced by the stochastic dependence of X_1 and X_2. In a Banach space, we could write $\|X_1 + X_2\| \le \|X_1\| + \|X_2\|$, but this bound would depend only on the marginal distributions of X_1 and X_2. In order to exploit and dependence structure (which in some applications will be crucial) this would have to be done 'within the norm' $\|X_1 + X_2\|$, so that the Banach space structure does not really offer any contribution in this regard.

Our interest in the Hilbert space will be precisely in its ability to capture forms of stochastic independence in its structure. This is expressed by the *inner product* $\langle \cdot, \cdot \rangle : \mathcal{V} \times \mathcal{V} \to \mathbb{R}$, which generalizes the 'dot product' $\langle a, b \rangle = \sum_i a_i b_i$ for Euclidean vectors $a = (a_1, \ldots, a_n)$ and $b = (b_1, \ldots, b_n)$.

Definition 6.13 *Suppose $\mathcal{V}$ is a vector space with a real valued scalar field, and suppose we have mapping $\langle \cdot, \cdot \rangle : \mathcal{V} \times \mathcal{V} \to \mathbb{R}$. Then $\langle \cdot, \cdot \rangle$ is an* inner product *(or* scalar product*), and $(\mathcal{V}, \langle \cdot, \cdot \rangle)$ is an* inner product space *(or* scalar product space*) if the following axioms hold:*

Symmetry *For any $x, y \in \mathcal{V}$ we have $\langle x, y \rangle = \langle y, x \rangle$,*
Scalar linearity *For any scalar a and vectors $x, y \in \mathcal{V}$ we have $\langle ax, y \rangle = a\langle x, y \rangle$,*
Vector linearity *For any vectors $x, y, z \in \mathcal{V}$ we have $\langle x + y, z \rangle = \langle x, z \rangle + \langle y, z \rangle$,*
Positive-definiteness *For any $x \in \mathcal{V}$ we have $\langle x, x \rangle \geq 0$,*
Identifiability *For any $x \in \mathcal{V}$ we have $\langle x, x \rangle = 0$ if and only if $x = \vec{0}$, the identity element of the vector group.*

Note that inner products may more generally assume complex values, but we will confine attention to *real inner product spaces*.

We will provisionally define the norm $\|x\| = \langle x, x \rangle^{1/2}$, then verify that it is a true norm. We first present a number of basic theorems.

Theorem 6.19 (Schwartz Inequality for Inner Products) *If $(\mathcal{V}, \langle \cdot, \cdot \rangle)$ is an inner product space, and $\|x\| = \langle x, x \rangle^{1/2}$, then $|\langle x, y \rangle| \leq \|x\| \|y\|$ for any $x, y \in \mathcal{V}$.*

Proof For any constant a set

$$0 \leq \langle x - ay, x - ay \rangle = \|x\|^2 - 2a\langle x, y \rangle + a^2 \|y\|^2.$$

If $y = 0$ the theorem holds directly. Otherwise, set $a = \langle x, y \rangle / \|y\|^2$, which completes the proof. ///

Theorem 6.20 (The Triangle Inequality for Inner Products) *If $(\mathcal{V}, \langle \cdot, \cdot \rangle)$ is an inner product space, and $\|x\| = \langle x, x \rangle^{1/2}$, then $\|x + y\| \leq \|x\| + \|y\|$ for any $x, y \in \mathcal{V}$.*

Proof Applying the Schwartz inequality we have

$$\begin{aligned} \|x + y\|^2 &= \langle x + y, x + y \rangle \\ &= \langle x, x \rangle + 2\langle x, y \rangle + \langle y, y \rangle \\ &\leq \|x\|^2 + 2\|x\| \|y\| + \|y\|^2 \\ &= \left(\|x\| + \|y\|\right)^2, \end{aligned}$$

which completes the proof. ///

Theorem 6.21 (Parallelogram Law for Inner Products) *If $(\mathcal{V}, \langle\cdot,\cdot\rangle)$ is an inner product space then $\|x+y\|^2 + \|x-y\|^2 = 2(\|x\|^2 + \|y\|^2)$.*

Proof Write

$$\begin{aligned}\|x+y\|^2 + \|x-y\|^2 &= \|x\|^2 + \|y\|^2 + 2\langle x,y\rangle + \|x\|^2 + \|y\|^2 + 2\langle x,-y\rangle \\ &= 2(\|x\|^2 + \|y\|^2).\end{aligned}$$

///

We may now verify that $\|x\| = \langle x,x\rangle^{1/2}$ is a true norm. That $\|\cdot\|$ satisfies the triangle inequality has just been shown. The identifiability of $\|x\|$ follows from the identifiability of $\langle\cdot,\cdot\rangle$, while the scalar homogeneity of $\|x\|$ follows from the scalar linearity, then the symmetry, of $\langle\cdot,\cdot\rangle^{1/2}$. This then implies that $d(x,y) = \langle x-y, x-y\rangle^{1/2}$ is a true metric, so that an inner product space is also a normed vector space and a metric space, which may be completed to define the *Hilbert space*:

Definition 6.14 A Hilbert space *is an inner product space $(\mathcal{V}, \langle\cdot,\cdot\rangle)$ which is complete with respect to the metric $d(x,y) = \langle x-y, x-y\rangle^{1/2}$.*

Then, any properties of a complete metric space or a Banach space hold also for a Hilbert space with respect to the metric or norm induced by the inner product.

6.8 EXAMPLES OF BANACH SPACES

In this section we discuss a number of Banach space structures we will make use of in later chapters.

6.8.1 Finite dimensional spaces

The Banach space structure of $\mathbb{R}^m$ is straightforward and intuitive. It is worth characterizing as an illustration of elementary Banach space methods, and also as an example of an important isomorphism, since any finite dimensional vector space has identical Banach space properties.

Convergence in $\mathbb{R}^m$ is usually taken to be equivalent to pointwise convergence. Since the dimension is finite, this is also uniform convergence. However, recall that the definition of convergence depends on the topology. We could decide that all sequences converge, or that only constant sequences do. We don't point this out to recommend such schemes, but rather to emphasize that the choice of topology is ours to make.

However, once we select a norm the notion of convergence is fixed, and in finite dimensional vector spaces all norms are equivalent. Recall that any finite dimensional vector space $\mathcal{V}$ may be represented by an m-dimensional basis $\tilde{V}$, that is, for each $V \in \mathcal{V}$ there is a unique vector $\tilde{a} = (a_1, \ldots, a_m)$ for which $V = \sum_{i=1}^m a_i V_i$. Any norm $\|\cdot\|$ on $\mathcal{V}$ must be a continuous function of $\tilde{a}$ in the Euclidean norm. This is because if $|\tilde{a}| = \epsilon$ we must also have $|a_i| \leq \epsilon$, so that

$$\|V\| \leq \sum_{i=1}^m |a_i| \|V_i\| \leq \epsilon \sum_{i=1}^m \|V_i\| = |\tilde{a}| \sum_{i=1}^m \|V_i\|.$$

The Euclidean norm on the coefficients $\tilde{a}$ is also a norm on $\mathcal{V}$, which dominates any other norm on $\mathcal{V}$, so that all norms on $\mathcal{V}$ form an equivalence class with the Euclidean norm.

6.8.2 Matrix norms and the submultiplicative property

Let M_n be the class of $n \times n$ real valued matrices A with elements $a_{i,j}$ in the ith row and jth column. This is clearly a finite dimensional vector space, so all norms are equivalent to the Euclidean norm defined on the n^2 matrix elements. However, when a matrix A is interpreted as a linear operator $A : \mathbb{R}^m \to \mathbb{R}^m$ a somewhat richer norm structure emerges. We provide a brief introduction to the topic, and refer the reader to, for example Horn and Johnson (1985) for a comprehensive treatment.

If a vector space admits a composition $AB \in \mathcal{V}$ for $A, B \in \mathcal{V}$ a norm $\|\cdot\|$ is *submultiplicative* if $\|AB\| \leq \|A\|\|B\|$. The composition of operators is always well defined, and an operator norm is always submultiplicative. However, despite the fact that M_n is a class of operators, and all norms on M_n are equivalent, not all norms are submultiplicative. For example, let $\|A\|_\infty = \max_{i,j} |a_{i,j}|$. This is a norm, but is not submultiplicative, as can be seen by considering matrix A with $a_{i,j} \equiv 1$, since in this case $\|A\|_\infty = 1$ but $\|AA\|_\infty = m$. Because of the importance of this property, the term *matrix norm* is sometimes reserved specifically for submultiplicative norms (as in Horn and Johnson (1985)), for which the notation $|||\cdot|||$ is reserved.

The operator norm is generated by a vector norm $\|\cdot\|_\alpha$ on $\mathbb{R}^n$ using the usual definition of the operator norm

$$|||A|||_\alpha = \sup_{\|x\| \neq 0} \frac{\|Ax\|_\alpha}{\|x\|_\alpha}. \tag{6.13}$$

Now, suppose $A \in M_n$ is taken as a linear operator on the Banach space $(\mathbb{R}^n, \|\cdot\|_\alpha)$. We have already established that multistage contraction is a norm equivalence property, and all norms on $\mathbb{R}^n$ are equivalent. An asymptotic contraction rate (Theorem 6.3) can be defined for operator A alone, and is in fact equal to the spectral radius $\rho(A)$ (see Section 2.3.2 for definition).

Theorem 6.22 *Suppose we are given a square real valued matrix* $A \in M_n$.
(*i*) $\lim_{k\to\infty} A^k = 0$ *if and only if* $\rho(A) < 1$.
(*ii*) *For any matrix norm* $|||\cdot|||$ *we have* $\lim_{k\to\infty} |||A^k|||^{1/k} = \rho(A)$.

Proof See Theorem 5.6.12 and Corollary 5.6.14 of Horn and Johnson (1985).

6.8.3 Weighted norms on function spaces

For the discussion of this section, it will suit our purposes best to let $\mathcal{F}(\mathcal{X})$ be the vector space of measurable functions on a Borel space $\mathcal{X}$. Then let $\mathcal{F}(\mathcal{X}, \|\cdot\|_{sup})$ be a vector space of bounded real valued functions on $\mathcal{X}$, which is a Banach space by Theorem 6.8. We also define the set of *weight functions* $\mathcal{W}(\mathcal{X}) = \{w \in \mathcal{F}(\mathcal{X}) \mid w(x) > 0\}$. For any weight function w we define the weighted supremum norm $\|\cdot\|_w$ on $\mathcal{F}(\mathcal{X})$ by $\|V\|_w = \sup_{x\in\mathcal{X}} w^{-1}(x)|V(x)|$. The space $\mathcal{F}(\mathcal{X}, \|\cdot\|_w) = \{wV \mid V \in \mathcal{F}(\mathcal{X}, \|\cdot\|_{sup})\}$ is clearly an isomorphism of $\mathcal{F}(\mathcal{X}, \|\cdot\|_{sup})$ and is therefore also a Banach space. If $w \equiv 1$ we recover the supremum norm $\|\cdot\|_w = \|\cdot\|_{sup}$.

Note that the weighted norm is monotone in the sense of Definition 6.11, that is, if $|V_1| \leq |V_2|$ then $\|V_1\|_w \leq \|V_2\|_w$.

For weight functions w_1, w_2 define the quantity:

$$d_w(w_1, w_2) = \|w_1\|_{w_2} \|w_2\|_{w_1}.$$

Definition 6.15 *Two weight functions w_1, w_2 are* equivalent, *written $w_1 \sim w_2$, if $d_w(w_1, w_2) < \infty$. An alterative definition is that $\sup_{x \in \mathcal{X}} d_s(w_1(x), w_2(x)) < \infty$.*

The analytical properties of weight functions and their equivalence relations will be of some importance, and so will be given in some detail in the following theorem.

Theorem 6.23 *The following statements hold for the weight functions $w_i \in \mathcal{W}(\mathcal{X})$:*

(*i*) *For any $V \in \mathcal{F}(\mathcal{X})$ we have $\|V\|_{w_1} \leq \|w_2\|_{w_1} \|V\|_{w_2}$. In addition, $\|\cdot\|_{w_1}$ and $\|\cdot\|_{w_2}$ are equivalent norms if and only if $w_1 \sim w_2$; and so $\mathcal{F}(\mathcal{X}, \|\cdot\|_{w_1}) = \mathcal{F}(\mathcal{X}, \|\cdot\|_{w_2})$ if and only if $w_1 \sim w_2$.*

(*ii*) *$d_w(w_1, w_2) \geq 1$, with equality if and only if $w_1 = a w_2$ for scalar $a > 0$ (hence $w_1 \sim a w_1$).*

(*iii*) *$w_1 \sim w_2$ if and only if there exist finite positive scalars a, b such that $a w_1 \leq w_2 \leq b w_1$.*

(*iv*) *Weight equivalence $w_1 \sim w_2$ is an equivalence relation in the sense of Definition 2.1.*

(*v*) *Suppose $w_2 \leq w_1$. Then $\|w_1 - w_2\|_{w_1} = 1 - \|w_1\|_{w_2}^{-1} \leq 1$, hence $w_1 \sim w_2$ if and only if $\|w_1 - w_2\|_{w_1} < 1$.*

(*vi*) *For general weight functions, $w_1 \sim w_2$ if and only if*

$$\min_{i=1,2} \|w_1 - a w_2\|_{w_i} < 1,$$

for some scalar $a > 0$.

(*vii*) *If weight functions w_1, w_2 satisfy $\inf_x w_1 > 0$ and $\inf_x w_2 > 0$ and $|w_1 - w_2| \leq a < \infty$ then $w_1 \sim w_2$.*

Proof

(*i*) We may write

$$\begin{aligned}\|V\|_{w_1} &= \sup_x (|V(x)|/w_1(x))(w_2(x)/w_2(x)) \\ &= \sup_x (|V(x)|/w_2(x))(w_2(x)/w_1(x)) \\ &\leq \sup_x (|V(x)|/w_2(x)) \sup_x (w_2(x)/w_1(x)) \\ &= \|V\|_{w_2} \|w_2\|_{w_1}.\end{aligned}$$

Then $\sup_V d_w(\|V\|_{w_1}, \|V\|_{w_2}) \leq \max(\|w_1\|_{w_2}, \|w_2\|_{w_1})$, which is a finite bound if $w_1 \sim w_2$. Conversely, suppose $\|w_1\|_{w_2} = \infty$. Then $\|\cdot\|_{w_1}$ and $\|\cdot\|_{w_2}$ cannot be equivalent, since $\|w_1\|_{w_1} = 1$.

(*ii*) $1 = \|w_1\|_{w_1} \leq \|w_1\|_{w_2}\|w_2\|_{w_1} = d_w(w_1, w_2)$. Suppose for x, y we have $w_1(x)/w_2(x) = a \neq w_1(y)/w_2(y) = b$. Then

$$\|w_1\|_{w_2}\|w_2\|_{w_1} \geq \max(a, b)\max(1/a, 1/b) > 1.$$

(*iii*) If $w_1 \sim w_2$ then set $a = \|w_1\|_{w_2}^{-1}$ and $b = \|w_2\|_{w_1}$. Conversely, we have $\|w_1\|_{w_2} \leq a^{-1}$ and $\|w_2\|_{w_1} \leq b$.

(*iv*) The equivalence relation is symmetric by construction, and reflexivity follows from the equality $\|w\|_w = 1$ for any weight function w. Transitivity follows by noting

$$\begin{aligned} d_w(w_1, w_3) &= \|w_1\|_{w_3}\|w_3\|_{w_1} \\ &\leq \|w_1\|_{w_2}\|w_2\|_{w_3}\|w_3\|_{w_2}\|w_2\|_{w_1} \\ &= d_w(w_1, w_2)d_w(w_2, w_3). \end{aligned}$$

(*v*) If $w_2 \leq w_1$ then $\|w_1 - w_2\|_{w_1} = 1 - \inf_x w_2(x)/w_1(x) = 1 - \|w_1\|_{w_2}^{-1}$. Then $\|w_1 - w_2\|_{w_1} < 1$ if and only if $\|w_1\|_{w_2} < \infty$ so the argument is completed by noting that $\|w_2\|_{w_1} \leq 1$.

(*vi*) We may write $\|w_1 - w_2\|_{w_1} = \max(1 - \|w_1\|_{w_2}^{-1}, \|w_2\|_{w_1} - 1)$. Then $\|w_1 - w_2\|_{w_1} < 1$ implies both $\|w_1\|_{w_2} < \infty$ and $\|w_2\|_{w_1} < \infty$. By exchanging w_1 and w_2 we conclude that $\|w_1 - w_2\|_{w_1} < 1$ implies $w_1 \sim w_2$. If $\min_{i=1,2}\|w_1 - aw_2\|_{w_i} < 1$ then $w_1 \sim aw_2 \sim w_2$. To prove the converse, note that if $w_1 \sim w_2$ then if $a = \|w_2\|_{w_1}^{-1}$ we have $aw_2 \leq w_1$ so that the remainder follows from (*v*).

(*vii*) Under the hypothesis $\|V\|_{w_i} = a/\inf_x w_i < \infty$ for any constant function $V \equiv a$. Then $w_1 \leq w_2 + a$, so that $\|w_1\|_{w_2} \leq 1 + \|a\|_{w_2} < \infty$, with the proof completed by exchanging w_1 and w_2. ///

The weighted norm can be particularly useful when dealing with positive linear operators. The following lemma will be particularly important.

Theorem 6.24 *If Q is a positive linear operator on $\mathcal{F}(\mathcal{X}, \|\cdot\|_w)$, then its operator norm is given by*

$$|||Q|||_w = \|Qw\|_w$$

Proof First note that since Q is positive $|QV| \leq Q|V|$. Then, if $\|V\|_w \leq 1$ we have $|V| \leq w$, consequently, $|QV| \leq Q|V| \leq Qw$. But $\|w\|_w = 1$, so by the monotonicity of $\|\cdot\|_w$ we conclude that $\|V\|_w$ is maximized for $\|V\|_w \leq 1$ by $\|Qw\|_w$. ///

6.8.4 Span seminorms

Suppose $\mathcal{V} \subset \mathcal{R}(\mathcal{X})$ is a vector space with identity $e \equiv \vec{0}$ and scalar field $\mathbb{R}$. In Section 4.8.3 the span seminorm was introduced:

$$\|f\|_{SP} = \sup_{x \in \mathcal{X}} f(x) - \inf_{x \in \mathcal{X}} f(x).$$

We also introduced the *span midpoint*:

$$mid(f) = 2^{-1}\left(\sup_{x\in\mathcal{X}} f(x) + \inf_{x\in\mathcal{X}} f(x)\right).$$

Scalar homogeneity is easily verified, while the triangle inequality follows from the inequality $\sup_{x\in\mathcal{X}}(f(x)+g(x)) \leq \sup_{x\in\mathcal{X}} f(x) + \sup_{x\in\mathcal{X}} g(x)$ and $\inf_{x\in\mathcal{X}}(f(x)+g(x)) \geq \inf_{x\in\mathcal{X}} f(x) + \inf_{x\in\mathcal{X}} g(x)$. Clearly, $\|\vec{0}\|_{SP} = 0$, but $\|f\|_{SP} = 0$ also for any constant function $f \equiv c$, so that $\|\cdot\|_{SP}$ is a seminorm. The equivalence classes are easily generated from the null subspace $\mathcal{N} = \{f \in \mathcal{V} \mid \|f\|_{SP} = 0\}$, which is the class of all constant functions, and they form the *span quotient space* $\mathcal{V}/\mathcal{N}$.

Suppose we have Banach space $(\mathcal{V}, \|\cdot\|_{sup})$, on which $\|\cdot\|_{SP}$ is defined. There is an interesting relationship between the suprememum norm and the span seminorm which will assume some importance in subsequent analysis, and should be noted. Assume $\mathcal{V}$ contains $\vec{1}$.

Theorem 6.25 *Suppose we have Banach space $(\mathcal{V}, \|\cdot\|_{sup})$ with real scalars, which contains $\vec{1}$ and identity $\vec{0}$. Then for all $f \in \mathcal{V}$ with $\|f\|_{sup} < \infty$,*

$$\|f\|_{SP} \leq 2\|f\|_{sup}, \text{ and} \tag{6.14}$$

$$\inf_{a\in\mathbb{R}} \|f - a\vec{1}\|_{sup} = \|f - mid(f)\vec{1}\|_{sup} = \frac{1}{2}\|f\|_{SP}. \tag{6.15}$$

In addition, the following inequalities hold:

$$\begin{aligned} &\|f\|_{SP} \leq \|f\|_{sup} \text{ if } \inf f \geq 0 \text{ or } \sup f \leq 0, \\ &\|f\|_{SP} \geq \|f\|_{sup} \text{ otherwise.} \end{aligned} \tag{6.16}$$

Proof Inequality (6.14) follows by noting $\|f\|_{SP} = \sup_{x\in\mathcal{X}} f(x) - \inf_{x\in\mathcal{X}} \leq |\sup_{x\in\mathcal{X}} f(x)| + |\inf_{x\in\mathcal{X}} f(x)|$. Then by the definition of $\|\cdot\|_{SP}$ and (6.14) we have

$$\frac{1}{2}\|f\|_{SP} = \frac{1}{2}\|f - a\vec{1}\|_{SP} \leq \|f - a\vec{1}\|_{sup},$$

with equality achieved by setting $a = mid(f)$, giving (6.15). Then (6.16) follows from the definition of $\|\cdot\|_{SP}$. ///

The definition of the weighted span seminorm follows from the weighted supremum norm, and Theorem 6.25 extends naturally to the weighted case:

$$\|f\|_{SP(\nu)} = \|\nu^{-1}f\|_{SP} = \sup_{x\in\mathcal{X}} \nu(x)^{-1}f(x) - \inf_{x\in\mathcal{X}} \nu(x)^{-1}f(x).$$

However, it is important to note that the equivalence classes will depend on the weight function, in particular we have,

$$\|f - g\|_{SP(\nu)} = 0 \quad \text{if and only if } f - g = a\nu \text{ and}$$
$$\|h\|_{SP(\nu)} = 0 \quad \text{if and only if } h = a\nu$$

for some constant a, and

$$\|f\|_{SP(\nu)} = \|f + a\nu\|_{SP(\nu)}$$

for any constant a. The equivalence classes may therefore be identified with the quotient space $\mathcal{V}/\mathcal{N}_\nu$, where we write the null space $\mathcal{N}_\nu = \{a\nu \mid a \in \mathbb{R}\}$.

As a matter of notation the symbol ν will be used to represent weight functions associated with weighted span seminorms, while w remains the weight function associated with weighted supremum norms. The two are the same type of object, but will often serve different purposes, and may have very different origins. It will sometimes be the case that $\nu = w$, but using distinct notation will give more clarity. Following this remark we have the following theorem:

Theorem 6.26 *If we have Banach space $(\mathcal{V}, \|\cdot\|_w)$ with real scalars, which contains w and identity $\vec{0}$, and $\nu \sim w$, then $(\mathcal{V}/\mathcal{N}_\nu, \|\cdot\|_{SP(\nu)})$ is also a Banach space.*

Proof From Theorem 6.25 we can see that if $\nu \sim w$ we have

$$\|V\|_w < \infty \iff \|V\|_\nu < \infty \iff \|V\|_{SP(\nu)} < \infty,$$

so that all elements of $\mathcal{V}/\mathcal{N}_\nu$ are of finite norm. Clearly, the null space $\mathcal{N}_\nu$ is closed *wrt* $\|\cdot\|_\nu \sim \|\cdot\|_w$, so by Theorem 6.18 $\mathcal{V}/\mathcal{N}_\nu$ is a Banach space *wrt* a norm calculable by (6.12). By Theorem 6.25 this norm is equivalent to $\|\cdot\|_{SP(\nu)}$. ///

6.8.5 Operators on span quotient spaces

We have seen that given Banach space $(\mathcal{V}, \|\cdot\|_w)$ the span seminorm $\|\cdot\|_{SP(\nu)}$ induces a Banach space $(\mathcal{V}/\mathcal{N}_\nu, \|\cdot\|_{SP(\nu)})$, provided $\nu \sim w$.

Next, consider a linear operator Q on $\mathcal{V}$. We may define the continuity of Q with respect to $\|\cdot\|_{SP(\nu)}$, with some caution, as the Lipschitz condition $\|QV\|_{SP(\nu)} \leq L\|V\|_{SP(\nu)}$ for all $V \in \mathcal{V}$ for some $L < \infty$. However, in doing so, we find that the essential condition is that Q be interpretable as an operator on the quotient space $\mathcal{V}/\mathcal{N}_\nu$.

Theorem 6.27 *Consider a linear operator Q on a vector space of functions $\mathcal{V}$, with weight function $\nu \in \mathcal{V}$, and define the following condition:*

$$Q\nu = \alpha\nu \quad \textit{for some scalar } \alpha. \tag{6.17}$$

(*i*) *If Q is Lipschitz continuous in $\|\cdot\|_{SP(\nu)}$ then (6.17) holds.*

(*ii*) *If Q is Lipschitz continuous in the weighted supremum norm $\|\cdot\|_\nu$ and (6.17) holds then Q is Lipschitz continuous in $\|\cdot\|_{SP(\nu)}$, with*

$$\|QV\|_{SP(\nu)} \leq \|\!|Q|\!\|_\nu \|V\|_{SP(\nu)}. \tag{6.18}$$

(*iii*) *If (6.17) holds, then Q is an operator on the quotient space $\mathcal{V}/\mathcal{N}_\nu$, where Q is understood to map subsets $E \subset \mathcal{V}$ to images $QE \subset \mathcal{V}$.*

Proof

(*i*) If Q is Lipschitz continuous in $\|\cdot\|_{SP(\nu)}$ then for some finite L we must have $\|Q\nu\|_{SP(\nu)} \leq L\|\nu\|_{SP(\nu)} = \|\vec{1}\|_{SP} = 0$, which implies (6.17).

(*ii*) Suppose $\|V\|_{sup} < \infty$. Then if (6.17) holds, by Theorem 6.25 we have

$$\begin{aligned}\|QV\|_{SP(\nu)} &= \|Q(V - \nu\, mid(V/\nu))\|_{SP(\nu)} \\ &\leq 2\|Q(V - \nu\, mid(V/\nu))\|_\nu \\ &\leq 2\|\!|Q|\!\|_\nu \|V - \nu\, mid(V/\nu)\|_\nu \\ &= \|\!|Q|\!\|_\nu \|V\|_{SP(\nu)},\end{aligned}$$

which completes part (*ii*).

(*iii*) Suppose $[U]$ is the coset of U modulo $\mathcal{N}_\nu$. If $V \in [U]$, then $V = U + a\nu$ for some scalar a. We then have $QV = Q(U + a\nu) = QU + \alpha a\nu$. Clearly, Q is a bijective mapping from $[U]$ to $[QU]$, and so is generally a bijective mapping between elements of $\mathcal{V}/\mathcal{N}_\nu$. If we then regard Q as a mapping from a subset $E \subset \mathcal{V}$ to its image $QE \subset \mathcal{V}$, then Q is well defined as an operator on $\mathcal{V}/\mathcal{N}_\nu$ since $E \in \mathcal{V}/\mathcal{N}_\nu$ implies $QE \in \mathcal{V}/\mathcal{N}_\nu$. ///

It is important to know that (6.17) is the essential condition for span continuity, and if it holds, that continuity in the supremum norm implies span continuity. The interesting consequence of (6.17) is that it permits the interpretation of Q as an operator on the Banach space $(\mathcal{V}/\mathcal{N}_\nu, \|\cdot\|_{SP(\nu)})$. We will see that this alternative Banach space construction can have considerable advantages.

Of course, the pair (α, ν) in (6.17) can be recognized as an eigenvalue and an associated eigenvector (see Section 2.3.2). It is well known that the existence of a real nonzero eigenvalue and an associated real eigenvector is an important property for operators. This is the subject of the Perron-Frobenius theorem (Theorem 2.13) when Q is a nonnegative matrix, and a central concern of *operator theory*, which considers the extension of the spectral properties of matrices to more general linear operators. We have discussed the implications of this for Markov chains, for which (6.17) will generally hold. For more general operators, it will suffice for our purposes to verify (6.17) by construction.

We will see that the Lipschitz constant given in (6.18) may not be sharp, so we next consider a more refined evaluation of $\|\!|Q|\!\|_{SP(\nu)}$.

6.9 MEASURE KERNELS AS LINEAR OPERATORS

We have introduced measure kernels in Section 4.9 as mappings of the form $Q:\mathcal{Y}\rightarrow \mathcal{M}_{\mathcal{X}}$, given Borel spaces $\mathcal{X},\mathcal{Y}$, where $\mathcal{M}_{\mathcal{X}}$ is the set of finite signed measures on $\mathcal{X}$. The range of Q may be restricted to $\mathcal{M}^1_{\mathcal{X}}$, $\mathcal{M}^{1^-}_{\mathcal{X}}$ or $\mathcal{M}^+_{\mathcal{X}}$, yielding stochastic, substochastic or positive kernels, respectively.

Under general assumptions, Q may be equivalently taken to be a linear operator between vector spaces $\mathcal{F}(\mathcal{X})$ and $\mathcal{F}(\mathcal{Y})$ using the evaluation

$$W(y)=Q(\cdot\mid y)V, \quad y\in\mathcal{Y}.$$

This assumes that the mapping preserves measurability, which will hold under Definition 4.7. We will assume that any measure kernel Q is measurable in this sense.

The purpose of this section is to clarify the role of measure kernels in the Banach space $\mathcal{F}(\mathcal{X},\|\cdot\|_w)$. If $\mathcal{Y}=\mathcal{X}$ then Q is an operator on $\mathcal{F}(\mathcal{X})$. If Q is a signed kernel the operator norm *wrt* $\|\cdot\|_w$ becomes,

$$\begin{aligned}
|\!|\!|Q|\!|\!|_w &= \sup_{\|V\|_w\leq 1}\|QV\|_w \\
&= \sup_{\|V\|_w\leq 1}\sup_{x\in\mathcal{X}} w(x)^{-1}\left|Q(\cdot\mid x)V\right|.
\end{aligned}$$

By Theorem 4.21 we may conclude the operator norm of Q is given by

$$|\!|\!|Q|\!|\!|_w=\|Qw\|_w=\sup_{x\in\mathcal{X}} w(x)^{-1}\|Q(\cdot\mid x)\|_{TV(w)}.$$

6.9.1 The contraction property of stochastic kernels

The operator norm $|\!|\!|Q|\!|\!|_w$ is a Lipschitz constant for Q on the Banach space $\mathcal{F}(\mathcal{X},\|\cdot\|_w)$, so there is considerable interest in determining whether or not $|\!|\!|Q|\!|\!|_w<1$, since this value will also determine the Lipschitz constant of various operators constructed from Q.

The contraction property of a positive operator Q in the supremum norm usually follows from the substochastic property. Possibly, $Q=\beta Q_0$, where Q_0 is a proper stochastic kernel and $\beta\in(0,1)$. Alternatively, Q may define a Markov process with an absorbing class Δ. Suppose we define the vector space $\mathcal{F}_\Delta(\mathcal{X})$ of measurable function for which $V(x)=0$ when $x\in\Delta$. Suppose we have $Q^J(\Delta\mid x)\geq\delta$ for all $x\in\mathcal{X}$ for some finite J (in particular, $Q^J(\Delta\mid x)=1$ when $x\in\Delta$). In this case, for any $V\in\mathcal{F}_\Delta(\mathcal{X})$ for which $\|V\|_{sup}\leq 1$ we must have $\|Q^J V\|_{TV}\leq 1-\delta$, from which the contraction property follows.

6.9.2 Stochastic kernels and the span seminorm

There will be considerable interest in examining the span seminorm properties of Q, first, in determining when it is span continuous, and then, in evaluating its span operator norm.

If Q is assumed continuous in $\|\cdot\|_\nu$, from Theorem 6.27, Q will be span continuous if and only if $Q\nu = \alpha\nu$ for some scalar α. This condition will be frequently encountered, but will also be violated for many interesting models.

First, we will assume that Q is span continuous, in which case we may calculate the span operator norm.

Theorem 6.28 *Suppose for Borel space $\mathcal{X}$, Q is a span continuous measure kernel with respect to weight function ν. Then the span operator norm is given by*

$$\|\|Q\|\|_{SP(\nu)} = \frac{1}{2} \sup_{x,y \in \mathcal{X}} \|(I_\nu^{-1}Q)(\cdot \mid x) - (I_\nu^{-1}Q)(\cdot \mid y)\|_{TV(\nu)} \leq \|\|Q\|\|_\nu. \tag{6.19}$$

Proof Suppose $V \in \mathcal{F}(\mathcal{X})$. Then

$$\|QV\|_{SP(\nu)} = \sup_{x,y \in \mathcal{X}} \nu(x)^{-1}Q(\cdot \mid x)V - \nu(y)^{-1}Q(\cdot \mid y)V.$$

Using the notation of Theorem 4.22 set

$$\mu = \nu(x)^{-1}Q(\cdot \mid x) - \nu(y)^{-1}Q(\cdot \mid y).$$

By hypothesis $\mu\nu = 0$, since $Q(\cdot \mid x)\nu = \alpha\nu(x)$, which gives

$$\|QV\|_{SP(\nu)} \leq \frac{1}{2} \sup_{x,y \in \mathcal{X}} \|(I_\nu^{-1}Q)(\cdot \mid x) - (I_\nu^{-1}Q)(\cdot \mid y)\|_{TV(\nu)} \|V\|_{SP(\nu)}. \tag{6.20}$$

Following the arguments of Section 4.8.3 we can see that equality in (6.20) can be attained within any $\epsilon > 0$ by selecting a suitable V. Then (6.19) follows directly, after noting that

$$\frac{1}{2}\|(I_\nu^{-1}Q)(\cdot \mid x) - (I_\nu^{-1}Q)(\cdot \mid y)\|_{TV(\nu)} \leq \sup_x \nu(x)^{-1}\|Q(\cdot \mid x)\|_{TV(\nu)} = \|\|Q\|\|_\nu.$$

///

If Q is span continuous, then the alternative Banach space $(\mathcal{V}/\mathcal{N}_\nu, \|\cdot\|_{SP(\nu)})$ may have quite different contraction properties than the underlying Banach space $(\mathcal{V}, \|\cdot\|_w)$. The following example will be quite relevant. Suppose Q_1 is a stochastic kernel with eigenpair $(1, 1)$. For $\beta < 1$ we have the uniformly substochastic model $Q = \beta Q_1$. Then the span operator norm is

$$\|\|Q\|\|_{SP} = \beta\|\|Q_1\|\|_{SP}. \tag{6.21}$$

Since Q_1 is strictly stochastic, we have

$$\|\|Q_1\|\|_{SP} = \frac{1}{2} \sup_{x,y \in \mathcal{X}} \|Q_1(\cdot \mid x) - Q_1(\cdot \mid y)\|_{TV} \leq 1.$$

so that the contraction constant for Q *wrt* the span seminorm is never larger than that *wrt* to the supremum norm. The next question is whether or not it is smaller. If there exists a pair $x, y \in \mathcal{X}$ for which

$$Q_1(\cdot \mid x) \perp Q_1(\cdot \mid y),$$

then the answer is no, since in this case

$$|||Q_1|||_{SP} \geq (1/2)\|Q_1(\cdot \mid x) - Q_1(\cdot \mid y)\|_{TV} = 1.$$

However, if the variation norm of all signed measures $Q_1(\cdot \mid x) - Q_1(\cdot \mid y)$ is bounded away from 2, then $|||Q_1|||_{SP} < 1$, and the contraction constant *wrt* the span seminorm is strictly smaller.

Chapter 7

Fixed point equations

Suppose we are given a metric space $(\mathcal{V}, d)$, on which is defined an operator T. The *fixed point equation* is simply $V = TV$, and is the common form of many important mathematical problems. The Banach fixed point theorem (Theorem 6.4) applies to complete metric spaces, and defines sufficient conditions under which is can be stated that

1. A solution to $V = TV$ exists,
2. The solution to $V = TV$ is unique,
3. The solution to V is the limit in the given metric of the sequence $T^k v_0$ for any $v_0 \in \mathcal{V}$.

When this holds we obtain the *fixed point algorithm* $V_k = T^k v_0$ for any $v_0 \in \mathcal{V}$, and we define a *fixed point iteration* as the step $V_{k+1} = TV_k$. The most obvious condition is that T is contractive in a metric space $(\mathcal{V}, d)$, but the theorem also implicitly assumes that T is closed on $\mathcal{V}$ in the sense that for any element $V \in \mathcal{V}$ we have $d(TV, V) < \infty$. If we are given a Banach space, the metric becomes $d(V_1, V_2) = \| V_1 - V_2 \|$, so that closure of T is stated as the condition that $\|TV - V\| < \infty$, so that by the triangle inequality $\|V\| < \infty$ implies $\|TV\| < \infty$. Under this condition, for each iterate $\|T^k v_0\| < \infty$ holds if $\|v_0\| < \infty$ holds. In this case the Banach fixed point theorem applies following the contraction property $\|TV_1 - TV_2\| = d(TV_1, TV_2) \leq \rho d(V_1, V_2) = \rho \|V_1 - V_2\|$ for some $\rho < 1$.

We summarize the discussion in the following theorem:

Theorem 7.1 *Suppose we are given operator T on Banach space $(\mathcal{V}, \|\cdot\|)$. Suppose any of the conditions hold:*

(*i*) *T is Lipshitz continuous, and for all $V, V' \in \mathcal{V}$ we have $\|T^J V - T^J V'\| \leq \rho \|V - V'\|$ for some $\rho < 1$ and finite J;*
(*ii*) *There exists $V' \in \mathcal{V}$ such that for all $V \in \mathcal{V}$ we have $\|T^J V - V'\| \leq \rho \|V - V'\|$ for some $\rho < 1$ and finite J.*

Then there exists a unique fixed point $V^* = TV^*$, which is the limit of the iterations $T^n V_0 \to V^*$ for any initial point $V_0 \in \mathcal{V}$. In the case of (ii) $V^* = V'$.

Proof The theorem is simply a restatament of Theorems 6.4 and 6.5 in the context of Banach spaces. ///

7.1 CONTRACTION AS A NORM EQUIVALENCE PROPERTY

When considering fixed point operators on a metric space, have seen that there are several versions of the contraction property, in particular pseudocontractivity or multistage contractivity. However by Theorem 6.3 we may associate a single asymptotic contraction rate ρ to an operator which gives a convergence rate of $O(\rho^k)$ for a fixed point algorithm. For this reason it is worth complementing Theorem 7.1 by restating Theorem 6.3 in terms of operators on Banach spaces (the proof would be identical):

Theorem 7.2 *Suppose we are given a Lipschitz continuous operator T on Banach space $(\mathcal{V}, \|\cdot\|)$. Then there exists a finite constant ρ such that*

$$\lim_{n\to\infty} |||T^n|||^{1/n} = \rho. \tag{7.1}$$

Furthermore, ρ is the best possible contraction rate in the sense that

$$|||T^n|||^{1/n} \geq \rho, \quad \textit{for all } n \geq 1, \tag{7.2}$$

and for any $\epsilon > 0$, there exists n_ϵ for which

$$|||T^n|||^{1/n} \leq (\rho + \epsilon), \quad \textit{for all } n \geq n_\epsilon. \tag{7.3}$$

The asymptotic contraction rate is determined by both the operator and the norm, so we define

$$\rho(T, \|\cdot\|) = \lim_{n\to\infty} |||T^n|||^{1/n}.$$

Since Lipschitz continuity guarantees the exists and finiteness of the limit $\rho(T, \|\cdot\|)$ we can adopt the convention of setting $\rho(T, \|\cdot\|) = \infty$ if and only if T is not Lipschitz continuous.

If T possesses a single stage contraction constant ρ_1 and asymptotic contraction rate ρ, then we have $\rho \leq |||T^n|||^{1/n} \leq \rho_1$. On the other hand, if we are only given $\rho(T, \|\cdot\|_\alpha) = \rho$, the most we can say about other contraction constants is that

$$|||T^n||| = b_n \rho^n, \quad \text{where } \lim_{n\to\infty} b_n^{1/n} = 1. \tag{7.4}$$

The sequence b_n may be unbounded, as long as its increase is subgeometric. Sometimes it will be useful to impose a contraction property more stringent than (7.4), but less stringent that the optimal single stage case. First, recall the notation of Section 6.1.1, which assigns to ρ_J the best Lipchitz constant of T^J, that is $\rho_J = |||T^J|||$. This is then standardized to a contraction rate $\bar{\rho}_J = \rho_J^{1/J}$. We have the following definition.

Definition 7.1 *Suppose we are given a Lipschitz continuous operator T on a normed vector space $(\mathcal{V}, \|\cdot\|)$. We say T is J-stage contractive with rate $\bar{\rho}_J$ if $|||T^J||| = \bar{\rho}_J^J < 1$. If there exists finite $K > 0$ and $\rho > 0$ for which $|||T^n||| \leq K\rho^n, n \geq 1$, then T is* uniformly contractive *with rate ρ.*

The relationship between asymptotic, J-stage and uniform contractivity is summarized in the following theorem.

Theorem 7.3 *Suppose we are given a normed vector space $(\mathcal{V}, \|\cdot\|)$, and a Lipschitz continuous operator T on $\mathcal{V}$. Then J-stage contractivity with rate ρ implies uniform contraction with rate ρ, which implies $\rho(T, \|\cdot\|) \leq \rho$.*

Conversely, for any $\epsilon > 0$ asymptotic (and therefore uniform) contraction with rate ρ implies J-stage contractivity with rate $(\rho + \epsilon)$ for some J.

Proof Suppose T is J-stage contractive with rate ρ, and that T possesses Lipschitz constant $L < \infty$. Write any integer $k = Jk_1 + k_2$ where $k_1 = \lfloor k/J \rfloor$ and $0 \leq k_2 \leq J - 1$. By the submultiplicative property of Lipschitz continuity we have

$$\begin{aligned}
\|T^k V - T^k V'\| &\leq \|T^{Jk_1 + k_2} V - T^{Jk_1 + k_2} V'\| \\
&\leq \|T^{Jk_1}(T^{k_2} V) - T^{Jk_1}(T^{k_2} V')\| \\
&\leq \|(T^J)^{k_1}(T^{k_2} V) - (T^J)^{k_1}(T^{k_2} V')\| \\
&\leq (\rho^J)^{k_1} \|T^{k_2} V - T^{k_2} V'\| \\
&\leq \rho^{Jk_1} L^{k_2} \|V - V'\| \\
&= \rho^{Jk_1 + k_2} (L/\rho)^{k_2} \|V - V'\| \\
&= K\rho^k
\end{aligned}$$

for all $k \geq 1$ for some finite constant K. Then, if uniform contraction holds for K, ρ we have

$$\limsup_{k \to \infty} \left[\|T^k V - T^k V'\| \right]^{1/k} \leq \lim_{k \to \infty} K^{1/k} \rho = \rho.$$

To show the converse, suppose $\rho(T, \|\cdot\|_\alpha) = \rho$. Then for any $\epsilon \in (0, 1 - \rho)$ there exists large enough J for which $\|T^J V - T^J V'\| \leq (\rho + \epsilon)^J < 1$. ///

We next show that uniform and asymptotic contraction rates are norm equivalence properties.

Theorem 7.4 *Suppose we are given a Lipschitz continuous operator T on Banach space $(\mathcal{V}, \|\cdot\|_\alpha)$. If $\|\cdot\|_\beta \leq \|\cdot\|_\alpha$, then $\rho(T, \|\cdot\|_\beta) \leq \rho(T, \|\cdot\|_\alpha)$. Therefore $\|\cdot\|_\beta \sim \|\cdot\|_\alpha$ implies $\rho(T, \|\cdot\|_\beta) = \rho(T, \|\cdot\|_\alpha)$.*

Similarly, if $\|\cdot\|_\beta \leq \|\cdot\|_\alpha$, and T possesses uniform contraction rate ρ wrt $\|\cdot\|_\alpha$ the same holds wrt $\|\cdot\|_\beta$. Therefore a uniform contraction rate is a norm equivalence property.

Proof If $\|\cdot\|_\beta \leq \|\cdot\|_\alpha$ then T is Lipschitz continuous *wrt* $\|\cdot\|_\beta$, and so we may write, for some finite constant $b > 0$

$$\rho(T, \|\cdot\|_\beta) = \lim_{n \to \infty} \|T^n\|_\beta^{1/n} \leq \liminf_{n \to \infty} b^{1/n} \|T^n\|_\alpha^{1/n} = \rho(T, \|\cdot\|_\alpha).$$

Norm equivalence follows by exchanging $\|\cdot\|_\alpha$ and $\|\cdot\|_\beta$ in the argument.

The remainder of the proof follows immediately from the definition of norm dominance. ///

As for operators on metric spaces, the theory of this section holds for pseudo-Lipschitz operators. We may define equivalent notions of J-stage, uniform and asymptotic contraction rates for pseudo-Lipschitz constants L satisfying $\|TV - V^*\| \leq L\|V - V^*\|$, since the submultiplicative property underlying Theorem 6.3 holds also for this definition.

To summarize, while a specific contraction property of an operator T may depend on a specific norm (for example, single stage contractivity) contraction in general, along with a single asymptotic contraction rate, is a norm equivalence property.

7.2 LINEAR FIXED POINT EQUATIONS

There will be special interest in fixed point equations based on bounded linear operators. Suppose on a Banach space $(\mathcal{V}, \|\cdot\|)$ we are given a fixed element $R \in \mathcal{V}$ and a linear operator $Q : \mathcal{V} \to \mathcal{V}$, and we are given the problem of finding a solution V to equation $V = R + QV$. The pair $\pi = (R, Q)$ will be refered to as a *model*, and the *fixed point operator* will be denoted $T_\pi V = R + QV$. The Jth iteration of T_π is given by

$$T_\pi^J V = T_\pi(T_\pi^{J-1} V) = R + QT_\pi^{J-1} V = \cdots = R_J + Q^J V, \tag{7.5}$$

where

$$R_J = \sum_{i=0}^{J-1} Q^i R.$$

Thus, T_π^J may be interpreted as the operator $T_{\pi[J]}$ for model $\pi[J] = (R_J, Q^J)$. Clearly, a fixed point of T_π is also a fixed point of T_π^J, and the converse holds under the conditions of Theorem 6.4.

Theorem 7.5 *Suppose we are given model $\pi = (R, Q)$ on a vector space $\mathcal{V}$. If $\|\cdot\|_\alpha$ completes $\mathcal{V}$, and we have $\|R\|_\alpha < \infty$, $\|\!|Q|\!\|_\alpha < \infty$ then T_π is an operator on Banach space $(\mathcal{V}, \|\cdot\|_\alpha)$.*

The iteration T_π^J possesses Lipschitz constant $\|\!|Q^J|\!\|_\alpha$. If $\|\!|Q^J|\!\|_\alpha < 1$ for some $J \geq 1$, then the conditions of the Banach fixed point theorem hold.

Proof Since Q is bounded, $\|V\|_\alpha < \infty$ implies $\|T_\pi V\|_\alpha \leq \|R\|_\alpha + \|QV\|_\alpha < \infty$. Thus, T_π is an operator on Banach space $(\mathcal{V}, \|\cdot\|_\alpha)$. The Lipschitz constant is given by $\|\!|Q|\!\|_\alpha$, since

$$\|TV_1 - TV_2\|_\alpha = \|Q(T_1 - T_2)\|_\alpha \leq \|\!|Q|\!\|_\alpha \|V_1 - V_2\|_\alpha.$$

Furthermore, refering to (7.5), we may interpret $T_\pi^J = T_{\pi[J]}$ as a linear operator on $(\mathcal{V}, \|\cdot\|_\alpha)$ with Lipschitz constant $\|\!|Q^J|\!\|_\alpha$, noting that $\|R_J\|_\alpha < \infty$. ///

7.3 THE GEOMETRIC SERIES THEOREM

The *geometric series theorem* gives an alternative approach to the Banach fixed point theorem which provides a more precise relationship between a fixed point solution $V = T_\pi V$ and the model π. The fixed point equation may be rewritten $(I - Q)V = R$, so that for any V in the domain of Q we may find an element R which generates a fixed point equation for which V is the solution. Of course, the converse problem is usually of more interest, that is, we would like to know that $(I - Q)$ possesses an inverse $(I - Q)^{-1}$, in which case fixed points could be obtained directly by $V = (I - Q)^{-1} R$. This leads to a fundamental characterization of the fixed point equation:

Theorem 7.6 (Geometric Series Theorem) *If Q is a bounded linear operator with $\|\|Q\|\| < 1$ then $I - Q$ is a bijection on $\mathcal{V}$, and its inverse is a bounded linear operator for which $\|\|(I - Q)^{-1}\|\| \leq (1 - \|\|Q\|\|)^{-1}$.*

In addition, if Q is a bounded linear operator with $\|\|Q^J\|\| < 1$ for some $J \geq 1$, then $I - Q$ is a bijection on $\mathcal{V}$, and its inverse is a bounded linear operator for which $\|\|(I - Q)^{-1}\|\| \leq (1 - \|\|Q^J\|\|)^{-1} \sum_{i=0}^{J-1} \|\|Q^i\|\|$.

Proof See, for example, Theorem 2.3.1 and Corollary 2.3.3 in Atkinson and Han (2001). ///

The geometric series theorem guarantees the existence of a solution to the fixed point equation for multistage contractive Q, but also relates this solution to R in a number of ways. First, V may be thought of as a continuous mapping of R, hence fixed point equations which are close in this sense will have close solutions. In addition, it provides the bound

$$\|V\| \leq \left[(1 - \|\|Q^J\|\|)^{-1} \sum_{i=0}^{J-1} \|\|Q^i\|\| \right] \|R\|.$$

Once its existence in given, the inverse map may be identified as

$$(I - Q)^{-1} = \sum_{i=0}^{\infty} Q^i$$

by substitution into $V = R + QV$, leading to fixed point

$$V = \sum_{i=0}^{\infty} Q^i R.$$

This also characterizes the convergence of the fixed point algorithm, since we have

$$T^J V_0 = R_J + Q^J V_0,$$

with $R_J \to V$ and $Q^J V_0 \to \vec{0}$, the vector of norm zero.

Something like a converse result can be obtained for positive linear operators on $\mathcal{F}(\mathcal{X}, \|\cdot\|_w)$.

Theorem 7.7 *Suppose for model $\pi = (R, Q)$ on vector space $\mathcal{V} \subset \mathcal{R}(\mathcal{X})$ we have $R \geq 0$ and Q is a positive bounded linear operator.*

(*i*) *If there exists a solution $V^* > 0$ to the fixed point equation $V = T_\pi V$ the following hold:*
  (*ia*) *T_π is nonexpansive.*
  (*ib*) *$R_J \sim V^* \iff \|\!|Q^J|\!\|_{V^*} = \| V^* - R_J\|_{V^*} = 1 - \| V^*\|_{R_J}^{-1} < 1$.*
  (*ic*) *$\|T_\pi^k V_0 - V^*\|_{V^*} \to_k 0$ for all $V_0 \in \mathcal{V}$ for which $\| V_0\|_{V^*} < \infty \iff R_J \sim V^*$ for some $J \geq 1$.*
  (*id*) *$R_J \sim V^* \Rightarrow R_{J+1} \sim V^*$.*

(*ii*) *If for any $J \geq 1$ we have $\|\!|Q|\!\|_{R_J} < \infty$ and $\|\!|Q^m|\!\|_{R_J} < 1$ for some $m \geq 1$ then there exists a fixed point solution V^* to $V = T_\pi V$ for which $V^* \sim R_J$.*

(*iii*) *Therefore, (i) and (ii) together imply that T_π possesses a unique fixed point V^* and is J-stage contractive wrt $\|\cdot\|_{V^*}$ if and only if condition (ii) holds.*

Proof First note that by hypothesis $Q^J V^* \geq 0$ and $R \geq 0$, therefore $V^* \geq R_J \geq 0$. To show (*ia*) note that

$$V^* - R = QV^*,$$

so that

$$\|\!|Q|\!\|_{V^*} = \| V^* - R\|_{V^*} \leq 1$$

since $V^* \geq R$ (apply Theorem 6.23 (*v*)).

To show (*ib*), note that by hypothesis, $V^* = R_J + Q^J V^* \geq 0$, so we have

$$V^* - R_J = Q^J V^*,$$

and therefore

$$\| V^* - R_J\|_{V^*} = \|\!|Q^J|\!\|_{V^*}. \tag{7.6}$$

Since $V^* \geq R_J$, by Theorem 6.23 (*v*) $V^* \sim R_J$ if and only if $\| V^* - R_J\|_{V^*} < 1$. Therefore, by (7.6) (*ib*) follows.

To show (*ic*) suppose $\|T_\pi^k V_0 - V^*\|_{V^*} \to_k 0$ for $V_0 = \vec{0}$. Since this gives $R_k = T_\pi^k V_0$, we must have $\|R_J - V^*\|_{V^*} < 1$ for some J, which we have shown implies $R^J \sim V^*$. Conversely, if $R^J \sim V^*$, by (*ib*) T_π is contractive, and the Banach fixed point theorem applies.

We obtain (*id*) by noting $R_J \leq R_{J+1} \leq V^*$.

Under the conditions of (*ii*) T_π is a contractive operator on Banach space $\mathcal{R}(\mathcal{X}, \|\cdot\|_{R^J})$ and therefore possess a fixed point V^* for which $\| V^*\|_{R_J} < \infty$. Since $V \geq R_J$ we must also have $V^* \sim R^J$.

Then (*iii*) follows directly. ///

We may refer to $\pi = (R, Q)$ as a *positive model* if $R \geq 0$ and Q is a positive operator. In this case R_k is an increasing sequence which, if Q is continuous, will converge to the fixed point. We would have pointwise convergence if $R_k(x) \rightarrow_k V^*(x)$ for each $x \in \mathcal{X}$. Theorem 7.7 describes uniform convergence weighted by V^*, that is,

$$\lim_{k\to\infty} \sup_{x\in\mathcal{X}} \left| \frac{V^*(x) - R_k(x)}{V^*(x)} \right| = 0.$$

It can be seen that while contractivity is not equivalent to the existence of a fixed point, it is equivalent in the positive model to uniform convergence of the fixed point algorithm. The importance of Theorem 7.7 is that *(ii)* defines a necessary and sufficient condition for this form of convergence which can be tested directly from the model elements defining $\pi = (R, Q)$, that is, there exists R_J for which T_π is multistage contractive *wrt* $\|\cdot\|_{R_J}$. Then, by Theorem 7.3, it suffices to verify contractivity of any form *wrt* any equivalent norm.

That a fixed point can exist without the contraction property is shown in the following counterexample.

Example 7.1 *Set $\mathcal{X} = (1, 2, \ldots)$ and define model $\pi = (R, Q)$ with stochastic kernel $Q(\{i+1\} \mid i) = 1$ and $R(i) = \alpha^{-i} i^{-2}$, with $\alpha < 1$. Then T_π possesses a fixed point V^* to which R_J converges pointwise, but not uniformly. It may be verfied that V^* is not weight equivalent to any R_J.*

7.4 INVARIANT TRANSFORMATIONS OF FIXED POINT EQUATIONS

We consider two forms of transformations of the model (R, Q) which yield corresponding transformations of the fixed point equations which are invariant in the sense that they retain exactly the same Banach space properties. In this case, we would have two iterative algorithms $V_k = TV_{k-1}$, $V'_k = T'V'_{k-1}$ which are connected by a well defined transformation $V'_k = H(V_k)$, which extends also to the respective fixed points. Furthermore, T and T' have the same Lipschitz properties, so the two algorithms are identical.

Theorem 7.8 *Suppose we are given Banach space $(\mathcal{V}, \|\cdot\|_w)$, $\mathcal{V} \subset \mathcal{R}(\mathcal{X})$, and a model $\pi = (R, Q)$ defining operator $T_\pi V = R + QV$ for $V \in \mathcal{V}$.*

(i) *Suppose V_π is a fixed point of T_π, and that for positive weight function w we have $Qw = \alpha w$ for some $\alpha \in (0, 1)$. Then for any $c \in \mathbb{R}$, if $\pi_c = (R + cw, Q)$ then T_{π_c} is an operator on $(\mathcal{V}, \|\cdot\|_w)$ with fixed point $V_{\pi_c} = V_\pi + (1-\alpha)^{-1} cw$. In addition, if*

$$H(V) = V + (1-\alpha)^{-1} cw,$$

then $V' = T_\pi V$ implies $H(V') = T_{\pi_c} H(V)$. Finally, T_π and T_{π_c} have identical Lipschitz constants.

(ii) *Suppose V_π is a fixed point of T_π. If we define transformed model $\pi_w = (I_w^{-1} R, I_w^{-1} Q I_w)$ then T_{π_w} is an operator on Banach space $(\mathcal{V}, \|\cdot\|_{sup})$ with fixed point $V_{\pi_w} = I_w^{-1} V_\pi$. In addition, if*

$$H(V) = I_w^{-1} V,$$

then $V' = T_\pi V$ implies $H(V') = T_{\pi_w} H(V)$. Finally, T_π and T_{π_w} have identical Lipschitz constants.

(*iii*) *Suppose T_π is an operator on Banach space $(\mathcal{V}/\mathcal{N}_\nu, \|\cdot\|_{SP(\nu)})$ with $Q\nu = \nu$, and that E is an equivalence class of solutions to the equation $V + h\nu = T_\pi V$ for some scalar h. Then for any $c \in \mathbb{R}$, if $\pi_c = (R + c\nu, Q)$ then E is also an equivalence class of solutions to $V + (h + c)\nu = T_{\pi_c} V_{\pi_c}$.*

Proof The statements are easily verified by direct substitution. ///

Statement (*i*) of Theorem 7.8 essentially states that a scalar multiple of w may be added to R without changing the fixed point equation in any important way. Thus, the span seminorm of R is a more important quantity of R than its supremum, since R may always be replaced by $R - \inf R$. For this reason, R is sometimes standardized into the range $[0, 1]$.

We may refer to the case $\|R\|_{sup} = \infty$ as an *unbounded model*. Clearly, the operator T_π cannot be defined on Banach space $(\mathcal{V}, \|\cdot\|_{sup})$, but can be defined on $(\mathcal{V}, \|\cdot\|_w)$ if $\|R\|_w < \infty$ and $\|\|Q\|\|_w < \infty$. However, the assumption that R is bounded often permits considerable simplification of the analysis, so it is important to know that an unbounded model may be transformed to the bounded model π_w given in statement (*ii*) of Theorem 7.8. The cost of doing this is that the transformation from operator Q to $I_w^{-1} Q I_w$ may introduce some complications in the eigenpair structure.

Case (*iii*) of Theorem 7.8 is of a somewhat different structure, since the fixed point does not change, and will be of relevance to the *average cost Markov decision process*, to be considered in Section 12.7.

7.5 FIXED POINT ALGORITHMS AND THE SPAN SEMINORM

Theorem 7.7 characterizes solutions to fixed point equations, as well as the contractive properties of T_π in a class of Banach spaces $\mathcal{F}(\mathcal{X}, \|\cdot\|_w)$. Suppose next that Q is span continuous *wrt* weight function ν, with $Q\nu = \alpha\nu$ for positive α and positive ν. Then T_π is well defined on the span quotient space $(\mathcal{V}/\mathcal{N}_\nu, \|\cdot\|_{SP(\nu)})$, mapping coset $[V]$ to $[T_\pi V]$. By Theorem 6.28 we have $\|\|Q\|\|_{SP(\nu)} \leq \|\|Q\|\|_\nu$, and by Theorem 6.25 we also have $\|R\|_{SP(\nu)} \leq 2\|R\|_\nu$, so that if T_π is an operator (or is contractive) on $(\mathcal{V}, \|\cdot\|_\nu)$ then it is also an operator (or is contractive) on $(\mathcal{V}/\mathcal{N}_\nu, \|\cdot\|_{SP(\nu)})$. We summarize this in the following theorem.

Theorem 7.9 *If $\|R\|_\nu < \infty$ and $\|\|Q\|\|_\nu < \infty$, and Q is continuous wrt $\|\cdot\|_{SP(\nu)}$, then T_π is an operator on $(\mathcal{V}/\mathcal{N}_\nu, \|\cdot\|_{SP(\nu)})$. If T_π is contractive wrt $\|\cdot\|_\nu$ it is also contractive wrt $\|\cdot\|_{SP(\nu)}$.*

The average cost Markov decision process of Section 12.7 gives an example of an operator which is usually contractive in the span seminorm but not the supremum norm. However, we will see that interest in the span seminorm need not be a matter only of necessity. Even when an operator is contractive in the supremum norm, it may still be worth considering convergence in $(\mathcal{V}/\mathcal{N}_\nu, \|\cdot\|_{SP(\nu)})$, since the contraction rate *wrt* $\|\cdot\|_{SP(\nu)}$ may be considerably smaller. Of course, this leaves open the problem of reconstructing a specific solution $V = T_\pi V$ in $\mathcal{V}$ from an equivalence class solution $[V] = T_\pi[V]$ in $\mathcal{V}/\mathcal{N}_\nu$. We next consider this question.

Assume, as in Theorem 7.9, that $|||Q|||_{SP(\nu)} < \infty$ and that Q is J-stage contractive on $(\mathcal{V}/\mathcal{N}_\nu, \|\cdot\|_{SP(\nu)})$. By the Banach fixed point theorem there exists a unique solution $E \in \mathcal{V}/\mathcal{N}_\nu$ to $E = T_\pi E$, that is, there exists a unique modulo $\mathcal{N}_\nu$ coset $[V^*]$ for which $T_\pi : [V^*] \to [V^*]$. This, by itself, does not guarantee the existence of a solution to $V = T_\pi V$ in the original Banach space $(\mathcal{V}, \|\cdot\|_w)$. If one exists, it must be in $[V^*]$, since for any other $E \in \mathcal{V}/\mathcal{N}_\nu$ we must have $T_\pi E \neq E$, since $[V^*]$ is the unique fixed point in $\mathcal{V}/\mathcal{N}_\nu$, and distinct cosets are disjoint.

Let us consider a somewhat more general problem. Since $T_\pi : [V^*] \to [V^*]$ is a unique coset fixed point, we have $T_\pi V - V \in \mathcal{N}_\nu$ if and only if $V \in [V^*]$, so that we might expect to find solutions to the equation

$$V + h\nu = T_\pi V, \tag{7.7}$$

for a fixed scalar h. Recall that we must have $Q\nu = \alpha\nu$, and in the applications we consider we will generally have $\alpha \in (0, 1]$. Suppose we may identify some $v^* \in [V^*]$. This means any solution to (7.7) must be of the form $v^* + a\nu$ for some scalar a. This gives

$$v^* + a\nu + h\nu = T_\pi v^* + \alpha a\nu$$

which, rearranged, is

$$T_\pi v^* - v^* = h\nu + a(1 - \alpha)\nu.$$

Suppose $\alpha < 1$. Then we may solve directly, so that

$$V_h = v^* + (1 - \alpha)^{-1}(T_\pi v^* - v^* - h\nu)$$

is a solution to (7.7). Furthermore, since all solutions may be written in the form $v^* + a\nu$, and a is uniquely determined, this solution must be unique, including that for the actual fixed point equation given by $h = 0$. This simple construction device will prove useful.

Next, suppose $\alpha = 1$. Then $a(1 - \alpha)\nu = \vec{0}$, and (7.7) is simply

$$T_\pi v^* - v^* = h\nu,$$

so that $v^* + a\nu$ is a solution for all scalars a. In fact, for any $V_1, V_2 \in [V^*]$ we must have $T_\pi V_1 - V_1 = T_\pi V_2 - V_2$, and we may equate this quantity with $h\nu$ in (7.7). This means that there is a unique h for which all $V \in [V^*]$ solve (7.7), and in this case solving the fixed point equation involves determining this value.

We summarize the discussion in the following theorem:

Theorem 7.10 *Suppose for model $\pi = (R, Q)$, T_π is an operator on $(\mathcal{V}/\mathcal{N}_\nu, \|\cdot\|_{SP(\nu)})$ with $Q\nu = \alpha\nu$, $\alpha \in (0, 1]$. Suppose the coset (modulo $\mathcal{N}_\nu$) $[v^*]$ is a unique fixed point of T_π.*

(*i*) *If* $\alpha < 1$ *then* $V_h^* = v + (1-\alpha)^{-1}(T_\pi v - v - hv)$ *is the unique fixed point solution to* $V + hv = T_\pi V$ *for any scalar* h *and* $v \in [v^*]$.

(*ii*) *If* $\alpha = 1$ *then there exists a unique scalar* h *for which all elements of* $[v^*]$ *solve* $V + hv = T_\pi V$. *This scalar is the solution to* $T_\pi v - v = hv$ *for any* $v \in [v^*]$.

7.5.1 Approximations in the span seminorm

The error of an approximation $\hat{V}$ to fixed point $V_\pi = T_\pi V_\pi$ is ideally expressed in the weighted supremum norm as a bound in the form $\| \hat{V} - V_\pi \|_w \leq \epsilon$, where $w \sim V_\pi$. We will see some advantage to measuring approximation error in the span seminorm, but this leaves the problem of reconstructing a single solution which is uniformly close to V_π.

Formally, if we have approximation $\hat{V}$ for which $\| \hat{V} - V_\pi \|_{SP(\nu)} \leq \epsilon$ we would like to use $\hat{V}$ to construct some $\hat{V}'$ for which $\| \hat{V}' - V_\pi \|_\nu = O(\epsilon)$. Under the conditions of Theorem 7.10, with $\alpha < 1$, if $\| \hat{V} - V_\pi \|_{SP(\nu)} = 0$ we may determine fixed point $V_\pi = \hat{V}'$ exactly with the equation

$$\hat{V}' = \hat{V} + (1-\alpha)^{-1}(T_\pi \hat{V} - \hat{V}). \tag{7.8}$$

We may therefore expect that an order $O(\epsilon)$ approximation to V_π may be based on $\hat{V}$ using a similar device. This will hold for nonlinear operators which share the same type of relationship to an eigenpair usually associated with linear operators.

Theorem 7.11 *Let* T *be a monotone operator on a vector space* $\mathcal{V}$. *Suppose there exists* $\nu \in \mathcal{V}$ *and constant* $\alpha < 1$ *for which* $T(V + a\nu) = TV + \alpha a\nu$ *for all* $V \in \mathcal{V}$. *Suppose* $V_\pi = TV_\pi$ *is the fixed point of* T. *Then*

$$\| \hat{V} - V_\pi \|_{SP(\nu)} \leq \epsilon \textit{ implies } \| \hat{V}' - V_\pi \|_\nu \leq \frac{1+\alpha/2}{(1-\alpha)}\epsilon \tag{7.9}$$

for $\hat{V}'$ *defined by (7.8).*

Proof By Theorem 6.25 there exists a (which can be identified as $mid(V_\pi - \hat{V})$) for which, given the assumption of (7.9),

$$\| V_\pi - (\hat{V} + a\nu) \|_\nu \leq \frac{1}{2}\epsilon,$$

from which we may write

$$\hat{V} + (a - \frac{1}{2}\epsilon)\nu \leq V_\pi \leq \hat{V} + (a + \frac{1}{2}\epsilon)\nu. \tag{7.10}$$

We may apply T_π to (7.10), which by monotonicity gives

$$T_\pi \hat{V} + \alpha(a - \frac{1}{2}\epsilon)\nu \leq V_\pi \leq T_\pi \hat{V} + \alpha(a + \frac{1}{2}\epsilon)\nu. \tag{7.11}$$

Then (7.10) and (7.11) may be combined to give

$$0 \leq T_\pi \hat{V} - \hat{V} - (1-\alpha)a\nu + \frac{1+\alpha}{2}\epsilon\nu$$

$$0 \geq T_\pi \hat{V} - \hat{V} - (1-\alpha)a\nu - \frac{1+\alpha}{2}\epsilon\nu,$$

from which (7.9) follows. ///

Thus, in Theorem 7.11 an approximation in the span seminorm of ϵ becomes order $O((1-\alpha)^{-1}\epsilon)$ in the supremum norm.

7.5.2 Magnitude of fixed points in the span seminorm

One of the consequences of the geometric series theorem (Theorem 7.6) is that the magnitude of the fixed point of T_π for model $\pi = (R, Q)$ follows from the expansion

$$V_\pi = \sum_{i=0}^{\infty} Q^i R,$$

so that in any suitable norm $\|\cdot\|$ we have

$$\| V_\pi \| \leq \sum_{i=0}^{\infty} \| Q^i \| \| R \|.$$

If Q is contractive, that is $\|Q\| < 1$ then,

$$\| V_\pi \| \leq \left(1 - \|Q\|\right)^{-1} \| R \|.$$

One situation commonly encountered is the model $Q = \beta Q_0$ where $\beta < 1$ and Q_0 is a proper stochastic kernel, for which $\|Q_0\|_{sup} = 1$, and so $\|Q\|_{sup} = \beta$. In this case

$$\| V_\pi \|_{sup} \leq (1-\beta)^{-1} \| R \|_{sup}. \tag{7.12}$$

Since $\vec{1}$ is a principal eigenvector of Q_0, it is also a principal eigenvector of Q, with principal eigenvalue β. Then

$$\| V_\pi \|_{SP} \leq \sum_{i=0}^{\infty} \beta^i \| Q_0^i \|_{SP} \| R \|_{SP}. \tag{7.13}$$

From Theorem 7.8 we may always replace $R \geq 0$ with $R - \inf R \geq 0$, so we lose no generality in assuming $\|R\|_{sup} = \|R\|_{SP}$. We have also seen that $\|Q_0^i\|_{SP} \leq 1$, so that $\| V_\pi \|_{SP} \leq \| V_\pi \|_{sup}$. However, as discussed in Section 5.2, under well defined ergodicity conditions Q_0^n approaches a steady state transition kernel Q_0^∞, in the sense that there exists a steady state distrbution P_0 for which $Q_0^\infty(\cdot \mid x) = P_0$ for all $x \in \mathcal{X}$. But this means $\|Q_0^\infty\|_{SP} = 0$, so it worth investing by what degree a bound for $\| V_\pi \|_{SP}$ can be improved over the order $O((1-\beta)^{-1})$ bound given in (7.12).

In fact, when Q_0 is a stochastic matrix, $\vert\vert\vert Q_0 \vert\vert\vert_{SP}$ is equivalent to *Dobrushin's ergodic coefficient*, and is an upper bound for $|\lambda_{SLEM}|$, where λ_{SLEM} is the second largest eigenvalue in magnitude (Section 2.3.4). See, for example, Theorem 7.2 of Brémaud (1999). *Dobrushin's inequality* states that $\vert\vert\vert P_1 P_2 \vert\vert\vert_{SP} \leq \vert\vert\vert P_1 \vert\vert\vert_{SP} \vert\vert\vert P_2 \vert\vert\vert_{SP}$ for any two stochastic matrices P_1, P_2 (Theorem 7.1 of Brémaud (1999)), and this is also a direct consequence of the interpretation of this coefficient as an operator norm, which possesses the submultiplicative property.

We have already seen that for finite state transition kernels $\vert\vert\vert Q_0 \vert\vert\vert_{SP} = 1$ if and only if there exists a pair of singular measures $Q_0(\cdot \mid x) \perp Q_0(\cdot \mid y)$ within Q_0. If this is true of Q_0, it need not be true of Q_0^J. If this is the case, suppose we have $\vert\vert\vert Q_0^J \vert\vert\vert_{SP} = \alpha < 1$ for some J. By the submultiplicative property we have $\vert\vert\vert Q_0^{mJ} \vert\vert\vert_{SP} \leq \alpha^m$, and since $\vert\vert\vert Q_0 \vert\vert\vert_{SP} = 1$ we have $\vert\vert\vert Q_0^i \vert\vert\vert_{SP} \leq \alpha^{\lfloor i/J \rfloor}$, and therefore $\vert\vert\vert Q^i \vert\vert\vert_{SP} \leq \beta^i \alpha^{\lfloor i/J \rfloor}$. We obtain the upper bound

$$\| V_\pi \|_{SP} \leq \min\left(J(1-\beta^J\alpha)^{-1}, (1-\beta)^{-1}\right) \|R\|_{SP}. \tag{7.14}$$

As we have discussed, as a consequence of the 'curse of the discount factor' it is important to consider the behavior of any bound as $\beta \uparrow 1$, and we can see that (7.14) implies

$$\| V_\pi \|_{SP} \leq J(1-\alpha)^{-1} \|R\|_{SP} \tag{7.15}$$

for any $\beta \leq 1$. The value of this magnitude, whether $\| V_\pi \|_{sup}$ or $\| V_\pi \|_{SP}$, will play a crucial role in our approximation theory, and so the ability to reduce dependence on the factor $(1-\beta)^{-1}$ will offer considerable practical advantage in many approximation methods.

7.6 STOPPING RULES FOR FIXED POINT ALGORITHMS

The practical implementation of a fixed point algorithm $V_k = T_\pi V_{k-1} = T_\pi^k V_0$ requires the ability to bound the approximation error after a finite number of iterations, and therefore to devise a stopping rule which, for any pretermined tolerance ϵ (or δ), is able to terminate the algorithm at iteration N so that $\| V^* - V_N \| \leq \epsilon$ (or $\| V^* - V_N \| / \| V^* \| \leq \delta$).

A fixed stopping time is based on any contraction bound of the form

$$\| V_N - V^* \| = \| T_\pi^N V_0 - V^* \| \leq K\rho^N \| V_0 - V^* \|.$$

It may be that the ratio $\| V_N - V^* \| / \| V_0 - V^* \|$ may be interpreted as a relative error, especially if $V_0 = \vec{0}$. Then the stopping time is simply

$$N^\delta = \min\{N \mid K\rho^N \leq \delta\} = \lceil \log(\delta/K)/\log(\rho) \rceil. \tag{7.16}$$

Then

$$\| V_{N^\delta} - V^* \| / \| V_0 - V^* \| \leq \delta.$$

From Theorem 6.6, a stopping rule based on the contraction constant $\rho_J = |||Q^J||| < 1$ of the Jth iterate T^J is given by

$$N_J^\epsilon = \min\{N \mid \rho_J(1-\rho_J)^{-1}\|V_N - V_{N-J}\| \leq \epsilon\}, \tag{7.17}$$

which guarantees absolute approximation bound

$$\|V_{N_J^\epsilon} - V^*\| \leq \epsilon.$$

As discussed in Section 6.2.1 it is also possible to devise a stopping rule based on the sequence of contraction constants ρ_n of the form

$$N_\infty^\epsilon = \min\{N \mid \rho_N(1-\rho_N)^{-1}\|V_N - V_0\| \leq \epsilon\}, \tag{7.18}$$

which guarantees absolute approximation bound

$$\|V_{N_\infty^\epsilon} - V^*\| \leq \epsilon.$$

This stopping rule may be difficult to implement in practice if the sequence ρ_n is not available, but does show that an approximation bound based on the best possible asymptotic contraction rate ρ discussed in Theorem 6.3 can be achieved.

7.6.1 Fixed point iteration in the span seminorm

Here, we consider the implemenation of fixed point algorithms for the operator T_π, $\pi = (R, Q)$ on the quotient space $(\mathcal{V}/\mathcal{N}_\nu, \|\cdot\|_{SP(\nu)})$. This assumes a positive eigenvalue and eigenvector pair $Q\nu = \rho\nu$, $\rho \in (0, 1]$.

We have already remarked that T_π may be contractive on $(\mathcal{V}/\mathcal{N}_\nu, \|\cdot\|_{SP(\nu)})$ but not $(\mathcal{V}, \|\cdot\|_w)$, so that $T_\pi^k V_0$ need not even possess a limit, and in many important cases will diverge pointwise. However, the operator is defined on an equivalence class, which may be represented by any member of that class. In this case divergent behavior may be avoided by using fixed point iterations of the form

$$V_k = T_\pi[V_{k-1} - (\inf_x V_{k-1}(x)/\nu(x))\nu] \tag{7.19}$$

or more simply

$$V_k = T_\pi[V_{k-1} - (V_{k-1}(x_0)/\nu(x_0))\nu] \tag{7.20}$$

for some fixed $x_0 \in \mathcal{X}$. Then any of the stopping rules (7.16) - (7.17) may be used based on the norm $\|\cdot\|_{SP(\nu)}$ and contraction constants $\rho_J = |||Q^J|||_{SP(\nu)}$.

7.7 PERTURBATIONS OF FIXED POINT EQUATIONS

An important concept in numerical analysis is the *perturbation* of a system of equations. The underlying notion is that a bounded perturbation of the equation parameters will yield a bounded perturbation of the solution. We have considered a family of models (R, Q), each element of which generates a fixed point equation

$V = R + QV$ possessing a unique solution. One advantage of the Banach space construction is that under general conditions we may establish continuous relationships between the three elements, since we may equivalently write $V = (I - Q)^{-1}R$ or $R = (I - Q)V$. Under the hypothesis of the geometric series theorem (Theorem 7.6) both $(I - Q)$ and $(I - Q)^{-1}$ are bounded linear operators, so a perturbation of either V or R while Q is fixed results in a Lipschitz continuous perturbation of the other. We may also consider a perturbation of Q. If V is fixed, then for fixed point equations $R_i = (I - Q_i)V$, $i = 1, 2$ we may write $R_2 - R_1 = -(Q_2 - Q_1)V$, so that a perturbation of Q results in a perturbation of R bounded by $\|R_2 - R_1\| \leq \|\!|Q_2 - Q_1|\!\| \|V\|$. Of course, the same holds when R is fixed, and we have bound $\|V_2 - V_1\| \leq \|\!|(I - Q_2)^{-1} - (I - Q_1)^{-1}|\!\| \|R\|$, which may be used in the same way. The practical problem is that the perturbation $Q_2 - Q_1$ will usually be more tractable than $(I - Q_2)^{-1} - (I - Q_1)^{-1}$, so we would usually prefer to characterize the effect of a perturbation of Q on V in terms of the former object, unless the form of the geometric expansion $(I - Q)^{-1} = \sum_{i \geq 0} Q^i$ is well understood.

We will briefly develop this idea in the context of fixed point equations. Suppose we are given fixed point operators $TV = R + QV$ and $\hat{T}V = \hat{R} + \hat{Q}V$. We are motivated by the idea that $(\hat{R}, \hat{Q})$ approximates (R, Q), so that we might claim approximation bounds $\|R - \hat{R}\| \leq \delta_R$ and $\|\!|Q - \hat{Q}|\!\| \leq \delta_Q$.

Theorem 7.12 *Suppose we are given Banach space $(\mathcal{V}, \|\cdot\|)$, elements $R, \hat{R} \in \mathcal{V}$ and bounded linear operators $Q, \hat{Q}$. Suppose solutions $V, \hat{V} \in \mathcal{V}$ exist to the fixed point equations*

$$\begin{aligned} V &= R + QV, \\ \hat{V} &= \hat{R} + \hat{Q}\hat{V}, \end{aligned} \tag{7.21}$$

and that $(I - \hat{Q})$ has a bounded inverse. Then

$$\|V - \hat{V}\| \leq \sum_{i=0}^{\infty} \|\!|\hat{Q}^i|\!\| \left[\|R - \hat{R}\| + \|QV - \hat{Q}V\| \right], \tag{7.22}$$

and if $\|\!|\hat{Q}|\!\| < 1$ then

$$\|V - \hat{V}\| \leq (1 - \|\!|\hat{Q}|\!\|)^{-1} \left[\|R - \hat{R}\| + \|QV - \hat{Q}V\| \right]. \tag{7.23}$$

Proof From (7.21), by subtraction we may obtain

$$(I - \hat{Q})(V - \hat{V}) = R - \hat{R} + (Q - \hat{Q})V.$$

Since $(I - \hat{Q})^{-1}$ is assumed to be a bounded operator we have

$$(V - \hat{V}) = \sum_{i=0}^{\infty} \hat{Q}^i \left[(R - \hat{R}) + (Q - \hat{Q})V \right],$$

from which (7.22) and (7.23) follow from the triangle inequality and the submuliplicative property of operator norms. ///

Suppose, to fix ideas, we assume $R = \hat{R}$, and $|||\hat{Q}||| = \beta < 1$. Then from (7.23) we obtain bound

$$\begin{aligned}\|V - \hat{V}\| &\leq (1-\beta)^{-1}\|QV - \hat{Q}V\| \\ &\leq (1-\beta)^{-1}|||Q - \hat{Q}|||\,\|V\| \\ &\leq (1-\beta)^{-1}\delta_Q\|V\|. \end{aligned} \tag{7.24}$$

In order for $\|V - \hat{V}\|$ to be small relative to $\|V\|$ (which is generally the goal) we must have δ_Q small relative to $(1-\beta)$. However, in some applications, including many MDP models, this is problematic. First, an approximation of Q may not involve β in any important way. This is especially true when β is a fixed 'discount factor', which is used to define operator $Q = \beta Q_0$, but which otherwise plays no role. Essentially, it would be Q_0 that is approximated, and so when β is close to 1, δ_Q would have little relationship to β. Therefore, requiring $\delta_Q \ll (1-\beta)$ would be an unrealistic burden, especially given the incentive in some applications to allow β to approach 1.

The situation improves considerably when there exists a positive eigenpair (α, ν) common to both Q and $\hat{Q}$. In this case $(Q - \hat{Q})V = (Q - \hat{Q})(V + c\nu)$ for any scalar c, so that

$$\|(Q - \hat{Q})V\|_\nu = \|(Q - \hat{Q})(V + c\nu)\|_\nu \leq |||Q - \hat{Q}|||_\nu\|V + c\nu\|_\nu.$$

However, we may minimize over c, so that by Theorem 6.25 we have

$$\begin{aligned}\|(Q - \hat{Q})V\|_\nu &\leq \inf_c |||Q - \hat{Q}|||_\nu\|V + c\nu\|_\nu \\ &\leq \frac{1}{2}|||Q - \hat{Q}|||_\nu\|V\|_{SP(\nu)}.\end{aligned}$$

This expression can be quite advantageous, since the quantity $\|V\|_{SP(\nu)}$ may be considerably smaller than $\|V\|_\nu$, so that (7.24) can be replaced by

$$\|V - \hat{V}\|_\nu \leq \frac{1}{2}(1-\beta)^{-1}\delta_Q\|V\|_{SP(\nu)}. \tag{7.25}$$

Suppose for some J we have $|||Q^J|||_{SP(\nu)} = \alpha\beta$ for fixed $\alpha < 1$ and all $\beta < 1$. We may conclude that $\|V\|_{SP(\nu)}$ is not of order $O((1-\beta)^{-1})$, remaining bounded as β approaches 1. Compare, for example, (7.12) and (7.15). In this case we have

$$\|V - \hat{V}\|_\nu = O\left((1-\beta)^{-2}\delta_Q\right)$$

using the bound (7.24), in contrast to

$$\|V - \hat{V}\|_\nu = O\left((1-\beta)^{-1}\delta_Q\right)$$

when using bound (7.25). We are thus able to reduce the approximation bound by a factor of $(1-\beta)^{-1}$.

Chapter 8

The distribution of a maximum

Approximations are often based on statistical estimation, and our reliance on the supremum norm as the measure of approximation will usually require the characterization of the maximum value of many random quantities. Suppose we are given a sequence of random variables $X_1, X_2, \ldots, X_m$, and assume that at least the means and variances $E[X_i] = \mu_i$, $var[X_i] = \sigma_i^2$ are finite. We will use the following notation:

$$\begin{aligned}
M &= \max_{1 \leq i \leq m} X_i, \\
\bar{M} &= E[M], \\
\mu_{max} &= \max_{1 \leq i \leq m} \mu_i, \\
\sigma_{max}^2 &= \max_{1 \leq i \leq m} \sigma_i^2, \\
\bar{\mu} &= m^{-1} \sum_{i=1}^{m} \mu_i, \\
\bar{\mu^2} &= m^{-1} \sum_{i=1}^{m} \mu_i^2 \\
\bar{\nu} &= \bar{\mu^2} - \bar{\mu}^2 \\
\bar{\sigma^2} &= m^{-1} \sum_{i=1}^{m} \sigma_i^2 \\
\bar{\sigma_j^2} &= m^{-1} \sum_{i=1}^{m} var[X_i - X_j], \quad j = 1, \ldots m.
\end{aligned}$$

When appropriate, subscripts will be used with these quantities to denote the number of random variables, say $M = M_m$ or $\bar{\mu} = \bar{\mu}_m$.

That $\bar{M} \geq \mu_{max}$ is clear by noting that $X_i \leq \max(X_1, X_2)$ for $i = 1, 2$, so that $E[X_i] \leq E[\max(X_1, X_2)]$. This means $\max(E[X_1], E[X_2]) \leq E[\max(X_1, X_2)]$. That $\bar{M} \geq \mu_{max}$ follows by induction. Furthermore, if $E[|X_i|] < \infty$ then since $M \leq \sum_i |X_i|$, we must have $\bar{M} \leq \sum_i E[|X_i|] < \infty$ since m is assumed finite.

8.1 GENERAL APPROACH

We will first discuss some general principles. We prefer not to rely on independence assumptions. One obvious reason is that, everything else being equal, it is always preferable to impose fewer restrictions, and under general depedendence assumptions only information about the marginal distributions F_{X_i} is used. Interestingly, we find that we end up losing little in this approach. For example, in the case of *iid* RVs approximations can yield precise limits, while the more general methods yield bounding inequalities. However, it turns out these limits are generally comparable to the upper bounds, the *iid* case thus representing a worst case scenario.

Next, it is important to note the role that stochastic ordering plays. Recall from Section 4.7 the following hierarchy:

$$X \geq_{st} Y \Rightarrow X \geq_{MGF} Y \Rightarrow X \geq_{var} Y.$$

A number of techniques for the approximation of M will be considered. We can classify the methods into three groups, according to whether the bounds are based on F_{X_i}, $m_{X_i}(t)$ or the moments (μ_i, σ_i^2). It will often be convenient to rely on stochastic ordering to develop bounds. Suppose we are given two collections of RVs X_i and X_i^*, $i=1,\ldots,m$, and we can verify that $X_i^* \geq_{st} X_i$, then a bound for $M^* = \max_i X_i^*$ based on CDFs, MGFs or the first two moments will also hold for $M = \max_i X_i$. Similarly, if $X_i^* \geq_{MGF} X_i$ then results for M^* based on MGFs or the first two moments will hold also for M, but since MGF ordering does not imply stochastic ordering, bounds based on the CDFs need not hold. It will sometimes be convenient to replace a sequence $X_1,\ldots,X_m$ with m RVs of identical distribution F_X (possibly $F_X = F_{X_j}$ for some X_j) under the assumption that X is stochastically larger than each X_i in some sense. This is a perfectly valid approach, provided the appropriate stochastic ordering is used.

This hierarchy is reflected thoughout the theory of stochastic maxima. We have already seen that bounds based directly on CDFs can be sharper than those constructed from MGFs, such as the Chernoff bound. We will also consider bounds based on the the first two moments alone. These are quite convenient to apply, but are generally far less efficient than those obtainable under assumptions we can expect to hold in the applications we consider.

We will first consider the problem of bounding $\bar{M}$, using an MGF based method similar to that employed in the Chernoff bound.

8.2 BOUNDS ON $\bar{M}$ BASED ON MGFs

A convenient but powerful method of bounding $\bar{M}$ can be based on the following theorem:

Theorem 8.1 *Suppose for random variables $X_1,\ldots,X_m$, $m > 1$, we are given functions $m_i(t)$, $i=1,\ldots,m$ for which $E[\exp(tX_i)] \leq m_i(t)$ for all $t \in \mathcal{T} \subset [0,\infty)$. Then*

$$\bar{M} \leq \inf_{t\in\mathcal{T}} t^{-1} \log\left(\sum_{i=1}^{m} m_i(t)\right). \tag{8.1}$$

In particular, if $m_i(t) \equiv m(t)$, then

$$\bar{M} \leq \inf_{t \in \mathcal{T}} t^{-1}\left(\log(m) + \log(m(t))\right). \tag{8.2}$$

Proof By Jensen's inequality, for any $t \in \mathcal{T}$,

$$\begin{aligned}
\exp\left\{tE\left[\max_{1\leq i\leq m} X_i\right]\right\} &\leq E\left[\exp\left\{t \max_{1\leq i\leq m} X_i\right\}\right] \\
&= E\left[\max_{1\leq i\leq m} \exp\{tX_i\}\right] \\
&\leq \sum_{i=1}^{m} E\left[\exp\{tX_i\}\right] \\
&\leq \sum_{i=1}^{m} m_i(t),
\end{aligned}$$

which gives (8.1) following a log transformation of the preceding inequality, and (8.2) after substituting $m(t) = m_i(t)$. ///

The CGF of $Z \sim N(\mu, \sigma^2)$ is $c_Z(t) = \mu t + \sigma^2 t^2/2$ for all $t \in \mathbb{R}$. Rather than apply Theorem 8.1 directly to this case, it will be worth considering a weaker condition.

Theorem 8.2 *Let $X_1, \ldots, X_m$ be a collection of RVs, and suppose there exists $t^* > 0$ for which*

$$c_{X_i}(t) \leq \mu t + \sigma^2 t^2/2 + \epsilon, \quad t \in [0, t^*], \text{ and } t^* \geq \sqrt{\frac{2(\log(m) + \epsilon)}{\sigma^2}}, \tag{8.3}$$

for each X_i. Then

$$\bar{M} \leq \mu + \sigma\sqrt{2(\log(m) + \epsilon)}. \tag{8.4}$$

In particular, if $X_i \leq_{\text{MGF}} Z$ for $1 \leq i \leq m$, where $Z \sim N(\mu, \sigma^2)$ then

$$\bar{M} \leq \mu + \sigma\sqrt{2\log(m)} \tag{8.5}$$

Proof From Theorem 8.1, using (8.3) we may write,

$$\begin{aligned}
\bar{M} &\leq \inf_{t \in [0, t^*]} t^{-1}\left(\log(m) + \epsilon + \mu t + \sigma^2 t^2/2\right) \\
&= \mu + \inf_{t \in [0, t^*]} t^{-1}\left(\log(m) + \epsilon + \sigma^2 t^2/2\right).
\end{aligned}$$

The preceding upper bound is minimized over $t > 0$ at stationary point $t_0 = \left(2(\log(m) + \epsilon)/\sigma^2\right)^{1/2}$. Under the hypothesis, $t_0 \leq t^*$, from which (8.4) follows. Then (8.5) is a direct application of (8.4). ///

We next present what is essentially a refinement of Theorem 8.2. Clearly, any CGF can be approximated by the normal CGF in a small enough neighborhood of $t=0$. The question is whether a well defined approximation holds uniformly over an interval large enough to include the stationary point used in the proof of the theorem.

Recall the following function defined in Section 4.6.1,

$$\kappa_2^*(t) = \begin{cases} \sup_{0 \le t' \le t} \left. \frac{d^2 c(s)}{ds^2} \right|_{s=t'} ; & t \ge 0, |c(t)| < \infty \\ \infty ; & t \ge 0, |c(t)| = \infty \end{cases}.$$

for any CGF $c(t)$.

Note that since $d^2 c_X(s)/ds^2 \big|_{s=0} = var[X]$ we must have $\kappa_2^*(t) \ge var[X] \ge 0$.

Theorem 8.3 *Given RVs $X_1, \ldots, X_m$ for which $E[X_i] \le \mu$, suppose for some CGF $c(t)$ we have $c_{X_i}(t) \le c(t)$ for $t \ge 0$ and for each X_i. Then*

$$\bar{M} \le \mu + \inf\{\sqrt{\kappa_2^*(t) 2 \log(m)} \mid t > 0, \ \ t(\kappa_2^*(t))^{1/2} \ge (2\log(m))^{1/2}\}. \tag{8.6}$$

Proof By Taylor's theorem, for $t \ge 0$

$$c(t) = \mu t + d^2 c_X(s)/ds^2 \big|_{s=\eta(t)} \frac{t^2}{2}$$

for some $\eta(t)$ satisfying $\eta(t) \in [0, t]$. This means that for any $s > 0$ we have

$$c(s) \le \mu s + \kappa_2^*(t) \frac{s^2}{2}, \quad s \in [0, t].$$

Using the notational conventions of Theorem 8.2 set $\sigma^2 = \kappa_2^*(0)$ and $K = \kappa_2^*(t)/\sigma^2$. Then if $t \ge (2\log(m)/\kappa_2^*(t))^{1/2}$ we conclude

$$E[M_m] \le \mu + \sqrt{\kappa_2^*(t) 2 \log(m)},$$

so that (8.6) follows by minimizing over all t satisfying the preceding constraint. ///

8.2.1 Sample means

Suppose for zero mean RV X the CGF exists for $t \in [0, t^*)$. Suppose we are given $S_1, \ldots, S_m$, where each S_i is a sum of n *iid* RVs equal in distribution to X. Then $c_{S_i}(t) = n c_X(t)$. Then let $\sigma^2 = var[X]$, and for X calculate $\kappa_2^*(t)$.

Applying Theorem 8.3 to the maximum M^S of $S_1, \ldots, S_m$ yields bounds on the expectation

$$\bar{M}^S \le \inf\{\sqrt{n\kappa_2^*(t) 2 \log(m)} \mid t(\kappa_2^*(t))^{1/2} \ge n^{-1/2}(2\log(m))^{1/2}\}.$$

For any $\epsilon > 0$ we many select $t_\epsilon > 0$ close enough to 0 to make

$$|\kappa_2^*(t_\epsilon) - \sigma^2| \leq \epsilon,$$

and then select large enough n_ϵ so that

$$t_\epsilon(\kappa_2^*(t_\epsilon))^{1/2} \geq n_\epsilon^{-1/2}(2\log(m))^{1/2}.$$

Hence

$$\bar{M}^S \leq \sqrt{n(\sigma^2 + \epsilon)2\log(m)},$$

for all $n \geq n_\epsilon$. At this point, we standardize the sums to obtain $Z_i = n^{-1/2}S_i$, and let M be the resulting maximum. We need only divide the preceding inequality by $n^{1/2}$, to obtain

$$\bar{M} \leq \sqrt{(\sigma^2 + \epsilon)2\log(m)}. \tag{8.7}$$

We may make ϵ arbitrarily small, to conclude that as $n \to \infty$ the bound approaches that equivalent to the normal distribution.

8.2.2 Gamma distribution

If $X \in gamma(\alpha, \lambda)$ then

$$c_X(t) = -\alpha\log(1 - t/\lambda), \quad \text{for } t \leq \lambda.$$

Then we may evaluate

$$\kappa_2^*(t) = \frac{\alpha/\lambda^2}{(1 - t/\lambda)^2}.$$

The condition of Theorem 8.3 becomes, for $0 \leq t < \lambda$,

$$t^2\kappa_2^*(t) > 2\log(m) \iff t/\lambda > \frac{\sqrt{2\log(m)/\alpha}}{1 + \sqrt{2\log(m)/\alpha}}.$$

Fortunately, the preceding lower bound is less than 1, so there exists $t \in [0, \lambda)$ satisfying this condition. Following Theorem 8.3 we take the infimum of all permissible bounds, yielding

$$\bar{M} \leq \frac{\alpha}{\lambda}\left[1 + \left(1 + \sqrt{2\log(m)/\alpha}\right)\sqrt{2\log(m)/\alpha}\right]. \tag{8.8}$$

8.3 BOUNDS FOR VARYING MARGINAL DISTRIBUTIONS

The strategy adopted up to this point has been to rely on ordering with respect to a single MGF, although Theorem 8.1 provides a bound for varying marginals. It is worth considering to what degree modeling distributional variation, as opposed to relying on a single dominating distribution, can improve bounds. We find that some improvement is possible, although dependence of the bound on the stochastically largest element of $X_1, \ldots, X_m$ remains strong.

We consider specifically the case of normal RVs.

Theorem 8.4 *Let $X_1, \ldots, X_m$, $m > 1$, be any random variables with MGFs $m_i(t)$ such that for some finite constant $K \geq 1$ we have $m_i(t) \leq K \exp(\sigma_i^2 t^2/2)$ for all $i = 1, \ldots, m$ and $t \geq 0$. Then the following bounds hold*

$$\bar{M} \leq \sigma_{max}\sqrt{2\log(Km)}, \tag{8.9}$$

and

$$\bar{M} \leq \frac{s}{\sqrt{2\log(Km)}} \log\left(K\sum_{i=1}^{m}(Km)^{\sigma_i^2/s^2}\right) \quad \textit{for any } s > 0. \tag{8.10}$$

The upper bound of (8.10) is no larger than that of (8.9) when $s = \sigma_{max}$, and is strictly smaller unless all variances σ_i^2 are equal. In addition, the upper bound of (8.10) is minimized in the range $s = (2^{-1}\sigma_{max}, 2\sigma_{max})$, and the sharpness of the bound is limited by

$$\frac{s}{\sqrt{2\log(Km)}} \log\left(K\sum_{i=1}^{m}(Km)^{\sigma_i^2/s^2}\right) > \frac{1}{4}\sigma_{max}\sqrt{2\log(Km)}. \tag{8.11}$$

Proof By Theorem 8.1 we may write

$$\bar{M} \leq t^{-1}\log(Km) + t\sigma_{max}^2/2, \quad t \geq 0.$$

The upper bound is minimized by $t = \left(2\log(Km)/\sigma_{max}^2\right)^{1/2}$, which gives (8.9).

Then (8.10) is obtained directly from (8.1) by the reparametrization $t = \left(2\log(Km)/s^2\right)^{1/2}$. If we set $s = \sigma_{max}$ in (8.10) we obtain

$$\bar{M} \leq \frac{\sigma_{max}}{\sqrt{2\log(Km)}} \log\left(K\sum_{i=1}^{m}(Km)^{\sigma_i^2/\sigma_{max}^2}\right). \tag{8.12}$$

However, we must have $K\sum_{i=1}^{m}(Km)^{\sigma_i^2/\sigma_{max}^2} \leq (Km)^2$, with strict inequality unless all variances σ_i^2 are equal. This forces our conclusion.

We know now that the sharpest upper bound of (8.10) is no larger than $\sigma_{max}\sqrt{2\log(Km)}$. First note that since $K \geq 1$ we have

$$K\sum_{i=1}^{n}(Km)^{\sigma_i^2/s^2} > (Km)^{\sigma_{max}^2/s^2},$$

and therefore

$$\frac{s}{\sqrt{2\log(Km)}}\log\left(K\sum_{i=1}^{m}(Km)^{\sigma_i^2/s^2}\right) > \frac{\sigma_{max}^2}{\sqrt{2}s}\sqrt{log(Km)}. \tag{8.13}$$

By substitution, this implies that the upper bound of (8.10) cannot be minimized by $s \leq \sigma_{max}/2$. We also may write

$$K\sum_{i=1}^{m}(Km)^{\sigma_i^2/s^2} > Km,$$

since $Km^{\sigma_i^2/s^2} > 1$. This implies

$$\frac{s}{\sqrt{2\log(Km)}}\log\left(K\sum_{i=1}^{m}(Km)^{\sigma_i^2/s^2}\right) > \frac{s}{\sqrt{2}}\sqrt{log(Kn)}, \tag{8.14}$$

which in turn implies that the upper bound of (8.10) cannot be minimized by $s \geq 2\sigma_{max}$. Finally, (8.11) is obtained by determining the infimum of either (8.13) or (8.14) over the range $s \in (\sigma_{max}/2, 2\sigma_{max})$. ///

We can do better than replacing all variances σ_i^2 with σ_{max}^2, but the bound will remain of order $O(\sigma_{max}(2\log(m))^{1/2})$. Therefore, $\sigma_{max}(2\log(m))^{1/2}$ can be used as a reasonable first approximation of the bound, which can always be improved by minimizing (8.10) over the range $s \in (\sigma_{max}/2, 2\sigma_{max})$.

Suppose the variances are given by $\sigma_i^2 = \sigma^2/n_i$, $i = 1, \ldots, m$, where n_i represents a sample size used in the calculation of X_i. Denote the total sample size $n = \sum_i n_i$. The harmonic mean of the variances is given by

$$HM[\sigma_i^2] = \left(\frac{\sum_{i=1}^{m} n_i/\sigma^2}{m}\right)^{-1} = \frac{m}{n}\sigma^2.$$

The quantity n/m is the average sample size, so that $HM[\sigma_i^2]$ represents the variance σ_i^2 obtained when the sample size is allocated uniformly to each RV X_i. In this sense, it represents a type of best case, and hence a lower bound for the approximation given in Theorem 8.4 which we verify next. We discuss this issue further in the next section.

Theorem 8.5 *Under the conditions of Theorem 8.4 the bound* (8.10) *of Theorem 8.4 satisfies*

$$\frac{s}{\sqrt{2\log(Km)}}\log\left(K\sum_{i=1}^{m}(Km)^{\sigma_i^2/s^2}\right)\geq HM[\sigma_i^2]\sqrt{2\log(Km)}. \tag{8.15}$$

Proof For convenience, set $HM[\sigma_i^2]=1/N$. Consider the function $g(x)=\exp(a/x)$ on $x\in(0,\infty)$ for $a>0$. It may be verified that $g(x)$ is convex, therefore $g(x)+g(y)>g((x+y)/2)+g((x+y)/2)$. Suppose $x_1,\ldots,x_m$ is a sequence of positive numbers with $\sum_i x_i=N$, for which $x_1\neq x_2$. Set $x_1'=x_2'=(x_1+x_2)/2$. Then

$$g(x_1')+g(x_2')+g(x_3)+\cdots+g(x_m)<g(x_1)+g(x_2)+g(x_3)+\cdots+g(x_m),$$

and

$$x_1'+x_2'+x_3+\cdots+x_m=x_1+x_2+x_3+\cdots+x_m=N.$$

We conclude that $\sum_i g(x_i)$ is minimized under the constraint $\sum_i x_i=N$, $x_i>0$ by solution $x_i=N/m$.

For any fixed s the minimization of (8.15) under constraint $HM[\sigma_i^2]=1/N$ is an example of the preceding minimization problem. The proof is completed by minimizing over s as in Theorem 8.4. ///

8.3.1 Example

We will further consider varying distributions generated by sample size allocation. Suppose we are given m zero mean normal RVs X_i with variances $\sigma_i^2=1/n_i$. This models the case in which each X_i is a estimator, such as a sample mean, based on a sample of size n_i. Suppose a total available sample size of n is to be allocated among the estimators. If the allocation was perfectly balanced, so that for each i, $n_i=n/m$ and $\sigma_i^2=m/n=\sigma_{max}^2$, we could use bound

$$E[M_m]\leq\sigma_{max}\sqrt{2\log(m)}=\sqrt{\frac{m}{n}}\sqrt{2\log(m)}, \tag{8.16}$$

following Theorem 8.4. Next, suppose n_i results from a uniformly random allocation of the total available sample, so that $n_i\sim bin(n,1/m)$. In this case $\sigma_{max}^2=1/\min_i n_i\geq m/n$, with strict inequality unless all sample sizes are equal by chance, which would usually have probability very close to zero. By Theorem 8.5 the best bound is obtained by a perfectly balanced allocation, since in this case $HM[\sigma_i^2]=(m/n)$. Therefore any randomness in the allocation of a fixed sample size will result in a larger bound. On the other hand, by Theorem 8.4 we know that minimizing (8.10) will result in a smaller bound than (8.9), which therefore defines the worst case bound.

Figure 8.1 demonstrates a simulation of this model with $m=100$ and $n=10000$. The left plot displays the variances $\sigma^2=1/n_i$, which range from 0.0079 to 0.0133. The right plot shows the upper bound (8.10) over the range $s\in[2^{-1}\sigma_{max},2\sigma_{max}]$. As

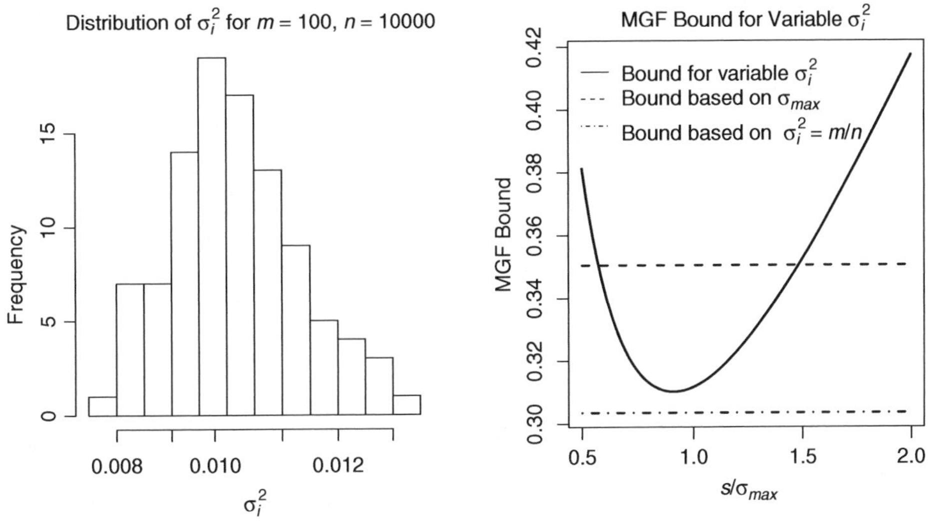

Figure 8.1 Summary of numerical example of Section 8.3.1.

required, a minimum is achieved in the given range, at a value of s slightly less than σ_{max}. For comparison, the bound (8.9) based on σ_{max} alone is also included in the plot, as well as the optimal bound based on balanced allocation given by Theorem 8.5. It is interesting to note that the bound obtained by minimizing (8.10) is reasonably close to the optimal, and is much closer to it than to the worst case bound.

8.4 TAIL PROBABILITIES OF MAXIMA

Reporting the expected value of a maximum M can be misleading unless the degree of deviation of M from $\bar{M}$ can also be predicted. We would therefore like to bound the tail probability $P(M/\bar{M} > \eta)$ for some value η which is greater than 1, but not larger than, say, 2 or 3. If $var[M]$ can be estimated, and $var[M]^{1/2} \ll \bar{M}$, then by Chebyshev's inequality

$$P\left(M/\bar{M} > \left[1 + t\frac{var[M]^{1/2}}{\bar{M}}\right]\right) = P\left(M - \bar{M} > t\,var[M]^{1/2}\right) \leq t^{-2}.$$

However, we may only have an upper bound for $\bar{M}$, say $\bar{M}^* \geq \bar{M}$, so instead we bound the tail probability $P(M/\bar{M}^* > \eta)$ (it may sometimes be convenient to use for $\bar{M}^*$ a quantity which is asymptotically equivalent to a formal upper bound). Using Markov's inequality we have

$$P\left(|M/\bar{M}^*| > \eta\right) = P\left(|M/\bar{M}^*|^k > \eta^k\right) \leq \frac{E[|M/\bar{M}^*|^k]}{\eta^k} \tag{8.17}$$

for any $k > 0$, and we will see an example of this strategy used to good effect.

However, when possible it is usually preferable to model tail probabilities directly, and the existence of a MGF for the random variables usually makes this possible.

8.4.1 Extreme value distributions

A rich asymptotic theory exists for the process $M_n = \max_{1 \leq i \leq n} X_i$ based on an indefinite *iid* sequence $X_1, X_2, \ldots$. In Section 4.12, it was pointed out that the CDF of M_n can be written

$$F_{M_n}(x) = P(M_n \leq x) = F_X(x)^n.$$

Since

$$F_{M_n}(Q_{1-1/n}) = (1 - 1/n)^n \to_n e^{-1},$$

the construction of a limiting distribution for a normalized M_n can be based on careful choices of quantiles near $Q_{1-1/n}$. In fact, the asymptotic theory for M_n, in the form of the *Fisher-Tippett-Gnedenko* theorem (Theorem 4.35), provides a ready method to do this. This theorem states that if a normalization for M_n in the form of $M'_n = (M_n - b_n)/a_n$ exists, in the sense that M'_n converges in distribution to a nondegenerate distribution G, then G must be one of the extreme value distributions (defined in Section 4.12).

If X_1 is the unit normal, then G is the Gumbel distribution, and it was shown in Hall (1979) that optimal norming constants are given by the solutions to

$$2\pi b_n^2 \exp(b_n^2) = n^2 \quad \text{and} \quad a_n = b_n^{-1}.$$

The solution is expressible by

$$b_n = (2 \log n)^{1/2} - \frac{\log \log n + \log 4\pi}{2(2 \log n)^{1/2}} + O(1/\log n), \tag{8.18}$$

and satisfies bounds

$$2 \log n - (\log \log n + \log 4\pi) < b_n^2 < 2 \log n, \quad n \geq 2. \tag{8.19}$$

Thus, $\bar{M} \approx \sqrt{2 \log(n)}$, and the bound of Theorem 8.4 can be seen to be sharp by considering the *iid* normal case.

8.4.2 Tail probabilities based on Boole's inequality

Although extreme value theory offers some insight into the problem of characterizing M, there are a number of drawbacks. First, and most obvious, we would rather not impose the restriction of independence if it is not needed. Second, we will see that there can be considerable advantage to eliminating the assumption of equal marginal distributions. A careful modeling of distributional variation can be a significant improvement over the strategy of selecting the most extreme distribution as the representational marginal. Finally, we cannot rely on the limiting distribution to approximate tail probabilities (consider, for example, bounds on tail probability approximations given in Hall (1979)).

If $X_1, \ldots, X_m$ are *iid*, $X_i \sim F_X$, then by the binomial theorem, if $\bar{F}_X(x) \ll 1/m$ we have

$$\bar{F}_M(x) = 1 - F_M(x) = 1 - (1 - \bar{F}_X(x))^m \approx m\bar{F}_X(x). \tag{8.20}$$

Suppose we take another approach which does not assume that the RVs are independent. Using Booles' inequality we have

$$\begin{aligned} \bar{F}_M(x) &= P(M > x) \\ &= P(\cup_i \{X_i > x\}) \\ &= \sum_i P(X_i > x) \\ &\leq m\bar{F}_X(x), \end{aligned} \tag{8.21}$$

which yields an upper bound close to (8.20), but with the considerable advantage that no assumption on dependence is needed in order to conclude that the inequality is an exact statement. Thus, although (8.21) appears to be a quite conservative bound, it often proves to yield accurate approximations. Essentially, the assumption of independence can be seen to represent a worst case.

In addition, the assumption of identical marginal distributions is no longer needed, since the application of Boole's inequality would be identical:

$$\bar{F}_M(x) \leq \sum_{i=1}^{m} \bar{F}_{X_i}(x). \tag{8.22}$$

8.4.3 The normal case

Recall the Chernoff bound for $X \sim N(0, \sigma^2)$ given in (4.20):

$$\bar{F}_X(x) \leq \exp(-2^{-1}x^2/\sigma^2).$$

Suppose we are given $X_1, \ldots, X_m$ satisfying $\bar{F}_{X_i}(x) \leq \exp(-2^{-1}x^2/\sigma_i^2)$. Then using Boole's inequality we have

$$\bar{F}_M(x) \leq \sum_{i=1}^{m} \exp(-2^{-1}x^2/\sigma_i^2) \leq m \exp(-2^{-1}x^2/\sigma_{max}^2). \tag{8.23}$$

The normalizing constant $b_n \approx \sqrt{2\log(m)}$ defined in (8.18) provides a natural candidate for a normalization method. Accordingly, set $x = \eta\sigma_{max}\sqrt{2\log(m)}$, yielding

$$\bar{F}_M(x) \leq m \exp(-\eta^2 \log(m)) = \exp((1 - \eta^2)\log(m)). \tag{8.24}$$

Thus, if $\eta > 1$ we obtain a natural form for the normalization of M, namely

$$P\left(\frac{M}{\sigma_{max}\sqrt{2\log(m)}} > \eta\right) \leq \exp((1-\eta^2)\log(m)) = \frac{1}{m^{(\eta^2-1)}}.$$

Thus, can expect M to be within a small multiple of bound $\bar{M}^* = \sigma_{max}\sqrt{2\log(m)}$ with a probability close to 1. Furthermore, the normalized tail probabilities decrease with m, so that stability increases with m, that is, for $\eta > 1$ we have limit $P(M/\bar{M}^* > \eta) \rightarrow_m 0$.

8.4.4 The *gamma*(α, λ) case

Suppose $X_i \sim gamma(\alpha_i, \lambda_i)$, and set $\mu_i = \alpha_i/\lambda_i$. Using (4.21) gives, for any $x \geq \mu_{max} = \max_i \mu_i$

$$\bar{F}_M(x) \leq \sum_{i=1}^{m} \exp\left(-\lambda_i x + \alpha_i + \alpha_i \log(x/\mu_i)\right),$$

or, for identical marginal distributions $gamma(\alpha, \lambda)$, $\mu = \alpha/\lambda$:

$$\bar{F}_M(x) \leq m \exp\left(-\lambda x + \alpha + \alpha \log(x/\mu)\right).$$

Suppose we use a normalizing constant of the form

$$x = \eta\mu(1+\delta), \quad \eta > 1, \quad \delta > 0,$$

then

$$\begin{aligned}\bar{F}_M(x) &\leq m \exp\left(-\alpha\eta(1+\delta) + \alpha + \alpha\log(\eta(1+\delta))\right)\\ &= m\exp\left(-\eta\alpha\delta + \alpha\log(1+\delta) + \alpha(1-\eta+\log(\eta)\right)\\ &\leq m\exp\left(-\eta\alpha\delta + \alpha\log(1+\delta)\right),\end{aligned}$$

since $1 - \eta + \log(\eta) \leq 0$ for any $\eta > 0$. Similarly, since $\eta > 1$ and $\delta > 0$, the exponent is negative, and is of order $O(\eta\alpha\delta)$ for large δ. More precisely, for $\delta \geq 3$ we always have $\log(1+\delta) < \delta/2$, and so

$$\bar{F}_M(x) \leq \exp\left(\log(m) - \eta\alpha\delta/2\right).$$

Based on (8.8) we choose $\delta \approx 2\log(m)/\alpha$, that is, we use

$$\bar{M}^* = \mu(1 + 2\log(m)/\alpha),$$

giving

$$\bar{F}_M(x) \leq \exp\left((1-\eta)\log(m)\right) = \frac{1}{m^{(\eta-1)}}.$$

Thus, similarly to the case of the normal, we may conclude that for $\eta > 1$, $P(M/\bar{M}^* > \eta) \rightarrow_m 0$. We also note that because we have used a Chernoff bound to estimate the CDFs, this result holds for RVs stochastically smaller than gamma RVs with respect to MGF ordering. The bound can be refined for the actual gamma distribution based on the hazard rate ordering discussed in Section 4.7.2.

8.5 VARIANCE MIXTURES BASED ON RANDOM SAMPLE SIZES

One type of RV which will be of interest results from a statistical estimate $\hat{\theta}$ of a parameter θ based on a sample size n. There will typically be a loss function $d(\theta, \hat{\theta})$. If we take $X = d(\theta, \hat{\theta})$, then statistical theory predicts under quite general conditions that $E[X] \approx \mu/n^{1/2}$, $var(X) \approx \sigma^2/n$.

Sometimes, however, it may be more appropriate to regard n as itself random. We have seen an example in Section 8.3.1. By Theorem 8.4, the upper bound obtained by MGF methods will always be sensitive to the maximum variance. When this is random, it will be important to estimate its distribution.

In some applications it may be useful to regard n as the realization of an ongoing counting process. A natural example would the cumulative number of transitions into a given state of a Markov chain, which would be a type of renewal process. If the renewal times were memoryless this counting process would be well approximated by a Poisson process. However, in Aldous (1989) it is shown that for Markov chains with relatively small stationary probabilities, such a counting process tends to be characterized by short periods of rapid arrivals separated by longer memoryless waiting times, refered to as *Poisson clumping*.

In this case, if N is the number of visits of a Markov chain path of T transitions to a state i with stationary probability π_i then $N \approx \kappa Y$, where κ is the average clump size, and $Y \sim pois(\lambda)$, where $E[N] = \pi_i T = E[\kappa]\lambda$. We then have

$$\lambda = E[\kappa]^{-1}\pi_i T.$$

The clumps are approximately independent, and of small duration relative to the waiting time between clumps. This means that κ is approximately a sample mean of an *iid* sample, and is independent of Y.

If N is the number of visits to state i, and we are interested in estimating some quantity associated with that state, then N will be the sample size for that estimate, which would have variance $\tau = \sigma^2/N$ for some constant σ^2.

This leads to the problem of bounding the maximum of $X_1, \ldots, X_m$, assuming that $var[X_i] = \sigma^2/n_i$ for random sample sizes n_i (we will content ourselves with modeling the average clump size κ as a constant). Suppose $n_i \sim pois(\lambda)$. We will make use of the following theorem, Proposition 1 of Glynn (1987).

Theorem 8.6 *Suppose $X \sim pois(\lambda)$ and $0 \leq n < \lambda$. Then*

$$P(X \leq n) \leq P(X = n)(1 - (n/\lambda))^{-1}. \tag{8.25}$$

Under completely uniform sampling $n_i = \lambda$, we would have the optimal allocation, with $var[X_i] = \sigma^2/\lambda$, and in particular, $\sigma^2_{max} = \sigma^2/\lambda$. Of course, we must anticipate that

$\sigma^2_{max} = \sigma^2/n_{min} > \sigma^2/\lambda$, where $n_{min} = \min_i n_i$. We then regard a bound on the expected maximum as conditional on the random quantity σ_{max}, that is,

$$E[M_m \mid \sigma_{max}] \leq \sigma_{max}\sqrt{2\log(m)}. \tag{8.26}$$

We would therefore like to develop a bound of the form

$$P\big(\sigma^2_{max} \geq \eta\sigma^2/\lambda\big) = P(1/n_{min} \geq \eta/\lambda),$$

for $\eta > 1$. Using Theorem 8.6 and Boole's inequality we have, for $\eta > 1$ (and assuming for convenience that λ/η is an integer),

$$\begin{aligned} P(1/n_{min} \geq \eta/\lambda) &= P(n_{min} \leq \lambda/\eta) \\ &\leq m\left(\frac{\eta}{\eta-1}\right)\frac{\lambda^{\lambda/\eta}}{(\lambda/\eta)!}\exp(-\lambda) \\ &\leq m\left(\frac{\eta}{\eta-1}\right)\frac{\lambda^{\lambda/\eta}}{\left(\frac{\lambda/\eta}{e}\right)^{\lambda/\eta}\sqrt{2\pi\lambda/\eta}}\exp(-\lambda) \\ &= m\left(\frac{\eta}{\eta-1}\right)\frac{1}{\sqrt{2\pi\lambda/\eta}}\exp(-\lambda+\lambda/\eta+(\lambda/\eta)\log(\eta)). \end{aligned}$$

We then note the inequality $\log(1+x) \leq x - \frac{x^2}{2(1+x)^2}$ for $x \geq 0$, which when applied to the preceding inequality yields

$$P(1/n_{min} \geq \eta/\lambda) \leq m\left(\frac{\eta}{\eta-1}\right)\frac{1}{\sqrt{2\pi\lambda/\eta}}\exp\left(-\frac{\lambda}{\eta}\frac{(\eta-1)^2}{2\eta^2}\right). \tag{8.27}$$

Thus, since λ increases proportionally with the total number of transitions T, the probability that the maximum variance σ^2_{max} exceeds σ^2/λ by a fixed ratio η decreases exponentially with T, so that the random variation of the sample size does not significantly affect the distribution of the maximum of $X_1, \ldots, X_m$ in comparison to a fixed sample size allocation. This will be demonstrated in Chapter 15.

8.6 BOUNDS FOR MAXIMA BASED ON THE FIRST TWO MOMENTS

We have seen that the existence of an MGF implies a bound on the rate of growth of the higher order moments, and so it is worth considering what bounds may be placed when this condition cannot be assumed. The case considered here will be of set of m RVs of general dependence structure possessing finite means and variances μ_i, σ_i^2, with no further restriction on higher order moments. In contrast to the order $O(\log(m)^{1/2})$ bound obtainable under for normal, or 'sufficiently normal', distributions, an order $O(m^{1/2})$ bound has been established, first for the *iid* case in Gumbel (1954) and Hartley and David (1954), and extended to dependent RVs in Arnold and Groeneveld (1979) and Aven (1985). A method of attaining tighter bounds from modifications of these earlier techniques is proposed in Bertsimas et al. (2006). We summarize below the approach taken in Aven (1985).

We will make use of the following lemma:

Lemma 8.1 *For any sequence of random variables $X_1, X_2, \ldots$ and $p \geq 1$*

$$E[M_n] \leq \sum_{i \leq n} E[|X_i|^p]^{1/p}, \tag{8.28}$$

where $M_n = \max_{i \leq n} X_i$.

Proof Clearly, $E[M_n] \leq \sum_{i=1}^{n} E[|X_i|]$. Recall that by Jensen's inequality $E[|X_i|^p]^{1/p}$ is increasing in p, which completes the proof. ///

The following theorem is due to Aven (1985) (Theorem 2.1), and we offer a proof for the $j=0$ case.

Theorem 8.7 *For any finite sequence of random variables $X_1, X_2, \ldots X_m$ possessing finite second moments the following inequalities hold:*

$$\bar{M} \leq \bar{\mu} + (m-1)^{1/2}\left(\bar{\sigma_j^2} + \nu\right)^{1/2}, \quad j = 0, 1, \ldots, m, \tag{8.29}$$

$$\bar{M} \leq \mu_{max} + (m-1)^{1/2}\left(\bar{\sigma_j^2}\right)^{1/2}, \quad j = 0, 1, \ldots, m. \tag{8.30}$$

Proof First, assume $\mu_i = 0$ for all i. Then for any real number c, by Lemma 8.1

$$E[M] = E[M + c] - c \leq \left(\sum_{i=1}^{m} E[(X_i + c)^2]\right)^{1/2} - c = m^{1/2}(\bar{\sigma_0^2} + c^2)^{1/2} - c.$$

It is easily verified that minimizing over c yields

$$E[M] \leq (m-1)^{1/2}(\bar{\sigma_0^2})^{1/2}.$$

The proof of (8.30) for general μ_i for $j = 0$ follows by noting

$$E[M] \leq \mu_{max} + E\left[\max_i (X_i - \mu_i)\right].$$

To prove (8.29) replace the sequence $\tilde{X}$ by a random permutation $\tilde{X}^*$ of $\tilde{X}$. The maximum of $\tilde{X}^*$ and of $\tilde{X}$ are identical. The elements of $\tilde{X}^*$ have identical marginal distributions with mean μ_{max} and variance $\bar{\sigma_0^2} + \nu$, so that (8.29) follows after applying (8.30) to $\tilde{X}^*$. ///

The proof of Theorem 8.7 for $j \geq 1$ may be found in Aven (1985) and follows similar arguments.

It is interesting to note that (8.29) follows from (8.30). However, this does not mean that (8.30) is uniformly sharper. The relative advantages of the two bounds will be discussed in more detail for a special case in Section 15.1. We will make a few general observations here.

8.6.1 Stability

The reliance of Theorem 8.7 on the first two moments limits stability results, but permits a power law to be obtained using Markov's inequality. Assume X_i are positive, and to fix ideas, assume the marginal distributions of X_i are equal. Using (8.30) we have

$$\bar{M} \leq \bar{M}^* = E[X_1] + (m-1)^{1/2} var[X_1]^{1/2}.$$

We may also apply (8.30) to bound the expected value of $M^k = \max_{i \leq m} X_i^k$, assuming the appropriate moments exist. Using the method of (8.17) we have

$$\begin{aligned} P(M/\bar{M}^* \geq \eta) &\leq \frac{1}{\eta^k} \frac{E\left[X_1^k\right] + (m-1)^{1/2} var[X_1^k]^{1/2}}{\left(E\left[X_1\right] + (m-1)^{1/2} var[X_1]^{1/2}\right)^k} \\ &\approx \frac{1}{\eta^k} \frac{1}{(m-1)^{(k-1)/2}} \frac{var[X_1^k]^{1/2}}{var[X_1]^{k/2}}, \end{aligned} \tag{8.31}$$

where the approximation holds if $\max(E[X_1], E[X_1^k]) \ll (m-1)^{1/2}$ for unit variance. This bound can be effective if $var[X_1^k]^{1/2}/var[X_1]^{k/2}$ is not too large. On interesting feature of (8.31) is that, as with the examples of Section 8.4, the tail probabilities decrease as the maximum order m increases when $k > 1$.

Part II

General theory of approximate iterative algorithms

Chapter 9

Background – linear convergence

The next three chapters form the central portion of this volume, in particular Chapter 10, which defines the approximate iterative algorithm (AIA) on Banach and Hilbert spaces, and establishes the main convergence results. Following this, Chapter 11 considers the problem of determining the optimal design of an AIA when approximation tolerances can be controlled. In such cases, it can be advantageous to start the AIA with a coarse approximation, which would reduce computation cost, gradually refining the approximation as the iterations proceed. The precise convergence rates derived in Chapter 10 can be used to determine near optimal rates of approximation refinement, the problem considered in Chapter 11.

The purpose of this chapter is to introduce some technical results relating to the properties of linearly convergent sequences. In some cases, in addition to verifying that a sequence α_i converges to 0 at a given rate, it may be necessary to establish that it does so 'smoothly', that is, the decrease of α_i must be regular and not too rapid. In a sense, this type of condition is analgous to the requirement that a real-valued function f on $\mathbb{R}$ possess a bounded derivate. In many applications of interest, this may not be realistic, for example, when α_i is stochastic. In such a case, it will suffice to construct an *envelope* $d_i \geq \alpha_i$, where d_i possesses the required smoothness properties. This is the subject of Sections 9.2 and 9.3.

The general theory of AIAs developed in Chapter 10 is largely a consequence of a straightforward idea presented in Section 9.4. In particular, the essential point of l'Hôpital's rule holds for series and sequences, as well as for real-valued functions and their derivatives. Viewed this way, it becomes natural to specify regularity conditions for sequences similar to those specified for real-valued functions in almost all theorems.

9.1 LINEAR CONVERGENCE

Suppose we are given a sequence of positive constants $\tilde{\rho} = \{\rho_k, k \geq 1\}$. Define the *product kernel*

$$\lambda^{k,j} = \begin{cases} 1 & ; \ j = 0 \\ \prod_{i=k-j+1}^{k} \rho_i & ; \ j \geq 1 \end{cases}$$

for any $k \geq j \geq 0$. In general, $\lambda^{k,j} = \lambda^{k,k} / \lambda^{k-j,k-j}$. Also, since $\lambda^{k,1} = \rho_k$, $\tilde{\rho}$ uniquely defines the kernel $\lambda^{k,j}$. For convenience, the superscripts may be omitted in a reference to a product kernel.

We have identified various modes of linear convergence in Section 2.1.7, namely

$$\lambda\{a_k\} = \lim_k a_{k+1}/a_k \quad \text{and} \quad \hat{\lambda}\{a_k\} = \lim_k a_k^{1/k},$$

with superscripts l, u denoting the respective limit infimum and limit supremum. For the kernel $\lambda^{k,k}$ these conditions are equivalently defined by the ρ_k sequences

$$\lambda\{\lambda^{k,k}\} = \rho \Leftrightarrow \lim_k \rho_k = \rho \quad \text{and}$$

$$\hat{\lambda}\{\lambda^{k,k}\} = \rho \Leftrightarrow \lim_k \left(\prod_{i=1}^{k} \rho_i\right)^{1/k} = \rho, \tag{9.1}$$

with analogous conditions for the upper and lower limits. Thus, the two modes of convergence are defined by convergence either of ρ_k to ρ or of the convergence of the cumulative geometric means of the sequence $\rho_1, \rho_2, \ldots$ to ρ, respectively.

The quantities $\lambda\{a_k\}$, $\hat{\lambda}\{a_k\}$ are interpretable as a rate of convergence. The following lemma is easily verified:

Lemma 9.1 *Let $\{a_k\}$ be a positive sequence. If a_k possess a finite positive limit then $\lambda\{a_k\} = \hat{\lambda}\{a_k\} = 1$. If a_k is nonincreasing then $\lambda^u\{a_k\}, \hat{\lambda}^u\{a_k\} \leq 1$, and if a_k is nondecreasing then $\lambda^l\{a_k\}, \hat{\lambda}^l\{a_k\} \geq 1$.*

Clearly, the first mode implies the second:

Lemma 9.2 *For $\rho > 0$ the implications*

$$\lambda^l\{\lambda^{k,k}\} = \rho \Rightarrow \hat{\lambda}^l\{\lambda^{k,k}\} \geq \rho \tag{9.2}$$

$$\lambda^u\{\lambda^{k,k}\} = \rho \Rightarrow \hat{\lambda}^u\{\lambda^{k,k}\} \leq \rho \tag{9.3}$$

hold, and therefore

$$\lambda\{\lambda^{k,k}\} = \rho \Rightarrow \hat{\lambda}\{\lambda^{k,k}\} = \rho.$$

Proof If $\lambda^l\{\lambda^{k,k}\} = \rho$ then for any positive $\epsilon < \rho$ we must have $k_\epsilon < \infty$ for which $\rho_k \geq (\rho - \epsilon)$ for all $k \geq k_\epsilon$. This implies that $\lambda^{k,k} \geq (\rho - \epsilon)^{k - k_\epsilon} \lambda^{k_\epsilon, k_\epsilon}$ for all $k \geq k_\epsilon$, in turn implying $\hat{\lambda}^l\{\lambda^{k,k}\} \geq \rho - \epsilon$ for all $\epsilon > 0$. Then (9.2) follows by allowing ϵ to approach 0, and (9.3) follows using a similar argument. ///

The distinction between statements $\lambda\{\lambda^{k,k}\} = \rho$ and $\hat{\lambda}\{\lambda^{k,k}\} = \rho$ is that the former assumes that the linear convergence rate of $\lambda^{k,k}$ approaches a constant ρ, while the latter states only that ρ is an average rate. For example, if ρ_k alternates between ρ and 1, then $\lambda\{\lambda^{k,k}\}$ would not exist, but we would have $\hat{\lambda}\{\lambda^{k,k}\} = \rho^{1/2}$. This latter case will sometimes occur in applications of interest.

Lemma 9.3 *Suppose $\{a_k\}$ is a sequence which converges to 0. Then*

$$\lim_k k^{-1} \sum_{i=1}^{k} \left|\log(1 + a_i)\right| = 0. \tag{9.4}$$

If in addition $\sum_{i\geq 1} |a_i| < \infty$ *then*

$$\sum_{i\geq 1} \left|\log(1+a_i)\right| < \infty. \tag{9.5}$$

Proof By Taylor's approximation theorem, for all $x \in (-1/2, 1/2)$ we have $|\log(1+x)| \leq 2|x|$. Therefore, for any $\epsilon \in (0, 1/2)$ there exists finite k_ϵ such that $a_k \leq \epsilon$ and therefore $|\log(1+a_k)| \leq 2\epsilon$ for all $k \geq k_\epsilon$. Setting $K_\epsilon = \sum_{i=1}^{k_\epsilon - 1} \left|\log(1+a_i)\right| < \infty$, we have $\sum_{i=1}^{k} \left|\log(1+a_i)\right| \leq K_\epsilon + k2\epsilon$, from which it follows $\limsup_k k^{-1} \sum_{i=1}^{k} \left|\log(1+a_i)\right| \leq 2\epsilon$. Then (9.4) follows by allowing ϵ to approach 0.

To complete the lemma, using the same constants we may write

$$\sum_{i=1}^{k} \left|\log(1+a_i)\right| \leq 2 \sum_{i=k_\epsilon}^{k} |a_i| + K_\epsilon \leq 2 \sum_{i\geq 1} |a_i| + K_\epsilon < \infty.$$

Then (9.5) follows by allowing $k \to \infty$. ///

We will sometimes need to know if a kernel $\lambda^{k,j}$ retains its convergence properties when vanishing terms δ_k are added to the sequence $\tilde{\rho}$.

Lemma 9.4 *Suppose product kernels* $\lambda^{k,j}$ *and* $\lambda_\delta^{k,j}$ *are defined by sequences* $(\rho_1, \rho_2, \ldots)$ *and* $(\rho_1 + \delta_1, \rho_2 + \delta_2, \ldots)$. *For* $\rho > 0$ *the following implications hold:*

(*i*) *If* $\delta_k \to 0$ *then*

$$\lambda^l\{\lambda^{k,k}\} = \rho \text{ or } \lambda^u\{\lambda^{k,k}\} = \rho \Rightarrow \lambda^l\{\lambda_\delta^{k,k}\} = \rho \text{ or } \lambda^u\{\lambda_\delta^{k,k}\} = \rho \tag{9.6}$$

(*ii*) *If* $\delta_k/\rho_k \to 0$ *then*

$$\hat{\lambda}^l\{\lambda^{k,k}\} = \rho \text{ or } \hat{\lambda}^u\{\lambda^{k,k}\} = \rho \Rightarrow \hat{\lambda}^l\{\lambda_\delta^{k,k}\} = \rho \text{ or } \hat{\lambda}^u\{\lambda_\delta^{k,k}\} = \rho \tag{9.7}$$

Proof First, (9.6) follows directly from (9.1). To verify (9.7) write

$$k^{-1} \log(\rho_k + \delta_k) = k^{-1}\left(\log(\rho_k) + \log(1 + \delta_k/\rho_k)\right)$$

then apply (9.4) of Lemma 9.3. ///

The condition for (*ii*) of Lemma 9.4 can be weakened to the constraint that the cumulative geometric mean of the sequence $1 + \delta_1/\rho_1, 1 + \delta_2/\rho_2, \ldots$ approaches 1, but we will be most interested in the case of vanishing δ_k.

Existence of a limit $\lambda\{\lambda^{k,k}\} = \rho$ does not imply the stronger limit $\lambda^{k,k} = \Omega(\rho^k)$. We may construct a sequence $\rho_k = a_k \rho$ for which $a_k \to 1$, but for which $\prod_{i=1}^{k} a_k$ is unbounded as $k \to \infty$, which provides a counterexample. Conditions for the equivalence of the two limits are given in the following lemma.

Lemma 9.5 *If $\rho_k = \rho + \delta_k$ where $\rho > 0$ and $\sum_{i\geq 1} |\delta_i| < \infty$ then $\lambda^{k,k} = \Omega(\rho^k)$.*

Proof We may write

$$\lambda^{k,k} = \exp\left(\sum_{i=1}^{k} \log(\rho + \delta_i)\right) = \exp(k\log(\rho) + \sum_{i=1}^{k} \log(1 + \delta_i/\rho)).$$

This gives

$$\lambda^{k,k}/\rho^k = \exp\left(\sum_{i=1}^{k} \log(1 + \delta_i/\rho)\right).$$

The lemma is proven after noting that by hypothesis, using Lemma 9.3 we may conclude that the partial sums $\sum_{i=1}^{k} \log(1 + \delta_i/\rho)$ posses a finite limit. ///

9.2 CONSTRUCTION OF ENVELOPES – THE NONSTOCHASTIC CASE

We next show that if we are given a kernel λ for which is convergent in the geometric mean, we may construct upper and lower envelopes with stronger linear convergence properties.

Theorem 9.1 *Consider a nonnegative sequence $\{a_k\}$.*

(*i*) *If for some $\rho \in (0,1]$ we have $\hat{\lambda}^u\{a_k\} = \rho$, then there exists a sequence $\{a^*_k\}$ for which $a^*_k \geq a_k$, $k \geq 1$, $\lambda^l\{a^*_k\} \geq \rho$ and $\hat{\lambda}\{a^*_k\} = \rho$.*

(*ii*) *If for some $\rho \in (0,1]$ we have $\hat{\lambda}^l\{a_k\} = \rho$, then there exists a sequence $\{a^*_k\}$ for which $a^*_k \leq a_k$, $k \geq 1$, $\lambda^u\{a^*_k\} \leq \rho$ and $\hat{\lambda}\{a^*_k\} = \rho$.*

Proof We first consider (*i*). First rewrite $a_k = c_k\rho^k$. If the sequence $\{c_k\}$ possesses a finite upper bound C then the objective is achieved by setting $a^*_k = C\rho^k$. Otherwise, set $c'_k = \max_{i\leq k} c_i$. We may assume without loss of generality that $c_1 \geq 1$. Set $\alpha_k = k^{-1}\log(c_k)$ and $\alpha'_k = k^{-1}\log(c'_k)$. Since $c'_1 = c_1 \geq 1$ and c'_k is nondecreasing we must have $\alpha'_k \geq 0$. Then for $k > 1$, α'_k may be equivalently written as

$$\alpha'_k = \max\left(k^{-1}\log(c_k), k^{-1}\log(c'_{k-1})\right),$$

and for $i > k > 1$ we have, similarly,

$$\alpha'_i = \max\left(i^{-1}\log(c_i), \ldots, i^{-1}\log(c_k), i^{-1}\log(c'_{k-1})\right).$$

This leads to the expression

$$\sup_{i\geq k} \alpha'_k = \max\left(\sup_{i\geq k} \alpha_i, (k-1)/k\alpha'_{k-1}\right). \tag{9.8}$$

By hypothesis, $\lim_{k\to\infty} \sup_{i\geq k} \alpha_i = 0$. We also note that the expression in (9.8) is nonincreasing in k, therefore α'_{k-1} possess a nonnegative limit K'. If $K' > 0$, then for some

$\rho' > 1$ we have $c'_k \geq (\rho')^k$ for all large enough k. However, since c_k is unbounded, there is an infinite subsequence of indices k' for which $c_{k'} = c'_{k'}$, which contradicts the hypothesis $\limsup_k k^{-1}\log(a_k) = \log(\rho)$, implying $K' = 0$. Hence, if we set $a^*_k = c'_k \rho^k$, we have $a^*_k \geq a_k$, where $\lim_k k^{-1}\log(a^*_k) = \log(\rho)$, and, since c'_k is increasing, we also must have $\lambda^l\{a^*_k\} \geq \rho$.

The proof of (*ii*) proceeds in a complementary manner. Again set $a_k = c_k\rho^k$. If the sequence $\{c_k\}$ possesses a positive finite lower bound C then the objective is achieved by setting $a^*_k = C\rho^k$. Otherwise, suppose c_k has a subsequence converging to zero, and set $c'_k = \min_{i\leq k} c_i$. We may assume without loss of generality that $c_1 \leq 1$. Set $\alpha_k = k^{-1}\log(c_k)$ and $\alpha'_k = k^{-1}\log(c'_k)$. Since $c'_1 = c_1 \leq 1$ and c'_k is nonincreasing we must have $\alpha'_k \leq 0$. Then for $k > 1$, α'_k may be equivalently written as

$$\alpha'_k = \min\left(k^{-1}\log(c_k), k^{-1}\log(c'_{k-1})\right),$$

and for $i > k > 1$ we have, similarly,

$$\alpha'_i = \min\left(i^{-1}\log(c_i), \ldots, i^{-1}\log(c_k), i^{-1}\log(c'_{k-1})\right).$$

This leads to the expression

$$\inf_{i\geq k} \alpha'_k = \min\left(\inf_{i\geq k} \alpha_i, (k-1)/k\alpha'_{k-1}\right). \tag{9.9}$$

By hypothesis, $\lim_{k\to\infty} \inf_{i\geq k} \alpha_i = 0$. We also note that the expression in (9.9) is nondecreasing in k, therefore α'_{k-1} possess a nonpositive limit K'. If $K' < 0$, then for some $\rho' < 1$ we have $c'_k \geq (\rho')^k$ for all large enough k. However, since c_k is not bounded away from 0, there is an infinite subsequence of indices k' for which $c_{k'} = c'_{k'}$, which contradicts the hypothesis $\liminf_k k^{-1}\log(a_k) = \log(\rho)$, implying $K' = 0$. Hence, if we set $a^*_k = c'_k\rho^k$, we have $a^*_k \leq a_k$, where $\lim_k k^{-1}\log(a^*_k) = \log(\rho)$, and, since c'_k is decreasing, we also must have $\lambda^u\{a^*_k\} \leq \rho$. ///

The following upper envelopes arise naturally. Relevant conditions are given in the following theorem.

Theorem 9.2 *The following bounds hold for $\rho > 0$:*

(*i*) *If $\rho_k \leq \rho + \delta_k$ where $\rho > 0$ and $\sum_{i=1}^k |\delta_i| < \infty$ then there exists a finite constant K for which $\lambda^{k,j} \leq K\rho^j$ for all k, j.*

(*ii*) *If $\lambda^u\{\lambda^{k,k}\} = \rho$ then for all $\epsilon > 0$ there exists finite K_ϵ such that $\lambda^{k,j} \leq K_\epsilon(\rho+\epsilon)^j$ for all k, j.*

(*iii*) *If $\hat{\lambda}^u\{\lambda^{k,k}\} = \rho$ then for all $\epsilon > 0$ there exists finite j_ϵ and $K_{\epsilon,j}$ and k_ϵ such that $\lambda^{k,k-j} \leq K_\epsilon(\rho+\epsilon)^{k-j}$ for all $k \geq j_\epsilon$.*

Proof (*i*) Following the proof of Lemma 9.5 we may write

$$\lambda^{k,j} = \rho^j \exp\left(\sum_{i=k-j+1}^{k} \log(1+\delta_i/\rho)\right).$$

If the hypothesis holds then by Lemma 9.4 the summation $\sum_{i=k-j+1}^{k} \log(1+\delta_i/\rho))$ has a finite upper bound over all $j \leq k$ so (i) holds.

To prove (ii) note that for any $\epsilon \in (0, 1-\rho)$ we may select finite index i_ϵ for which $|\delta_i| < \epsilon$ for all $i > i_\epsilon$. Define $\rho_\epsilon^u = \max\{\rho + \epsilon, \max_{i=1,\ldots,i_\epsilon} \rho_i\}$. The upper bound follows by setting $K_\epsilon = [\rho_\epsilon/(\rho+\epsilon)]^{i_\epsilon}$.

To prove (iii), note that under the hypothesis we may write $\lambda^{k,k} = (\rho + \epsilon_k)^k$, where $\epsilon_k \to 0$. ///

Example 9.1 *Consider for some* $\rho \in (0,1)$ *the sequence* $\tilde{\alpha} = \rho, \rho, \rho^2, \rho^2, \ldots$, *that is,* $\alpha_k = \rho^{\lfloor (k+1)/2 \rfloor}$. *We have* $\lambda^l\{\tilde{\alpha}\} = \rho < \lim_k k^{-1}\log(\alpha_k) = \rho^{1/2} < \lambda^u\{\tilde{\alpha}\} = 1$. *Following Theorem 9.1 we have representation* $\alpha_k = c_k(\rho^{1/2})^k$, *where* $c_k \in \{\rho^{1/2}, 1\}$, *so that* $\alpha_k^* = (\rho^{1/2})^k$ *defines an envelope of* $\tilde{\alpha}$.

9.3 CONSTRUCTION OF ENVELOPES – THE STOCHASTIC CASE

If a sequence $\{z_n\}$ is a mapping of *iid* sequences we may be able to deduce from the law of the iterated logarithm $\|z_n\| \leq d_n = Kn^{-1/2}\log\log n$ for some finite K, in which case $\lambda^l\{z_n\} = 1$. See Section 4.11.

The following approach can be used for more general stochastic sequences. Suppose $z_n \geq 0$, and $\tau_n = E[z_n]$. We will set $\sup_n E[(z_n/\tau_n)^p] = m_p$. Suppose we then have another positive sequence $\{\alpha_n\}$ which converges to 0. From Lemma 8.1 we have for $q \geq 1$

$$E\left[\max_{i \leq n} [\alpha_i(z_i/\tau_i)]^q\right] \leq m_q \sum_{i \leq n} \alpha_i^q.$$

If $\sum_{i \geq 1} \alpha_i^q < \infty$ and $m_q < \infty$ we may conclude that the random variable $Z^* = \sup_i [\alpha_i(z_i/\tau_i)]^q$ has a finite first moment, and so does $(Z^*)^{1/q}$, and we may write

$$z_n \leq \alpha_n^{-1}\tau_n(Z^*)^{1/q}, \qquad (9.10)$$

so that we may assert that if, for example, $\lambda^l\{\alpha_n^{-1}\tau_n\} = r$ then $wp1$ $\{z_n\}$ has an envelope $\tilde{d}$ for which $\lambda^l\{\tilde{d}\} = r$. We further note that if $m_q < \infty$ for all $q \geq 1$ we may set the convergence rate of α_n as slow as we wish, noting that $\alpha_n = n^{-1/(q+\epsilon)}$, for any $\epsilon > 0$, would permit a bound of the form (9.10) for which $(Z^*)^{1/q}$ has finite first moment.

9.4 A VERSION OF L'HÔPITAL'S RULE FOR SERIES

Much of the convergence theory of Chapter 10 depends on a version of l'Hôpital's rule due to Fischer (1983) (Theorem 9.1). A slightly modified version is given here.

Lemma 9.6 *Suppose* $\{a_n\}$, $\{b_n\}$ *are two real valued sequences such that* $b_{n+1} > b_n > 0$ *for all* n, *and* $\lim_{n\to\infty} b_n = \infty$. *Then*

$$\liminf_{n\to\infty} \frac{a_{n+1}-a_n}{b_{n+1}-b_n} \leq \liminf_{n\to\infty} \frac{a_n}{b_n} \leq \limsup_{n\to\infty} \frac{a_n}{b_n} \leq \limsup_{n\to\infty} \frac{a_{n+1}-a_n}{b_{n+1}-b_n}. \qquad (9.11)$$

Proof We first consider the first inequality of (9.11). This holds trivially if $\liminf_{n\to\infty} \frac{a_{n+1}-a_n}{b_{n+1}-b_n} = -\infty$. Otherwise assume $\liminf_{n\to\infty} \frac{a_{n+1}-a_n}{b_{n+1}-b_n} \geq L$ where $|L| < \infty$. Then for any $\epsilon > 0$ there exists N_ϵ such that $(a_{n+1} - a_n)/(b_{n+1} - b_n) > L - \epsilon$ for $n \geq N_\epsilon$. It follows that, for $k > 0$,

$$\frac{a_{N_\epsilon+k} - a_{N_\epsilon}}{b_{N_\epsilon+k} - b_{N_\epsilon}} > L - \epsilon.$$

We may then write

$$\begin{aligned}
a_{N_\epsilon+k} - a_{N_\epsilon+k-1} &> (L-\epsilon)(b_{N_\epsilon+k} - b_{N_\epsilon+k-1}) \\
a_{N_\epsilon+k-1} - a_{N_\epsilon+k-2} &> (L-\epsilon)(b_{N_\epsilon+k-1} - b_{N_\epsilon+k-2}) \\
&\vdots \\
a_{N_\epsilon+1} - a_{N_\epsilon} &> (L-\epsilon)(b_{N_\epsilon+1} - b_{N_\epsilon})
\end{aligned}$$

which after summing the inequalities is equivalent to

$$\frac{a_{N_\epsilon+k}}{b_{N_\epsilon+k}} > \frac{a_{N_\epsilon}}{b_{N_\epsilon+k}} + (1 - \frac{b_{N_\epsilon}}{b_{N_\epsilon+k}})(L - \epsilon).$$

Letting $k \to \infty$ gives

$$\liminf_{k\to\infty} \frac{a_{N_\epsilon+k}}{b_{N_\epsilon+k}} > L - \epsilon,$$

so that $\liminf_{n\to\infty} \frac{a_{n+1}-a_n}{b_{n+1}-b_n} \geq L$ implies $\liminf_{n\to\infty} \frac{a_n}{b_n} \geq L$ by letting ϵ approach 0. If $\liminf_{n\to\infty} \frac{a_{n+1}-a_n}{b_{n+1}-b_n} = L$, then the first inequality of (9.11) holds. If $\liminf_{n\to\infty} \frac{a_{n+1}-a_n}{b_{n+1}-b_n} = \infty$ then $\liminf_{n\to\infty} \frac{a_{n+1}-a_n}{b_{n+1}-b_n} \geq L$, and therefore $\liminf_{n\to\infty} \frac{a_n}{b_n} \geq L$ for all finite L, which completes the argument.

The final inequality of (9.11) holds using essentially the same argument. ///

The essential point of Lemma 9.6 is that for partial sums $s_n = \sum_{i=1}^n c_i$, the term $c_n = s_n - s_{n-1}$ is analogous to a derivative, and as for the convential l'Hôpital's rule, the limit of the ratio of partial sums is the same as the limit of the ratio of their 'derivatives'. Our interest in Lemma 9.6 will be in the evaluation of a type of convolution which will appear in several applications.

Lemma 9.7 *Let $s_n = \sum_{i=1}^n c_i$ be the partial sums for a nonnegative sequence c_n, and let τ_n be any positive sequence. Suppose $s_n \to \infty$. Then*

$$\begin{aligned}
\liminf_{n\to\infty} \tau_n \left[1 - (\tau_{n-1}c_{n-1})/(\tau_n c_n)\right] &\leq \liminf_{n\to\infty} \tau_n c_n/s_n \\
&\leq \limsup_{n\to\infty} \tau_n c_n/s_n \\
&\leq \limsup_{n\to\infty} \tau_n \left[1 - \frac{\tau_{n-1}c_{n-1}}{\tau_n c_n}\right]. \qquad (9.12)
\end{aligned}$$

Proof If we set $a_n = \tau_n c_n$ and $b_n = s_n$ then the hypothesis of Lemma 9.6 is satisfied, from which (9.12) follows after noting that $(a_n - a_{n-1})/(b_n - b_{n-1}) = \tau_n [1 - (\tau_{n-1}c_{n-1})/(\tau_n c_n)]$ and $a_n/b_n = \tau_n c_n/s_n$. ///

Two convolution limit bounds will be of particular importance.

Lemma 9.8 *Suppose we are given a product kernel λ, and a positive sequence of constants d_n with $\lambda^u\{\lambda^{n,n}\} = \rho$ and $\lambda^l\{d_n\} = r > \rho$. Then the convolution*

$$S_n = \sum_{i=1}^{n} \lambda^{n,n-i} d_i$$

possess the limit

$$1 - \rho/r \leq \liminf_{n\to\infty} d_n/S_n \leq \limsup_{n\to\infty} d_n/S_n \leq 1 - \lambda^l\{\lambda^{n,n}\}/\lambda^u\{d_n\}.$$

Proof In Lemma 9.7 set $\tau_n = 1$, $c_n = d_n/\lambda^{n,n}$ and therefore $s_n = S_n/\lambda^{n,n}$. Under the hypothesis we may conclude that $s_n \to \infty$, so that the hypothesis is satisfied, the result following froma direct application of (9.12). ///

Theorem 9.3 *Suppose we are given a product kernel λ defined by ρ_n, and a positive sequence of constants d_n. Define*

$$\alpha_n = (d_{n-1}/d_n)\rho_n, \quad n \geq 1,$$

and suppose for all large enough n α_n is nondecreasing with $\alpha_n < 1$. Then the convolution

$$S_n = \sum_{i=1}^{n} \lambda^{n,n-i} d_i$$

possess the limit

$$\liminf_{n\to\infty} \frac{d_n}{S_n(1-\alpha_n)} \geq 1. \tag{9.13}$$

Proof Define c_n and s_n as in Lemma 9.8, and set $\tau_n = (1 - \alpha_n)^{-1}$. By hypothesis α_n is positive and increasing for large enough n, at which point the lower bound of (9.12) is no smaller than 1. In this case $c_{n-1}/c_n = \alpha_n < 1$, so that c_n is increasing, therefore $s_n \to \infty$, from which (9.13) follows directly. ///

Chapter 10

A general theory of approximate iterative algorithms (AIA)

We are given a seminormed linear space $(\mathcal{V}, \|\cdot\|)$ on which a sequence of operators $\tilde{T} = T_1, T_2, \ldots$ is defined. We may take norms to be a special case of seminorms, the theory of this chapter generally applying in the same manner to each. We usually expect each operator to have a common fixed point $V^* = T_k V^*$, or an equivalence class of fixed points. At the very least, interest is in evaluating or approximating some fixed point V^* using the iteration algorithm:

$$\begin{aligned} V_0 &= v_0 \\ V_k &= T_k V_{k-1}, \quad k = 1, 2, \ldots, \end{aligned} \tag{10.1}$$

for a given initial solution $v_0 \in \mathcal{V}$. We always assume $\|v_0\| < \infty$ and $\|V^*\| < \infty$. The intention is that $V_k \in \mathcal{V}$ for all $k \geq 1$, and that the sequence converges to V^* in the given seminorm, that is, $\lim_k \|V_k - V^*\| = 0$. In many cases, the algorithm will be homogenous in the sense that $T_k = T$, so that $V_k = T^k v_0$, but it turns out that results obtainable for homogenous algorithms are extendible to the nonhomgenous case with little loss of generality, and much gain in applicability.

Our interest is in cases for which evaluation of T_k is not feasible, can only be evaluated with error or can be approximated with some advantage. In this case T_k in (10.1) may be replaced by approximate operator $\hat{T}_k$, giving *approximate iteration algorithm* (AIA)

$$\begin{aligned} V_0 &= v_0 \\ V_k &= \hat{T}_k V_{k-1} \\ V_k &= T_k V_{k-1} + U_k, \quad k = 1, 2, \ldots, \end{aligned} \tag{10.2}$$

setting $U_k = \hat{T}_k V_{k-1} - T_k V_{k-1}$, which we take to be the error resulting from the approximation. We may therefore define an *exact iterative algorithm* (EIA) as the duple $\mathcal{Q}_e = (\tilde{T}, V_0)$, where $V_0 \subset \mathcal{V}$ is the starting point. An approximate iterative algorithm will be based on an EIA $\mathcal{Q}_e$, but will also incorporate the approximate operators $\hat{T}_k$.

However, it is important to stress that we are not taking the point of view that an AIA is to be analyzed as an EIA after new operators have been substituted. This approach is certainly not uncommon, but has the disadvantage that any convergence properties of the AIA tend to depend on specific details of the approximation method. In contrast, the formulation implied by (10.2) would appear to permit the convergence properties of an AIA to be established using only properties of an approximation method which could be easily generalized.

How then do we characterize $\hat{T}_k$? Certainly, it must map $\mathcal{V}$ into $\mathcal{V}$, and we also expect that $\|\hat{T}_k V\| < \infty$ when $\|V\| < \infty$, as in the EIA. In addition, as a practical matter we must permit $\hat{T}_k$ to depend on process history, as would be the case in adaptive control applications. To do this define the *history process* $\tilde{H} = H_1, H_2, \ldots$ of an AIA as $H_k = (V_0, U_1, \ldots, U_{k-1})$ for $k > 1$ and $H_1 = (V_0)$, and let $\mathcal{H}_k$ be the set of all possible histories H_k. Then we may consider the mapping $\hat{T}_k : \mathcal{V} \times \mathcal{H}_k \to \mathcal{V}$, or alternatively, $\hat{T}_k$ represents a set of operators indexed by H_k.

It may also be the case that $\hat{T}_k$ must be regarded as stochastic. In this case, T_k would remain deterministic (if unknown), so that random outcomes would be expressible entirely in the history process. Accordingly, we construct a probability space $\mathcal{P} = (\Omega, \mathcal{F}, P)$, with the σ-fields $\mathcal{F}_k = \sigma(H_k)$ so that we may set $\mathcal{F} = \cup_i \mathcal{F}_i$. Then $\tilde{H}$ is a filtration process (Definition 4.2), with filtration $\mathcal{F}_1 \subset \mathcal{F}_2 \ldots$. Since H_k is $\mathcal{F}_k$-measurable, it follows that $V_k = \hat{T}_k V_{k-1} = T_k V_{k-1} + U_k$ is $\mathcal{F}_{k+1}$-measurable. There are two approachs to take at this point. It may be possible to identify an event $E \in \mathcal{F}$ which occurs with a probability of 1, or close to 1, in which case the AIA can be analyzed as a deterministic algorithm with a history satisfying those conditions implied by E. Alternatively, $\mathcal{V}$ itself may be defined on a probability space, with seminorm $\|\cdot\|$ based on stochastic L^p norms. Thus, while the specific design of the approximate operator $\hat{T}_k$ would naturally be the central concern for any particular AIA, from the point of view of developing a general theory for the characterization of AIAs, the more fruitful approach will be to regard an AIA as an EIA coupled with an error history process, that is, $\mathcal{Q}_a = (\tilde{T}, \tilde{H})$.

The material in this and the subsequent chapter is organized around three questions:

(Q1) If an EIA converges to a fixed point, under what conditions does an AIA also converge to the same fixed point?

(Q2) How can the approximation error of an AIA be expressed in terms of the sequence U_k, $k \geq 1$?

(Q3) If a range of approximate operators T_ϵ exists, and the iterations of an AIA are successively refined, can a rate of refinement be determined which minimizes computation time?

(Q1) has been discussed in the literature, and it is been long established (as in Ostrowski (1964) or Ortega and Rheinboldt (1967)) that contractive algorithms evaluated with vanishing rounding error converge to the intended fixed point. In our notation, $T \equiv T_k$ is contractive, and $\lim_k \|U_k\| = 0$. A general theory for (Q2) is not as well developed, and results tend to be available for specific models only. To the best of the author's knowledge, very little theory exists with which to answer (Q3).

In this chapter, a general theory of the convergence properties of (10.2) will be developed. When $\tilde{T}$ is collectively contractive, this theory unites (Q1) and (Q2), in the sense that convergence of $\|U_k\|$ to zero suffices to establish convergence of (10.2) to V^*, at a rate directly determined by that of $\|U_k\|$. It will also be possible to characterize convergence in terms of relative errors $\|U_k\| / \|V_{k-1}\|$, which will often be the most natural expression for the error of approximate operator evaluation. In the subsequent chapter, the theory will be used to determine general principles with which to answer (Q3).

10.1 A GENERAL TOLERANCE MODEL

The definition of an AIA is closely tied to an EIA, the convergence properties of which are assumed known, and are incorporated into the analysis. The AIA is then considered to be a noisy implementation of an EIA. The question is how the errors modify the EIA convergence properties. It will therefore be useful to define a *general tolerance model* which will summarize the approach.

Usually, we are concerned with the convergence of the quantity $\| V_k - V^* \|$, which we refer to as the *algorithm error*. Within the kth iteration the quantity $\| \hat{T}_k V_{k-1} - T_k V_{k-1} \|$ is referred to as the *operator error*. If a bound on operator error $\epsilon_k \geq \| \hat{T}_k V_{k-1} - T_k V_{k-1} \|$ exists, then ϵ_k is referred to as the *operator tolerance*. Similarly, any bound on algorithm error, denoted $\eta_k \geq \| V_k - V^* \|$ is referred to as *algorithm tolerance*. We can also imagine the evolution $V'_0, V'_1, \ldots$ of the EIA (assuming it shares the same initial solution $V'_0 = V_0$ as the AIA), which has its' own algorithm error $\| V'_k - V^* \|$. We refer to any bound $B\alpha_k \geq \| V'_k - V^* \|$, $B > 0$, as the *exact algorithm tolerance*. It may also be convenient to refer to α_k independently as the *exact algorithm tolerance rate* (our intention is to reserve the term *tolerance* for any bound imposed or guaranteed by a specific algorithm on a corresponding error).

It will be useful to express the algorithm tolerance using the form

$$\| V_k - V^* \| \leq \eta_k = B\alpha_k + u_k, \quad k \geq 1 \tag{10.3}$$

where u_k is interpretable as a bound on the cumulative effect of the operator errors. We then refer to u_k as the *approximation tolerance*. In this way, the algorithm tolerance η_k can be compared to the exact algorithm tolerance $B\alpha_k$. Clearly, we expect that the convergence rate of an AIA can never be strictly better than that of the EIA, but it might be asymptotically equivalent. In fact, the main conclusion of Chapter 11 is that this should generally be a goal in the design of AIAs.

The model (10.3) becomes useful when the approximation tolerance u_k can be expressed in terms of the operator tolerance, since this quantity is often tractible, and is in fact widely used in the literature. That this is generally possible is one of the important consequences of the theory developed below.

10.2 EXAMPLE: A PRELIMINARY MODEL

The general approach will be illustrated using a simple approximation model for a homogenous algorithm based on ρ-contractive operator T. First, suppose we may set a constant operator tolerance $\epsilon_k = \epsilon > 0$, for all $k \geq 1$. Following Isaacson and Keller (1966), we have

$$\| V_k - V^* \| \leq \| TV_{k-1} - V^* \| + \| V_k - TV_{k-1} \| \leq \rho \| V_{k-1} - V^* \| + \epsilon,$$

which applied iteratively gives

$$\|V_k - V^*\| \leq \rho^k \|V_0 - V^*\| + \epsilon(1-\rho)^{-1}, \tag{10.4}$$

so that the algorithm tolerance is dominated by ϵ. In particular, we achieve algorithm tolerance (10.3) with $u_k = \epsilon(1-\rho)^{-1}$ and $\alpha_k = \rho^k$.

Now suppose ϵ_k vanishes as $k \to \infty$. This gives an algorithm tolerance

$$\|V_k - V^*\| \leq \rho^k \|V_0 - V^*\| + \rho^k \sum_{i=1}^{k} \rho^{-i}\epsilon_i, \tag{10.5}$$

using an argument similar to (10.4). To fix ideas, consider the special case $\epsilon_k = Kr^k$ for some $K > 0$, $r \in (0,1)$. Then (10.5) becomes

$$\|V_k - V^*\| \leq \begin{cases} \rho^k \|V_0 - V^*\| + Kr\frac{\rho^k - r^k}{\rho - r}; & r \neq \rho \\ \rho^k \|V_0 - V^*\| + Kk\rho^k; & r = \rho \end{cases}.$$

Note that the algorithm tolerance corresponds to the structure of (10.3) by setting approximation tolerance $u_k = Kr(\rho^k - r^k)/(\rho - r)$ or $u_k = Kk\rho^k$ for $r \neq \rho$ or $r = \rho$ respectively. Thus, we have $u_k = \Omega(\max(\epsilon_k, \rho^k))$ when $\epsilon_k \neq \Omega(\rho^k)$. When $\epsilon_k = \Omega(\rho^k)$ we have $\log(u_k) = \Omega(k\log(\rho))$, but it is important to note that $\epsilon_k = o(u_k)$. Stated more directly, we have algorithm tolerance of order

$$\|V_k - V^*\| \leq \begin{cases} O(\max(\epsilon_k, \rho^k)); & r \neq \rho \\ O(k\rho^k); & r = \rho \end{cases}$$

for the range of algorithms explicitly considered. It turns out that this form is representative of the general case, particularly with respect to its direct relationship to the operator tolerance. This fact forms the basis of our theory of AIAs.

10.3 MODEL ELEMENTS OF AN AIA

As suggested by the inequality (10.5) there are three elements required to characterize the behavior of an AIA. The first we generally assume to be given, that is, the convergent behavior of the EIA (the first term of the upper bound of (10.5)). The remaining elements contribute to the second term. These are the sequence of operator tolerances for the operator errors U_k, and then the Lipschitz properties of the sequence $\tilde{T}$, which we discuss in the next section.

10.3.1 Lipschitz kernels

If we are given an operator sequence $\tilde{T}$, this will define an *operator kernel* $T^{k,j} = T_k \cdots T_{k-j+1}$, from which the EIA $V_k = T^{k,k}V_0$ follows. We say $\bar{\lambda} = \{\bar{\lambda}^{k,j}; k \geq j \geq 0\}$ is a *Lipschitz kernel* for $\tilde{T}$ if $T^{k,j}$ possesses Lipschitz constant $\bar{\lambda}^{k,j}$ (the definition extends in the obvious way to a pseudo-Lipschitz kernel). We generally set $\bar{\lambda}^{k,0} = 1$. By convention, we write $\bar{\lambda}_1 \le \bar{\lambda}_2$ if $\bar{\lambda}_1^{k,j} \le \bar{\lambda}_2^{k,j}$ for all k,j (in which case we say $\bar{\lambda}_2$ *dominates* $\bar{\lambda}_1$). We also write $\bar{\lambda} < \infty$ if each $\bar{\lambda}^{k,j} < \infty$. A Lipschitz kernel $\bar{\lambda}_1^{k,j}$ of $\tilde{T}$ is *sharp* if $\bar{\lambda}_1^{k,j} \le \bar{\lambda}_2^{k,j}$ for any other Lipschitz kernel $\bar{\lambda}_2^{k,j}$.

A Lipschitz kernel has a *product form* (or, is a product kernel, Section 9.1), if there is a sequence $\tilde{\lambda} = \lambda_1, \lambda_2, \ldots$ for which $\bar{\lambda}^{k,j} = \lambda^{k,j}$ for all k,j. In this case we say that $\bar{\lambda}^{k,j}$ is the (product form) kernel induced by $\tilde{\lambda}$. In this case, for convenience, the notation $\tilde{\lambda}$ will refer equivalently to the sequence and to the kernel itself. Suppose $\bar{\lambda}$ is a Lipschitz kernel for $\tilde{T}$. Write the sequence $\tilde{\lambda}_{pf} = \bar{\lambda}^{1,1}, \bar{\lambda}^{2,1}, \ldots$. If $\bar{\lambda}_{pf}$ is the product form kernel induced by $\tilde{\lambda}_{pf}$ then it is also a Lipschitz kernel for $\tilde{T}$. It is important to note that if $\bar{\lambda}$ is sharp, $\bar{\lambda}_{pf}$ need not be, as would be expected in, for example, multistage contractive models. However, even in such cases it may be useful to construct a product form kernel which dominates $\bar{\lambda}$.

One other form of kernel will be of interest. Suppose the EIA is based on a single operator T, for which the asymptotic contraction rate may be considerably smaller than a J-stage contraction rate (see Section 7.1). Then we may set $\bar{\lambda}^{k,i} = \beta_i$ for all $k \ge i \ge 0$, where β_J is a contraction constant for T^J.

10.3.2 Lipschitz convolutions

We next generalize the second term of the upper bound in (10.5) by defining an *Lipschitz convolution* of order p:

$$\mathcal{I}_k^p(\tilde{d}, \bar{\lambda}) = \sum_{i=1}^{k} [\bar{\lambda}^{k,k-i} d_i]^p, \quad k \ge 1,$$

for positive constant p, Lipschitz kernel $\bar{\lambda}$ and sequence $\tilde{d} = (d_1, d_2, \ldots)$. Our objective will be quite precise, to construct an algorithm tolerance model of the form

$$\eta_k^p = B\alpha_k^p + \mathcal{I}_k^p(\tilde{d}, \bar{\lambda}), \tag{10.6}$$

where $\bar{\lambda}$ is a Lipschitz kernel for the EIA, $B\alpha_k$ is the tolerance for the EIA and $\tilde{d}$ is a sequence of operator tolerances. Note that α_k often will, but need not, be related to $\bar{\lambda}^{k,1}$. We will generally be able to set $\tilde{d} = (\|U_1\|, \|U_2\|, \ldots)$, but it will sometimes be necessary to use a smooth upper envelope of $\|U_k\|$, so the model is best generalized in this way. See the discussion of envelopes in Chapter 9.

Once (10.6) is established, the next step is to evaluate $\mathcal{I}_k^p(\tilde{d}, \bar{\lambda})$. This turns out to be relatively simple to do using the extension of l'Hôpital's rule for series given in Section 9.4. The comparison is then made to the exact algorithm tolerance rate α_k to resolve the convergence properties of η_k itself.

Generally, an AIA will be defined on a Banach space. There will sometimes be a natural Hilbert space structure to the AIA, which, as will be seen, can generally yield sharper approximation bounds. This will typically arise when the operator error can be modeled as a martingale (see Section 4.4). In this case, approximation errors are signed, so that the resulting averaging effect can be exploited to improve the approximation bound. In either case, in (10.6) $p=1$ will apply to the Banach space model, while $p=2$ is used for the Hilbert space model.

To study the asymptotic properties of $\mathcal{I}_k^p(\tilde{d},\bar{\lambda})$ we will make use of the following quantities:

$$I_k^p(\tilde{d},\bar{\lambda}) = d_k^{-p}\mathcal{I}_k^p(\tilde{d},\bar{\lambda}), \quad k \geq 1,$$
$$\bar{I}^p(\tilde{d},\bar{\lambda}) = \limsup_{k\to\infty} I_k^p(\tilde{d},\bar{\lambda}),$$
$$\hat{I}^p(\tilde{d},\bar{\lambda}) = \sup_k I_k^p(\tilde{d},\bar{\lambda}).$$

It is also worth noting the simple iteration rule for product form Lipschitz kernels:

$$\mathcal{I}_k^p(\tilde{d},\bar{\lambda}) = d_k + \lambda_k \mathcal{I}_{k-1}^p(\tilde{d},\bar{\lambda}). \tag{10.7}$$

10.4 A CLASSIFICATION SYSTEM FOR AIAs

As is well known, the Lipschitz properties of an EIA are crucial to determining its convergence properties. The same is obviously true for an AIA, but we have defined it as $\mathcal{Q}_a = (\tilde{T},\tilde{H})$, a composition of the original EIA and an error history. The point of this is that the Lipschitz properties of the original EIA are sufficient to establish the convergence properties of the AIA, provided the error terms can be suitably bounded. Accordingly, we have the following definitions:

Definition 10.1 (Pseudo-Lipschitz EIA/AIA) *A sequence of operators $\tilde{T}$ on vector space $\mathcal{V}$ is* pseudo-Lipschitz *if all T_k possess a common set of fixed points V^* for which $\|V^*\| < \infty$; and each T_k possesses a pseudo-Lipschitz constant λ_k for all $V \in \mathcal{V}$. An EIA $\mathcal{Q}_e = (\tilde{T}, V_0)$ is pseudo-Lipschitz if $\tilde{T}$ is pseudo-Lipschitz and $\|V_0\| < \infty$. An AIA $\mathcal{Q}_a = (\tilde{T},\tilde{H})$ is pseudo-Lipschitz if $\tilde{T}$ is pseudo-Lipschitz, $\|V_0\| < \infty$, $U_k \in \mathcal{V}$, and $\|U_k\| < \infty$ for all k.*

Definition 10.2 (Lipschitz EIA/AIA) *An sequence of operators $\tilde{T}$ on vector space $\mathcal{V}$ is* Lipschitz *if for each T_k there exists at least one $V' \in \mathcal{V}$ for which $\|V'\| < \infty$ and $\|T_k V'\| < \infty$; and there is a Lipschitz kernel $\bar{\lambda} < \infty$ such that each compound operator $T^{k,j}$ possesses Lipschitz constant $\bar{\lambda}^{k,j}$. An EIA $\mathcal{Q}_e = (\tilde{T}, V_0)$ is Lipschitz if $\tilde{T}$ is Lipschitz and $\|V_0\| < \infty$. An AIA $\mathcal{Q}_a = (\tilde{T},\tilde{H})$ is Lipschitz if $\tilde{T}$ is Lipschitz, $\|V_0\| < \infty$, $U_k \in \mathcal{V}$, and $\|U_k\| < \infty$ for all k.*

The assumptions in Definitions 10.1 and 10.2 are enough to guarantee that for an AIA, $\|V_k\| < \infty$. For Definition 10.1, if $\|V_{k-1}\| < \infty$, then

$$\begin{aligned}
\|V_k\| &= \|T_k V_{k-1} - V^* + V^* + U_k\| \\
&\leq \lambda_k \|V_{k-1} - V^*\| + \|V^*\| + \|U_k\| \\
&\leq \lambda_k \|V_{k-1}\| + (1+\lambda_k)\|V^*\| + \|U_k\| < \infty.
\end{aligned}$$

For Definition 10.2, there exists V' with $\|V'\| < \infty$ and $\|T_k V'\| < \infty$, so that if $\|V_{k-1}\| < \infty$, then

$$\begin{aligned}
\|V_k\| &= \|T_k V_{k-1} - T_k V' + T_k V' + U_k\| \\
&\leq \bar{\lambda}^{k,1}\|V_{k-1} - V'\| + \|T_k V'\| + \|U_k\| \\
&\leq \bar{\lambda}^{k,1}\|V_{k-1}\| + \bar{\lambda}^{k,1}\|V'\| + \|T_k V'\| + \|U_k\| < \infty.
\end{aligned}$$

The definitions assume that $\|V_0\| < \infty$, so that the assertion that $\|V_k\| < \infty$ follows by induction. The same remark of course applies to EIAs by omitting the $\|U_k\|$ terms.

It is of some interest to note that while Definition 10.1 relies on the notion of a fixed point, Definition 10.2 does not. In fact, our foundational theory does not assume that an EIA satisfying that definition is formally a fixed point algorithm, merely that it converges to some solution. The analysis and conclusions are the same whether or not that solution happens to be some fixed point of interest.

We may also classify EIAs and AIAs in terms of their contraction properties.

Definition 10.3 (Contractive EIA/AIA) *An EIA is contractive if (1) it is pseudo-Lipschitz and $\bar{\lambda}_{k,1} \leq \rho$ for some $\rho < 1$, for all $k \geq 1$; or (2) it is Lipschitz and $\bar{\lambda}^{N,k} \leq K\rho^k$ for some finite K and $\rho < 1$, for all $N \geq k \geq 0$. A pseudo-Lipschitz or Lipschitz AIA is contractive if its associated EIA is contractive.*

Definition 10.4 (Weakly Contractive EIA/AIA) *An EIA is weakly contractive if it is pseudo-Lipschitz or Lipschitz and $\lim_{k\to\infty} \bar{\lambda}^{N+k,k} = 0$ for all $N \geq 0$. A pseudo-Lipschitz or Lipschitz AIA is weakly contractive if its associated EIA is weakly contractive.*

Definition 10.5 (Nonexpansive EIA/AIA) *An AIA is nonexpansize if it is pseudo-Lipschitz or Lipschitz and $\bar{\lambda}^{N,k} \leq 1$ for all $N \geq k \geq 0$. A pseudo-Lipschitz or Lipschitz AIA is nonexpansize if its associated EIA is nonexpansize.*

For greater clarity, the contraction property of Definition 10.3 will sometimes be referred to as *strong contraction*, in contrast with the weak contraction property of Definition 10.4.

A few points are worth noting. A contractive pseudo-Lipschitz EIA is assumed to be single stage contractive, while a contractive Lipschitz EIA may be J-stage contractive. The reason for this is that for pseudocontractive EIAs the product form representation of $\bar{\lambda}$ becomes important. Of course, if we are given an operator that is J-stage pseudo-contractive, it would always be possible to consider the EIA defined by iterations of the J-step operator T^J, yielding a pseudo-Lipschitz contractive EIA according to Definition 10.3.

In addition, to say that an AIA is contractive, weakly contractive or nonexpansive does not imply that these properties hold for the approximate operators $\hat{T}_k$, and in many interesting cases this implication will not hold. Of course, it will sometimes be useful to be able to state that this implication does hold, but in general the important point is to study how the contractive properties of the EIAs interact with the error terms, so these definitions formally apply to the original operators $\tilde{T}$ only.

10.4.1 Relative error model

The objective is to resolve the convergence properties of an AIA by imposing sufficiently tight bounds on the sequence $\|U_k\|$. However, the assumption that an absolute operator tolerance $d_k \geq \|U_k\|$ may be imposed will sometimes be quite restrictive, since in many applications we cannot expect the quantity $\|T_k V_{k-1} - TV_{k-1}\|$ to be uniformly bounded over all possible V_{k-1}. In Almudevar (2008) an approximation model was developed which assumed a relative bound $\|U_k\| \leq b_k \|V_{k-1}\|$ for some sequence b_k. Accordingly, we develop a general error model which incorporates both absolute and relative errors, expressed in the following assumption. Here, the model is generalized to permit the use of alternative seminorms $\|\cdot\|_0$ which are dominated by $\|\cdot\|$.

(ARE) In AIA $\mathcal{Q}_a = (\tilde{T}, \tilde{H})$ the following bound holds

$$\|U_k\| \leq a_k + b_k \|V_{k-1}\|_0, \quad k \geq 1 \tag{10.8}$$

for some sequence of nonnegative finite constants $\{a_k; k \geq 1\}$, $\{b_k; k \geq 1\}$, where $\|\cdot\|_0$ is a seminorm for which $\|V\|_0 \leq \kappa_0 \|V\|$ for all $V \in \mathcal{V}$ for some finite constant κ_0.

The use of a seminorm in (10.8) is motivated by the following observation. Suppose there exists $\nu_0, \nu_1 \in \mathcal{V}$ and constant β such that for all $V \in \mathcal{V}$ and scalars a we have

$$T_k(V + a\nu_0) = T_k V + \beta a \nu_1 \quad \text{and} \quad \hat{T}_k(V + a\nu_0) = \hat{T}_k V + \beta a \nu_1, \quad k \geq 1. \tag{10.9}$$

Suppose we may assert

$$\left\| T_k V - \hat{T}_k V \right\| \leq L \|V\|, \quad V \in \mathcal{V}.$$

If (10.9) holds, then for any scalar a

$$\left\| T_k V - \hat{T}_k V \right\| = \left\| T_k(V + a\nu_0) - \hat{T}_k(V + a\nu_0) \right\| \leq L \inf_a \|V + a\nu_0\|.$$

Then set $\|V\|_0 = \inf_a \|V + a\nu_0\|$, which is a seminorm (see Section 6.3.1), which, by construction, is dominated by $\|\cdot\|$. Possibly, ν_0 is an eigenvector with associated eigenvalue β, in which case $\nu_0 = \nu_1$, but we may encounter models for which (10.9) holds without this assumption.

It is important to note that as long as the boundedness alone of an AIA satisfying (ARE) can be established there need not be an important difference between the relative and absolute operator tolerance models, since if $\sup_k \|V_k\|_0 \leq M$ we may claim absolute bound $\|U_k\| \leq a_k + b_k M$, and this observation will sometimes be quite useful.

If we may claim that the AIA converges to V^* it may be useful to define the sequence $\tilde{d}^* = (d_1^*, d_2^*, \ldots)$ by

$$d_k^* = a_k + b_k \|V^*\|_0.$$

If we first suppose that $\|V^*\|_0 > 0$, then d_k^* is, asymptotically, an absolute operator tolerance, since $a_k + b_k \|V_{k-1}\|_0 \leq a_k + b_k \|V^*\|_0 + b_k \|V_{k-1} - V^*\|_0 = d_k^* + o(b_k)$. On the other hand, if $\| V^* \|_0 = 0$, the bound will depend on the relative convergence rates of the various components, but will be resolvable with the theory developed here.

We may directly establish an important principle, that if an AIA is contractive, convergence to fixed point V^* follows if the operator tolerance converges to zero either absolutely, or relative to the iterates of the algorithm itself. In terms of the relative error model, this follows if $a_k \to_k 0$ and $b_k \to 0$. This is summarized in the next theorem.

Theorem 10.1 *Suppose a pseudo-Lipschitz AIA $\mathcal{Q}_a = (\tilde{T}, \tilde{H})$ is contractive, with product kernel $\tilde{\lambda}$ satisfying $\lim_{k\to\infty} \lambda_k = \rho < 1$. If for any $s \geq 0$ and $\delta \in [0, 1-\rho)$ the limit*

$$\limsup_{n\to\infty} \| U_n \| / \max(s, \| V_{n-1} \|) = \delta$$

holds, then

$$\limsup_{n\to\infty} \| V_n - V^* \| \leq \delta(1-\rho-\delta)^{-1} \max(s, \| V^* \|).$$

Proof Fix $\epsilon > 0$ such that $\rho + \delta + \epsilon < 1$. Then there exists N_ϵ such that $\| U_n \| \leq (\delta + \epsilon/2) \max(s, \| V_{n-1} \|)$ and $\lambda_n \leq \rho + \epsilon/2$ for all $n \geq N_\epsilon$. Then

$$\begin{aligned}
\| V_n - V^* \| &\leq \| T^{n,1} V_{n-1} - T^{n,1} V^* \| + \| U_n \| \\
&\leq (\rho + \epsilon/2) \| V_{n-1} - V^* \| + (\delta + \epsilon/2) \max(s, \| V_{n-1} \|) \\
&\leq (\rho + \epsilon/2) \| V_{n-1} - V^* \| + (\delta + \epsilon/2) \max(s, \| V_{n-1} - V^* \| + \| V^* \|) \\
&= \max((\rho + \epsilon/2) \| V_{n-1} - V^* \| + (\delta + \epsilon/2) s \\
&\quad + (\rho + \delta + \epsilon) \| V_{n-1} - V^* \| + (\delta + \epsilon/2) \| V^* \|) \\
&\leq (\rho + \delta + \epsilon) \| V_{n-1} - V^* \| + (\delta + \epsilon/2) \max(s, \| V^* \|),
\end{aligned} \tag{10.10}$$

for all $n \geq N_\epsilon$. Applying (10.10) iteratively,

$$\begin{aligned}
\| V_{N_\epsilon + n} - V^* \| &\leq (\rho + \delta + \epsilon)^n \| V_{N_\epsilon} - V^* \| + (\delta + \epsilon) \sum_{i=0}^{n-1} (\rho + \delta + \epsilon)^i \max(s, \| V^* \|) \\
&\leq (\rho + \delta + \epsilon)^n \| V_{N_\epsilon} - V^* \| + (\delta + \epsilon)(1 - \rho - \delta - \epsilon)^{-1} \max(s, \| V^* \|)
\end{aligned}$$

hence

$$\limsup_{n\to\infty} \| V_n - V^* \| \leq (\delta + \epsilon)(1 - \rho - \delta - \epsilon)^{-1} \max(s, \| V^* \|)$$

which proves the theorem by making ϵ arbitrarily small. ///

10.5 GENERAL INEQUALITIES

From Theorem 10.1 the convergence of a contractive AIA to solution V^* follows from the convergence to zero in $\|\cdot\|$ of the error terms U_k, both in the absolute and relative sense. The remaining task is to estimate this rate of convergence, or to bound the algorithm tolerance when $\|U_k\|$ does not vanish. We consider separately pseudo-Lipschitz and Lipschitz models, each in Banach and Hilbert spaces. The pseudo-Lipschitz model is further specialized to the relative error model.

Let $\tilde{U} = (U_1, U_2, \ldots)$ and $\tilde{d}_U = \tilde{U} = (\|U_1\|, \|U_2\|, \ldots)$. We consider each case in turn.

Theorem 10.2 (Pseudo-Lipschitz AIAs) *For any pseudo-Lipschitz AIA,*

$$\|V_k - V^*\| \leq \lambda^{k,k}\|V_0 - V^*\| + \mathcal{I}_k^1(\tilde{d}_U, \tilde{\lambda}), \quad k \geq 1. \tag{10.11}$$

Proof We may write

$$\begin{aligned}\|V_k - V^*\| &\leq \|T_k V_{k-1} - V^* + U_k\| \\ &\leq \lambda_k \|V_{k-1} - V^*\| + \|U_k\|.\end{aligned}$$

The argument is then applied to the quantity $\|V_{k-1} - V^*\|$ of the upper bound, after which sufficient iterations yield (10.11). ///

Theorem 10.3 (Lipschitz AIAs) *For any Lipschitz AIA,*

$$\|V_k - V'\| \leq \|T^{k,k} V_0 - V'\| + \mathcal{I}_k^1(\tilde{d}_U, \bar{\lambda}), \quad k \geq 1.$$

Proof For any operator T' with Lipschitz constant L, and any fixed $V_0 \in \mathcal{V}$ we may write

$$\begin{aligned}\|T'(W + U) - V'\| &\leq \|T'(W + U) - T'W + T'W - V'\| \\ &\leq \|T'W - V'\| + \|T'(W + U) - T'W\| \\ &\leq \|T'W - V'\| + L\|W + U - W\| \\ &\leq \|T'W - V'\| + L\|U\|.\end{aligned} \tag{10.12}$$

Then for any $k \geq 3$ we may write

$$\begin{aligned}\|V_k - V'\| &\leq \|T_k V_{k-1} - V'\| + \|U_k\| \\ &= \|T_k(T_{k-1} V_{k-2} + U_{k-1}) - V'\| + \|U_k\|.\end{aligned}$$

Applying (10.12) to the first term of the upper bound gives

$$\begin{aligned}\|V_k - V'\| &\leq \|T_k V_{k-1} - V'\| + \|U_k\| \\ &\leq \|T^{k,2} V_{k-2} - V'\| + \|U_k\| + \bar{\lambda}^{k,1}\|U_{k-1}\|.\end{aligned}$$

A second application gives

$$\begin{aligned}\|V_k - V'\| &\leq \|T^{k,2}(T_{k-2}V_{k-3} + U_{k-2}) - V'\| + \|U_k\| + \bar{\lambda}^{k,1}\|U_{k-1}\| \\ &\leq \|T^{k,3}V_{k-3} - V'\| + \|U_k\| + \bar{\lambda}^{k,1}\|U_{k-1}\| + \bar{\lambda}^{k,2}\|U_{k-2}\|,\end{aligned}$$

and repeating the argument yields (10.12). ///

It is interesting to note that for contractive pseudo-Lipshitz AIAs, under relative error model (ARE) if $b_k \to_k 0$, then the constant b_k can be incorporated into the kth contraction constant λ_k . To see this, given a product form Lipschitz kernel $\tilde{\lambda}$ we may introduce the adjusted kernel:

$$\lambda_b^{k,j} = \begin{cases} 1 & ; j=0 \\ \prod_{i=k-j+1}^{k}(\lambda_i + b_i) & ; j \geq 1 \end{cases}$$

denoted $\tilde{\lambda}_b$.

This is summarized in the following theorem:

Theorem 10.4 (Pseudo-Lipschitz AIAs with Relative Error) *For any pseudo-Lipschitz AIA, under assumption (ARE)*

$$\|V_k - V^*\| \leq \lambda_b^{k,k}\|V_0 - V^*\| + \mathcal{I}_k^1(\tilde{d}^*, \tilde{\lambda}_b), \quad k \geq 1. \tag{10.13}$$

where $\lambda_b^{k,k}$ is the product kernel generated by $(\lambda_1 + \kappa_0 b_1, \lambda_2 + \kappa_0 b_2, \ldots)$.

Proof Following Theorem 10.2 we write

$$\begin{aligned}\|V_k - V^*\| &\leq \lambda_k\|V_{k-1} - V^*\| + \|U_k\| \\ &\leq \lambda_k\|V_{k-1} - V^*\| + a_k + b_k \left\|V_{k-1}\right\|_0 \\ &\leq \lambda_k\|V_{k-1} - V^*\| + a_k + b_k \left\|V^*\right\|_0 + b_k \left\|V_{k-1} - V^*\right\|_0 \\ &\leq \lambda_k\|V_{k-1} - V^*\| + a_k + b_k \left\|V^*\right\|_0 + b_k \left\|V_{k-1} - V^*\right\|_0 \\ &\leq \lambda_k\|V_{k-1} - V^*\| + d_k^* + b_k\kappa_0 \left\|V_{k-1} - V^*\right\|_0 \\ &\leq (\lambda_k + \kappa_0 b_k)\|V_{k-1} - V^*\| + d_k^*.\end{aligned}$$

An iterative argument similar to that used in Theorem 10.2 yields (10.13). ///

10.5.1 Hilbert space models of AIAs

Within the structure permitted by normed linear spaces, the methods discussed here depend on the ability to bound the terms $\|U_k\|$. As will be seen, in some algorithms which are naturally expressed as AIAs, this will be too restrictive. However, when Hilbert space structure is present, a strengthening of this bound may be possible. In practice, this will occur when we may rely on the error terms U_k to fluctuate in sign about a central quantity, in a manner permitting their cumulative effect to obey a law of large numbers. This can be expected to result in weaker convergence criteria than would be predicted from the bounds $\|U_i\|$ alone. Martingale theory will play an important role here (Section 4.4).

The averaging requirement on the error terms will take the form of the following nested conditions

(H1) For some V', for all $k \geq 1$ we have $\langle U_k, T_k V_{k-1} - V' \rangle \leq 0$.
(H2) For some V', for all $k \geq 1$ and $m \geq 1$ we have:

$$\langle T^{k+m,m}(T_k V_{k-1} + U_k) - T^{k+m,m}(T_k V_{k-1}), T^{k+m,m}(T_k V_{k-1}) - V' \rangle \leq 0.$$

Note that conditions (H1)–(H2) are specified for a particular V'. This will usually be a fixed point V^* (but see next section for an exception). The intention is that V' not depend on $\tilde{U}$. These conditions represent precisely the conditions required for the subsequent lemmas. However, they would follow from a more intuitive set of conditions which we now discuss. Recall the AIA history process $H_k = (V_0, U_1, \ldots, U_{k-1})$, $k \geq 1$. For a fixed operator sequence we may consider V_{k-1} or $T_k V_{k-1}$ to be a mapping of H_k. It may be the case that in the iteration $V_k = T_k V_{k-1} + U_k$ certain interesting properties of U_k may not depend on history H_k. Accordingly, we let $\mathcal{V}[\mathcal{H}_k]$ denote all mappings from $\mathcal{H}_k$ to $\mathcal{V}$.

We will motivate a new set of conditions by considering a simple linear space $\mathcal{V}$ of real valued random variables with inner product $\langle X, Y \rangle = E[XY]$ for $X, Y \in \mathcal{V}$, which induces the L^2 norm $\|X\|_2 = E[X^2]^{1/2}$. Suppose U_k is a martingale adapted to H_k, in particular $E[U_k \mid H_k] = 0$, so that

$$\langle U_k, X \rangle = E[U_k X] = E[E[U_k X \mid H_k]] = E[X E[U_k \mid H_k]] = 0$$

whenever $X \in \mathcal{V}[\mathcal{H}_k]$, so that (H1) will hold, assuming V' is fixed, which holds if $V' \in \mathcal{V}[\mathcal{H}_1]$.

In considering (H2) suppose $T(x)$ is a linear function, and that $X, Y \in \mathcal{V}[\mathcal{H}_k]$. Then by a similar argument

$$\langle T(X + U_k) - T(X), Y \rangle = 0$$

so that (H2) will hold.

Thus (H1)–(H2) may be restated in a somewhat more restricted but also more intuitive way. These conditions will be used in the subsequent development, but may always be replaced by (H1a)–(H2a):

(H1a) For all $k \geq 1$ we have $\langle U_k, V \rangle \leq 0$ for any $V \in \mathcal{V}[\mathcal{H}_k]$.
(H2a) For all $k \geq 1$ and $m \geq 1$ we have $\langle T^{k+m,m}(V' + U_k) - T^{k+m,m}(V'), V'' \rangle \leq 0$ for any $V', V'' \in \mathcal{V}[\mathcal{H}_k]$.

Versions of Theorems 10.2 and 10.3 for Hilbert space may now developed using analagous arguments. We will assume that any norm $\|\cdot\|$ is induced by an inner product $\langle \cdot, \cdot \rangle$.

Theorem 10.5 (Pseudo-Lipschitz AIAs on Hilbert Spaces) *For any pseudo-Lipschitz AIA with an inner product satisfying (H1) with $V' = V^*$,*

$$\| V_k - V^* \|^2 = [\lambda^{k,k}]^2 \| V_0 - V^* \|^2 + \mathcal{I}_k^2(\tilde{d}_U, \tilde{\lambda}), \quad k \geq 1. \tag{10.14}$$

Proof We may write

$$\begin{aligned} \| V_k - V^* \|^2 &= \| T_k V_{k-1} - V^* + U_k \|^2 \\ &= \| T_k V_{k-1} - V^* \|^2 + \| U_k \|^2 + 2\langle U_k, T_k V_{k-1} - V^* \rangle \\ &\leq \lambda_k^2 \| V_{k-1} - V^* \|^2 + \| U_k \|^2. \end{aligned}$$

Then (10.14) holds after sufficient iterations. ///

Theorem 10.6 (Lipschitz AIAs on Hilbert Spaces) *For any Lipschitz AIA with an inner product satisfying (H1)–(H2) for some V'*

$$\| V_k - V' \|^2 = \| T^{k,k} V_0 - V' \|^2 + \mathcal{I}_k^2(\tilde{d}_U, \bar{\lambda}), \quad k \geq 1. \tag{10.15}$$

Proof For any operator T' with Lipschitz constant L, and any fixed $V_0 \in \mathcal{V}$ we may write

$$\begin{aligned} &\| T'(W + U) - V_0 \|^2 \\ &\quad = \| T'(W + U) - T'W + T'W - V_0 \|^2 \\ &\quad \leq \| T'W - V_0 \|^2 + L^2 \| U \|^2 + 2\langle T'(W + U) \\ &\quad - T'W, T'(W + U) - T'W - V_0 \rangle \|. \end{aligned} \tag{10.16}$$

Then for any $k \geq 3$ we may write

$$\begin{aligned} &\| V_k - V' \|^2 \\ &\quad = \| T_k V_{k-1} - V' + U_k \|^2 \\ &\quad = \| T_k V_{k-1} - V' \|^2 + \| U_k \|^2 + 2\langle U_k, T_k V_{k-1} - V' \rangle \\ &\quad \leq \| T_k (T_{k-1} V_{k-2} + U_{k-1}) - T_k T_{k-1} V_{k-2} + T_k T_{k-1} V_{k-2} - V' \|^2 + \| U_k \|^2 \\ &\quad \leq \| T_k T_{k-1} V_{k-2} - V' \| + \| T_k (T_{k-1} V_{k-2} + U_{k-1}) - T_k T_{k-1} V_{k-2}) \|^2 + \| U_k \|^2 \\ &\qquad + 2\langle T_k (T_{k-1} V_{k-2} + U_{k-1}) - T_k T_{k-1} V_{k-2}, T_k T_{k-1} V_{k-2} - V' \rangle \\ &\quad \leq \| T^{k,2} V_{k-2} - V' \| + \| U_k \|^2 + [\bar{\lambda}^{k,1}]^2 \| U_{k-1} \|^2. \end{aligned}$$

A second application of (10.16) gives

$$\begin{aligned}\|V_k - V'\|^2 &\leq \|T^{k,2}V_{k-2} - V'\| + \|U_k\|^2 + [\bar{\lambda}^{k,1}]^2\|U_{k-1}\|^2 \\ &\leq \|T^{k,2} - V'\| + \|U_k\|^2 + [\bar{\lambda}^{k,1}]^2\|U_{k-1}\|^2 + [\bar{\lambda}^{k,2}]^2\|U_{k-2}\|^2.\end{aligned}$$

Sufficient iterations of the argument yields (10.15). ///

Finally, we consider the relative error model in Hilbert spaces.

Theorem 10.7 (Pseudo-Lipschitz AIAs on Hilbert Spaces with Relative Error) *Suppose we are given a pseudo-Lipschitz AIA with an inner product satisfying (H1) and product form Lipschitz kernel* $\bar{\lambda}$. *Suppose assumption (ARE) holds. Define the product form Lipschitz kernels* $\tilde{\lambda}_b^{(1)} = (\lambda_1 + b_1\kappa_0, \lambda_2 + b_2\kappa_0, \ldots)$ *and* $\tilde{\lambda}_b^{(2)} = (\lambda_1^2 + 2b_1^2\kappa_0^2, \lambda_2^2 + 2b_2^2\kappa_0^2, \ldots)$. *Then if* $b_k\kappa_0 \leq 2\lambda_k$ *for all* $k \geq 1$

$$\|V_k - V^*\|^2 \leq [\lambda_b^{k,k}]^2\|V_0 - V^*\|^2 + 2\mathcal{I}_k^2(\tilde{d}^*, \tilde{\lambda}_b^{(1)}) \quad k \geq 1. \tag{10.17}$$

and in general

$$\|V_k - V^*\|^2 \leq \left\{\prod_{i=1}^{k}(\lambda_i^2 + 2b_i^2)\right\}\|V_0 - V^*\|^2 + 2\mathcal{I}_k^1([\tilde{d}^*]^2, \tilde{\lambda}_b^{(2)}) \quad k \geq 1, \tag{10.18}$$

where $[\tilde{d}^*]^2$ *is obtained by raising each element of* $\tilde{d}^*$ *to the power 2.*

Proof Following Lemma 10.5 we may write:

$$\begin{aligned}\|V_k - V^*\|^2 &= \|T_kV_{k-1} - V^* + U_k\|^2 \\ &= \|T_kV_{k-1} - V^*\|^2 + \|U_k\|^2 + 2\langle U_k, T_kV_{k-1} - V^*\rangle \\ &\leq \lambda_k^2\|V_{k-1} - V^*\|^2 + (a_k + b_k\|V_{k-1}\|_0)^2 \\ &\leq \lambda_k^2\|V_{k-1} - V^*\|^2 + (d_k^* + b_k\|V_{k-1} - V^*\|_0)^2 \\ &\leq \lambda_k^2\|V_{k-1} - V^*\|^2 + 2b_k^2\|V_{k-1} - V^*\|_0^2 + 2(d_k^*)^2 \\ &\leq (\lambda_k^2 + 2b_k^2\kappa_0^2)\,\|V_{k-1} - V^*\|^2 + 2(d_k^*)^2.\end{aligned}$$

Then (10.18) is obtained after sufficient iteration, while (10.17) follows after noting that $(\lambda_k^2 + 2b_k^2\kappa_0^2) \leq (\lambda_k + b_k\kappa_0)^2$ under the given assumption. ///

It is not necessary to assume that the seminorm $\|\cdot\|_0$ used in definition (ARE) is induced by an inner product, or otherwise has Hilbert space properties.

The inequality (10.18) of Theorem 10.7 is uniformly sharper than (10.17) and does not require the additional assumption. The latter is included to emphasize the point that for both Banach and Hilbert space models accomodating a relative error model is largely a matter of adding the constants b_k to the original Lipschitz kernel.

The condition $b_k \leq 2\lambda_k$ will be met by almost any model we would consider for large enough k.

10.6 NONEXPANSIVE OPERATORS

Referring to a tolerance model of the form (10.6) it is clearly ideal that $\mathcal{I}^{1,k} \to_k 0$, since in this case if the EIA converges, then convergence of the AIA follows, with bounds on the convergence rate following naturally. As we will see, this will generally be the case when the AIA is contractive. However, convergence properties of the AIA may follow from those of the EIA under weaker conditions. To see this, we start with the following definition.

Suppose we are given a Banach space $(\mathcal{V}, \|\cdot\|)$. One important property of the contractive operator T on $\mathcal{V}$ is that there exists exactly one fixed point V^*, to which the sequence $T^n V_0$ converges for all $V_0 \in \mathcal{V}$. If we are given any general operator kernel $\tilde{T}$, it will be important to establish that $T^{n,n} V_0$ converges (in the norm) for all $V_0 \in \mathcal{V}$, and to characterize the set of all limits (which may be greater than 1). An additional issue arises with AIAs. The introduction of perturbations means that a larger class of sequences may need to be considered if $\tilde{T}$ is nonhomogeneous. Therefore, we say $\tilde{T}$ is *universally convergent* if the sequence $T^{n+N-1,n} V$ converges in the norm for all $N \geq 1$, $V \in \mathcal{V}$. If we were concerned with EIAs it may suffice for this condition to hold for $N = 1$ alone.

We define the *offset* operation on a sequence: if $\tilde{a} = (a_1, a_2, \ldots)$ then $\tilde{a}^{(N)} = (a_N, a_{N+1}, \ldots)$ (so that $\tilde{a}^{(1)} = \tilde{a}$). In particular, this can be applied to $\tilde{T}$, $\tilde{d}$ or $\tilde{d}_U$. This will be used in the following way. Suppose we consider a particular iteration of an AIA:

$$V_N = T_N V_{N-1} + U_N.$$

we may regard the sequence $V_N, V_{N+1}, \ldots$ as the N-*offset* AIA with starting point V_{N-1}, operator kernel $\tilde{T}^{(N)}$ and operator tolerances $\tilde{U}^{(N)}$. Of course, the offset AIA has the same limit as the original, but any of the general inequalities may be applied.

For a universally convergent operator kernel define the set

$$\mathcal{V}^*(\tilde{T}) = \left\{ \lim_{n \to \infty} T^{n+N-1,n} V : V \in \mathcal{V},\ N \geq 1 \right\},$$

and define the distance from any subset $\mathcal{V}' \subset \mathcal{V}$

$$D(V, \mathcal{V}') = \inf_{V' \in \mathcal{V}'} \|V - V'\|$$

Next, for any $V \in \mathcal{V}$ let $\bar{V}^{(N)} = \lim_{n \to \infty} T^{n+N-1,n} V \in \mathcal{V}^*(\tilde{T})$ be the limit associated with starting point V and operator sequence $T_N, T_{N+1}, \ldots$. Then define the uniform convergence bound

$$\delta_n = \sup_{V \in \mathcal{V}, N \geq 1} \|T^{n+N-1,n} V - \bar{V}^{(N)}\|.$$

If δ_n is finite, then the tolerance for any N-offset AIA from any starting point V is bounded by δ_n after n iterations.

10.6.1 Application of general inequalities to nonexpansive AIAs

A suitable general inequality may be applied to nonexpansive operator kernels, which possesses Lipschitz kernel $\bar{\lambda}^{k,j} \leq 1$ for all $k \geq j \geq 0$. Define

$$E_1^{N:n} = \sum_{i=1}^{n} \|U_{N+i-1}\|,$$

$$E_1^{N} = \sum_{i \geq N} \|U_i\|,$$

$$E_2^{N:n} = \sum_{i=1}^{n} \|U_{N+i-1}\|^2,$$

$$E_2^{N} = \sum_{i \geq N} \|U_i\|^2.$$

In this case the Lipschitz convolutions are bounded by

$$\mathcal{I}_n^1(\tilde{d}_U, \bar{\lambda}) \leq E_1^{1:n},$$

$$\mathcal{I}_n^2(\tilde{d}_U, \bar{\lambda}) \leq E_2^{1:n}.$$

We begin with the following lemma:

Lemma 10.1 *For a nonexpansive Lipschitz AIA,*

$$\|V_{n+N-1} - T^{n+N-1,n} V_{N-1}\| \leq E_1^{N:n} \tag{10.19}$$

for any $N, m \geq 1$. *For a nonexpansive Lipschitz AIA on an inner product space satisfying (H1a) and (H2a) we have*

$$\|V_{n+N-1} - T^{n+N-1,n} V_{N-1}\|^2 \leq E_2^{N:n}. \tag{10.20}$$

Proof We may apply Theorem 10.3 to the N-offset AIA, yielding

$$\|V_{n+N-1} - V'\| \leq \|T^{n+N-1,n} V_{N-1} - V'\| + E_1^{N:n}. \tag{10.21}$$

Then (10.19) follows by setting $V' = T^{n+N-1,n} V_{N-1}$.

It remains to apply Theorem 10.6 to the N-offset AIA with $V' = T^{n+N-1,n} V_{N-1}$. Note that (in the original AIA) $V_{N-1} \in \mathcal{V}[\mathcal{H}_k]$ for all $k \geq N$ and therefore so is

$T^{n+N-1,n}V_{N-1}$, for all $n \geq 1$. This means that (H1)–(H2) hold for each N-offset AIA with $V' = T^{n+N-1,n}V_{N-1}$, and (10.20) follows. ///

We now give the main theorem.

Theorem 10.8 *Suppose we are given a nonexpansive Lipschitz AIA $\mathcal{Q}_a = (\tilde{T}, \tilde{H})$, where $\tilde{T}$ is universally convergent. Then*

(*i*) *If $E_1^N \to_N 0$ then $D(V_n, \mathcal{V}^*(\tilde{T})) \to_n 0$.*
(*ii*) *The finite bound holds:*

$$D(V_{2n}, \mathcal{V}^*(\tilde{T})) \leq E_1^n + \delta_n.$$

(*iii*) *If $E_1^N \to_N 0$ and $\mathcal{V}^*(\tilde{T})$ is finite, the AIA possesses a limit.*

For a nonexpansive Lipschitz AIA on an inner product space satisfying (H1a) and (H2a) we have

(*i*)′ *If $\sqrt{E_2^N} \to_N 0$ then $D(V_n, \mathcal{V}^*) \to_n 0$.*
(*ii*)′ *The finite bound holds:*

$$D(V_{2n}, \mathcal{V}^*(\tilde{T})) \leq \sqrt{E_2^N} + \delta_n.$$

(*iii*)′ *If $\sqrt{E_2^N} \to_N 0$ and $\mathcal{V}^*(\tilde{T})$ is finite, the AIA possesses a limit.*

Proof Fix N. By Lemma 10.1

$$\|V_{n+N-1} - T^{n+N-1,n}V_{N-1}\| \leq E_1^{N:n}. \tag{10.22}$$

Then

$$\|V_{n+N-1} - \bar{V}^{(N)}\| \leq E_1^{N:n} + \|T^{n+N-1,n}V_{N-1} - \bar{V}^{(N)}\|. \tag{10.23}$$

Letting $n \to \infty$ gives

$$\limsup_{n\to\infty} \|V_n - \bar{V}^{(N)}\| \leq E_1^N.$$

Then part (*i*) is proven by noting that $D(V_n, \mathcal{V}^*(\tilde{T})) \leq \|V_n - \bar{V}^{(N)}\|$, then making ϵ_N arbitrarily small.

Part (*ii*) follows directly from (10.23) by substituting $N = n+1$.

To prove part (*iii*), let $\epsilon^* = \min_{V_1, V_2 \in^* \mathcal{V}^*(\tilde{T})} \|V_1 - V_2\|$. Then select N for which $E_1^N < \epsilon^*$, so that V_n must converge to $\bar{V}^{(N)}$.

To prove (*i*)′, (*ii*)′ and (*iii*)′ we note that by Lemma 10.1 the upper bound of (10.22) may be replaced by $\sqrt{E_2^{N:n}}$, then the remaining argument is identical. ///

10.6.2 Weakly contractive AIAs

The weakly contractive model is something of a hybrid. Nominally, it is nonexpansive. However, the convergence properties of the EIA rely on the convergence of $\lambda^{k,k}$ to zero. Essentially, the sequence of operators $\tilde{T}$ are contractive, but the contraction constant approaches 1, with $\lambda^{k,k}$ vanishing if the approach to 1 is at a slow enough rate. In a manner similar to Theorem 10.8 we may also conclude that a bound $\sum_{i\geq 1} \|U_i\| < \infty$ or $\sum_{i\geq 1} \|U_i\|^2 < \infty$ suffices for the convergence of an AIA. However, we will see that the weak contraction property can be exploited within the Lipschitz convolution itself to provide more general results, and in some cases precise convergence rates.

Theorem 10.9 *Suppose an AIA is pseudo-Lipschitz and weakly contractive. Suppose further that* $E_1^N \to_N 0$. *Then* $\lim_k \|V_k - V^*\| = 0$. *If in addition the AIA is defined an inner product space satisfying (H1) for* $V' = V^*$ *then* $E_2^N \to_N 0$ *implies* $\lim_k \|V_k - V^*\| = 0$.

Proof Let $E_N = \sum_{i>N} \|U_i\|$. Applying Theorem 10.2 to the N-offset AIA yields the bound

$$\|V_{N+k} - V^*\| = \lambda^{N+k,k}\|V_N - V^*\| + E_1^N$$

so that under the assumptions, $\limsup_{k\geq 1}\|V_{N+k} - V^*\| \leq E_1^N$. However, this holds for all N, so the lemma holds by noting $\lim_{N\to\infty} E_1^N = 0$. Under the hypothesis, we may apply Theorem 10.5 to complete the proof in a similar manner. ///

There exists exactly one fixed point for a weakly contractive EIA and this holds for the AIA. When only the nonexpansive property is given, a fixed point either need not exist, or may not be unique.

10.6.3 Examples

A number of EIAs are commonly used for determination of a fixed point $V^* = TV^*$ when T is nonexpansive. Some of the most commonly studied algorithms are defined by the following update rules:

$$\begin{aligned}
V_{k+1} &= TV_k && \text{(Picard iteration),}\\
V_{k+1} &= (1-\alpha)V_k + \alpha TV_k && \text{(Krasnoselskij iteration),}\\
V_{k+1} &= (1-\alpha_k)V_k + \alpha_k TV_k && \text{(Mann iteration),}\\
V_{k+1} &= (1-\alpha_k)V_k + \alpha_k TY_k, && \text{where}\\
Y_k &= (1-\beta_k)V_k + \beta_k TV_k && \text{(Ishikawa iteration).}
\end{aligned}$$

See Berinde (2007) for a source on this subject.

All the employed parameters may be, at least provisionally, assumed to be in the interval $[0,1]$, so that each iteration listed includes all the preceding ones as special cases. The distinctions are important with respect to the conditions on the parameters, operator T and on the structure of $(\mathcal{V}, \|\cdot\|)$, and also on the relative convergence rates. For this reason, there is an advantage to studying the various algorithms separately, even though they may be regarded as special cases of the Ishikawa iteration.

Define the compound operator $T_\alpha = (1-\alpha)I + \alpha T$, which defines a mapping $T_\alpha : \mathcal{V} \to \mathcal{V}$. Then all but the Ishikawa iteration can be written in the form:

$$\begin{aligned} V_{k+1} &= T_k V_k, \quad \text{where} \\ T_k &= T_{\alpha_k} \end{aligned} \tag{10.24}$$

for some sequence $\alpha_k \in [0,1]$. Clearly, T_α has the same fixed points as T. Also, if T possesses (pseudo-) Lipschitz constant λ, then T_α is easily shown to be (pseudo-) Lipschitz with constant $(1-\alpha) + \alpha\lambda$.

The Ishikawa iteration is based on operators written explicitly as

$$T_{\alpha,\beta} V = (1-\alpha)V + \alpha T[(1-\beta)V + \beta TV].$$

If T possesses (pseudo-) Lipschitz constant λ, then $T_{\alpha,\beta}$ possesses (pseudo-) Lipschitz constant $(1-\alpha) + \alpha(1-\beta)\lambda + \alpha\beta\lambda^2$.

Since a single evaluation of $T_{\alpha,\beta}$ involves two evaluations of T, it is appropriate to decompose the operator error into two sources, yielding AIA:

$$\begin{aligned} V_{k+1} &= (1-\alpha_k)V_k + \alpha_k T Y_k + \alpha_k U_k, \\ Y_k &= (1-\beta_k)V_k + \beta_k T V_k + \beta_k W_k. \end{aligned} \tag{10.25}$$

We may therefore represent an approximation evaluation of $T_{\alpha,\beta} V$ as

$$\hat{T}_{\alpha,\beta} V = (1-\alpha)V + \alpha T[(1-\beta)V + \beta TV + \beta W] + \alpha U, \tag{10.26}$$

leading to operator tolerance

$$\begin{aligned} &\left\| \hat{T}_{\alpha,\beta} V - T_{\alpha,\beta} V \right\| \\ &\quad \leq \alpha \|U\| + \alpha \|T[(1-\beta)V + \beta TV + \beta W] - T[(1-\beta)V + \beta TV]\| \\ &\quad \leq \alpha \|U\| + \beta \|W\|. \end{aligned} \tag{10.27}$$

Theorem 10.8 then applies directly, and in the notation of the hypothesis,

$$\begin{aligned} E_1^n &= \sum_{i \geq n} (\alpha_i \|U_i\| + \beta_i \|W_i\|), \\ E_2^n &= \sum_{i \geq n} (\alpha_i \|U_i\| + \beta_i \|W_i\|)^2. \end{aligned}$$

Suppose we are given fixed series α_k, β_k, $k \geq 1$. Construction of an AIA based on the approximate operator $\hat{T}_{\alpha,\beta}$ given in (10.26) leads to iterations of the form (10.25). We may compare the AIA to the EIA based on operator sequence $\tilde{T} = (T_{\alpha_1,\beta_1}, T_{\alpha_2,\beta_2}, \ldots)$. Under the conditions of Theorem 10.8 if the operator sequence $\tilde{T}$ is universally convergent, then the AIA converges to the solution space of the EIA for the Banach space, or Hilbert space, model if $E_1^n \to_n 0$, or $E_2^n \to_n 0$, respectively.

The requirement that $E_n^1 \to_n 0$ is a standard one (see Definition 6.1 of Berinde (2007), also Liu (1995), Osilike (1997), Deng and Li (2000), Liu (2001)). Requirements on the constants α_k, β_k for the convergence of the EIA were originally reported in Ishikawa (1974) as

$$0 \le \alpha_k \le \beta_k \le 1, \quad k \ge 1$$

$$\lim_{k\to\infty} \beta_k = 0$$

$$\sum_{k\ge 1} \alpha_k \beta_k = \infty,$$

under the stated hypothesis, that T is a Lipschitz map on a convex compact subset of a Hilbert space, satisfying the 'pseudocontractive' property (using another terminological convention):

$$\|V - W\| \le \|(1+r)(V - W) - r(TV - TW)\|, \quad \text{for all } V, W \in \mathcal{V}, r > 0.$$

As a general principle, when an EIA is convergent, under the contraction property an associated AIA is also convergent if the operator tolerance vanishes, while under the nonexpansive property the same holds if the summation of the operator tolerances is finite. We next examine an important intermediate case.

10.6.4 Stochastic approximation (Robbins-Monro algorithm)

The *Robbins-Monro algorithm* (RMA), generally known as *stochastic approximation*, is used to find the solution t_0 to an equation $g(t) = g_0$ in $\mathbb{R}$, where $g(y)$ is an increasing function which can be evaluated only with random errors (Robbins and Monro (1951)). Soon after its introduction, the algorithm was extended to the optimization problem, in this form known as the *Kiefer-Wolfowitz algorithm* (Kiefer and Wolfowitz (1952)). The RMA is useful when it is easier to simulate a random variable with mean $g(y)$ then to evaluate $g(y)$ itself. Like other simulation-based computational tools the RMA is very general in its applicability, and straightforward to implement.

First, assume $g_0 = 0$ (if necessary, replace $g(t)$ with $g(t) - g_0$). Suppose for any t we can obtain (by simulation or otherwise) a noisy evaluation of $g(t)$:

$$G_t = g(t) + \epsilon_t, \quad E[\epsilon_t] = 0.$$

Let $a_n \to 0$ be a sequence of positive constants. The RMA is defined by

$$Y_n = Y_{n-1} - a_n Z_n, \quad n \ge 1, \tag{10.28}$$

where $Z_n \sim G_{Y_{n-1}}$, and the simulation is independent of process history conditional on Y_{n-1}. Conditions under which $Y_n \to t_0$ in the L^2 norm were originally reported in Robbins and Monro (1951):

$$\sum_{n\ge 1} a_n = \infty, \quad (RM1)$$

$$\sum_{n\ge 1} a_n^2 < \infty. \quad (RM2)$$

Almost sure convergence under (RM1) and (RM2) was verified in Ljung (1978).

The RMA can be expressed as an AIA. For $a \in \mathbb{R}$ define the operator $T_a y = y - ag(y)$. The fixed point equation $y = T_a y$ is solved by $y = t_0$. Then set

$$Y_n = T_n Y_{n-1} + U_n, \quad \text{where}$$
$$T_n = T_{a_n}, \quad \text{and}$$
$$U_n = a_n\,[g(Y_{n-1}) - Z_n].$$

Under most implementations we may expect U_n to possess the martingale property $E[U_n \mid \mathcal{F}_{n-1}] = 0$, where $\mathcal{F}_{n-1} = \sigma(U_1, \ldots, U_{n-1})$ represents process history. Also, assume $E[U_n^2 \mid \mathcal{F}_{n-1}] \leq a_n^2 \sigma^2$ for all n. We impose further conditions on $g(y)$, first, that it possesses pseudo-Lipschitz constant L, and that there is a nonzero constant η for which $g(y)/y \geq \eta$ for all $y \neq 0$. Otherwise, g need be continuous only at $y = 0$. Thus, for all $a < L^{-1}$, T_a is pseudocontractive with constant $1 - a\eta$. To see this, fix $y > 0$, and suppose $a < L^{-1}$. This means $y - ag(y) > 0$, so that $|y - ag(y)| = y - ag(y) \leq (1 - a\eta)y$ (note that we must have $\eta \leq L$, and therefore $(1 - a\eta) > 0$). The same argument applies for $y < 0$.

We will now apply Theorem 10.5. The Hilbert space will be the space of random variables with inner product $E[XY]$. Because $a_k \to 0$, the condition $a_k < L^{-1}$ will eventually be met. By restarting the algorithm at this point, we may equivalently assume that $a_k < L^{-1}$ for all k from an abitrary starting point y_0. For this reason, implementation of the RMA does not require the knowledge of the actual values η and L. So, we have contraction constants $\rho_k = 1 - \eta a_k$. Furthermore, the condition $\langle U_k, T_k Y_{k-1} \rangle = E[U_k T_k Y_{k-1}] = 0$ follows from the martingale property. We therefore have (assuming, without loss of generality, $\rho_k \leq 1$),

$$\begin{aligned} E[(Y_k - t_0)^2] &\leq \left[\prod_{i=1}^{k} \rho_i^2\right] E[(Y_0 - t_0)^2] + \sum_{i=1}^{k} \rho_k^2 E[U_i^2] \\ &\leq \left[\prod_{i=1}^{k} (1 - \eta a_i)\right]^2 E[(Y_0 - t_0)^2] + \sigma^2 \sum_{i=1}^{k} a_i^2. \end{aligned} \tag{10.29}$$

In general, for $a \in [0,1)$ we have $\log(1 - a) \leq -a$, so that the product in (10.29) is bounded by

$$\prod_{i=1}^{k} (1 - \eta a_i) = \exp\left(\sum_{i=1}^{k} \log(1 - \eta a_i)\right) \leq \exp\left(-\eta \sum_{i=1}^{k} a_i\right)$$

By condition (RM1) the preceding bound approaches 0 as $k \to \infty$, and by condition (RM2) the summation in the bound given in (10.29) is bounded as $k \to \infty$, which together imply that $E[(Y_k - t_0)^2]$ remains bounded as $k \to \infty$.

Next, by condition (RM2) for any ϵ we may select N for which $\sum_{i \geq N} a_i^2 \leq \epsilon$. Then, considering the offset algorithm we have

$$E[(Y_{N+k} - t_0)^2] \leq \left[\prod_{i=N}^{N+k} (1 - \eta a_i) \right]^2 E[(Y_N - t_0)^2] + \sigma^2 \epsilon.$$

Letting $k \to \infty$ we have, following a similar argument, $\limsup_k E[(Y_k - t_0)^2] \leq \sigma^2 \epsilon$. This holds for any ϵ, so we conclude that Y_n converges to t_0 in the L^2 norm. Thus, the RMA can be seen as an AIA, with EIA based on contraction constants ρ_k. This argument is, in fact, a special case of Theorem 10.9.

Note that in deriving the upper bound in (10.29) we simply accepted the nonexpansive assumption $\rho_k \leq 1$ in the summation. We will see in Section 10.8 below that exploiting condition (RM1) within this term will lead to a stronger result, in that we can obtain explicit convergence rates, and can dispense with condition (RM2).

10.7 RATES OF CONVERGENCE FOR AIAs

In this section we obtain asymptotic rates or bounds for AIAs satisfying a general inequality with a product form Lipschitz kernel. The main task is to resolve the asymptotic properties of the Lipschitz convolution in the tolerance model (10.6), the final step being to compare this rate to the convergence rate of the EIA α_n, thus resolving the convergence rate of the AIA. Clearly, the tolerance η_k can never be smaller than α_k (the question of sharpness will be considered below). We will find, however, that the contribution of the approximation error, $\mathcal{I}_k^p(\tilde{d}, \bar{\lambda})$ can be asymptotically smaller than the exact tolerance, and we may predict when this will occur.

In this section we will assume that an AIA satisfies the general inequality

$$\| V_k - V^* \|^p = [\bar{\lambda}^{k,k}]^p \| V_0 - V^* \|^p + \mathcal{I}_k^p(\tilde{d}, \bar{\lambda}) \quad k \geq 1, \tag{10.30}$$

where $\bar{\lambda}$ is a product Lipschitz kernel generated by sequence $(\rho_1, \rho_2, \ldots)$, and $\tilde{d}$ is some positive sequence. General inequalities of the form (10.30) can be directly verified from Theorems 10.2, 10.4, 10.5 or 10.7.

It is important to note that the general inequality (10.30) may be used with a Lipschitz kernel other than that of the AIA itself. For example, it may be the case that (10.30) holds for some $\bar{\lambda}$ which does not satisfy some regularity condition, in which case it may be possible to replace $\bar{\lambda}$ with an upper envelope $\bar{\lambda}' \geq \bar{\lambda}$ which does. This is discussed below in Section 10.7.1. Alternatively, we may be able to establish (10.30) only for the adjusted kernel $\bar{\lambda}_b$ (using Theorems 10.4 or 10.7) when the relative error model (ARE) holds. In either case, the following theory does not require that the Lipschitz kernel satisfying (10.30) be the actual Lipschitz kernel for the AIA. However, in the case of the relative error model, there may be some value in decomposing the bound into components corresponding to the Lipshitz kernel of the AIA and the sequences defining the relative error model (ARE). This discussion is deferred to Section 10.7.7 below.

10.7.1 Monotonicity of the Lipschitz kernel

The Lipschitz convolution is clearly monotone, so that $\mathcal{I}_k^p(\tilde{d}, \bar{\lambda}) \leq \mathcal{I}_k^p(\tilde{d}', \bar{\lambda})$ or $\mathcal{I}_k^p(\tilde{d}, \bar{\lambda}) \leq \mathcal{I}_k^p(\tilde{d}, \bar{\lambda}')$ if $\tilde{d} \leq \tilde{d}'$ or $\bar{\lambda} \leq \bar{\lambda}'$. In addition, it will be natural to use a product form Lipschitz kernel generated by sequence λ_k. When $\bar{\lambda}$ is not product form, we may substitute any other bounding kernel. We may, for example, always define a product form kernel from the sequence $\lambda_k = \bar{\lambda}^{k,1}$. For multistage contraction operators, we can not expect that $\bar{\lambda}^{k,1} < 1$. However, if the operator sequence is J-stage contractive, that is, $T^{k,J}$ possesses contraction constant ρ for each $k \geq J$, then we must have $\bar{\lambda}^{k,j} \leq (\rho^{1/J})^{j-J+1}$, which can serve as a bounding product form kernel. See Chapter 9 for an extensive discussion of this issue.

10.7.2 Case I – strongly contractive models with nonvanishing bounds

We first establish that the asymptotic behavior of an AIA is dominated by d_k under strong contraction.

Theorem 10.10 *Suppose an AIA satisfies general inequality (10.30).*

(*i*) *If $\bar{\lambda}$ is a product form Lipschitz kernel for which $\lambda^u\{\bar{\lambda}^{k,k}\} = \rho < 1$ then:*

$$\limsup_{k\to\infty} \mathcal{I}_k^p(\tilde{d}, \bar{\lambda}) \leq (1-\rho^p)^{-1} \limsup_{k\to\infty} d_k^p, \quad \textit{and therefore}$$

$$\limsup_{k\to\infty} \| V_k - V^* \| \leq (1-\rho^p)^{-1/p} \limsup_{k\to\infty} d_k. \tag{10.31}$$

(*ii*) *If for some nonincreasing sequence β_i, $i \geq 0$, for which $\lim_i \beta_i = 0$, we have $\bar{\lambda}^{k,i} = \beta_i$ for all $k \geq i \geq 0$ then*

$$\limsup_{k\to\infty} \mathcal{I}_k^p(\tilde{d}, \bar{\lambda}) \leq \left[\sum_{i=0}^{\infty} \beta_i^p\right] \limsup_{k\to\infty} d_k^p, \quad \textit{and therefore}$$

$$\limsup_{k\to\infty} \| V_k - V^* \| \leq \left[\sum_{i=0}^{\infty} \beta_i^p\right]^{1/p} \limsup_{k\to\infty} d_k. \tag{10.32}$$

Proof Set $d' = \limsup_k d_k$. First take the $p=1$ case. We may select k_ϵ large enough such that $\sup_{k \geq k_\epsilon} d_k \leq d' + \epsilon$. Then, for $k > k_\epsilon$ we may write

$$\begin{aligned}
\sum_{i=1}^{k} \bar{\lambda}^{k,k-i} d_i &= \sum_{i=1}^{k_\epsilon} \bar{\lambda}^{k,k-i} d_i + \sum_{i=k_\epsilon+1}^{k} \bar{\lambda}^{k,k-i} d_i \\
&\leq \sum_{i=1}^{k_\epsilon} \bar{\lambda}^{k,k-i} d_i + (d' + \epsilon) \sum_{i=k_\epsilon+1}^{k} \bar{\lambda}^{k,k-i}.
\end{aligned} \tag{10.33}$$

For part (*i*) we may also select k_ϵ large enough that $\rho_k \leq \rho + \epsilon$ for all $k \geq k_\epsilon$. Then by (10.33) we have

$$\sum_{i=1}^{k} \bar{\lambda}^{k,k-i} d_i \leq \sum_{i=1}^{k_\epsilon} \bar{\lambda}^{k,k-i} d_i + (d' + \epsilon) \sum_{i=0}^{k} (\rho + \epsilon)^i.$$

Each term in the first summation of the preceding upper bound approaches 0 as $k \to \infty$. The argument is completed by allowing ϵ to approach 0.

Similarly, for part (*ii*) we have

$$\sum_{i=1}^{k} \bar{\lambda}^{k,k-i} d_i \leq \sum_{i=1}^{k_\epsilon} \bar{\lambda}^{k,k-i} d_i + (d' + \epsilon) \sum_{i=0}^{k} \beta_i,$$

with the remaining argument the same as for part (*i*). The argument for general $p > 0$ is essentially the same. ///

10.7.3 Case II – rapidly vanishing approximation error

We say the approximation errors *vanish rapidly* if the following condition holds:

$$\sum_{i \geq 1} \left(d_i / \bar{\lambda}^{i,i}\right)^p < \infty. \tag{10.34}$$

It is easy to prove that when this condition holds the AIA converges at the same rate as the EIA.

Theorem 10.11 *Suppose an AIA satisfies general inequality (10.30). If assumption (10.34) holds then*

$$\mathcal{I}_k^p(\tilde{d}, \bar{\lambda}) = \Omega\left([\bar{\lambda}^{k,k}]^p\right) \text{ and therefore } \| V_k - V^* \| \leq \Omega(\bar{\lambda}^{k,k}). \tag{10.35}$$

Proof The result holds directly by noting $\mathcal{I}_k(\tilde{d}, \bar{\lambda}) = \bar{\lambda}^{k,k} \sum_{i=1}^{k} d_i / \bar{\lambda}^{i,i}$. ///

A convenient special case of Case II is available.

Theorem 10.12 *If $\hat{\lambda}^l\{\bar{\lambda}^{k,k}\} = \rho$ and $\hat{\lambda}^u\{\tilde{d}\} = r < \rho$ then (10.34) holds for any $p > 0$.*

Proof Select any $\epsilon > 0$ for which $r + \epsilon < \rho - \epsilon$. There exists finite k_ϵ for which $\bar{\lambda}^{k,k} \geq (\rho - \epsilon)^k$ and $d_k \leq (r + \epsilon)^k$ for all $k > k_\epsilon$. We may write for large enough k and any $p > 0$

$$\begin{aligned}
\sum_{i=1}^{k} (d_i / \bar{\lambda}^{i,i})^p &\leq \sum_{i=1}^{k_\epsilon} (d_i / \bar{\lambda}^{i,i})^p + \sum_{i=k_\epsilon+1}^{k} \left(\left[\frac{r+\epsilon}{\rho-\epsilon}\right]^p\right)^i \\
&\leq K_\epsilon + \left(1 - \left[\frac{r+\epsilon}{\rho-\epsilon}\right]^p\right)^{-1}
\end{aligned}$$

where K_ϵ is a finite constant which does not depend on k. By hypothesis the upper bound is finite, therefore (10.34) holds. ///

10.7.4 Case III – approximation error decreasing at contraction rate

We say the approximation errors *decrease at the contraction rate* if the following condition holds:

$$\hat{\lambda}\left\{\sum_{i=1}^{k}(d_i/\bar{\lambda}^{i,i})^p\right\} = 1. \tag{10.36}$$

The simplest such case occurs when $d_k = \bar{\lambda}^{k,k} = \rho^k$, in which case $\sum_{i=1}^{k}(d_i/\bar{\lambda}^{i,i})^p = k$, so that (10.36) holds. Note that if $\tilde{d}$ is positive, which we assume within the general inequality (10.30), we must have $\hat{\lambda}\left\{\sum_{i=1}^{k}(d_i/\bar{\lambda}^{i,i})^p\right\} \geq 1$ by Lemma 9.1.

Theorem 10.13 *Suppose an AIA satisfies general inequality (10.30). If assumption (10.36) holds then*

$$\hat{\lambda}^u\left\{\mathcal{I}_k^p(\tilde{d},\bar{\lambda})\right\} = \hat{\lambda}^u\{[\bar{\lambda}^{k,k}]^p\} \text{ and therefore } \hat{\lambda}^u\{\|V_k - V^*\|\} \leq \hat{\lambda}^u\{\bar{\lambda}^{k,k}\}. \tag{10.37}$$

Proof The result holds directly by noting

$$k^{-1}\log\left(\mathcal{I}_k^p(\tilde{d},\bar{\lambda})\right) = k^{-1}\left(\log\left([\bar{\lambda}^{k,k}]^p\right) + \log\left(\sum_{i=1}^{k}(d_i/\bar{\lambda}^{i,i})^p\right)\right)$$

and applying assumption (10.36) directly. ///

A result analagous to Theorem 10.12 for Case II is available.

Theorem 10.14 *If $\tilde{d}$ is positive, $\hat{\lambda}^l\{\bar{\lambda}^{k,k}\} = \rho$ and $\hat{\lambda}^u\{\tilde{d}\} = \rho$ then (10.36) holds for any $p > 0$.*

Proof Select any $\epsilon > 0$ for which $0 < \rho - \epsilon$ and $\rho + \epsilon < 1$. There exists finite k_ϵ for which $\bar{\lambda}^{k,k} \geq (\rho - \epsilon)^k$ and $d_k \leq (\rho + \epsilon)^k$ for all $k > k_\epsilon$. We may write for large enough k and any $p > 0$

$$\begin{aligned}\sum_{i=1}^{k}(d_i/\bar{\lambda}^{i,i})^p &\leq \sum_{i=1}^{k_\epsilon}(d_i/\bar{\lambda}^{i,i})^p + \sum_{i=k_\epsilon+1}^{k}\left(\left[\frac{\rho+\epsilon}{\rho-\epsilon}\right]^p\right)^i \\ &\leq K_\epsilon + k\left(\left[\frac{\rho+\epsilon}{\rho-\epsilon}\right]^p\right)^k \end{aligned} \tag{10.38}$$

where K_ϵ is a finite constant which does not depend on k. The inequality (10.38) then leads to

$$\limsup_{k\to\infty} k^{-1}\sum_{i=1}^{k}(d_i/\bar{\lambda}^{i,i})^p \leq \log\left(\left(\left[\frac{\rho+\epsilon}{\rho-\epsilon}\right]^p\right)\right),$$

which completes the proof after noting that the upper bound approaches 0 as ϵ approaches 0. ///

10.7.5 Case IV – Approximation error greater than contraction rate

Case IV covers an important special case, in particular, $d_k = k^{-q}$ for $q > 0$. For example, when approximations are based on statistical estimation we will often have $q = 1/2$. The point here is that d_k approaches zero more slowly than the EIA. In this case we conclude, in general, that $\| V_k - V^* \| = O(d_k)$. The proof makes use of the l'Hôpital's rule analog for discrete series of Section 9.4. Of course, the original l'Hôpital's rule makes use of derivatives. In the discrete case, this manifests itself in a smoothness requirement for the sequence d_k. If this does not hold, then d_k can be replaced by a smooth upper envelope (see Chapter 9 for some relevant techniques).

Theorem 10.15 *Suppose an AIA satisfies general inequality (10.30) and* $\lambda^u\{\bar{\lambda}^{k,k}\} = \rho < 1$.

(*i*) *If* $\lambda^l\{\tilde{d}\} = r > \rho$ *then*

$$\limsup_{k\to\infty} d_k^{-p} \mathcal{I}_k^p(\tilde{d}, \bar{\lambda}) \leq \left(1 - [\rho/r]^p\right)^{-1} \text{ and therefore}$$

$$\limsup_{k} d_k^{-1} \| V_k - V^* \| \leq (1 - [\rho/r]^p)^{-1/p}. \tag{10.39}$$

(*ii*) *If* $\hat{\lambda}^u\{\tilde{d}\} = r > \rho$ *then*

$$\hat{\lambda}^u\{\mathcal{I}_k^p(\tilde{d}, \bar{\lambda})\} \leq r^p \text{ and therefore}$$

$$\hat{\lambda}^u\{\| V_k - V^* \|\} \leq r. \tag{10.40}$$

Proof Part (*i*) follows directly from Lemma 9.8. For part (*ii*) from Theorem 9.1 there exists a sequence $\tilde{d}^* = \{d_k^*\}$ for which $d_k^* \geq d_k$, $\lambda^l\{d_k^*\} \geq \rho$ and $\hat{\lambda}\{a_k^*\} = \rho$. Since $\mathcal{I}_k^p(\tilde{d}, \bar{\lambda}) \leq \mathcal{I}_k^p(\tilde{d}^*, \bar{\lambda})$, we may apply (10.39) to $\tilde{d}^*$ obtain (10.40). ///

10.7.6 Case V – Contraction rates approaching 1

We next consider a class of algorithms based on contractive operators, but with contractive constants approaching 1 as $k \to \infty$. In such cases, while the convergence properties do depend on the contractivity property, the overall convergence rate is no longer linear. However, convergence rates may be obtained using much the same analysis as for the strongly contractive case.

The essential assumption is that $\lim_{k\to\infty} \rho_k = 1$. It will additionally be convenient to assume that ρ_k is increasing. If it isn't, it may be replaced by the envelope $\rho_k^* = \sup_{i\leq k} \rho_i$.

First consider the exact algorithm (that is, $\|U_k\| = 0$). Directly from Theorem 10.2 we then have

$$\|V_k - V^*\| \leq \lambda^{k,k} \|V_0 - V^*\|, \quad k \geq 1, \tag{10.41}$$

so that convergence of the AIA requires, at least, that $\lim_k \lambda^{k,k} = 0$ (compare to the RMA, Section 10.6.4). Following Section 9.1, a necessary and sufficient condition is easily stated. If $\rho_k = 1 - \delta_k$, then $\lim_k \lambda^{k,k} = 0$ if and only if $\sum_k \delta_k = \infty$. Intuitively, this requirement forces ρ_k to approach 1 slowly enough for the contractive property to force convergence to the fixed point. However, it is important to note that even without this condition, approximation theory is still relevant. It may be that the bound (10.41) suffices for a particular application even if it does not converge to 0, in which case we may still predict the effect of approximation in the same way.

First note that Case II (Section 10.7.3) does not require the strong contraction assumption, so we also have

$$\|V_k - V^*\| \leq \Omega\left(\bar{\lambda}^{k,k}\right), \tag{10.42}$$

for the Case II condition. For example, if $\tilde{d}$ convervges to 0 linearly, and $\rho_k \to 1$, then $\bar{\lambda}^{k,k}$ either possesses a nonzero limit, or converges to 0 sublinearly, in which case (10.42) clearly holds. A similar remark holds for Case III.

With some additional assumptions, a more direct comparison with Case IV can be made. In particular, if for case IV we have $\|V_k - V^*\| \approx (1 - \rho^p)^{-1/p} d_k$, we might expect that for Case V we have $\|V_k - V^*\| \approx (1 - \rho_k^p)^{-1/p} d_k$, which is in fact the case.

Theorem 10.16 *Suppose an AIA satisfies general inequality (10.30). Define*

$$\alpha_k = (d_{k-1}/d_k)\rho_k, \quad k \geq 1.$$

Suppose the hypothesis of Theorem 9.3 is satisfied, that is, α_n is nondecreasing with $\alpha_n < 1$. Then

$$\limsup_{k\to\infty} \frac{\mathcal{I}_k^p(\tilde{d}, \bar{\lambda})}{d_k^p (1 - \alpha_k^p)^{-1}} \leq 1 \quad \textit{and therefore}$$

$$\limsup_{k\to\infty} \frac{\|V_k - V^*\|}{d_k (1 - \alpha_k^p)^{-1/p}} \leq 1. \tag{10.43}$$

In addition, if $\tilde{d}$ is nonincreasing and $\rho_k \to 1$ then

$$\limsup_{k\to\infty} \frac{\left(\frac{d_{k-1}}{d_k}\right)^p - 1}{1 - \rho_k^p} = r < 1 \tag{10.44}$$

and

$$\limsup_{k\to\infty} \frac{\mathcal{I}_k^p(\tilde{d},\bar{\lambda})}{d_k^p(1-\rho_k^p)^{-1}(1-r)^{-1}} \leq 1 \quad \textit{and therefore}$$

$$\limsup_{k\to\infty} \frac{\| V_k - V^* \|}{d_k(1-\rho_k^p)^{-1/p}(1-r)^{-1/p}} \leq 1. \tag{10.45}$$

Proof The first statement of (10.43) is a direct application of Theorem 9.3, raising ρ_n and d_n to the pth power. Within the argument of Theorem 9.3 it is shown that $\min_{k\to\infty} (\bar{\lambda}^{k,k})^{-p}\mathcal{I}_k^p(\tilde{d},\bar{\lambda}) = \infty$, so that in (10.30) the term $\mathcal{I}_k^p(\tilde{d},\bar{\lambda})$ is asymptotically dominant, which completes the proof.

To prove (10.45) write

$$1 - \alpha_k^p = (1-\rho_k^p) + \rho_k^p(1 - \frac{d_{k-1}}{d_k}),$$

then from (10.44), and the fact that $\rho_k \to 1$, we have

$$\liminf_{k\to\infty} \frac{1-\alpha_k^p}{1-\rho_k^p} = 1 - r, \tag{10.46}$$

then (10.45) follows by substitution of (10.46) into (10.43). ///

In order to clarify the assumption in Theorem 10.16 that α_n is nondecreasing, and $\alpha_n < 1$ for all large enough n, consider a simple example in which $d_n = an^{-s}$ and $\delta_n = bn^{-t}$ for positive constants s,t,a and b, so that $\rho_n = 1 - \delta_n$. As discussed earlier, to obtain convergence we need to have $\sum_n \delta_n = \infty$, so we assume that $t \leq 1$. This gives

$$d_{n-1}/d_n = (1-(1/n))^{-s} \quad \text{and} \quad \rho_n = 1 - b(1/n)^t$$

It will be convenient to define the functions $D(x) = (1-x)^{-s}$ and $\rho(x) = 1 - b(x)^t$ on $x \in [0,\infty)$ so that $d_{n-1}/d_n = D(1/n)$ and $\rho_n = \rho(1/n)$, and we investigate the behavior of the model as $n \to \infty$ by allowing $x \to 0$. We may use Taylor's expansions to obtain

$$\begin{aligned} D(x)^p &= 1 + spx + sp(sp+1)x^2/2 + O(x^3) \quad \text{and} \\ \rho(x)^p &= 1 - p(bx^t) + p(p-1)(bx^t)^2/2 + O(x^{3t}). \end{aligned}$$

Then set $\alpha(x) = D(x)\rho(x)$ so that $\alpha_n = \alpha(1/n)$. If $t < 1$ it can be verified that the first derivative $\alpha'(0) < 0$, and since $\alpha(0) = 1$, it follows that $\alpha_n \uparrow 1$. If $t = 1$ then

$$\frac{d\alpha(x)}{dx} = \frac{(s-b) + bx(1-s)}{(1-x)^{s+1}},$$

so that $\alpha'(0) < 0$ if $b > s$ or $b = s > 1$.

Then consider the rate of convergence of $\| V_n - V^* \|^p$ given in in (10.43), in particular $d_n/(1-\alpha_n^p)^{1/p}$. Taking $(1-\alpha_n^p)$ separately, for $t < 1$ we have

$$1 - \alpha(x)^p = 1 - D(x)^p \rho(x)^p \approx pbx^t$$

so that $1 - \alpha_n^p \approx pbn^{-t}$, giving

$$\frac{d_n}{(1-\alpha_n^p)^{1/p}} \approx \frac{a}{(pb)^{1/p}} \frac{1}{n^{s-t/p}}.$$

Thus, we have convergence if $s > t/p$, with the Hilbert space model resulting in a strict improvement in the convergence rate.

Next, suppose $t = 1$. We have two cases, $b > s$ and $b = s > 1$. If the first holds we have

$$1 - \alpha(x)^p \approx (b - s)px,$$

so that

$$\frac{d_n}{(1-\alpha_n^p)^{1/p}} \approx \frac{a}{(p(b-s))^{1/p}} \frac{1}{n^{s-t/p}},$$

obtaining the same convergence rate as the $t < 1$ case, but with a different constant.

Next, suppose $t = 1$ and $b = s > 1$. we have

$$1 - \alpha(x)^p \approx ps(s-1)/2x^2,$$

so that

$$\frac{d_n}{(1-\alpha_n^p)^{1/p}} \approx \frac{a}{(ps(s-1)/2)^{1/p}} \frac{1}{n^{s-2/p}}$$

so that the convergence rate now becomes $O(1/n^{s-2/p})$.

10.7.7 Adjustments for relative error models

The relative error model (ARE) is accomodated by verifying that the general inequality (10.30) holds with the adjusted kernel $\bar{\lambda}_b$. This will generally be adequate if $b_k \to 0$, otherwise it is best to use a bound which explicitly separates the Lipshitz kernel of the AIA from the sequences defining (ARE). In this case the Banach space and Hilbert space models can be are treated separately to some advantage as will be discussed in the next section.

Theorem 10.17 *Suppose we are given an AIA with product form Lipschitz kernel* $\bar{\lambda} = (\rho_1, \rho_2, \ldots)$ *and (ARE) holds with* $\limsup_k \kappa_0 b_k = \delta$ *for some* $\delta \in [0, 1-\rho)$. *Then*

$$\limsup_{k\to\infty} \| V_k - V^* \| \leq (1 - \rho - \delta)^{-1} \limsup_{k\to\infty} d_k^*. \tag{10.47}$$

If in addition the assumptions of Theorem 10.7 hold, then if $\kappa_0 b_k \leq \delta \in \left[0, 2^{-1/2}\sqrt{1-\rho^2}\right)$,

$$\limsup_{k\to\infty} \| V_k - V^* \| \leq 2^{1/2}(1-\rho^2-2\delta^2)^{-1/2} \limsup_{k\to\infty} d_k^*. \tag{10.48}$$

Proof The bound (10.47) follows directly from Theorem 10.10 by substituting $\bar{\lambda}_b$ for $\bar{\lambda}$. The bound (10.48) is obtained from (10.18) of Theorem 10.7 by applying Theorem 10.10. ///

10.7.8 A comparison of Banach space and Hilbert space models

There are two issues on which to elaborate. First, the bound (10.31) of Theorem 10.10 depends on the quantity $(1-\rho^p)^{-1/p}$, while the bound (10.39) of Theorem 10.15 depends on the quantity $(1-[\rho/r]^p)^{-1/p}$. In the following discussion, we lose nothing by considering the quantity $(1-\rho^p)^{-1/p}$ alone. In both the Banach and Hilbert space cases (that is, $p=1$ or $p=2$ respectively), the dependence of the convergence of the AIA on the asymptotic properties of $\tilde{d}$ is exactly the same. However, there is an important difference with respect to the constant $(1-\rho^p)^{-1/p}$. Although this value

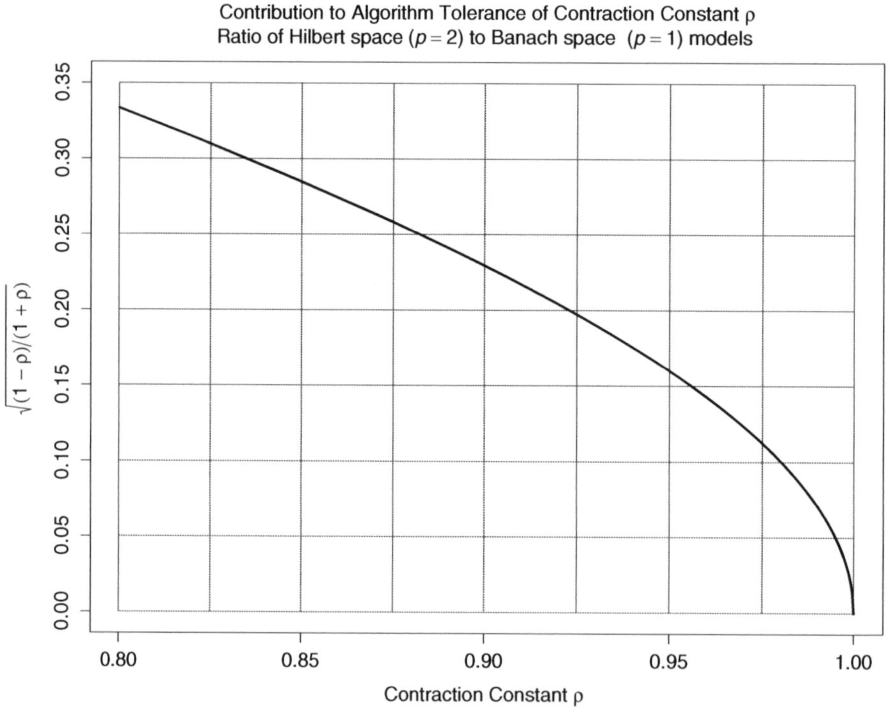

Figure 10.1 The graph representes the ratio of the contributions to the algorithm tolerance of the contraction constant for the Hilbert space model (numerator) to the Banach space model (denominator) (Section 10.7.8).

does not affect the convergence properties of the AIA, it is still an important quantity with respect to the algorithm tolerance, particularly because the contraction constant will, in practice, often be close to 1. To assess the impact of the power p, take the ration of the quantities:

$$\frac{(1-\rho^2)^{-1/2}}{(1-\rho)^{-1}} = \sqrt{\frac{1-\rho}{1+\rho}}.$$

This quantity is plotted in Figure 10.1. Interestingly, this quantity approaches 0 as $\rho \to 1$, so that if the Hilbert space model can be used, a much stronger bound on the algorithm tolerance may be possible.

To deal with the second issue note that (10.47) of Theorem 10.17 states that $\limsup_{k\to\infty} \| V_k - V^* \| \, / \, \| V^* \| \leq \delta(1-\rho-\delta)^{-1}$. The tolerance thus becomes relative to solution $\| V^* \|$, and to attain a useful bound we would need to have δ small in relation to $(1-\rho)^{-1}$. This means that the constraint $\delta < 1-\rho$ of Theorem 10.17 is a necessary one. It is thus interesting to note that use of the Hilbert space model brings another advantage, namely that the bound on δ can be relaxed to $2^{-1/2}\sqrt{1-\rho^2}$, which is strictly larger over $\rho \in (1/3, 1)$ (bound (10.47) can always be used should we have $\rho \leq 1/3$). See Figure 10.2.

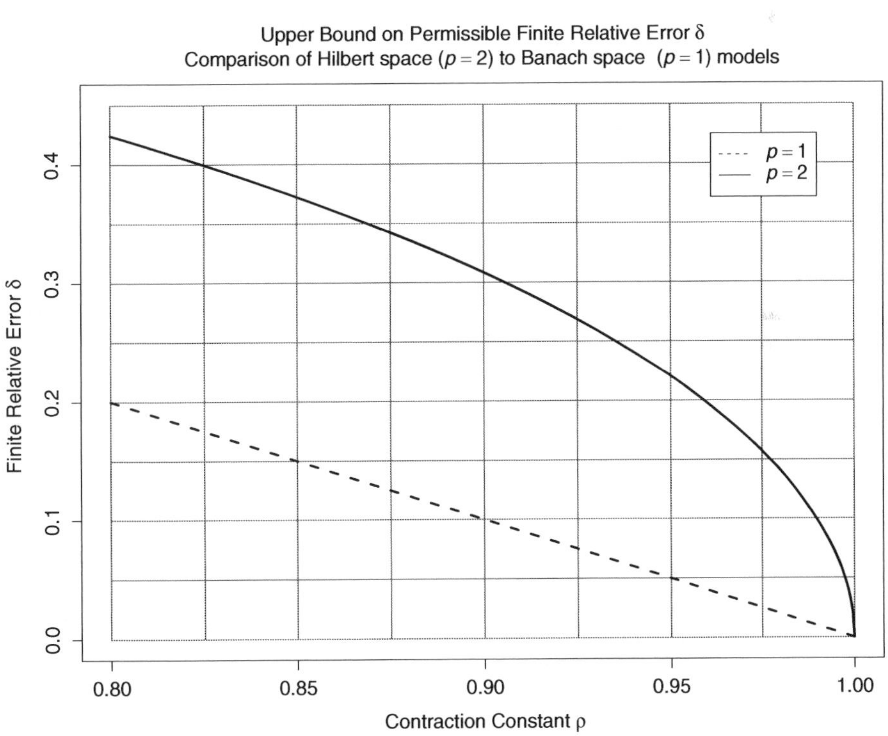

Figure 10.2 Permissible bounds for relative error for Banach space and Hilbert space model. See Theorem 10.17.

10.8 STOCHASTIC APPROXIMATION AS A WEAKLY CONTRACTIVE ALGORITHM

We continue the discussion of the Robbins-Monro algorithm introduced in Section 10.6.4. The kth operator is contractive with constant $\rho_k = 1 - \eta a_k$. We have already shown that $\lambda^{k,k} \to_k 0$ if and only if $\sum_k a_k = \infty$, which is precisely condition (RM1). It is important to note at this point that if were able to use the RMA to determine a root of $g(t)$ using noise-free evaluations, we could do so, as long as (RM1) held. In other words, if RMA is interpreted as an AIA, then (RM1) would be the sole condition needed to ensure convergence of the EIA on which it is based.

We will make use of Theorems 9.3 and 10.16, and the notation used the respective proofs. Because we are using a Hilbert space model, for convenience we may set $\lambda_k = \rho_k^2$ and $d_k = \sigma^2 a_k^2$. Then

$$\tau_k = (1 - [a_{k-1}/a_k]^2 (1 - \eta a_k)^2)^{-1}.$$

A common practice in the analysis of the RMA is to set $a_k \propto k^{-q}$. The standard regularity conditions force $q \in (1/2, 1]$. To analyze τ_k define function

$$f(x \mid \eta, q) = (1 - \eta x^q)(1 - x)^{-q}, \quad \text{so that}$$
$$\tau_k = (1 - f(1/k \mid \eta, q)^2)^{-1}.$$

The derivative with respect to x is

$$\frac{df(x \mid \eta, q)}{dx} = \frac{q(1 - q\eta x^{q-1})}{(1 - x)^{q+1}}. \tag{10.49}$$

Then, for $q \in (0, 1)$ and $\eta > 0$ we have $f(0 \mid \eta, q) = 1$, and from (10.49) $df(x \mid \eta, q)/dx < 0$ for all small enough $x > 0$, since $q < 1$. It follows that τ_k satisfies the required conditions.

Next, consider the problem of determining constants A and p for which

$$\lim_{x \downarrow 0} A x^p (1 - f(x \mid \eta, q)^2)^{-1} = 1.$$

An application of l'Hôpital's rule gives $p = q$ and $A = 2q\eta$, so that

$$\limsup_{k \to \infty} \frac{\| V_k - V^* \|^2}{2q\eta\sigma^2 k^{-q}} \le 1, \tag{10.50}$$

so that $\| V_k - V^* \| = O(k^{-q/2})$. Thus, in contrast with the strongly contractive algorithm, for which $\| V_k - V^* \| = O(d_k)$ for the RMA we have $\| V_k - V^* \| = O(d_k^{1/2})$.

The case $q=1$ requires a separate analysis, and is in fact the recommendation given in Robbins and Monro (1951). First note that

$$\lambda^{k,k} = \exp\left(\sum_{i=1}^{k} \log(1-\eta a_k)\right)$$
$$\approx k^{-c_0\eta}$$

where c_0 is a finite positive constant independent of all parameters. We then have

$$\|V_k - V^*\|^2 \leq (\rho^{k,k})^2 \|V_0 - V^*\|^2 + \sigma^2(\rho^{k,k})^2 \sum_{i=1}^{k} (d_i/\rho^{i,i})^2$$
$$\approx k^{-2c_0\eta}\left(\|V_0 - V^*\|^2 + \sigma^2 \sum_{i=1}^{k} i^{2c_0\eta-2}\right)$$
$$\approx \begin{cases} k^{-2c_0\eta}\left(\|V_0 - V^*\|^2 + \frac{\sigma^2 c_1}{2c_0\eta-1} k^{2c_0\eta-1}\right); & c_0\eta \neq 1/2 \\ k^{-2c_0\eta}\left(\|V_0 - V^*\|^2 + \sigma^2 c_2 \log(k)\right); & c_0\eta = 1/2 \end{cases},$$

where c_1, c_2 are finite positive constants independent of all parameters. This leads to bounds

$$\|V_k - V^*\| \leq \begin{cases} \|V_0 - V^*\| k^{-c_0\eta} + O(k^{-1/2}); & c_0\eta < 1/2 \\ \sigma c_2^{1/2}[k^{-1}\log(k)]^{1/2} + O(k^{-1/2}); & c_0\eta = 1/2 \\ \sqrt{\frac{\sigma^2 c_1}{2c_0\eta-1}}[k^{-1/2}] + O(k^{-c_0\eta}); & c_0\eta > 1/2 \end{cases}. \tag{10.51}$$

For the $q=1$ case, the algorithm tolerance rate depends on the value of η. From (10.51) we may conclude that for all large enough η the algorithm tolerance is $\|V_k - V^*\| \leq O(k^{-1/2})$, with smaller values yielding slower convergence rates.

With this caveat, we have a general form of the algorithm tolerance $\|V_k - V^*\| \leq O(k^{-q/2})$ for any $q \in (0,1)$. It will be of some interest to note that this holds for values of $q < 1/2$ which violate (RM2), thus we conclude that this assumption is not necessary. A number of alternative conditions for the convergence of the RMA which do not rely on (RM2) can be found in the literature, for example, Kushner and Yin (2003) (see Chapters 1 and 5).

10.9 TIGHTNESS OF ALGORITHM TOLERANCE

Clearly, an AIA will not converge more quickly than the error term. This is expressed formally in the next theorem, in the sense that the algorithm error $\|V_k - V^*\|$ will be of order $\|U_k\|$ infinitely often. For simplicity we consider the homogenous EIA based on T, which is assumed to be Lipschitz, but not necessarily contractive.

Theorem 10.18 *In algorithm* (10.2) *let*

$$d_n = \sup_{n' \geq n} \|U_{n'}\|$$

for $n \geq 1$. *If* V^* *is a fixed point of* T, *and there exists a finite constant* L *such that* $\|TV - TV^*\| \leq L\|V - V^*\|$ *for all* $V \in \mathcal{V}$, *then*

$$\limsup_{n\to\infty} \frac{\|V_{n-1} - V^*\|}{d_n} > 0. \tag{10.52}$$

Proof Suppose (10.52) does not hold. Fix $\epsilon > 0$. Then there exists N such that

$$\|V_{n-1} - V^*\| \leq \epsilon d_n \;\; \forall n \geq N. \tag{10.53}$$

Let N_1 be the smallest integer not less than N such that $\|U_{N_1}\| \geq (1-\epsilon)d_N$, which exists by the definition of d_n. This implies $\|U_{N_1}\| \geq (1-\epsilon)d_{N_1}$. Applying (10.53) gives

$$\begin{aligned}
\|U_{N_1}\| &= \|\left(V_{N_1} - V^*\right) - \left(TV_{N_1-1} - TV^*\right)\| \\
&\leq \|V_{N_1} - V^*\| + \|TV_{N_1-1} - TV^*\| \\
&\leq \|V_{N_1} - V^*\| + L\|V_{N_1-1} - V^*\| \\
&\leq \epsilon d_{N_1+1} + L\epsilon d_{N_1} \\
&\leq (1+L)\epsilon d_{N_1}.
\end{aligned}$$

We can always set ϵ small enough to force $(1+L)\epsilon d_{N_1} < (1-\epsilon)d_{N_1}$, in which case (10.53) leads to a contradiction, since $\|U_{N_1}\| \geq (1-\epsilon)d_{N_1}$. Hence (10.52) follows. ///

10.10 FINITE BOUNDS

The convergence results given above are based on asymptotic bounds. However, inequality (10.30) provides a means to calculate finite upper bounds on $\|V_k - V^*\|$. In this section we suppose we may bound errors with constants $\|U_k\| \leq d_k$, where we assume $d_k > 0$. We will consider the single stage ρ-contractive operator T, setting $p = 1$ in (10.30), which will be easily extended to more general models.

The first term in the upper bound is composed of the exponentially decreasing error of the exact algorithm. The second term may be bounded by

$$\sum_{i=1}^{n} \rho^{n-i}\|U_i\| \leq d_n I_n^1(\tilde{d}, \bar{\lambda}), \quad n \geq 1$$

using the notation introduced in Section 10.3.2. For convenience we may write $I_n = I_n^1(\tilde{d}, \bar{\lambda})$. We then have

$$\|V_n - V^*\| \leq \rho^n\|V_0 - V^*\| + d_n I_n, \quad n \geq 1. \tag{10.54}$$

Under the conditions of Theorem 10.15 we have

$$\limsup_{n\to\infty} I_n \leq (1 - \rho/r)^{-1}. \tag{10.55}$$

When a simpler finite bound is required we may write

$$\| V_n - V^* \| \leq \rho^n \| V_0 - V^* \| + d_n \hat{I}^1(\tilde{d}, \bar{\lambda}), \quad n \geq 1. \tag{10.56}$$

If d_n has a tractable form, we may simply calculate $I_n^1(\tilde{d}, \bar{\lambda})$ numerically. In this case, it will be useful to know something of the iterative properties of $I_n^1(\tilde{d}, \bar{\lambda})$. We show in the following lemma that under a type of convexity assumption on the sequence d_n once $I_n^1(\tilde{d}, \bar{\lambda})$ decreases in n, it decreases indefinitely.

Lemma 10.2 *If a positive sequence $\{d_n; n \geq 1\}$ satisfies*

$$d_{n+1}/d_{n+2} \leq d_n/d_{n+1}, \quad n \geq 1, \tag{10.57}$$

then $I_{n+1}^1(\tilde{d}, \bar{\lambda}) < I_n^1(\tilde{d}, \bar{\lambda})$ implies $I_{n+2}^1(\tilde{d}, \bar{\lambda}) < I_{n+1}^1(\tilde{d}, \bar{\lambda})$.

Proof We may write

$$I_{n+1} - I_n = 1 + \left(\rho \frac{d_n}{d_{n+1}} - 1\right) I_n, \quad n \geq 1$$

from which it follows

$$I_{n+1} - I_n < 0 \text{ if and only if } \left(1 - \rho \frac{d_n}{d_{n+1}}\right) I_n > 1. \tag{10.58}$$

Then, if $I_{n+1} - I_n < 0$ we have $\left(1 - \rho \frac{d_n}{d_{n+1}}\right) > 0$ and hence $\left(1 - \rho \frac{d_{n+1}}{d_{n+2}}\right) > 0$ from condition (10.57). This in turn implies by (10.58)

$$\begin{aligned}
\left(1 - \rho \frac{d_{n+1}}{d_{n+2}}\right) I_{n+1} &= \left(1 - \rho \frac{d_{n+1}}{d_{n+2}}\right) \left(1 + \rho \frac{d_n}{d_{n+1}} I_n\right) \\
&> \left(1 - \rho \frac{d_{n+1}}{d_{n+2}}\right) \left(1 - \rho \frac{d_n}{d_{n+1}}\right)^{-1} \geq 1
\end{aligned}$$

which proves the lemma. ///

Condition (10.57) is quite general, and is satisfied by any polynomially decreasing bound $1/n^k$. Denote any maximum $I^* = \max_n I_n$. As an example, for $k = 2$, $\rho = 0.9$, it is easily determined that I_n increases up to $n = 21$, at which $I_{21} = 104.67$. However, $I_{22} = 104.39$, so we conclude $I^* = I_{21}$. Note that this is considerably larger than the limiting bound for I_n given in (10.55), in this case 10.

10.10.1 Numerical example

There are two approaches to obtaining a finite bound. First we can simply accept (10.56). Otherwise, we may apply Lemma 10.2, assuming the hypothesis holds, which it will for $d_n = (1/n)^q$, $q > 0$. To do this, we identify n^* satisfying $I_{n^*} = I^*$. By Lemma 10.2 this is the minimum n for which $I_{n+1} < I_n$ (if the maximum I^* is not unique, this procedure identifies the largest maximizing n, which is suitable for our purpose). If the quantities ρ and $\tilde{d}$ are known, we may determine n^* by calculating $I_1, I_2, \ldots$ until I_n is observed to decrease. We then have the finite bound

$$\|V_n - V^*\| \leq \rho^n \|V_0 - V^*\| + d_n I_{n^{**}} \quad n \geq n^{**} \tag{10.59}$$

for any $n^{**} \geq n^*$, which holds since $I_n \leq I_{n^{**}}$. Since eventually I_n is decreasing, from (10.55) we have $I_n \downarrow (1 - \rho/r)^{-1}$. Therefore, $I_n \geq (1 - \rho/r)^{-1}$ for $n \geq n^*$, and a reasonable strategy would be to select n^{**} for which $I_{n^{**}}(1 - \rho/r) = 1 + \epsilon$ is slightly larger than 1. Then for $n \geq n^{**}$ the asymptotic bound will be within this tolerance of the finite bound.

To illustrate the procedure, we take $d_n = n^{-q}$, with $q = 1/2$ and $\rho = 0.9$, so that $r = 1$, and we expect $I_n \to_n (1 - \rho)^{-1}$. To standardize the comparison define the quantity $c(\rho, q) = I_n(1 - \rho)$. Figure 10.3 indicates the values for $n = 1, \ldots, 500$. The symbols A, B and C indicate $n = 26, 63, 465$, at which $I_n = I^*$, $I_n(1 - \rho) = 1.1$ and $I_n(1 - \rho) = 1.01$, respectively.

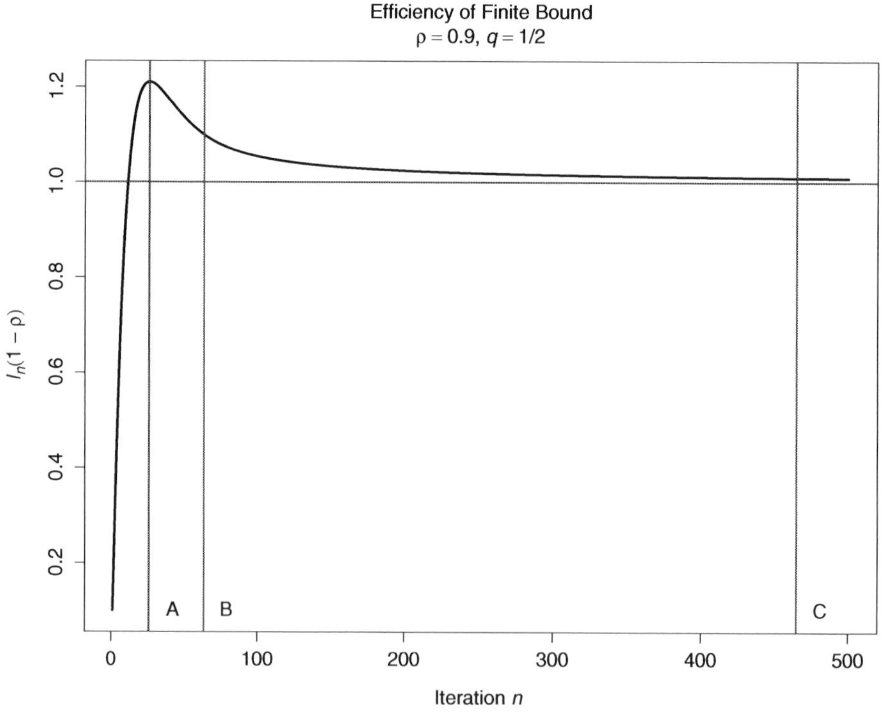

Figure 10.3 Example of finite bound method of Section 10.10.1. Parameters are set to $\rho = 0.9$, $q = 1/2$. Symbols A, B and C indicate $n = 26, 63, 465$, at which $I_n = I^*$, $I_n(1 - \rho) = 1.1$ and $I_n(1 - \rho) = 1.01$, respectively.

10.11 SUMMARY OF CONVERGENCE RATES FOR STRONGLY CONTRACTIVE MODELS

The analysis of this chapter is based on ideas presented in Almudevar (2008), but which concerned algorithms for which the EIA was defined by a single operator $T_n \equiv T$. It is therefore worth noting that extensions to nonhomogenous operator sequences, including nonexpansive algorithms, may be developed using essentially the same techniques. Furthermore, introducing Hilbert space structure can lead to considerable refinement of approximation bounds, and in the case of the Robbins-Monro algorithm allows a resolution of convergence not obtainable using Banach space structure alone.

However, the single contractive operator case remains quite central to many applications, including those considered here, so we will present a specialized summary.

First, for $\rho \in (0,1)$ define the following families of sequences:

$$\mathcal{F}_\rho^L = \left\{ \{d_k\} \in \mathcal{S} : \sum_{k=1}^{\infty} \rho^{-k} d_k < \infty \right\},$$

$$\mathcal{F}_\rho = \left\{ \{d_k\} \in \mathcal{S} : \hat{\lambda}\{d_k\} = \rho \right\}, \text{ and}$$

$$\mathcal{F}_\rho^U = \left\{ \{d_k\} \in \mathcal{S} : \lambda^l\{d_k\} > \rho \right\}. \tag{10.60}$$

Theorem 10.19 *Suppose $(\mathcal{V}, \|\cdot\|)$ is a Banach space on which T is a ρ-contractive operator (which therefore possesses fixed point $V^* = TV^* \in \mathcal{V}$). Suppose T_k is a sequence of operators on $(\mathcal{V}, \|\cdot\|)$ which defines AIA*

$$V_k = T_k V_{k-1}, \quad k \geq 1,$$
$$V_0 = v_0,$$

for which for which relative model (ARE) holds, in particular,

$$\|T_k V_{k-1} - TV_{k-1}\| \leq a_k + b_k \|V_{k-1}\|_0, \quad k \geq 1,$$

for sequences a_k, b_k, where $\|\cdot\|_0$ is a seminorm for which $\|V\|_0 \leq \kappa_0 \|V\|$ for all $V \in \mathcal{V}$ for some finite constant κ_0. Define

$$d_k^* = a_k + b_k \left\| V^* \right\|_0, \tag{10.61}$$

and

$$\lambda_k = \prod_{i=1}^{k} (\rho + \kappa_0 b_i). \tag{10.62}$$

The the following statements hold:

(*i*) *If* $\limsup_k \kappa_0 b_k = \delta \in [0, 1-\rho)$ *then*

$$\limsup_k \|V_k - V^*\| \leq (1-\rho-\delta)^{-1} \limsup_k d_k^* \tag{10.63}$$

(*ii*) *If* $\{d_k^*\} \in \mathcal{F}_\rho^L$ *then*

$$\|V_k - V^*\| \leq \Omega(\lambda_k). \tag{10.64}$$

(*iii*) *If* $\{d_k^*\} \in \mathcal{F}_\rho$ *then*

$$\hat{\lambda}^u \left\{\|V_k - V^*\|\right\} \leq \rho. \tag{10.65}$$

(*iv*) *If* $\{d_k^*\} \in \mathcal{F}_\rho^U$ *and* $d_k^* \to 0$ *then*

$$\limsup_k [d_k^*]^{-1} \|V_k - V^*\| \leq (1 - \rho/\lambda^l\{d_k^*\})^{-1}. \tag{10.66}$$

Proof We first note that the model satisfies general inequality (10.30) for $p = 1$ with sequence $\tilde{d} = \{d_k^*\}$ defined in (10.61) and product kernel (10.62). Then statements (*i*)–(*iv*) are applications of Cases I–IV.

First, note that if $b_k \to_k 0$ then $\lambda\{\lambda_k\} = \hat{\lambda}\{\lambda_k\} = \rho$ (by Lemma 9.3).

(i) is a direct application of Theorem 10.17.
(ii) follows from Theorem 10.11, after noting that if $\{d_k^*\} \in \mathcal{F}_\rho^L$, then the assumption (10.34) follows from the fact that $\lambda^k \geq \rho^k$.
(iii) follows from Lemma 10.14 and then Theorem 10.13, after noting that $b_k \to 0$, so that $\hat{\lambda}\{\lambda_k\} = \rho$.
(iv) follows directly from Theorem 10.15, after noting that $b_k \to 0$ and therefore $\hat{\lambda}\{\lambda_k\} = \rho$. ///

More informally, we have

$$\|V_k - V^*\| \leq \Omega(\max(d_k^*, (\rho+\epsilon)^k)) \tag{10.67}$$

for all $\epsilon > 0$ if $d_k^* \to 0$ at a linear rate of at least ρ, and if $d_k^* \to 0$ at a linear rate less than ρ, then $\|V_k - V^*\| \to 0$ at the best possible linear rate ρ. Furthermore, these rates are sharp by Theorem 10.18.

While it is crucial to understand the convergence rate of these algorithms, there will still be much to gain by considering the proportionality constant for the rate, particularly its dependence on $(1-\rho)^{-1}$. We have seen that improvements are possible when Hilbert space structure can be exploited. We will also see that the use of various seminorms in place of $\|\cdot\|$ in the relative error model (ARE) may also improve the bounds.

At the very least, it should be emphasized that the general inequality (10.30) may be used as it is. For the relative error model (ARE) this becomes

$$\| V_k - V^* \|^p \leq [\rho^k]^p \| V_0 - V^* \|^p + \mathcal{I}_k^p(\tilde{d}, \bar{\lambda}) \tag{10.68}$$

$$= [\rho^k]^p \| V_0 - V^* \|^p + \sum_{i=1}^{k} \rho^{k-i} (a_k + b_k \left\| V_{k-1} \right\|_0), \quad k \geq 1.$$

That the bound converges to 0 may be verified by Theorem 10.19, and the bound (10.68) is no larger than the bounds on which those results are based. Applications of the triangle inequality needed to obtain convergence rates can only weaken the bound. The only unknown quantity here is V^*, but it will be possible to conclude in many cases that the first term of (10.68) will converge to zero more quickly than the second.

It is also worth noting that the iteration formula (10.7) for $\mathcal{I}_k^1(\tilde{d}, \bar{\lambda})$ reduces to

$$\mathcal{I}_k^1(\tilde{d}, \bar{\lambda}) = a_k + b_k \left\| V_{k-1} \right\|_0 + \rho \mathcal{I}_{k-1}^1(\tilde{d}, \bar{\lambda}),$$

$$\mathcal{I}_1^1(\tilde{d}, \bar{\lambda}) = a_1 + b_1 \| V_0 \|_0 . \tag{10.69}$$

Finally, we note that the multistage contractive case can be handled by bounding the Lipschitz kernel, as discussed in Section 10.7.1. If for the underlying operator T, T^J has contraction constant ρ, then ρ in Theorem 10.19 would be replaced by $\rho^{1/J}$, and the inequalities adjusted by a multiplicative constant as discussed in Section 10.7.1, Chapter 9 or in Almudevar (2008).

Chapter 11

Selection of approximation schedules for coarse-to-fine AIAs

Suppose we are given a homogeneous EIA $V_k = TV_{k-1}$, $k \geq 1$, and that the contractive operator T is to be approximated. Suppose next that there exists an indexed family of approximate operators T_τ, where τ is some natural approximation parameter, such as grid size or sample size. In this case any AIA is defined by a sequence τ_k by defining a sequence of approximate operators $\hat{T}_k = T_{\tau_k}$. If operator tolerance can be expressed in terms of τ, then the convergence properties of the AIA may be determined by the theory of Chaper 10.

However, a new issue arises in algorithm design when the sequence τ_k may be selected by the designer with the objective of optimizing algorithmic efficiency. Clearly, if computational cost is relatively insensitive to the choice of τ, then a satisfactory resolution is achieved by any sequence τ_k which forces a convergence rate on the operator tolerance no slower than that of the EIA (see, for example, Equation (10.67)). Of course, we usually expect that the computational cost of evaluating T_τ will increase significantly as operator tolerance approaches zero. In fact, we will assume that this limiting cost is unbounded.

The material in this chapter is based on Almudevar and de Arruda (2012).

11.1 EXTENDING THE TOLERANCE MODEL

Recall equation (10.3) from Section 10.1:

$$\| V_k - V^* \| \leq \eta_k = B\alpha_k + u_k, \quad k \geq 1 \tag{11.1}$$

where V_k are the iterates of an AIA with limit V^*, η_k is the algorithm tolerance, $B\alpha_k$ is the tolerance of the EIA, and u_k is the approximation tolerance. We have already seen that this model usefully bounds a contractive AIA, both for the preliminary model of Section 10.2, and for the more general contractive model of Section 10.4. We have also seen that under general conditions we will have $u_k = O(\epsilon_k)$ where ϵ_k is the operator tolerance, thus providing a direct relationship between the latter quantity, which can usually be precisely estimated, and the overall performance of the algorithm.

Now suppose we add a cost structure to the model. In particular, let g_k be the computation cost for the evaluation of the kth iteration. If g_k is approximately constant, then η_k is an appropriate metric with which to evaluate algorithmic efficiency. On the other hand, suppose this does not hold, that is, the *cumulative cost*

$$\bar{G}_k = \sum_{i=1}^{k} g_i$$

does not statisfy $\bar{G}_k = \Omega(k)$. The efficiency of the algorithm is now appropriately evaluated relative to $\bar{G}_k$. In effect, we need to transform η_k to some function which behaves more like $\hat{\eta}_t \approx \eta_k$ when $t = \bar{G}_k$, which we refer to as the *computational algorithm tolerance*, and is interpretable as the algorithm tolerance achieved after a computational effort of t.

Now, suppose we have some choice regarding the approximation parameter τ, and this can be varied within an AIA. For example, evaluation of an operator may involve approximation on a grid, where τ is the grid size. The positive relationship between τ and the operator tolerance is well understood, and we also expect computation cost to be proportional to τ^{-d} for dimension d. Clearly for this, and almost any other approximation technique, there will be a decreasing relationship between ϵ_k and g_k, parametetrized by an approximation parameter τ_k.

We have already seen that allowing ϵ_k to approach zero will yield an AIA which converges to V^*. However, in our example, we would expect g_k to increase unboundedly, therefore η_k will no longer be a suitable tolerance metric. It would be more appropriate to compare algorithms by estimating η_k and g_k for any sequence τ_k, then transforming η_k into $\hat{\eta}_t$ as just discussed.

Once this is done, we may consider the problem of choosing a *tolerance schedule* τ_k, $k \geq 1$, which yields the best possible computational algorithm tolerance $\hat{\eta}_t$.

We proceed in the following way. Because we are able to develop a precise relationship between the computation cost, approximation tolerance and algorithm tolerance, we are able to reach some quite general conclusions about the optimal approximation tolerance rate. Examining the algorithm tolerance bound, we can see that η_k is always at least order $O(\alpha_k)$. We expect that forcing a decrease in u_k must always be at the expense of greater cumulative computation cost. However, if we already have $u_k = o(\alpha_k)$, then the algorithm tolerance cannot be improved. Clearly, if we increase $\bar{G}_k$ without significantly changing η_k, we must increase $\hat{\eta}_t$, so we conclude that there can be no advantage to decreasing u_k in this case. We can be more precise for the contractive model. In this case $\alpha_k \approx \alpha^k$. We also know that if $\epsilon_k \approx r^k$ for $r < \alpha$ we would have $u_k = o(\alpha^k)$, and hence $\eta_k \approx \alpha^k$. Again, we should expect $\hat{\eta}_t$ to decrease as r approaches α from below. Interestingly, this argument holds for any type of computation cost that increases with decreasing operator tolerance.

The case of $u_k \geq \Omega(\alpha_k)$ presents more of an apparent trade-off, since increasing u_k increases η_k while decreasing computation cost. However, the theory proposed in Almudevar and de Arruda (2012) may be used to show that for models with linearly convergent EIAs the situation is not much different than for the lower bound case just discussed. In particular, it turns out that increasing u_k cannot reduce the cumulative computation cost by an amount sufficient to force a strict improvement in the rate of $\hat{\eta}_t$. Furthermore, under certain well defined conditions doing so will yield a strictly larger $\hat{\eta}_t$. In particular, if u_k converges to 0 sublinearly, then $\hat{\eta}_t$ will be strictly larger than an algorithm for which u_k converges linearly at a rate $r > \alpha$. Again, this holds for quite general cost structures.

This makes a recommendation for a sequence of tolerance parameters τ_k quite straightforward. Noting that $u_k = \Omega(\epsilon_k)$ when $\epsilon_k > \Omega(\alpha^k)$, and $u_k = o(\alpha^k)$ when $\epsilon_k = o(\alpha^k)$, we may confine attention to approximation schedules for which $\epsilon_k \approx r^k$ for some $r \in [\alpha, 1)$, and as will be seen, in some cases further resolution is possible. However, we note an important omission, the case that $\epsilon_k \approx \alpha^k$, noting that in this case $u_k > \Omega(\epsilon_k)$. Pending further analysis, we may only conjecture that $\epsilon_k \approx \alpha^k$ would be an effective choice, as calculations in Almudevar and de Arruda (2012) suggest.

11.1.1 Comparison model for tolerance schedules

We are given an EIA with linear congergence rate $\alpha \in (0, 1)$, so that the algorithm tolerance is

$$\| V_k - V^* \| \leq \eta_k = B\alpha^k + u_k, \quad k \geq 1. \tag{11.2}$$

We also assume there is a nonincreasing nonnegative *computation function* G on $[0, \infty)$. We then refer to a *tolerance model* as the pair $M = (\alpha, G)$. We define an *approximation schedule*, or simply *schedule*, to be any sequence $S = \{u_k\} \in \mathcal{S}^-$. We have assumed that $u_k > 0$, since it will be natural to interpret $u_k \equiv 0$ as being equivalent to the exact algorithm. It will be convenient to regard u_0 as a dummy value of arbitrarily large value. The basis for the comparison of schedules is given in the following definition:

Definition 11.1 *A set $\mathcal{A}^S$ of AIAs conforms to a* tolerance model $M = (\alpha, G)$ *if a schedule $S = \{u_k\}$ may be associated with each algorithm for which (i) (11.2) holds for some finite B and for which (ii) the computation cost of the kth iteration is given by $g_k = G(u_k)$.*

It is important to note that the definition of G as a function of the approximation tolerance might seem problematic, since u_k may depend on previous iterations, while we expect $g_k = G(u_k)$ to depend only on iteration k. It would seem to be more natural to define the cost function as a function of the tolerance parameter, say $G^{app}(\tau)$, in which case $g_k = G^{app}(\tau_k)$. However, the relevant analysis concerns the interaction of cost with algorithm tolerance, and therefore with the approximation tolerance u_k. In effect, Definition 11.1 represents the hypothesis that u_k and G can interact in this direct manner, and we have shown earlier that this is the case. We can generally deduce a function $\epsilon_k = h(\tau_k)$ relating the tolerance parameter to the operator tolerance (we will see examples in later chapters). We have already shown that relationships such as $u_k \approx (1 - \alpha)^{-1}\epsilon_k$ may be established, so that the cost function

$$G(u) \approx G^{app}(h^{-1}((1 - \alpha)u)) \tag{11.3}$$

can be used for Definition 11.1. Given these relationships, an approximation schedule is equivalently expressed as a tolerance schedule $\{\tau_k\}$, which is of course our ultimate objective.

11.1.2 Regularity conditions for the computation function

It will be useful to define the following condition on the computation function G.

(A) The function $G:[0,\infty)\to[0,\infty]$ is finite on $(0,\infty)$, and satisfies the following properties for any two sequences $\{u_k\},\{u'_k\}\in\mathcal{S}$:

  (i) if $u'_k=o(u_k)$ then $\liminf_{k\to\infty}\left(G(u'_k)-G(u_k)\right)=d_G$ for some $d_G>0$ (possibly, $d_G=\infty$),

  (ii) if $\lim_{k\to\infty} G(u'_k)/G(u_k)=\infty$ then $u'_k=o(u_k)$.

Note that condition (A) implies that $G(u)$ is unbounded, since $G(u)$ cannot have a finite limit as $u\to 0$ (in effect, $G(0)=\infty$, so that the EIA, at least regarded as the limiting case $\tau_k\to 0$, is not tractable). We also anticipate that arbitrarily low approximation tolerance requires arbitrarily high computation time, otherwise no meaningful tradeoff exists.

We will also employ the following growth conditions on G and schedule S:

(B1) There exists a finite constant r_G for which the pair (G,S) satisfies

$$\limsup_{k\to\infty} G(u_{k+1})/G(u_k)=r_G.$$

(B2) The pair (G,S) satisfies

$$\lim_{k\to\infty} G(u_k)/\sum_{j=1}^{k} G(u_j)=0.$$

We will sometimes need to strengthen an ordering $u'_k=o(u_k)$ of two schedules $S=\{u_k\}$ and $S'=\{u'_k\}$. It will suffice to impose the following condition:

(C) Given two schedules S,S' we have $u'_k=o(u_{k+d})$ for all integers $d\geq 0$.

It is easily verified that (C) holds when $u'_k=o(u_k)$ and $\lambda^l\{u_k\}>0$.

11.2 MAIN RESULT

Theorems 11.1, 11.2 and 11.3 together establish a general principle. Suppose we are given two schedules $\{u_k\}$ and $\{u'_k\}$, with respective computational algorithm tolerances $\hat{\eta}_t$ and $\hat{\eta}'_t$. If schedule $\{u_k\}$ is closer than $\{u'_k\}$ to the linear rate of convergence α^k, then $\hat{\eta}_t$ is not worse, and may be strictly better than, $\hat{\eta}'_t$.

The first theorem deals with the lower bound case. Suppose $u_k=O(\alpha^k)$. If $u'_k=o(u_k)$ then $\hat{\eta}_t$ cannot be improved by $\hat{\eta}'_t$, and if in addition condition (C) holds, $\hat{\eta}_t$ will be strictly better.

Theorem 11.1 *Given tolerance model $M=(\alpha,G)$, with G satisfying condition (A), suppose we are given two schedules S and S', with constants B, B'. Suppose $u_k=O(\alpha^k)$.*

Then

(*i*) *if* $u'_k = o(u_k)$ *then* $\hat{\eta}_t = O(\hat{\eta}'_t)$,
(*ii*) *if* S, S' *satisfy condition (C) then* $\hat{\eta}_t = o(\hat{\eta}'_t)$.

Proof See Theorem 2 in Almudevar and de Arruda (2012) for the proof. ///

For the upper bound case it is assumed that $u_k \geq \Omega(\alpha^k)$ and that $u_k = o_\ell(u'_k)$. Under the weaker condition (B1) $\hat{\eta}_t$ cannot be improved by $\hat{\eta}'_t$, and if u'_k converges sublinearly then $\hat{\eta}_t$ is strictly better (Theorem 11.2). Under condition (B2) $\hat{\eta}_t$ is strictly better in either case (Theorem 11.3).

Theorem 11.2 *Suppose a tolerance model* $M = (\alpha, G)$ *and two schedules* S, S' *satisfy* (*B1*) *with* $u_k = o_\ell(u'_k)$. *Suppose. also, that schedule* S *satisfies*

$$\limsup_{k\to\infty} B\alpha^k/u_k = \kappa < \infty. \tag{11.4}$$

Then for any positive constant $r < 1$ *we have*

(*i*) $\liminf_{v\to 0} \bar{G}'_\eta\left(r(1+\kappa)^{-1}v\right)/\bar{G}_\eta(v) \geq r_G^{-1}$,
(*ii*) *for any finite constant* γ *there exists* $\beta_\gamma > 0$ *such that if* $\beta' \leq \beta_\gamma$ *then*

$$\liminf_{v\to 0} \bar{G}'_\eta\left(r(1+\kappa)^{-1}v\right)/\bar{G}_\eta(v) \geq \gamma,$$

(*iii*) *furthermore, if* S' *converges sublinearly then*

$$\lim_{v\to 0} \bar{G}'_\eta\left(r(1+\kappa)^{-1}v\right)/\bar{G}_\eta(v) = \infty.$$

Proof See Theorem 3 in Almudevar and de Arruda (2012) for the proof. ///

Theorem 11.3 *Suppose a tolerance model* $M = (\alpha, G)$ *and two schedules* S, S' *satisfy* (*B2*) *with* $u_k = o_\ell\left(u'_k\right)$. *Suppose, in addition, that* $\liminf_{i\to\infty} u_i/\alpha^i > 0$. *Then* $\hat{\eta}_t = o(\hat{\eta}'_t)$.

Proof See Theorem 4 in Almudevar and de Arruda (2012) for the proof. ///

11.3 EXAMPLES OF COST FUNCTIONS

We will find that cost functions will often assume some general form, so we will consider here two, in order to investigate conditions under which regularity conditions (A), (B1) or (B2) hold. Consider the general form of $G(u)$ in (11.3). Suppose an approximate operator T_τ is constructed by numerical integration on an interval $[a, b] \subset \mathbb{R}$, using step sizes τ (see, for example, Isaacson and Keller (1966)). Under general conditions we might establish an operator tolerance $\epsilon = O(\tau^p)$, where the power p depends on the particular method. The computation function would then be $G^{app}(\tau) = O(\tau^{-1})$, and given (11.3) we would have $G(u) = O(u^{-1/p})$.

First, we will give a stricter form of (A), based on part (*ii*) of the definition.

Lemma 11.1 *If* $\lim_{k\to\infty} G(u'_k)/G(u_k) = \infty$ *holds if and only if* $u'_k = o(u_k)$, *then condition (A) holds.*

Proof Part (*ii*) of (A) holds by hypothesis. Then if $\lim_{k\to\infty} G(u'_k)/G(u_k) = \infty$, we must have $\liminf_{k\to\infty} (G(u'_k) - G(u_k)) = \liminf_{k\to\infty} G(u'_k)\,(1 - G(u_k)/G(u'_k)) = \liminf_{k\to\infty} G(u'_k) > 0$, so that (*i*) holds. ///

We can expect many cost functions to be of the form $G(u) = \Omega(u^{-d})$. The following theorem characterizes this case.

Theorem 11.4 *If a computation function assumes form $G(u) = \Omega(u^{-d})$, $d > 0$, then (A) holds. In addition, for any schedule $S = \{u_k\}$, $\lambda_l\{u_k\} > 0$ implies condition (B1) and $\lambda_l\{u_k\} = 1$ implies condition (B2).*

Proof Given two schedules $\{u_k\}, \{u'_k\}$ we have $G(u'_k)/G(u_k) = (u'_k/u_k)^{-d}$, so that by Lemma 11.1 condition (A) holds.

Now suppose $\lambda_l\{u_k\} > 0$. Directly we have $\limsup_{k\to\infty} G(u_{k+1})/G(u_k) = \limsup_{k\to\infty} (u_k/u_{k+1})^d = [\lambda_l\{u_k\}]^{-d} < \infty$, so that (B1) holds.

Next, suppose $\lambda_l\{u_k\} = 1$. We may write

$$G(u_k)/\sum_{j=1}^{k} G(u_j) = \left[\sum_{j=1}^{k} \left(\frac{u_k}{u_j}\right)^d\right]^{-1}. \tag{11.5}$$

Fix small $\epsilon > 0$, and let N be any integer. We may select k' large enough so that $u_k/u_{k-1} > (1-\epsilon)^{1/dN}$ for all $k \geq k' - N$, in which case, for any $j = 1, \ldots, N$, and $k \geq k'$

$$\begin{aligned}\left(\frac{u_k}{u_{k-j}}\right)^d &= \prod_{i=1}^{j} \left(\frac{u_{k-i+1}}{u_{k-i}}\right)^d \\ &\geq \prod_{i=1}^{j} \left((1-\epsilon)^{1/dN}\right)^d \\ &= (1-\epsilon)^{j/N} \\ &\geq (1-\epsilon).\end{aligned} \tag{11.6}$$

The sum in the denominator in (11.5) may be bounded in the following manner, for all $k \geq k'$:

$$\sum_{j=1}^{k} \left(\frac{u_j}{u_k}\right)^d \geq \sum_{j=k-N+1}^{k} \left(\frac{u_k}{u_j}\right)^d \geq (1-\epsilon)N$$

using inequality (11.6), from which we may conclude $\limsup_{k\to\infty} G(u_k)/\sum_{j=1}^{k} G(u_j) \leq [(1-\epsilon)N]^{-1}$. Then (B2) holds, since N may be made arbitrarily large. ///

Another cost function which will arise naturally is $G(u) = -b\log(cu)$, for positive constants b, c. The hypothesis of Lemma 11.1 does not hold (part (*ii*) of condition (A) holds, but not the converse).

Theorem 11.5 *If a computation function assumes form* $G(u) = \Omega(-b\log(cu))$, $d > 0$, *then (A) holds. In addition, for any schedule* $S = \{u_k\}$, $\lambda_l\{u_k\} > 0$ *condition (B2) holds.*

Proof Given two schedules $\{u_k\}, \{u'_k\}$ we test part (*i*) of (A) by writing $\liminf_{k\to\infty} \left(G(u'_k) - G(u_k)\right) = \liminf_{k\to\infty} -b\log(cu'_k/u_k)$. This limit is unbounded if $u'_k = o(u_k)$, so part (*i*) holds.

Now suppose $\lambda_l\{u_k\} = \rho > 0$, and set $\rho_k = u_{k+1}/u_k$, so that $\liminf_{k\to\infty} \rho_k = \rho$. Then write

$$\frac{\sum_{j=1}^{k} G(u_j)}{G(u_k)} = \sum_{j=1}^{k} \frac{G(u_j)}{G(u_k)}. \tag{11.7}$$

For any $K < \infty$, we may choose k' and $1 > \delta > 0$ such that $u_k/u_j \geq \delta$ whenever $k' \leq j \leq k \leq k' + K$. Additionally, for any finite positive constant M k' may also be chosen so that $-\log(cu_j) > M$ when $j \geq k'$. Then, if $k \geq k'$ we may write

$$\begin{aligned}
\frac{\sum_{j=1}^{k+K} G(u_j)}{G(u_{k+K})} &\geq \sum_{j=k+1}^{k+K} \frac{G(u_j)}{G(u_{k+K})} \\
&= 1 + \sum_{j=k+1}^{k+K-1} \frac{-\log(cu_j)}{-\log(u_{k+K}/u_j) - \log(cu_j)} \\
&\geq K/(-\log(\delta)M^{-1} + 1).
\end{aligned}$$

Consequently,

$$\limsup_{k\to\infty} G(u_k)/\sum_{j=1}^{k} G(u_j) \leq K^{-1}(-\log(\delta)M^{-1} + 1).$$

This holds for all K, which, following a suitable choice for M, completes the proof. ///

11.4 A GENERAL PRINCIPLE FOR AIAs

The simplicity of the tolerance-cost model permits the formulation of a general principle for AIAs. Suppose our intention is to replace α-contractive operator T with a single approximate operator $\hat{T}$ with operator tolerance ϵ (note that it is not necessary to verify any contractivity properties of $\hat{T}$ once the operator tolerance is known to be bounded). We can achieve an algorithm tolerance of order $\epsilon(1-\alpha)^{-1}$. On the other hand, if we are given access to a class of approximate operators T_τ indexed by τ with cost function G^{app}, we may devise instead an AIA based on a sequence of approximate operators $\hat{T}_k = T_{\tau_k}$. The tolerance schedule τ_k is selected to yield an approximation schedule $u_k \approx \alpha^k$, which may be done using the methods described earlier. This coarse-to-fine strategy can be recommended for any suitable approximation scheme. We will consider two examples in Chapters 16 and 17.

Part III

Application to Markov decision processes

Chapter 12

Markov decision processes (MDP) – background

A *Markov decision process* (MDP) is a controlled Markov chain (Section 5.2), the objective of the control being to minimize a sum or average of (possibly discounted) costs associated with each *stage* (the term used to describe all aspects of a single time index). The set of admissible controls is precisely given, so we have a well defined optimization problem. Formally, the MDP itself is a selection, through the control process, from a class of Markov chains, and so may not itself be a Markov chain. This depends on whether or not the control itself is Markovian, in the sense that it depends only on the system's current state. It can be shown that under general conditions the optimal control can be chosen to be Markovian, assuming perfect system identification. But even when this holds, there may be formidable computational challenges to overcome, so, for a number of reasons, there has has been considerable development of approximation methods for MDPs, for example, Hernández-Lerma (1989b), Bertsekas and Tsitsiklis (1996), Chang et al. (2007), Si et al. (2004), Buşoniu et al. (2010), Powell (2011).

Given complete model identification, and a feasible computation method, the theory of MDPs permits the calculation of an optimal control. Obviously, it is ideal that the optimal control is available thoughout the entire system history, otherwise *regret*, or the cost in excess of the minimum attainable due to the application of suboptimal control, is accrued.

An *off-line* approximation method mimics this ideal by replacing the optimal control with an approximation, then using this approximation in the actual system as though it were the optimal control. An *on-line* approximation method does not make this distinction. The control itself incorporates approximation methods, and therefore has two objectives, minimizing current regret and refining the approximation so as to minimize future regret. As will be seen, these two objectives may be conflicting, forming a new optimization problem.

Approximations may take many forms, but are generally of two types. The first type arises when model estimates are used in place of the true but unknown model. This typically involves statistical estimation and sampling. The second type involves functional approximation. In this case the model is assumed known, but calculation of an optimal control is problematic, possibly because of high computational complexity, or because of dependence on numerical approximation methods.

The AIA theory developed earlier will allow a general method of relating the tolerance of an approximation method to regret, usually expressible in finite bounds.

Once this is done, we may determine the direct relationship between, for example, the sample size of a statistical model estimate, or the grid size of a numerical integration method, to the achieved regret of the resulting control. It will then be possible to exploit these bounds in the design of approximate algorithms, determining, for example, the sample size needed to achieve regret within a fixed tolerance, the most efficient allocation of finite computing resources, or the optimal rate at which an on-line control engages in exploratory behavior at the expense of cost minimizing behavior.

In this chapter, we define the MDP, and develop the theory necessary for the determination of optimal controls, and for the estimation of regret associated with approximate controls. *Value iteration* (VI) is discussed in Chapter 13, which is an iterative algorithm commonly used to calculate the optimal control of MDPs. This will become the exact iterative algorithm (EIA) described in Chapter 10, for which an approximate iteration algorithm (AIA) is to be constructed. Some general principles for the construction of AIAs for MDPs are introduced in Chapter 14.

The remaining chapters discuss specific AIAs. In Chapter 15 model approximation by statistical estimation is considered, both from the point of view of estimation methods and of sample size determination. Two examples of functional approximation follow. Chapter 16 discusses approximation via truncation of the probability distributions within the stochastic kernel defining an MDP. Chapter 17 discusses discrete grid approximations of continuous state MDP models. Both of these methods exploit the optimal tolerance scheduling of Chapter 11. We finally consider the problem of adaptive control in Chapter 18, using the general AIA theory to deduce optimal exploration rates, where exploration may be defined as suboptimal control intended to yield data for improved model estimation.

12.1 MODEL DEFINITION

In the standard formulation, a MDP navigates indefinitely through a *state space* $\mathcal{X}$ by discrete stages. Each stage consists of a state x from which an *action* a from *action space* $\mathcal{A}$ is taken, resulting in *cost* $R(x,a)$. The process transfers to a subsequent state according to probability measure $Q(\cdot \mid x,a)$, which begins a new stage. The state/action space $\mathcal{K} \subset \mathcal{X} \times \mathcal{A}$ consists of all state/actions pairs (x,a) for which action a is available from state x. If β is a cost *discount factor*, we refer to the object $\pi = (\mathcal{K}, Q, R, \beta)$ as a Markov control model (MCM). A *control policy* Φ is a rule (which may be probabilistic) for choosing an action based on the current state and process history. A MCM π with a control policy Φ together define a MDP.

The total discounted cost is $\sum_{n\geq 1} \beta^{n-1} R(X_n, A_n)$, where (X_n, A_n) is the state/action pair for the nth stage. We will consider other forms of cost, but it turns out that developing the theory for this case first is a useful approach.

Calculation of an optimal control usually proceeds by considering the *value function* $\bar{V}_\pi$, which gives the lowest achievable expected discounted cost $\bar{V}_\pi(x)$ for initial state x under MCM π. Under general conditions, $\bar{V}_\pi$ is the fixed point of the *Bellman operator* $\bar{T}_\pi$, which is defined for a specific MCM π. Then $\bar{V}_\pi$ may be calculated by the iterative algorithm $V_n = \bar{T}_\pi V_{n-1}$, $n \geq 1$, that is, value iteration, and it may be established that the algorithm converges to $\bar{V}_\pi$ under general conditions. The optimal control can then be derived by first calculating $\bar{V}_\pi$, then implicitly through an evaluation of $\bar{T}_\pi \bar{V}_\pi$. Therefore, π must be known in order to optimize cost.

We adopt the following conventions. If $\mathcal{U}$ is a Borel space, the class of Borel sets is denoted $\mathcal{B}(\mathcal{U})$. Then $\mathcal{M}(\mathcal{U})$ is the set of probability measures on $\mathcal{U}$. In addition, $\mathcal{F}(\mathcal{U})$ is the vector space of all measurable functions $V : \mathcal{U} \to \mathbb{R}$. As usual, if $\|\cdot\|$ is a seminorm defined on $\mathcal{F}(\mathcal{U})$ then $\mathcal{F}(\mathcal{U}, \|\cdot\|) = \{V \in \mathcal{F}(\mathcal{U}) : \|V\| < \infty\}$.

A Markov decision process will be made of the following elements (see, for example, Hernández-Lerma and Lasserre (1996, 1999)):

(M1) A Borel space $\mathcal{X}$. We refer to $\mathcal{X}$ as the *state space*.
(M2) A Borel space $\mathcal{A}$. We refer to $\mathcal{A}$ as the *action space*.
(M3) With each $x \in \mathcal{X}$ associate $\mathcal{K}_x \in \mathcal{B}(\mathcal{A})$, with $\mathcal{K}_x \neq \emptyset$. The *state/action space* $\mathcal{K} = \{(x, a) \in \mathcal{X} \times \mathcal{A} : a \in \mathcal{K}_x\}$ is assumed to be a measurable subset of $\mathcal{X} \times \mathcal{A}$.
(M4) A *measurable stochastic kernel* $Q : \mathcal{K} \to \mathcal{M}(\mathcal{X})$ (see Definition 4.7).
(M5) A measurable mapping $R : \mathcal{K} \to \mathbb{R}$, referred to as the *cost function.*
(M6) A *discount factor* $\beta \geq 0$. ///

A reference to state/action space $\mathcal{K}$ will implicitly include $(\mathcal{X}, \mathcal{A})$. We have already defined a Markov control model (MCM) as the object $\pi = (\mathcal{K}, Q, R, \beta)$. We will use $H_n^x = (X_1, A_1, \ldots, X_n)$ and $H_n^a = (X_1, A_1, \ldots, X_n, A_n)$ to denote vectors in $\mathcal{K}^{n-1} \times \mathcal{X}$ and $\mathcal{K}^n$, respectively, with H_0^x, H_0^a set to a null vector when needed. These objects represent system history.

Let $\mathcal{K}_f$ be the set of all measurable mappings $f : \mathcal{X} \to \mathcal{A}$ for which $f(x) \in \mathcal{K}_x$ for all $x \in \mathcal{X}$. We will assume $\mathcal{K}$ is constructed so that $\mathcal{K}_f$ is not empty. We have already seen that a kernel $Q : \mathcal{K} \to \mathcal{M}(\mathcal{X})$ can be considered to be a family of stochastic kernels $Q^\phi : \mathcal{X} \to \mathcal{M}(\mathcal{X})$ indexed by $\phi \in \mathcal{K}_f$, evaluated for each $x \in \mathcal{X}$ by $Q^\phi(\cdot \mid x) = Q(\cdot \mid x, \phi(x))$. Similarly, $R^\phi \in \mathcal{F}(\mathcal{X})$ is defined by $R^\phi(x) = R(x, \phi(x))$.

A Markov decision process will consist of a MCM $\pi = (\mathcal{K}, Q, R, \beta)$ coupled with a control policy:

Definition 12.1 *A* control policy *consists of a sequence of stochastic kernels* $\Phi = \{\Phi_n, n \geq 1\}$ *of the form* $\Phi_n : (\mathcal{K})^{n-1} \times \mathcal{X} \to \mathcal{M}(\mathcal{A})$. *We assume* $\Phi_n(\mathcal{K}_{x_n} \mid H_n^x = h_n^x) = 1$ *for* $n \geq 1$. *We refer to* Φ_n *as a* stage n control function. *We say control function* Φ_n *is* Markovian *if* $\Phi_n(\cdot \mid H_n^x) = \Phi_n(\cdot \mid X_n)$, *and policy* Φ *is Markovian if each control function is Markovian. A control function* Φ_n *is* deterministic *if there exists a measurable mapping* $\phi_n : (\mathcal{K})^{n-1} \times \mathcal{X} \to \mathcal{A}$ *such that* $\Phi_n(E_a \mid H_n^x) = I\{\phi_n(H_n^x) \in E_a\}$. *A policy* Φ *is deterministic if each control function is determinstic. A control policy is* stationary *if* $\Phi_n(\cdot \mid H_n^x) = \Phi_1(\cdot \mid X_n)$ *for each stage n. A stationary policy is necessarily Markovian.*

We will alternatively define a Markovian deterministic policy using $\Phi = \{\phi_1, \phi_2, \ldots\}$ *where* $\phi_n \in \mathcal{K}_f$, *and a stationary Markovian determinstic policy may be written simply as* $\Phi = \phi \in \mathcal{K}_f$.

The class of all admissible policies will be denoted Π, *and the class of all Markovian deterministic policies will be denoted* Π_{MD}.

Intuitively, the MDP is realized as a random process $(X_1, A_1, X_2, A_2, \ldots)$ on the product space $\mathcal{K}^\infty$. Stage n is taken to refer to (X_n, A_n), at which a cost of $R(X_n, A_n)$ is realized. It has been shown (see Hernández-Lerma (1989)) that for each $x \in \mathcal{X}$ there exists a unique probability measure P_x^Φ on $\mathcal{K}^\infty$ which satisfies

$$\begin{aligned} P_x^\Phi(X_1 = x) &= 1, \\ P_x^\Phi(A_n \in E_a \mid H_n^x) &= \Phi_n(E_a \mid H_n^x), \;\; \forall E_a \in \mathcal{B}(\mathcal{A}), H_n^x \in \mathcal{K}^{n-1} \times \mathcal{X}, \\ P_x^\Phi(X_{n+1} \in E_x \mid H_n^a) &= Q(E_x \mid X_n, A_n), \;\; \forall E_x \in \mathcal{B}(\mathcal{X}), H_n^a \in \mathcal{K}^n, \end{aligned} \tag{12.1}$$

for $n \geq 1$ and each admissible history H_n^x, H_n^a.

The criterion for choosing a policy Φ is given as a β-discounted cost from initial point $X_1 = x$;

$$V_\pi^\Phi(x) = E_x^\Phi\left[\sum_{n=1}^{\infty} \beta^{n-1} R(X_n, A_n)\right], \quad x \in \mathcal{X}, \tag{12.2}$$

where E_x^Φ is the expectation operator of P_x^Φ. We refer to V_π^Φ as the *policy value function* for Φ.

It will also be useful to consider remaining cost at stage n, precisely,

$$\Lambda_n^\Phi(H_n^x) = E_x^\Phi\left[\sum_{i=0}^{\infty} \beta^i R(X_{n+i}, A_{n+i}) \mid H_n^x\right], \quad n \geq 1, \tag{12.3}$$

which may be intepreted as the discounted cost of a MDP which has been 'restarted' at state X_n at stage n. Anticipating nonMarkovian control policies, this value is allowed to depend on H_{n-1}^a as well as X_n, so that $\Lambda_n^\Phi(H_n^x)$ may not be expressible as a function $V_\pi^{\Phi'}(x)$ for some policy Φ'. Of course, the special case

$$\Lambda_1^\Phi(h_1^x) = V_\pi^\Phi(x_1)$$

holds.

Next, suppose we are given control policy $\Phi = \{\Phi_n, n \geq 1\}$. Fix both $n \geq 1$ and history $H_{n-1}^a = h_{n-1}^a$. Then for $j \geq 1$, define $\Phi_j[n, h_{n-1}^a] = \Phi_{j+n-1}(\cdot \mid H_j^x \times h_{n-1}^a)$ as a mapping from H_j^x to $\mathcal{M}(\mathcal{A})$. If we then set $\Phi_{(n)}[h_{n-1}^a] = (\Phi_1[n, h_{n-1}^a], \Phi_2[n, h_{n-1}^a], \ldots)$ this satisfies the definition of a control policy, which we can intepret as the *nth stage offset policy* conditional on history $H_{n-1}^a = h_{n-1}^a$.

If $\Phi = (\phi_1, \phi_2, \ldots) \in \Pi_{MD}$, then the offset policy is simply $\Phi_{(n)}[H_{n-1}^a] = \Phi_{(n)} = (\phi_n, \phi_{n+1}, \ldots) \in \Pi_{MD}$.

Then, given the construction of P_x^Φ according to (12.2) we have

$$\Lambda_n^\Phi(H_n^x) = V_\pi^{\Phi_{(n)}[H_{n-1}^a]}(X_n), \tag{12.4}$$

since the distribution of $X_n, A_n, X_{n+1}, A_{n+1}, \ldots$ depends on H_{n-1}^a only through the control policy.

12.2 THE OPTIMAL CONTROL PROBLEM

Equation (12.2) defines the problem of minimizing $V_\pi^\Phi(x)$ over the family of admissible policies Φ. If we are given model π, we have enough information to evaluate $V_\pi^\Phi(x)$ for any candidate Φ, so the problem is well defined. This can be conceived of as a separate problem for each $x \in \mathcal{X}$ but, for very good reasons, the usual approach is to consolidate all optimization tasks into one.

Ideally, there is a control policy Φ^* which achieves this minimum, that is $V_\pi^{\Phi^*}(x) = \inf_\Phi V_\pi^\Phi(x)$ for each x. However, as in other optimization problems involving a continuum, this is not guaranteed, at least without certain continuity assumptions. This can occur in a quite intuitive and practical models, for example, the *capacity expansion problem* (see, for example, Davis et al. (1987) or Almudevar (2001)). Suppose there does exist some Φ' for which $V_\pi^{\Phi'}(x) < \infty$. Then for any $\epsilon > 0$ there must exist an *ϵ-optimal* control policy Φ^ϵ for which $V_\pi^{\Phi^\epsilon}(x) \leq \inf_\Phi V_\pi^\Phi(x) + \epsilon$.

12.2.1 Adaptive control policies

The structure of a control policy is potentially quite complex, allowing the control decision applied at state $X_n = x$ to depend on the accumulated history H_n^x. There exist very good reasons for allowing this. Even if we accept the form of an MCM π as a good model, we may not have sufficient model identification to determine the optimal policy. Of course, H_n^x will contain information with which the model could be estimated. Furthermore, as $n \to \infty$ the model estimate can be refined as more information becomes available. The more accurate the estimate, the closer to optimal the resulting control can be. In this way, new information can always be converted to lower costs, and so there is no reason to terminate this type of *adaptive* updating process. In this case, we expect that the control Φ will not be Markovian. An example of this type of control policy is discussed in Chapter 18.

12.2.2 Optimal control policies

Of course, we first need to consider the problem of determining an optimal control under the assumption that π is known. The need for complex history dependence is not so apparent in this case, so it would be useful to know if the optimization problem can be confined to the simplest type of control, that is, the Markovian deterministic policy.

That this should be the case seems reasonable. Suppose we are at state X_n at stage n. We have observed history H_{n-1}^a, and so will subsequently adopt a policy given by $\Phi_{(n)}[H_{n-1}^a]$, as intended by our adoption of the policy Φ. However, we would also be able to follow policies given by $\Phi_{(n)}[h_{n-1}^a]$ for *any* admissible history h_{n-1}^a. We would also be able to calculate the resulting expected future cost by evaluating $V_\pi^{\Phi_{(n)}[h_{n-1}^a]}(X_n)$. Thus, rather than proceed according to the original policy, we may simply choose instead whatever policy $\Phi_{(n)}[h_{n-1}^a]$ minimizes $V_\pi^{\Phi_{(n)}[h_{n-1}^a)]}(X_n)$ over h_{n-1}^a.

An interesting technical issue arises at this point. Possibly, no particular h_{n-1}^a achieves the infimum of $V_\pi^{\Phi_{(n)}[h_{n-1}^a]}(X_n)$. In this case, we may always improve over the observed H_{n-1}^a, but the control would still depend on this observed value. In this case interpret $E_x^{\Phi}[V_\pi^{\Phi_{(n)}[H_{n-1}^a]}(X_n) \mid H_{n-1}^a] = \psi(H_{n-1}^a)$ as a random function of H_{n-1}^a. Clearly, there exists some h_{n-1}^a for which $\psi(h_{n-1}^a) \leq E_x^{\Phi}[\psi(H_{n-1}^a)]$, then we may reduce the expected discounted cost by using policy $\Phi_{(n)}[h_{n-1}^a]$ rather than $\Phi_{(n)}[H_{n-1}^a]$ from stage n onwards. In either case, the control function at n under the modified policy is Markovian. We may therefore construct a Markovian policy that is at least as good as the original by sequentially carrying out this process at stages $2, 3, \ldots$.

A similar argument can be used to show that nondeterminstic policies may be improved by deterministic ones. Suppose Φ is a Markovian policy. We may write

$$\begin{aligned} V_\pi^{\Phi}(x) &= E_x^{\Phi}\left[R(X_1, A_1) + \beta V_\pi^{\Phi(2)}(X_2)\right] \\ &= E_x^{\Phi}\left[E_x^{\Phi}\left[R(x, A_1) + \beta V_\pi^{\Phi(2)}(X_2) \mid H_1^a\right]\right] \\ &= E_x^{\Phi}\left[\psi(x, A_1)\right] \end{aligned} \tag{12.5}$$

where the distribution of A_1 is given by $\Phi_1(\cdot \mid x)$. We may similarly improve Φ by replacing Φ_1 with any $\phi_1(x) = a_1$ which minimizes $\psi(x, a_1)$ for fixed x, or a_1 for which $E_x^{\Phi}[\psi(x, a_1)] \leq E_x^{\Phi}[\psi(x, A_1)]$. A determinstic policy may then be constructed sequentially as just described. There will therefore be special interest in the class of Markovian deterministic policies Π_{MD}, and we may summarize our argument by the following theorem:

Theorem 12.1 *For any policy Φ and initial state $X_1 = x$ there is a Markovian deterministic policy $\Phi' \in \Pi_{MD}$ for which $V_\pi^{\Phi'}(x) \leq V_\pi^{\Phi}(x)$.*

The value function is formally defined elementwise on $\mathcal{X}$ by

$$V_\pi^{\Pi}(x) = \inf_{\Phi \in \Pi} V_\pi^{\Phi}(x).$$

This term is reserved for the best possible cost (recall that the term 'policy value' function refers to the cost attained by a specific policy Φ). If there is a control policy Φ for which $V_\pi^{\Phi} \leq V_\pi^{\Pi} + \epsilon$ for $\epsilon > 0$ we say Φ is *ϵ-optimal*, and if $V_\pi^{\Phi} \leq (1+\epsilon) V_\pi^{\Pi}$, we say Φ is *scalar ϵ-optimal.*

This value function is well defined, but because the infimum in V_π^{Π} is taken pointwise there is no guarantee that there is any single V_π^{Φ}. However, it is easy to verify that the class of policies Π is rich enough to guarantee that this is the case.

Theorem 12.2 *If $V_\pi^{\Pi} < \infty$ then there exists ϵ-optimal and scalar ϵ-optimal policies.*

Proof For each x select Φ_x for which either $V_\pi^{\Phi_x} < V_\pi^{\Pi}(x) + \epsilon$ or $V^{\Phi_x} < (1+\epsilon) V_\pi^{\Pi}(x)$. Then let Φ be the (possibly nonMarkovian) policy which assumes policy Φ_x given starting point $X_1 = x$. Then Φ has value value function $V_\pi^{\Phi}(x) = V_\pi^{\Phi_x}(x)$, and so is ϵ-optimal and scalar ϵ-optimal accordingly. *///*

For continuous state/action spaces the issue of measurability arises twice. First, the calculation of any policy value function V_π^{Φ} assumes the ability to evaluate the expected value of $R(X_n, A_n)$. Second, the value iteration methods we will introduce require also that V_π^{Φ} or $\bar{V}_\pi$ are measurable functions. The two conditions are obviously related, and we will consider the question of measurability in more detail in Section 12.3.5. However, it is worth stating at this point the essential measurability requirement:

Assumption 12.1 *For every control Φ for which V_π^Φ exists, there exists $\Phi' \in \Pi_{MD}$ for which $V_\pi^{\Phi'} \in \mathcal{F}(\mathcal{X})$ and $V_\pi^{\Phi'} \leq V_\pi^\Phi$. This implies that $V_\pi^\Pi \in \mathcal{F}(\mathcal{X})$.*

We will assume Assumption 12.1 holds throughout.

12.3 DYNAMIC PROGRAMMING AND LINEAR OPERATORS

We first consider the problem of analyzing the policy value function. By definition we have

$$\begin{aligned} V_\pi^\Phi(x) &= E_x^\Phi\left[\sum_{n=1}^\infty \beta^{n-1} R(X_n, A_n)\right] \\ &= E_x^\Phi\left[R(X_1, A_1)\right] + \beta E_x^\Phi\left[\sum_{n=1}^\infty \beta^{n-1} R(X_{n+1}, A_{n+1})\right] \\ &= E_x^\Phi\left[R(X_1, A_1)\right] + \beta E_x^\Phi\left[\Lambda_2^\Phi(H_2^x)\right] \\ &= E_x^\Phi\left[R(X_1, A_1)\right] + \beta E_x^\Phi\left[V_\pi^{\Phi_{(2)}[H_1^a]}(X_2)\right]. \end{aligned} \tag{12.6}$$

Suppose we are given Markovian deterministic policy $\Phi = (\phi_1, \phi_2, \ldots) \in \Pi_{MD}$. This means $A_n = \phi_n(X_n)$ for $n \geq 1$ $wp1$, and that $\Phi_{(2)}[H_1^a] = \Phi_{(2)} = (\phi_2, \phi_3, \ldots) \in \Pi_{MD}$. Then by (12.6)

$$\begin{aligned} V_\pi^\Phi(x) &= R(x, \phi_1(x)) + \beta E_x^\Phi\left[V_\pi^{\Phi_{(2)}}(X_2)\right] \\ &= R(x, \phi_1(x)) + \beta \int_{y \in \mathcal{X}} V_\pi^{\Phi_{(2)}}(y) dQ(y \mid x, \phi_1(x)). \end{aligned} \tag{12.7}$$

This expression is quite intuitive. First, from initial state $X_1 = x$ assume cost $R(x, \phi_1(x))$. Transfer randomly to the next state X_2 according to distribution $Q(\cdot \mid x, \phi_1(x))$. From this point add to the total, after discounting by β, the expected remaining cost, which is $V_\pi^{\Phi_{(2)}}(X_2)$, given subsequent state X_2.

We can also see in (12.7) the type of linear operator considered in Section 7.2, that is, the *policy operator*

$$T_\pi^\phi V = R^\phi + \beta Q^\phi V,$$

so that (12.7) becomes

$$V_\pi^\Phi = T_\pi^{\phi_1} V_\pi^{\Phi_{(2)}}. \tag{12.8}$$

This point of view offers considerable flexibility. For example, the operation defined in (12.8) may be iterated indefinitely,

$$V_\pi^\Phi = T_\pi^{\phi_1} V_\pi^{\Phi(2)} = T_\pi^{\phi_1} T_\pi^{\phi_2} V_\pi^{\Phi(3)} = \cdots = T_\pi^{\phi_1} \ldots T_\pi^{\phi_{n-1}} V_\pi^{\Phi(n)},$$

where evaluation of the operators satisfies the associative property.

The operator may act on functions other than policy value functions. For example, $T_\pi^\phi \vec{0} = R^\phi$. Suppose we wish to evaluate the cost assumed by an MDP for the first N stages only. We may do this with the expression

$$E_x^\Phi\left[\sum_{n=1}^{N} \beta^{n-1} R(X_n, A_n)\right] = T_\pi^{\phi_1} \ldots T_\pi^{\phi_N} \vec{0}. \tag{12.9}$$

Next, suppose $\Phi = \phi$ is a stationary control. Then $\Phi_{(2)} = \Phi$, and (12.8) becomes

$$V_\pi^\phi = T_\pi^\phi V_\pi^\phi, \tag{12.10}$$

that is, the policy value function for stationary policy $\Phi = \phi$ must be a fixed point of T_π^ϕ.

12.3.1 The dynamic programming operator (DPO)

The operator T_π^ϕ provides a convenient way of evaluating policy value functions, but also provides a framework for the optimal control problem. Suppose our intention is to use control policy $\Phi = (\phi_1, \phi_2, \ldots) \in \Pi_{MD}$.

Define mapping $T_\pi^a : \mathcal{F}(\mathcal{X}) \to \mathcal{F}(\mathcal{K})$ elementwise for each $(x, a) \in \mathcal{K}$

$$(T_\pi^a V)(x, a) = R(x, a) + \beta \int_{y \in \mathcal{X}} V(y) dQ(y \mid x, a), \tag{12.11}$$

and write, following (12.8) and (12.11),

$$V_\pi^\Phi(x) = (T_\pi^{\phi_1} V_\pi^{\Phi(2)})(x) = (T_\pi^a V_\pi^{\Phi(2)})(x, \phi_1(x)). \tag{12.12}$$

Of course, this formulation gives us an opportunity to improve the control. At the first stage we may select instead of $\phi_1(x)$ any action $a \in \mathcal{K}_x$ yielding a smaller value, for example,

$$a_x^* = \operatorname{argmin}_{a \in \mathcal{K}_x} (T_\pi^a V_\pi^{\Phi(2)})(x, a), \tag{12.13}$$

so that

$$(T_\pi^a V_\pi^{\Phi(2)})(x, a_x^*) \le V_\pi^\Phi(x). \tag{12.14}$$

If this is done for each $x \in \mathcal{X}$, we may devise a new policy function $\phi_1^*(x) = a_x^*$. If the original policy $\Phi = (\phi_1, \phi_2, \dots)$ is modified by replacing ϕ_1 with ϕ_1^*, yielding $\Phi' = (\phi_1^*, \phi_2, \dots)$, then the comparison is straightforward,

$$V_\pi^{\Phi'} = T_\pi^{\phi_1'} V_\pi^{\Phi(2)} \le T_\pi^{\phi_1} V_\pi^{\Phi(2)} = V_\pi^{\Phi},$$

and so we have either improved the policy, or verified that it is the best available under the given constraints. The minimization operations defined in (12.13) (one for each $x \in \mathcal{X}$) have an intuitive meaning. It results in the optimal cost policy for stage 1 under the constraint that policy $\Phi_{(2)}$ will be used from stage 2 onwards. More generally, it is the optimal stage 1 policy given that the remaining cost (before discounting) given a transition to state $X_2 = x$ is $V(x) = V_\pi^{\Phi(2)}(x)$ for any x. Of course, we may substitute any measurable V. This minimization procedure may be expressed as a new operator, defined elementwise by

$$\begin{aligned} \bar{T}_\pi V(x) &= \inf_{a \in \mathcal{K}_x} (T_\pi^a V)(x, a) \\ &= \inf_{a \in \mathcal{K}_x} R(x, a) + \beta \int_{y \in \mathcal{X}} V(y) dQ(y \mid x, a), \end{aligned} \tag{12.15}$$

which is the infimum of all attainable costs when the remaining cost before discounting from state $X_2 = x$ is $V(x)$. The operator $\bar{T}_\pi$ defined in (12.15) is known as the *Bellman operator*, or the *dynamic programming operator* (DPO).

12.3.2 Finite horizon dynamic programming

We have introduced the idea of a *finite horizon* process, that is, one in which costs are assumed only for a finite number of stages N (in which case, the discount factor β may or may not play a role). It is necessary only to define a control function for the first N stages, but the reasoning behind Theorem 12.1 still applies. This means that (12.9) gives a well defined optimization problem, that is, to minimize the expected cost over the finite stage control policy $(\phi_1, \dots, \phi_N)$.

As is well known, effective optimization strategies are often based on the decomposition of a problem into simpler subproblems. *Dynamic programming* is such an approach to problems with a natural sequential decomposition. Suppose we write

$$\begin{aligned} T_\pi^{\phi_1} \dots T_\pi^{\phi_N} \vec{0} &= T_\pi^{\phi_1} \dots T_\pi^{\phi_{N-1}} (T_\pi^{\phi_N} \vec{0}) \\ &= T_\pi^{\phi_1} \dots T_\pi^{\phi_{N-1}} V_N, \end{aligned}$$

setting $V_N = T_\pi^{\phi_N} \vec{0}$. Note that V_N depends only on the MCM π and the final control function ϕ_N. Also, since each Q^ϕ is a positive linear operator, each operator T_π^ϕ is monotone (that is, $V \le V'$ implies $T_\pi^\phi V \le T_\pi^\phi V'$), so a necessary condition for an optimal solution is that ϕ_N minimize $T_\pi^{\phi_N} \vec{0}$, over all *single* control functions. This means the optimal selection of ϕ_N can be made independently of the rest of the problem. Clearly, this is achieved by $V_N^*(x) = \bar{T}_\pi \vec{0} = \inf_{a \in \mathcal{K}_x} R(x, a)$, and the optimal control function ϕ_N^* (if it exists) solves $T_\pi^{\phi_N^*} \vec{0} = \bar{T}_\pi \vec{0}$.

We then move back a step, writing

$$T_\pi^{\phi_1} \dots T_\pi^{\phi_N^*} \vec{0} = T_\pi^{\phi_1} \dots T_\pi^{\phi_{N-2}} (T_\pi^{\phi_{N-1}} V_N^*),$$

so that the next step is to minimize $V_{N-1} = T_\pi^{\phi_{N-1}} V_N^*$ *wrt* policy function ϕ_{N-1}. Fortunately, this problem is almost as simple as the previous one, since V_N^* is known from the last step, and we proceed by selecting ϕ_{N-1}^* which satisfies $T_\pi^{\phi_{N-1}^*} V_N^* = \bar{T}_\pi V_N^*$. This procedure, known as *backwards recursion*, is repeated to determine $\phi_N^*, \phi_{N-1}^*, \dots, \phi_1^*$ in order, which gives the solution. Formally, we have defined a form of dynamic programming algorithm:

$$\begin{aligned} V_N^* &= \inf_{a \in \mathcal{K}_x} R(x,a), \\ V_n^* &= \bar{T}_\pi V_{n+1}^*, \quad n = N-1, N-2, \dots, 1, \text{ with policy} \\ R(x, \phi_N^*(x)) &= \inf_{a \in \mathcal{K}_x} R(x,a), \\ T_\pi^{\phi_n^*} V_{n+1}^* &= \bar{T}_\pi V_{n+1}^*, \quad n = N-1, N-2, \dots, 1. \end{aligned} \tag{12.16}$$

The optimal cost function is then

$$V^* = \bar{T}_\pi^N \vec{0}.$$

Note that the existence of optimal control functions is not needed to calculate V^*. When these don't exist, we may always determine instead an ϵ-optimal control policy.

The advantage here is that a single optimization problem over an N-dimensional space of control functions has been replaced with N sequentially independent optimization problems, each over a single control function. If the state/action space $\mathcal{K}$ is finite, the search space has been reduced in size from $|\mathcal{K}_f|^N$ to $N|\mathcal{K}_f|$.

12.3.3 Infinite horizon problem

Next consider the *infinite horizon* problem, that is, the determination of a control policy Φ which minimizes (12.2). By Theorem 12.1 we may confine attention to policies in Π_{MD}. Suppose $\Phi = (\phi_1, \phi_2, \dots) \in \Pi_{MD}$ is optimal. This means $V_\pi^\Phi = V_\pi^\Pi$. It also means (for any $\Phi \in \Pi_{MD}$),

$$V_\pi^\Phi = T_\pi^{\phi_1} V_\pi^{\Phi(2)}.$$

First, note that since $V_\pi^\Phi = V_\pi^\Pi$ we must have $V_\pi^{\Phi(2)} \geq V_\pi^\Phi$. Optimality does not require strict equality, since $V_\pi^\Phi(x)$ is the expected total discounted cost given initial state $X_1 = x$. It would be possible to have $V_\pi^{\Phi(2)}(x') > V_\pi^\Phi(x')$ for some x' if $P(X_2 = x') = 0$. However, by the optimality of Φ and the monotonicity of $T_\pi^{\phi_1}$ we must have

$$V_\pi^\Phi = T_\pi^{\phi_1} V_\pi^{\Phi(2)} \geq T_\pi^{\phi_1} V_\pi^\Phi = V_\pi^{\Phi'},$$

for policy $\Phi' = (\phi_1, \phi_1, \phi_2, \ldots)$. By optimality $V_\pi^\Phi \le V_\pi^{\Phi'}$, so we must have $V_\pi^\Phi = V_\pi^{\Phi'}$, so that Φ' is also optimal. This also means

$$V_\pi^\Phi = T_\pi^{\phi_1} V_\pi^\Phi. \tag{12.17}$$

We may iterate as many times as we wish, so that if Φ is optimal, then any other policy in which the first $J < \infty$ control functions are ϕ_1, followed by the control functions of Φ, is also optimal.

We also have a necessary condition for ϕ_1. The value function V_π^Π can be defined independently of any specific policy, so ϕ_1 is the solution to:

$$V_\pi^\Pi = T_\pi^{\phi_1} V_\pi^\Pi. \tag{12.18}$$

Interestingly, we have so far not made use of the DPO $\bar{T}_\pi$, which we now do. While the equation (12.18) seems objective enough, it incorporates the problem of determining V_π^Π as well. By construction of the DPO, we must have $\bar{T}_\pi V_\pi^\Phi \le T_\pi^{\phi_1} V_\pi^\Phi$. However, if we had strict inequality $\bar{T}_\pi V_\pi^\Phi < T_\pi^{\phi_1} V_\pi^\Phi$ then ϕ_1 would not be optimal. Therefore, $V_\pi^\Pi = V_\pi^\Phi = \bar{T}_\pi V_\pi^\Phi = T_\pi^{\phi_1} V_\pi^\Phi$, and we have

$$V_\pi^\Phi = \bar{T}_\pi V_\pi^\Phi, \text{ or equivalently } V_\pi^\Pi = \bar{T}_\pi V_\pi^\Pi.$$

So, the value function is a fixed point of $\bar{T}_\pi$. If an optimal policy exists, and $\bar{T}_\pi$ has a unique fixed point, that fixed point must be V_π^Π. Then we may solve $T_\pi^{\phi_1} V_\pi^\Pi = V_\pi^\Pi$ to determine ϕ_1. An optimal policy can be constructed by applying control function ϕ_1 indefinitely. At this point, the original optimal policy Φ plays no role, and our optimal control follows entirely from a single optimal control function, denoted $\phi^* = \phi_1$. In fact, by comparing $T_\pi^{\phi_1} V_\pi^\Pi = V_\pi^\Pi$ with (12.10), V_π^Π can be equated with the policy value function of stationary policy $\Phi^* = \phi^*$.

This, in summary, is the solution to the infinite horizon discounted cost MDP control problem, although the reader is no doubt aware that a number of technical issues must be considered in order to make the preceding ideas mathematically valid.

12.3.4 Classes of MDP

We originally conceived of the MDP as an infinite horizon problem, in which cost is assumed indefinitely. The infinite horizon problem optimizes either the total discounted cost (as in (12.2)) or the long run average cost (see Section 12.7 below).

For the finite horizon problem, the cost will be finite, so there is no need for a discount factor $\beta < 1$, although we are free to use one when cost is appropriately discounted. In addition, there is no need for the transition kernel Q and cost function R to be the same for each stage. If they differ, the finite horizon dynamic programming algorithm given in (12.16) is easily modified by employing the appropriate operator (based on the appropriate cost function and transition kernel) at each iteration.

We may also also define a *shortest path problem*, in which there exists at least one terminal state from which no further cost is assumed, and in which the process may stay indefinitely. The stage structure is otherwise the same, as is the goal of minimizing total cost. This means that the process balances the goals of ending the process as soon as possible with the cost accrued in achieving this goal. The cost is usually assumed to be undiscounted.

12.3.5 Measurability of the DPO

A number of measure theoretic issues arise from the use of stochastic kernels and of the infimum operation in $\bar{T}_\pi$. Evaluation of $V' = \bar{T}_\pi V$ or $V' = T_\pi^\phi V$ assumes that V is Borel measurable. We need to know that V' is also measurable. This poses no particular problem for T_π^ϕ under the assumption that Q^ϕ is a measurable stochastic kernel by Definition 4.7 (see discussion in Section 4.9.1).

The situation is somewhat more complicated for $\bar{T}_\pi$ due to the infimum operation when $\mathcal{K}$ is uncountable (the reader can refer to Section 3.4 for more background on this issue). To see this suppose $\mathcal{U}, \mathcal{V}$ are two Borel spaces, and $f(u,v)$ is a measurable function on a product space $\mathcal{U} \times \mathcal{V}$. Evaluation of $\bar{T}_\pi$ is equivalent to evaluation of $f^*(u) = \inf_{v \in \mathcal{V}} f(u,v)$ (see the definition in (12.11)). However, without further assumptions $f^*(u)$ is not necessarily a Borel measurable function on $\mathcal{U}$. This problem is considered in Bertsekas and Shreve (1978) (Section 7.5), from which we cite the following two theorems:

Theorem 12.3 *Suppose $\mathcal{U}$ is a metrizable Borel space, $\mathcal{V}$ is a compact metrizable Borel space, D is a closed subset of $\mathcal{U} \times \mathcal{V}$ and f is a lsc function on $\mathcal{U} \times \mathcal{V}$. Evaluate*

$$f^*(u) = \inf_{v \in D_u} f(u,v), \quad D_u = \{v \mid (u,v) \in D\},$$

for $u \in proj_{\mathcal{U}} D$. Then $proj_{\mathcal{U}} D$ is closed in $\mathcal{U}$, f^* is lsc and there exists a Borel measurable function $\phi : proj_{\mathcal{U}} D \to \mathcal{V}$ such that $f(u, \phi(u)) = f^*(u)$.

Theorem 12.4 *Suppose $\mathcal{U}$ is a metrizable Borel space, $\mathcal{V}$ is a separable metrizable Borel space, D is an open subset of $\mathcal{U} \times \mathcal{V}$ and f is an usc function on $\mathcal{U} \times \mathcal{V}$. Evaluate*

$$f^*(u) = \inf_{v \in D_u} f(u,v), \quad D_u = \{v \mid (u,v) \in D\},$$

for $u \in proj_{\mathcal{U}} D$. Then $proj_{\mathcal{U}} D$ is open in $\mathcal{U}$, f^* is usc and for every $\epsilon > 0$ there exists a Borel measurable function $\phi_\epsilon : proj_{\mathcal{U}} D \to \mathcal{V}$ such that $f(u, \phi_\epsilon(u)) \le f^*(u) + \epsilon$.

We may, for example, use Theorem 4.24 to verify that if V is *lsc* on $\mathcal{X}$ and R is *lsc* on $\mathcal{K}$ then $(T_\pi^a V)(x,a)$ is *lsc* on $\mathcal{K}$. Then by Theorem 12.3 $\bar{T}_\pi V$ is *lsc* on $\mathcal{X}$.

In addition to Bertsekas and Shreve (1978) we may also recommend Hernández-Lerma (1989b) for a rigorous treatment of this issue. Otherwise, we rely on Assumption 12.1.

12.4 DYNAMIC PROGRAMMING AND VALUE ITERATION

We first introduce some notation and a preliminary lemma. Various bounds on R will need to be imposed. Accordingly, we define the following quantities:

$$R^{\phi}(x) = R(x,\phi(x)),$$
$$R^{inf}(x) = \inf_{a\in\mathcal{K}_x} R(x,a) \quad \text{and} \quad \bar{R}^{inf} = \inf_{x\in\mathcal{X}} R^{inf}(x),$$
$$R^{sup}(x) = \sup_{a\in\mathcal{K}_x} R(x,a) \quad \text{and} \quad \bar{R}^{sup} = \sup_{x\in\mathcal{X}} R^{sup}(x).$$

Clearly, if $\bar{R}^{inf} \geq 0$ we must have $V_\pi^\Phi \geq 0$ for any policy Φ, directly from (12.2).

Suppose we are given a control policy $\Phi = (\phi_1, \phi_2, \ldots) \in \Pi_{MD}$. We have already seen the iterative formula

$$V_\pi^\Phi = T_\pi^{\phi_1} \ldots T_\pi^{\phi_J} V_\pi^{\Phi^{(J+1)}}. \tag{12.19}$$

A J-step policy can be taken as $\tilde{\phi} = (\phi_1, \ldots, \phi_J) \in \mathcal{K}_f^J$. If we define the operator $T_\pi^{\tilde{\phi}} = T_\pi^{\phi_1} \ldots T_\pi^{\phi_J}$ then (12.19) can be written in the form $V_\pi^\Phi = T_\pi^{\tilde{\phi}} V_\pi^{\Phi^{(J+1)}}$. Of course, $T_\pi^{\tilde{\phi}}$ remains a linear operator,

$$\begin{aligned}
T_\pi^{\tilde{\phi}} V &= R^{\tilde{\phi}} + \beta^J Q^{\tilde{\phi}} V, \quad \text{where} \\
R^{\tilde{\phi}} &= R^{\phi_1} + \beta Q^{\phi_1} R^{\phi_2} + \beta^2 Q^{\phi_1} Q^{\phi_2} R^{\phi_3} + \cdots + \beta^{J-1} Q^{\phi_1} \ldots Q^{\phi_{J-1}} R^{\phi_J}, \quad \text{and} \\
Q^{\tilde{\phi}} &= Q^{\phi_1} \ldots Q^{\phi_J},
\end{aligned} \tag{12.20}$$

which follows from the iterations

$$\begin{aligned}
T_\pi^{\tilde{\phi}} V &= R^{\phi_1} + \beta Q^{\phi_1} T_\pi^{\phi_2} \ldots T_\pi^{\phi_J} V \\
&= R^{\phi_1} + \beta Q^{\phi_1} [R^{\phi_2} + \beta Q^{\phi_2} (T_\pi^{\phi_3} \ldots T_\pi^{\phi_J} V)] \\
&= R^{\phi_1} + \beta Q^{\phi_1} R^{\phi_2} + \beta^2 Q^{\phi_1} Q^{\phi_2} T_\pi^{\phi_3} \ldots T_\pi^{\phi_J} V \\
&\ \ \vdots
\end{aligned}$$

The definition of $R^{\tilde{\phi}}$ is consistent with R^{ϕ} for the special case of the 1-step policy.

So that the notation will not become too cumbersome, we will append the symbol $[n]$ to an operator to signify its nth iteration, that is, $T[n] = T^n$. In particular,

$$\bar{T}_\pi[n]V = (\bar{T}_\pi)^n V, \quad T_\pi^{\tilde{\phi}}[n]V = (T_\pi^{\tilde{\phi}})^n V, \quad \text{and} \quad Q^{\tilde{\phi}}[n] = \left(Q^{\tilde{\phi}}\right)^n.$$

We also set

$$\bar{R}[n] = \bar{T}_\pi[n]\vec{0} \quad \text{and} \quad R^{\tilde{\phi}}[n] = T_\pi^{\tilde{\phi}}[n]\vec{0},$$

and in particular, $\bar{R}[1] = R^{inf}$ and $R^{\tilde{\phi}}[1] = R^{\tilde{\phi}}$.

In a manner similar to (12.20) we note that for J-step policy function $\tilde{\phi}$, $T_\pi^{\tilde{\phi}}[n]$ is a linear operator equal to

$$T_\pi^{\tilde{\phi}}[n]V = R^{\tilde{\phi}}[n] + \beta^{nJ} Q^{\tilde{\phi}}[n]V. \tag{12.21}$$

Note that the operator $T_\pi^{\tilde{\phi}}[n]$ nodels nJ stages, as indicated by the power of the discount factor appearing in (12.21).

Some properties of $\bar{T}_\pi$ and $T_\pi^{\tilde{\phi}}$ are readily apparent, and summarized in the following lemma:

Lemma 12.1 *The following statements hold for DPO operators:*

(*i*) *For any* $V \in F(\mathcal{X})$ *and* $\tilde{\phi} \in \mathcal{K}_f^J$ *we have* $\bar{T}_\pi[J]V \leq T_\pi^{\tilde{\phi}} V$.

(*ii*) $\bar{T}_\pi$ *and* $T_\pi^{\tilde{\phi}}$ *are monotone operators.*

(*iii*) *If* $\bar{R}^{inf} \geq 0$ *and* $T = \bar{T}_\pi$ *or* $T = T_\pi^{\tilde{\phi}}$ *then:*

$$\begin{aligned} T(V_1 + V_2) &\leq TV_1 + TV_2 \\ TrV &\leq rTV, \quad \textit{for scalar } r \geq 1 \\ TrV &\geq rTV, \quad \textit{for scalar } r \in [0,1]. \end{aligned}$$

(*iv*) *If* $\bar{R}^{inf} \geq 0$ *then* $\bar{R}[n]$ *and* $R^{\tilde{\phi}}[n]$ *are increasing sequences.*

We will usually assume that $\bar{R}^{inf} \geq 0$, and sometimes some simplification follows from the assumption that $\bar{R}^{inf} > 0$. This constraint is intuitively reasonable in a minimum cost problem, since if costs are allowed to approach $-\infty$, the problem assumes a very different character. It is important to note that adding a constant c to R does not alter the problem, since the change to the total expected cost will be exactly the same for all policies (this idea is made precise in Section 7.4). The essential requirement is that $\bar{R}^{inf}$ is finite. For this reason it usually suffices to assume that R is nonnegative.

A distinction in the literature is often between bounded and unbounded costs, defined by the constraint $\bar{R}^{sup} < \infty$. If R is bounded the object would be to construct an algorithm which is convergent to the value function in the supremum norm.

If R is unbounded, the next question is whether or not there exists a policy Φ for which V_π^Φ is bounded. If so, the unboundedness of R is not decisive, and convergence in the supremum norm would be the goal. On the other hand, if all V_π^Φ are unbounded (with at least one policy value function being finite everywhere), then we would need to employ the weighted supremum norm, based on a weight function satisfying contraints imposed by both R and Q.

In our approach, employing the more general weighted supremum norm does not greatly complicate the analysis, and offers advantages even for the bounded cost model, so we will generally do so.

12.4.1 Value iteration and optimality

We are now in a position to make precise the analysis introduced in Section 12.3.3. Validation of the role of the DPO is based on establishing the equivalence of the value function V_π^Π with the objects

$$
\begin{aligned}
V_\pi^* &= \bar{T}_\pi V_\pi^* \\
\bar{V}_\pi &= \lim_{n\to\infty} \bar{R}[n],
\end{aligned} \tag{12.22}
$$

that is, the fixed point of $\bar{T}_\pi$, if it exists, and the limit $\bar{V}_\pi$, which by Lemma 12.1 (*iv*) exists when $\bar{R}^{inf} \geq 0$, which we generally assume. Following the discussion of Section 12.3.3 we will be able to define conditions under which $V_\pi^* = V_\pi^\Pi$. It is also important to identify $\bar{V}_\pi$ with V_π^Π, since this provides the algorithm with which to solve the optimization problem. In fact, $\bar{V}_\pi$ is the limit of an iterative algorithm based on operator $\bar{T}_\pi$ and initial solution $V_0 = \vec{0}$.

More generally, the goal is to achieve a *Value Iteration Algorithm* (VIA) which possesses the following elements:

Definition 12.2 *A* Value Iteration Algorithm *(VIA) consists of spaces $\mathcal{V} \subset \mathcal{F}(\mathcal{X})$, $\mathcal{V}_0 \subset \mathcal{V}$ and operator $T : \mathcal{V} \to \mathcal{V}$ for which* (*i*) *a unique fixed point $V^* = TV^* \in \mathcal{V}$ exists, and* (*ii*) $\lim_{n\to\infty} T^n V_0 = V^*$ *for all $V_0 \in \mathcal{V}_0$.*

In many cases it will be possible to set $\mathcal{V}_0 = \mathcal{V}$, which offers the possiblity of reducing the computation time by selecting V_0 to be close to V^* based on some initial estimate.

The central fact of the MDP optimization problem is the equivalence of V_π^Π, V_π^* and $\bar{V}_\pi$, which have nomimally very different derivations. Of course the value function always exists, as long as at least one policy may be evaluated, but its definition makes analysis somewhat difficult.

The quantity $V_\pi^* = \bar{T}_\pi V_\pi^*$ will prove to be more directly tractable, but before we use it, we must verify its existence and uniqueness. In most analyses, this is achieved by verifying that $\bar{T}_\pi$ is a contraction mapping.

First, we generally expect V_π^Π to be a fixed point of $\bar{T}_\pi$.

Theorem 12.5 *Under Assumption 12.1 we have $V_\pi^\Pi = \bar{T}_\pi V_\pi^\Pi$.*

Proof For any $\delta > 0$ there is $\Phi \in \Pi_{MD}$ for which $V_\pi^\Phi \leq (1+\delta) V_\pi^\Pi$. By appending any other $\phi \in \mathcal{K}_f$ to Φ we have the optimality conditions

$$
V_\pi^\Pi \leq T_\pi^\phi V_\pi^\Phi \leq \bar{T}_\pi V_\pi^\Phi \leq (1+\delta) \bar{T}_\pi V_\pi^\Pi
$$

so we may conclude that $V_\pi^\Pi \leq \bar{T}_\pi V_\pi^\Pi$ by letting δ vanish. Conversly, for any $\phi \in \mathcal{K}_f$

$$
\bar{T}_\pi V_\pi^\Pi \leq T_\pi^\phi V_\pi^\Pi \leq T_\pi^\phi V_\pi^\Phi.
$$

for any control policy Φ. The preceding upper bound may be set equal to V_π^Φ for any $\Phi \in \Pi_{MD}$. In particular, we may choose Φ for which $V_\pi^\Phi \leq (1+\delta) V_\pi^\Pi$. The proof is completed by letting δ vanish. ///

The next step is to establish the relationship between $\bar{V}_\pi$ and V_π^Π.

Theorem 12.6 *If* $\bar{R}^{inf} \geq 0$ *then* $\lim_{n\to\infty} \bar{R}[n] = \bar{V}_\pi \leq \bar{T}_\pi \bar{V}_\pi \leq V_\pi^\Pi$.

Proof If $\bar{R}^{inf} \geq 0$ then $\bar{R}[n]$ is increasing, with $\bar{R}[n] \leq \bar{V}_\pi$. In addition, $\bar{T}_\pi$ is monotone, so that $\bar{T}_\pi \bar{V}_\pi \geq \limsup_{n\to\infty} \bar{T}_\pi \bar{R}[n] = \limsup_{n\to\infty} \bar{R}[n+1] = \bar{V}_\pi$.

Next, suppose $\Phi \in \Pi_{MD}$. Then since $\bar{R}^{inf} \geq 0$, using (12.19) and Lemma 12.1 (*i*) we may write

$$V_\pi^\Phi = T_\pi^{\phi_1} \dots T_\pi^{\phi_n} V_\pi^{\Phi(n+1)} \geq T_\pi^{\phi_1} \dots T_\pi^{\phi_n} \vec{0} \geq \bar{T}_\pi^n \vec{0} = \bar{R}[n], \quad n \geq 1, \tag{12.23}$$

so, from (12.23) we may conclude $\bar{V}_\pi \leq V_\pi^\Phi$ for all $\Phi \in \Pi_{MD}$. By Theorem 12.1 it follows that $\bar{V}_\pi \leq V_\pi^\Pi$. Next, suppose we are given $\Phi = (\phi_1, \phi_2, \dots) \in \Pi_{MD}$. We must have $\bar{V}_\pi \leq V_\pi^{\Phi(2)}$, and by monotonicity of $\bar{T}_\pi$ also $\bar{T}_\pi \bar{V}_\pi \leq \bar{T}_\pi V_\pi^{\Phi(2)} \leq T_\pi^{\phi_1} V_\pi^{\Phi(2)} = V_\pi^\Phi$. We may similarly argue that $\bar{T}_\pi \bar{V}_\pi \leq V_\pi^\Pi$, which completes the proof. ///

After Theorem 12.6 two steps remain. First, we wish to know when $\bar{V}_\pi = \bar{T}_\pi \bar{V}_\pi$. If we examine the proof of the Banach fixed point theorem (Theorem 6.4) it can be seen that the existence of a fixed point for some operator T follows from its Lipschitz continuity on a metric space. Essentially, if $\bar{R}[n] \to \bar{V}_\pi$ and $\bar{T}_\pi$ is continuous we have $\bar{T}_\pi \bar{V}_\pi = \bar{T}_\pi \lim_n \bar{R}[n] = \lim_n \bar{T}_\pi \bar{R}[n] = \lim_n \bar{R}[n+1] = \bar{V}_\pi$, where convergence is taken with respect to the metric. This can be verified, and usually is, by defining a Banach space on which $\bar{T}_\pi$ is a Lipschitz operator under the supremum norm, so that $\bar{R}[n]$ converges uniformly to $\bar{V}_\pi$ and therefore also to $\bar{T}_\pi \bar{V}_\pi$.

It is worth noting, however, that we need only verify continuity of $\bar{T}_\pi$ with respect to the sequence $\bar{R}[n]$, and continuity with respect to pointwise convergence will suffice. The following theorem, an application of Dini's theorem (Theorem 3.16) gives an example of this approach.

Theorem 12.7 *Define*

$$\begin{aligned} v_{n+1}(x,a) &= (T_\pi^a \bar{R}[n])(x,a) \\ &= R(x,a) + \beta \int_{y\in\mathcal{X}} \bar{R}[n](y) dQ(y \mid x,a), \quad n \geq 1, \textit{ and} \\ \bar{v}(x,a) &= (T_\pi^a \bar{V}_\pi)(x,a) \\ &= R(x,a) + \beta \int_{y\in\mathcal{X}} \bar{V}_\pi(y) dQ(y \mid x,a). \end{aligned}$$

Suppose for each $x \in \mathcal{X}$ *and* $n \geq 1$ *the function* $g_n^x(a) = \bar{v}(x,a) - v_n(x,a)$ *is usc over* $a \in \mathcal{K}_x$, *and* $\mathcal{K}_x$ *is countably compact. Then* $\bar{V}_\pi = \bar{T}_\pi \bar{V}_\pi$

Proof For fixed x we have $\bar{R}[n](x) = \inf_a v_n(x,a)$ and $\bar{T}_\pi \bar{V}_\pi(x) = \inf_a \bar{v}(x,a)$, with the infimum taken over $a \in \mathcal{K}_x$. By the monotone convergence theorem $v_n(x,a)$ converges monotonically to $\bar{v}(x,a)$ for each (x,a).

By hypothesis, for each n and fixed x the function $g_n^x(a) = \bar{v}(x,a) - v_n(x,a)$ is *usc* over $a \in \mathcal{K}_x$. Since $\mathcal{K}_x$ is assumed countably compact, Theorem 3.16 implies that

$\bar{v}(x,a) - v_n(x,a)$ converges to 0 uniformly over $a \in \mathcal{K}_x$, from which is follows that $\bar{V}_\pi(x) = \lim_{n \lim \infty} \bar{R}[n](x) = \bar{T}_\pi \bar{V}_\pi(x)$. ///

The requirement that $\bar{v}(x,a) - v_n(x,a)$ be *usc* in the hypothesis of Theorem 12.7 would typically be resolved by, for example, Theorem 4.24.

That $\bar{V}_\pi = V_\pi^\Pi$ if and only if $\bar{V}_\pi = \bar{T}_\pi \bar{V}_\pi$ holds under quite general conditions, as stated in Proposition 9.16 of Bertsekas and Shreve (1978) (Definitions 8.1 and 9.1 define the precise model). However, as discussed in Section 9.5 of Bertsekas and Shreve (1978), deriving general conditions under which either statement holds is not simple (see, for example, Proposition 9.17 of Bertsekas and Shreve (1978)).

In the context of the present development, we have one half of the equivalence statement, since under the conditions of Theorem 12.6, $\bar{V}_\pi = V_\pi^\Pi$ implies $\bar{V}_\pi = \bar{T}_\pi \bar{V}_\pi$. That $\bar{V}_\pi = \bar{T}_\pi \bar{V}_\pi$ implies $\bar{V}_\pi = V_\pi^\Pi$ will be established in the next section. We therefore have two strategies.

The first approach is to use Theorem 12.6. For any Φ we must have $\bar{V}_\pi \leq V_\pi^\Phi$. Suppose $\bar{V}_\pi$ may be approximated arbitrarily well by some Φ, in the sense that $\bar{V}_\pi \leq V_\pi^\Pi \leq V_\pi^\Phi \leq \bar{V}_\pi + \epsilon$. If $\epsilon > 0$ can be made arbitrarily small then we also have $V_\pi^\Pi = \bar{V}_\pi$. Note that we have not assumed that there exists Φ^* for which $V_\pi^{\Phi^*} = V_\pi^\Pi$.

The second approach is to verify that $\bar{V}_\pi = \bar{T}_\pi \bar{V}_\pi$. If it can be further verified that any fixed point of $\bar{T}_\pi$ is unique, then by Theorem 12.5 we must have $\bar{V}_\pi = V_\pi^\Pi$. In fact, in most MDP models the DPO $\bar{T}_\pi$ is contractive and possesses a unique fixed point, so this suffices. However, we will be able to establish equivalence without relying on this structure, so that more general models may be considered.

We take the point of view that the question of the existence of an optimal control should not dominate the analysis, especially when the reader of this book is motivated by the difficulty in determining it even when it exists. In some developments, the analysis is simplified by its existence, but we argue that this need not be the case, by giving a simple method of constructing ϵ-optimal solutions. In the process, we will establish the other half of the equivalence statement, that is, $\bar{V}_\pi = \bar{T}_\pi \bar{V}_\pi$ implies $\bar{V}_\pi = V_\pi^\Pi$.

12.5 REGRET AND ε-OPTIMAL SOLUTIONS

The material of this section will be important in resolving optimality conditions, but will also play a central role in the design of approximate algorithms. The concept of regret can be intuitively defined as the amount by which the realized cost of a policy exceeds the optimal achievable. This definition does not rely on the existence of an optimal policy, as long as we have a sharp lower bound on achievable costs. This role will be played by $\bar{T}_\pi \bar{V}_\pi$.

We will make considerable use of the following simple quantity. For any MCM π and $(x,a) \in \mathcal{K}$ define

$$\lambda_\pi(x,a) = (T_\pi^a \bar{V}_\pi)(x,a) - (\bar{T}_\pi \bar{V}_\pi)(x). \tag{12.24}$$

Essentially the same quantity is employed in Schäl (1987).

Recall $\Lambda_n^\Phi(H_n^x)$ defined by (12.3), which represents the expected remaining discounted cost calculated from stage n given history H_n^x. The importance of this quantity is given in the following theorem:

Theorem 12.8 *If $\bar{R}^{inf} \geq 0$ and $\bar{V}_\pi = \bar{T}_\pi \bar{V}_\pi$ then*

$$\Lambda_n^\Phi(H_n^x) = (\bar{V}_\pi)(X_n) + E_x^\Phi\left[\sum_{i=0}^{\infty} \beta^i \lambda_\pi(X_{n+i}, A_{n+i}) \mid H_n^x\right]. \tag{12.25}$$

Proof First, write

$$\begin{aligned}\lambda_\pi(X_{n+i}, A_{n+i}) &= R(X_{n+i}, A_{n+i}) + \beta \int_{y\in\mathcal{X}} \bar{V}_\pi(y) dQ(y \mid X_{n+i}, A_{n+i}) - \bar{V}_\pi(X_{n+i}) \\ &= R(X_{n+i}, A_{n+i}) + \beta E_x^\Phi\left[\bar{V}_\pi(X_{n+i+1}) \mid H_{n+i}^a\right] - \bar{V}_\pi(X_{n+i}).\end{aligned}$$

Noting that $\sigma(H_n^x) \subset \sigma(H_{n+i}^a)$ for $i \geq 0$, we may write

$$\begin{aligned}&E_x^\Phi\left[\sum_{i=0}^{N} \beta^i \lambda_\pi(X_{n+i}, A_{n+i}) \mid H_n^x\right] \\ &\quad = E_x^\Phi\left[\sum_{i=0}^{N} \beta^i R(X_{n+i}, A_{n+i}) + \beta^{N+1} \bar{V}_\pi(X_{n+N+1}) \mid H_n^x\right] - \bar{V}_\pi(X_n).\end{aligned} \tag{12.26}$$

The expected value on the right side of equation (12.26) can be bounded above and below as follows:

$$\begin{aligned}\Lambda_n^\Phi(H_n^x) &= E_x^\Phi\left[\sum_{i=0}^{N} \beta^i R(X_n, A_n) + \beta^{N+1} V_\pi^{\Phi^{-(n+N+1)}}(X_{n+N+1}) \mid H_n^x\right] \\ &\geq E_x^\Phi\left[\sum_{i=0}^{N} \beta^i R(X_{n+i}, A_{n+i}) + \beta^{N+1} \bar{V}_\pi(X_{n+N+1}) \mid H_n^x\right] \\ &\geq E_x^\Phi\left[\sum_{i=0}^{N} \beta^i R(X_{n+i}, A_{n+i}) \mid H_n^x\right],\end{aligned}$$

since by Theorem 12.6 $\bar{V}_\pi$ is a lower bound on any achievable cost, so that letting $N \to \infty$ completes the proof. ///

An immediate application of Theorem 12.8 is the identification $\bar{V}_\pi = V_\pi^\Pi$, as well as the existence of various forms of ϵ-optimal policies in Π_{MD}.

Theorem 12.9 *If $\bar{V}_\pi = \bar{T}_\pi \bar{V}_\pi$ and $\bar{R}^{inf} \geq 0$ then for any $\epsilon > 0$ there exists policy $\Phi \in \Pi_{MD}$ for which*

$$V_\pi^\Phi(x) - \epsilon \leq \bar{V}_\pi(x) \leq V_\pi^\Phi(x), \quad \textit{for each } x \in \mathcal{X}. \tag{12.27}$$

Consequently, $\bar{V}_\pi = V_\pi^\Pi$.

In addition, the following hold:

(*i*) *If $\beta < 1$ there exists a stationary ϵ-optimal policy ϕ.*
(*ii*) *If $\bar{R}^{inf} > 0$ there exists a stationary scalar ϵ-optimal policy ϕ.*

Proof First note that $\Lambda_1^{\Phi}(H_1^x) = V_\pi^{\Phi}(x_1)$, so that Theorem 12.8 may be used directly to bound policy value functions.

To prove (12.27) select an nth stage control function ϕ_n for $\Phi \in \Pi_{MD}$ which satisfies $\lambda_\pi(x, \phi_n(x)) \leq \epsilon_n$ for all $x \in \mathcal{X}$. Each ϵ_n may be chosen as small as needed, in particular, we may force $\sum_n \beta^n \epsilon_n \leq \epsilon$ from which (12.27) follows after applying Theorem 12.8. Since we may construct an example of the inequality $\bar{V}_\pi \leq V_\pi^\Pi \leq V_\pi^\Phi \leq \bar{V}_\pi + \epsilon$ for any $\epsilon > 0$, we conclude that $\bar{V}_\pi = V_\pi^\Pi$.

To prove (*i*) select $\phi \in \mathcal{K}_f$ such that $\lambda_\pi(x, \phi(x)) \leq \epsilon$ for all $x \in \mathcal{X}$. Then by Theorem 12.8 we have $V^\phi(x) \leq (\bar{T}_\pi \bar{V}_\pi)(x) + \epsilon(1-\beta)^{-1}$. We then make ϵ as small as needed. The results follows from the fact that $V_\pi^\Pi = \bar{V}_\pi$.

To prove (*ii*) select $\phi \in \mathcal{K}_f$ such that $\lambda_\pi(x, \phi(x)) \leq \epsilon R^{inf}(x)$ for all $x \in \mathcal{X}$. Set $\epsilon < 1$. Then letting $\Phi = \phi$, we have

$$E_x^\Phi\left[\sum_{i=0}^{\infty} \beta^i \lambda_\pi(X_{n+i}, A_{n+i}) \mid H_n^x\right] \leq E_x^\Phi\left[\sum_{i=0}^{\infty} \beta^i \epsilon R^{inf}(x) \mid H_n^x\right]$$
$$\leq \epsilon \Lambda_n^\Phi(H_n^x).$$

An application of Theorem 12.8 yields $V_\pi^\Phi \leq (1-\epsilon)^{-1}\bar{V}_\pi = (1-\epsilon)^{-1}V_\pi^\Pi$. Then make ϵ as small as needed. ///

12.6 BANACH SPACE STRUCTURE OF DYNAMIC PROGRAMMING

We have seen that for the nonnegative cost model continuity properties of $\bar{T}_\pi$ (equivalently, the existence of a fixed point), coupled with the existence of a single policy yielding a finite value function, yields an iterative algorithm $\bar{R}[n]$ which converges to $\bar{V}_\pi = V_\pi^\Pi$. Before continuing to a Banach space construction, we show that multiple starting points are permitted.

We first consider policy evaluation.

Theorem 12.10 *If $\bar{R}^{inf} \geq 0$ then for any policy $\tilde{\phi} \in \mathcal{K}_f^J$*

$$\lim_{n\to\infty} (T_\pi^{\tilde{\phi}}[n]\vec{0})(x) = \lim_{n\to\infty} R^{\tilde{\phi}}[n](x) = V_\pi^{\tilde{\phi}} \tag{12.28}$$

for each $x \in \mathcal{X}$. If $V_\pi^{\tilde{\phi}} = T_\pi^{\tilde{\phi}} V_\pi^{\tilde{\phi}}$ then for any V_0 for which $\|V_0\|_{V_\pi^{\tilde{\phi}}} < \infty$ we also have

$$\lim_{n\to\infty} (T_\pi^{\tilde{\phi}}[n]V_0)(x) = V_\pi^{\tilde{\phi}}(x). \tag{12.29}$$

Proof A direct calculation gives, by the monotone convergence theorem,

$$V_\pi^{\bar{\phi}}(x) = \lim_{n\to\infty} E_x^{\bar{\phi}}\left[\sum_{i=1}^{n} \beta^{i-1} R(X_i, A_i)\right] = \lim_{n\to\infty} R^{\bar{\phi}}[n],$$

so that the pointwise limit (12.28) holds.

Next, suppose $V_\pi^{\bar{\phi}} = T_\pi^{\bar{\phi}} V_\pi^{\bar{\phi}}$. An indefinite number of iterations yields

$$V_\pi^{\bar{\phi}} = R^{\bar{\phi}}[n] + \beta^{nJ} Q^{\bar{\phi}}[n] V_\pi^{\bar{\phi}}.$$

From (12.28) we conclude that $\beta^{nJ} Q^{\bar{\phi}}[n] V_\pi^{\bar{\phi}} \to_n \vec{0}$ pointwise. Then

$$T_\pi^{\bar{\phi}}[n] V_0 = R^{\bar{\phi}}[n] + \beta^{nJ} Q^{\bar{\phi}}[n] V_0,$$

and

$$\left|\beta^{nJ} Q^{\bar{\phi}}[n] V_0\right| \leq \| V_0 \|_{V_\pi^{\bar{\phi}}} \beta^{nJ} Q^{\bar{\phi}}[n] V_\pi^{\bar{\phi}},$$

the upper bound of which converges pointwise to 0, which completes the proof. ///

For the DPO operator we have already identified $\bar{V}_\pi$ as the limit of a monotone sequence $\bar{R}[n]$, which is also an iterative algorithm based on $\bar{T}_\pi$ and starting point $V_0 = \vec{0}$. The task here is to identify the limit of $\bar{T}_\pi[n] V_0$ for a suitable class of general starting points V_0, under the assumption that $\bar{V}_\pi = \bar{T}_\pi \bar{V}_\pi$.

Theorem 12.11 *If $\bar{R}^{inf} \geq 0$ and $\bar{V}_\pi = \bar{T}_\pi \bar{V}_\pi$ then*

(*i*) *If $\vec{0} \leq V_0 \leq \bar{V}_\pi$ then* $\lim_{n\to\infty} \bar{T}_\pi[n] V_0(x) = \bar{V}_\pi(x)$

(*ii*) *In addition, if either $\beta < 1$ or $\bar{R}^{inf} > 0$ then* $\lim_{n\to\infty} \bar{T}_\pi[n] V_0(x) = \bar{V}_\pi(x)$ *for each $x \in \mathcal{X}$ where $V_0 \geq \vec{0}$ and $\| V_0 \|_{\bar{V}_\pi} < \infty$.*

Proof To prove part (*i*), by monotonicity we have

$$\vec{0} \leq V_0 \leq \bar{V}_\pi \Rightarrow \bar{T}_\pi[n]\vec{0} \leq \bar{T}_\pi[n] V_0 \leq \bar{T}_\pi[n] \bar{V}_\pi = \bar{V}_\pi \quad \text{for all } n \geq 1,$$

which completes the proof by noting $\bar{T}_\pi[n]\vec{0} \to_n \bar{V}_\pi$.

First suppose $V_0 \leq \bar{V}_\pi$. Then $\limsup_{n\to\infty} \bar{T}_\pi[n] V_0(x) \leq \bar{V}_\pi(x)$. On the other hand, by monotonicity we also have $\bar{T}_\pi[n] V_0(x) \geq \bar{T}_\pi[n] v_0(x) = \bar{R}[n]$ so that by we may conclude $\lim_{n\to\infty} \bar{T}_\pi[n] V_0(x) = \bar{V}_\pi(x)$.

To prove (*ii*) first suppose $V_0 \geq \bar{V}_\pi$. If $\beta < 1$ then by Theorem 12.9 for any $\epsilon > 0$ there is an ϵ-optimal policy ϕ. Since $V_\pi^{\phi} \geq \bar{V}_\pi$ by hypothesis $\| V_0 \|_{V_\pi^{\phi}} < \infty$. We may therefore apply Theorem 12.10

$$\bar{V}_\pi \leq \limsup_{n\to\infty} \bar{T}_\pi[n] V_0 \leq \lim_{n\to\infty} T_\pi^{\phi}[n] V_0 = V_\pi^{\phi} \leq \bar{V}_\pi + \epsilon.$$

Since this holds for any ϵ we must have $\lim_{n\to\infty} \bar{T}_\pi[n] V_0(x) = \bar{V}_\pi(x)$. Essentially the same argument is made if $\bar{R}^{inf} > 0$, based on Theorem 12.9.

The general case is then resolved by constructing an envelope, defined pointwise by $V_0^-(x) = \min(V_0(x), V_\pi(x)) \leq V_0(x) \leq \max(V_0(x), V_\pi(x)) = V_0^+(x)$. The cases already resolved may be used to verify pointwise convergence of both $\bar{T}_\pi[n]V_0^-$ and $\bar{T}_\pi[n]V_0^+$ to $\bar{V}_\pi$ and therefore also of $\bar{T}_\pi[n]V_0$ by monotonicity. ///

12.6.1 The contraction property

It is worth considering conditions under which we expect $\bar{T}_\pi$ to be contractive. If we first suppose that $R(x,a)$ is bounded and $\beta < 1$ then it is easily shown that $\bar{T}_\pi$ and T_π^ϕ, when calculable, are β-contractive. However, it is important to realize that contractivity need not be explicitly expressed using the discount parameter β. A unity of approach becomes apparent when the discount factor is interpreted as a 'kill rate', so that at each stage, there is a probability $1-\beta$ that the system moves to a 'kill state' Δ, at which the system stops, or in the context of an infinite horizon model, remains indefinitely without assuming any more cost. It is possible to incorporate this transition probability into the kernel, setting $Q(\{\Delta\} \mid x,a) = 1-\beta$ for each $(x,a) \in \mathcal{K}$, and renormalizing all other probability by a factor of β. In this case, there is no need to incorporate the discount factor in the DPO (so we would set $\beta = 1$), but the effect is exactly the same.

In more complex models, the kill rate may vary across (x,a). This is particularly true of *semi-Markov processes*, which are continuous-time processes in which a Markov chain is embedded at time points $t_1 < t_2 < \ldots$. In this case, it would usually be appropriate to regard discounting as being applied over continuous time, so that the discount factor for a stage may be taken to be $\exp(-\lambda T)$, where $\lambda > 0$ is a discount rate and T is the time length of the stage (presumed to be a function of state $x \in \mathcal{X}$). In this case, the contraction constant would be bounded above by $\exp(-\lambda T_{min})$, where T_{min} is the minimum time length for a stage. To establish the contraction property it would be enough to verify that $T_{min} > 0$, but the actual asymptotic contraction rate of any resulting operator may be strictly smaller.

Finally, we note that the contraction properties of shortest path MDPs (Section 12.3.4) can be established in this way, with the terminal state set as the kill state. In this case we would generally not expect single stage contration to hold, but would require instead that $Q^J(\{\Delta\} \mid x,a)$ be uniformly bounded from zero over $\mathcal{K}$, meaning that Δ is reachable from all states under all controls within J transitions.

12.6.2 Contraction properties of the DPO

Since $T_\pi^{\bar{\phi}}$ is a linear operator its contraction properties can be precisely resolved by Theorems 7.6 and 7.7 and the subsequent discussion. It is therefore straightforward to define a Banach space $\mathcal{F}(\mathcal{X}, \|\cdot\|_w)$ on which $T_\pi^{\bar{\phi}}$ is a contractive operator.

The DPO $\bar{T}_\pi$ is of course not a linear operator, but may be regarded as a type of composition of linear operators, in the sense that for any evaluation $\bar{T}_\pi V$ there is a linear operator $T_\pi^\phi V \approx \bar{T}_\pi V$, with the approximation as accurate as we like, and possibly exact under well defined continuity conditions. Therefore, it seems reasonable to suppose that certain properties that hold uniformly over all T_π^ϕ also hold for $\bar{T}_\pi$.

This, in fact, is the approach commonly taken. However, we will find that it may suffice to find a single operator T_π^ϕ that sufficiently resembles $\bar{T}_\pi$, the properties of the former then holding for the latter.

The first step is to verify that $\bar{T}_\pi$ is an operator on a Banach space $\mathcal{F}(\mathcal{X}, \|\cdot\|_w)$ for some $w \in \mathcal{W}$.

For an MCM we may define, for model elements R, Q

$$\eta_Q^w = \sup_{(x,a)\in\mathcal{K}} w(x)^{-1} \left\|\left| Q(\cdot \mid x,a) \right|\right\|_w$$
$$\eta_R^w = \sup_{(x,a)\in\mathcal{K}} w(x)^{-1} |R(x,a)|. \tag{12.30}$$

We can see from the definition that for any $\phi \in \mathcal{K}_k$ we must have

$$\left\|\left| Q^\phi \right|\right\|_w \leq \eta_Q^w \quad \text{and} \quad \left\| R^\phi \right\|_w \leq \eta_R^w.$$

For multistep policies $\tilde{\phi}$ it follows that $\left\|\left| Q^{\tilde{\phi}} \right|\right\|_w < \infty$ and $\left\| R^{\tilde{\phi}} \right\|_w < \infty$ when $\eta_Q^w, \eta_R^w < \infty$ (more precise bounds can follow from (12.20)). To verify that $\bar{T}_\pi$ is an operator on $\mathcal{F}(\mathcal{X}, \|\cdot\|_w)$ it will suffice to bound η_Q^w, along with related conditions on R. In contrast, contractivity may require consideration of J-step operators. Accordingly, we define

$$\eta_Q^w[J] = \sup_{\tilde{\phi}\in\mathcal{K}_f^J} \left\|\left| Q^{\tilde{\phi}} \right|\right\|_w. \tag{12.31}$$

We will impose the restriction $\eta_Q^w < \infty$, but the minimization procedure defining $\bar{T}_\pi$ permits a weaker assumption than $\eta_R^w < \infty$. This is summarized in the following theorem:

Theorem 12.12 *If for MCM π and weight function $w \in \mathcal{W}$ we have $\eta_Q^w < \infty$ and there exists finite positive constant M_R and at least one $\phi' \in \mathcal{K}_f$ for which*

$$\inf_{(x,a)\in\mathcal{K}} w(x)^{-1} R(x,a) \geq -M_R \quad \textit{and} \quad \left\| R^{\phi'} \right\|_w \leq M_R,$$

then

$$\left\| \bar{T}_\pi V \right\|_w \leq M_R + \beta\eta_Q^w \|V\|_w,$$

so that $\bar{T}_\pi$ is an operator on $\mathcal{F}(\mathcal{X}, \|\cdot\|_w)$.

Proof We always have

$$\begin{aligned}
w(x)^{-1}(\bar{T}_\pi V)(x) &\leq w(x)^{-1}(T_\pi^{\phi'} V)(x) \\
&= w(x)^{-1}R^{\phi}(x) + \beta w(x)^{-1} \int_{y \in \mathcal{X}} V(y)dQ(y \mid x, \phi(x)) \\
&\leq \left\| R^{\phi'} \right\|_w + \beta \left\| Q^{\phi'} \right\|_w \| V \|_w \\
&\leq M_R + \beta \eta_Q^w \| V \|_w \,.
\end{aligned}$$

Similarly,

$$\begin{aligned}
w(x)^{-1}(\bar{T}_\pi V)(x) &= \inf_{a \in \mathcal{K}_x} w(x)^{-1}R(x,a) + \beta w(x)^{-1} \int_{y \in \mathcal{X}} V(y)dQ(y \mid x, a) \\
&\geq -M_R - \beta \left\| Q(\cdot \mid x, a) \right\|_w \| V \|_w \\
&\geq -M_R - \beta \eta_Q^w,
\end{aligned}$$

which completes the proof. ///

That contractivity of $\bar{T}_\pi$ follows from the uniform contractivity of the family T_π^ϕ is a consequence of the following elementary theorem, proposed for this application in Hinderer (1970). We will have several occasions to use this theorem, given here in a slightly different form.

Theorem 12.13 *Let f_1, f_2 be real valued functions on a set E. Suppose* $\inf_x f_2(x) > -\infty$ *and* $|f_2(x)| < \infty$ *for all $x \in E$. Then* $| \inf_x f_1(x) - \inf_x f_2(x)| \leq \sup_x |f_1(x) - f_2(x)|$.

Proof First note that since f_2 is everywhere finite, $\sup_x |f_1(x) - f_2(x)|$ is well defined. If $| \inf_x f_1(x)| = \infty$ it follows that $\sup_x |f_1(x) - f_2(x)| = \infty$. Then suppose $\inf_x f_1(x)$ is finite. For any $\epsilon > 0$ there exists $x^* \in E$ such that $f_2(x^*) \leq \inf_x f_2(x) + \epsilon$. Then

$$\inf_x f_1(x) - \inf_x f_2(x) \leq f_1(x^*) - f_2(x^*) + \epsilon \leq \sup_x |f_1(x) - f_2(x)| + \epsilon.$$

A similar argument gives $\inf_x f_2(x) - \inf_x f_1(x) \leq \sup_x |f_1(x) - f_2(x)| + \epsilon$. The theorem follows by letting ϵ approach 0. ///

The application of uniform contractivity follows directly.

Theorem 12.14 *The J-step DPO $\bar{T}_\pi[J]$ possesses Lipschitz constant $\eta_Q^w[J]$, defined in* (12.31).

Proof From Theorem 12.13 we may write

$$
\begin{aligned}
& w(x)^{-1} \left| \bar{T}_\pi^J V_1 - \bar{T}_\pi^J V_2 \right| \\
& = w(x)^{-1} \left| \inf_{\bar{\phi}} \left[R^{\bar{\phi}}(x) + \beta \int_{\mathcal{X}} V_1 dQ^{\bar{\phi}}(x) \right] - \inf_{\bar{\phi}} \left[R^{\bar{\phi}}(x) + \beta \int_{\mathcal{X}} V_2 dQ^{\bar{\phi}}(x) \right] \right| \\
& \leq w(x)^{-1} \sup_{\bar{\phi}} \left| R^{\bar{\phi}}(x) + \beta \int_{\mathcal{X}} V_1 dQ^{\bar{\phi}}(x) - R^{\bar{\phi}}(x) + \beta \int_{\mathcal{X}} V_2 dQ^{\bar{\phi}}(x) \right| \\
& = w(x)^{-1} \sup_{\bar{\phi}} \left| \beta \int_{\mathcal{X}} V_1 - V_2 dQ^{\bar{\phi}}(x) \right| \\
& \leq \eta_Q^w[J] \left\| V_1 - V_2 \right\|_w ,
\end{aligned}
$$

which completes the proof. ///

Theorems 12.12 and 12.14 together may be used to verify that $\bar{T}_\pi$ is a contraction operator. They are in fact somewhat weaker than those proposed in Van Nunen and Wessels (1978) (following earlier work in Lippman (1975)) which can be summarized as $\eta_Q^w < 1$ and $\eta_R^w < \infty$.

We conclude with the following theorem.

Theorem 12.15 *Suppose for MCM π the assumptions of Theorem 12.12 hold for weight function $w \in \mathcal{W}$. Then $\bar{T}_\pi$ is an operator on $\mathcal{F}(\mathcal{X}, \|\cdot\|_w)$ for which $\bar{T}_\pi[J]$ possesses Lipschitz constant $\eta_Q^w[J]$.*

Proof The result follows directly from Theorems 12.12 and 12.14. ///

12.6.3 The equivalence of uniform convergence and contraction for the DPO

From the discussion of Section 7.3 we concluded that the contraction property of a positive linear operator T was equivalent to uniform convergence of the sequence $T^n\vec{0}$ to its limit. This applies directly to $T_\pi^{\bar{\phi}}$ under our assumptions. As an alternative to the approach underlying Theorem 12.15, we show how this equivalence can be extended to the DPO.

It will clarify the matter to introduce the definition

$$
\zeta_n = \inf_{x \in \mathcal{X}} \bar{R}[n](x) / \bar{V}_\pi(x) = \left\| \bar{V}_\pi \right\|_{\bar{R}[n]}^{-1} ,
$$

so that uniform convergence of $\bar{R}[n]$ to $\bar{V}_\pi$ is equivalent to $\zeta_n \uparrow 1$. In fact, J-stage contraction of any operator $T_\pi^{\bar{\phi}}$ follows if $\left\| V_\pi^{\bar{\phi}} \right\|_{\bar{V}_\pi} < \infty$ and $\zeta_n > 0$ for any $n \geq 1$. This is shown in the following theorem.

Theorem 12.16 *For any $\tilde{\phi} \in \mathcal{K}_f^m$ we have*

$$\left\| Q^{\tilde{\phi}}[n] \right\|_{V_\pi^{\tilde{\phi}}} \leq 1 - \left\| V_\pi^{\tilde{\phi}} \right\|_{R^{\tilde{\phi}}[n]}^{-1} \leq 1 - \zeta_{nm} \left\| V_\pi^{\tilde{\phi}} \right\|_{\bar{V}_\pi}^{-1}. \tag{12.32}$$

In addition, the operators $Q^{\tilde{\phi}}[J]$ for all policies $\tilde{\phi} \in \mathcal{K}_f^m$ satisfying $\left\| V_\pi^{\tilde{\phi}} \right\|_{\bar{V}_\pi} \leq 1 + \delta$ satisfy the uniform bound

$$\left\| Q^{\tilde{\phi}}[n]^J \right\|_{\bar{V}_\pi} \leq \left[1 - \zeta_{nm}(1+\delta)^{-1}\right]^J (1+\delta). \tag{12.33}$$

Proof By Theorem 7.7 we have directly

$$\left\| Q^{\tilde{\phi}}[n] \right\|_{V_\pi^{\tilde{\phi}}} \leq 1 - \left\| V_\pi^{\tilde{\phi}} \right\|_{R^{\tilde{\phi}}[n]}^{-1}.$$

Also, $R^{\tilde{\phi}}[n] \geq \bar{R}[nm]$ and $\bar{V}_\pi \leq V_\pi^{\tilde{\phi}}$, hence $\left\| V_\pi^{\tilde{\phi}} \right\|_{R^{\tilde{\phi}}[n]} \leq \left\| V_\pi^{\tilde{\phi}} \right\|_{\bar{V}_\pi} \| \bar{V}_\pi \|_{\bar{R}[nm]}$, which completes the proof. ///

The inequality (12.33) suggests that the family of operators $T_\pi^{\tilde{\phi}}$ for which $\left\| V_\pi^{\tilde{\phi}} \right\|_{\bar{V}_\pi} \leq 1 + \delta$ forms a type of neighborhood of DPO $\bar{V}_\pi$, over which a contraction constant may be uniformly bounded. This suggests the idea of contructing an approximate DPO based on this neighborhood. To do this we rely on the idea of a restricted policy set $\mathcal{K}^\delta$ defined by,

$$\mathcal{K}_x^\delta = \{a \in \mathcal{K}_x \mid \lambda_\pi(x,a) \leq \delta \bar{V}_\pi(x)\},$$

the set of restricted policy functions $\phi \in \mathcal{K}_f^\delta$ constructed from $\mathcal{K}^\delta$, and the resulting restricted DPO, defined for $x \in \mathcal{X}$ by

$$\bar{T}_\pi^\delta V(x) = \inf_{a \in \mathcal{K}_x^\delta} T_\pi^a V(x).$$

Clearly, we have

$$\bar{V}_\pi = \bar{T}_\pi^\delta \bar{V}_\pi = \bar{T}_\pi \bar{V}_\pi,$$

and we will need to show that this equality holds for V near $\bar{V}_\pi$. In addition, from Theorem 12.8 we may may conclude that $\left\| V_\pi^{\tilde{\phi}} \right\|_{\bar{V}_\pi} \leq (1+\delta)$ for any $\tilde{\phi} \in \mathcal{K}_f^\delta$. The contraction property for $\bar{T}_\pi^\delta$ follws directly:

Theorem 12.17 *If there exists n for which $\zeta_n > 0$ then $\bar{T}_\pi^\delta$ is J-stage contractive in $\|\cdot\|_{\bar{V}_\pi}$ for all $\delta > 0$.*

Proof Since $\phi \in \mathcal{K}_f^\delta$ implies $\| V_\pi^\phi \|_{\bar{V}_\pi} \leq (1+\delta)$, by Theorem 12.16 we have for any J the uniform bound

$$\| Q^\phi[n]^J \|_{\bar{V}_\pi} \leq \left[1 - \zeta_n(1+\delta)^{-1}\right]^J (1+\delta), \tag{12.34}$$

over $\phi \in \mathcal{K}_f^\delta$. By hypothesis J may be selected large enough for force the upper bound to be less than one. That $\bar{T}_\pi^\delta$ is multistage contractive follows from Theorem 12.14. ///

The remaining step is to determine when $\bar{T}_\pi^\delta$ and $\bar{T}_\pi$ are interchangeable.

Theorem 12.18 *Suppose $\bar{R}^{inf} \geq 0$. Given two positive constants $b < 1 < c$, if $(1+\delta) > c/b$ then $b\bar{V}_\pi \leq V \leq c\bar{V}_\pi$ implies $\bar{T}_\pi^\delta V = \bar{T}_\pi V$.*

Proof We must have $\bar{T}_\pi^\delta V \geq \bar{T}_\pi V$. If $\bar{T}_\pi^\delta V(x') > \bar{T}_\pi V(x')$ for some x' then there exists $a' \notin \mathcal{K}_x^\delta$ for which $(T_\pi^a V)(x', a') < \bar{T}_\pi^\delta V(x')$. Suppose $b\bar{V}_\pi \leq V \leq c\bar{V}_\pi$, so that using Lemma 12.1 we may write

$$(T_\pi^a V)(x', a') < T_\pi^\delta V(x') \leq c\bar{T}_\pi^\delta \bar{V}_\pi(x') = c\bar{V}_\pi(x'). \tag{12.35}$$

Conversely, since $a' \notin \mathcal{K}_{x'}^\delta$ we have

$$\begin{aligned}(T_\pi^a V)(x', a') &\geq R(x', a') + b\beta \int_{\mathcal{X}} \bar{V}_\pi(y) dQ(y \mid x', a') \\ &\geq b(\bar{T}_\pi \bar{V}_\pi(x') + \delta \bar{V}_\pi(x')) \\ &= b(1+\delta)\bar{V}_\pi(x'). \end{aligned} \tag{12.36}$$

However (12.35) and (12.36) contradict the hypothetical constraint on the constants δ, δ', so the theorem must hold. ///

The preceding theorems may be used in the following way. First suppose $\bar{R}^{inf} \geq 0$. If we can find some policy ϕ and $m \geq 1$ for which $R^\phi[m] \sim V_\pi^\phi$ we know that T_π^ϕ is contractive. If in addition $\bar{R}[m] \sim R^\phi[m]$ we may conclude

$$\| \bar{V}_\pi \|_{\bar{R}[m]} \leq \| \bar{V}_\pi \|_{R^{\bar{\phi}}[m]} \left\| R^{\bar{\phi}}[m] \right\|_{\bar{R}[m]} \leq \left\| V_\pi^{\bar{\phi}} \right\|_{R^{\bar{\phi}}[m]} \left\| R^{\bar{\phi}} \right\|_{\bar{R}[m]} < \infty,$$

so that $\zeta_m > 0$. Theorem 12.17 therefore holds.

Then, suppose we construct a VIA from starting point $V_0 \geq 0$. We must have $V_n = \bar{T}_\pi[n]V_0 \geq \bar{R}[n] \geq \zeta_m \bar{V}_\pi$ for all $n \geq m$. Furthermore, since $\bar{T}_\pi$ is nonexpansive, we have $\| V_{n+1} - \bar{V}_\pi \|_{\bar{V}_\pi} \leq C$ when $\| V_n - \bar{V}_\pi \|_{\bar{V}_\pi} \leq C$. Thus, we may find a single pair of constants b, c for which the hypothesis of Theorem 12.18 holds for all V_n, for $n \geq m$. Thus, for some δ the VIA is equivalent to $V_{n+1} = \bar{T}_\pi^\delta V_n$, that is, the VIA satisfies the contraction property.

12.7 AVERAGE COST CRITERION FOR MDP

There are two motivations for introducing a discount factor $\beta < 1$. There may be economic reasons for discounting future costs. The practical consequence of this is to assign greater weight to short term costs, so that as $\beta \to 0$ we would minimize the single stage cost $R(x, a)$ only.

If economic criterion play no role, the purpose of $\beta < 1$ is to ensure that the total cost is finite. In this case, it might seem unsatisfactory that the attained cost, and hence the optimal policy, depends on this value, which has no other meaning to the analyst.

At this point, recall that β equivalently defines a 'kill' process, in the sense that the MDP stops assuming cost after a random number of stages which is geometrically distributed with mean $(1-\beta)^{-1}$.

We can predict the magnitude of a value function to be within $(1-\beta)^{-1}\bar{R}^{sup}$, which is the maximum cost per stage multiplied by the expected number of stages. In this sense, $(1-\beta)\bar{V}_\pi$ represents the resulting 'average' cost per stage. If we let $\beta \to 1$ we should expect to attain an optimal *undiscounted* cost, as an long term average.

Calculated directly, this becomes

$$V_\pi^\Phi(x) = \lim_{N\to\infty} N^{-1} E_x^\Phi \left[\sum_{n=1}^{N} R(X_n, A_n) \right], \quad x \in \mathcal{X}. \tag{12.37}$$

The expectation can be evaluated as before using the measure P_x^Φ, yielding a well defined optimization problem over the space of controls. Of course, it would be helpful if the problem could be formulated so as to exploit the theory developed for the total discounted cost. One idea which presents itself is to calculate a sequence of solutions as β approaches 1, then accept the limit as the optimal average cost model. This is a viable approach, but would be quite cumbersome, particularly since the complexity of the computation can be very sensitive to the factor $(1-\beta)^{-1}$.

We propose instead to develop an optimality criterion by reformulating the problem as a shortest path MDP (following the approach discussed in Bertsekas (1995b), Chapter 4).

Assume that a controlled MDP is ergodic (Section 5.2), and that $\mathcal{K}$ is countable. In particular, we can take any state, say $x = 1 \in \mathcal{X}$, and decompose the process into renewal periods, which begin from state $X_n = 1$ and end when the system first returns to state $X_{n+j} = 1$ following $j \geq 1$ transitions. Since we have a formal renewal process, the long term behavior can be deduced from the properties of a single renewal period.

We do this in the following way. Consider a shortest path MDP which begins as state $X_1 = 1$ and which upon the next return to state 1 transitions instead to a kill state Δ, accruing no further cost.

This results in a modification of the original model π. We have appended a kill state Δ to $\mathcal{X}$, and the original stochastic kernel Q has been replaced by Q_Δ, for which

$$\begin{aligned} Q_\Delta(\Delta \mid x, a) &= Q(\{1\} \mid x, a) \\ Q_\Delta(\{1\} \mid x, a) &= 0 \\ Q_\Delta(\{y\} \mid x, a) &= Q(\{y\} \mid x, a), \quad y \in \mathcal{X} - \{1\}, \end{aligned}$$

for all $(x, a) \in \mathcal{K}$. If, for some $\phi \in \mathcal{K}_f$, Q^ϕ defines the transition kernel for an ergodic Markov chain, then all states communicate with state 1, so that, following the discussion in Section 12.6.1, the operator Q_Δ^ϕ will be multistage contractive. The cost function R remains the same, except that it is extended to $R(\Delta, a) \equiv 0$ for any dummy action a. Denote the new model $\pi[\Delta]$.

The total cost, using control policy Φ, and starting from state 1 is

$$R[\Phi,\Delta]=E_1^{\Phi}\left[\sum_{n=1}^{N^{tot}} R(X_n,A_n)\right]$$

where $N^{tot}\geq 1$ is the number of states visited *before* the kill state (including the initial state $X_1=1$).

Next, note that we may always find a constant λ for which

$$R[\Phi,\Delta]=E_1^{\Phi}\left[\sum_{n=1}^{N^{tot}} \lambda\right]=\lambda E_1^{\Phi}\left[N^{tot}\right],$$

which may be interpreted as an average cost per stage.

The MCM will be modified in one more way. Return to the original shortest path MDP, but now use cost function $R_\lambda(x,a)=R(x,a)-\lambda$, with $R_\lambda(\Delta,a)\equiv 0$, leaving the exact value of λ open for the moment. The new model is denoted $\pi[\Delta,\lambda]$.

Suppose we use stationary policy $\phi\in\mathcal{K}_f$ for model $\pi[\Delta,\lambda]$. The resulting policy value function satisfies the equation

$$V^{\phi}_{\pi[\Delta,\lambda]}(x)=R^{\phi}(x)-\lambda+\int_{y\in\mathcal{X}-\{1\}} V^{\phi}_{\pi[\Delta,\lambda]}dQ^{\phi}(y\,|\,x),\quad x\neq\Delta, \tag{12.38}$$

where we may force $V^{\phi}_{\pi[\Delta,\lambda]}(\Delta)=0$. This can be reexpressed in terms of the operator $T^{\phi}_{\pi[\Delta]}$ as

$$V^{\phi}_{\pi[\Delta,\lambda]}+\lambda=T^{\phi}_{\pi[\Delta]}V^{\phi}_{\pi[\Delta,\lambda]}. \tag{12.39}$$

Now, return to the original interpretation of λ as the solution, denoted $\lambda=\lambda_\phi$, to $R[\Phi,\Delta]=\lambda_\phi E_1^{\Phi}\left[N^{tot}\right]$, with $\Phi=\phi$. Then

$$V^{\phi}_{\pi[\Delta,\lambda_\phi]}(1)=E_1^{\Phi}\left[\sum_{n=1}^{N^{tot}}(R(X_n,A_n)-\lambda_\phi)\right]=R[\Phi,\Delta]-\lambda_\phi E_1^{\Phi}\left[N^{tot}\right]=0,$$

so that

$$\lambda_\phi=\left(T^{\phi}_{\pi[\Delta]}V^{\phi}_{\pi[\Delta,\lambda_\phi]}\right)(1).$$

Finally, consider the original problem of minimizing λ_ϕ. It is tempting at this point to conclude directly that an optimality equation can be obtained from (12.39) simply by replacing the linear operator with the DPO, that is, $V=\lambda+\bar{T}_\pi V$. This is close to being correct, but it must be remembered that the shortest path problem is not the problem we are considering. To complete the argument, suppose we can somehow identify a policy ϕ^* which achieves the minimum average cost λ^*. Now, we consider optimizing the shortest path MDP under model $\pi[\Delta,\lambda^*]$. Possibly, ϕ^* is not optimal for this problem, and we may be able to find $\hat{\phi}$ for which $V^{\hat{\phi}}_{\pi[\Delta,\lambda^*]} \leq V^{\phi^*}_{\pi[\Delta,\lambda^*]}$. From (12.39), since λ^* is common to both policy value functions, we also have $T^{\hat{\phi}}_{\pi[\Delta]} V^{\hat{\phi}}_{\pi[\Delta,\lambda^*]} \leq T^{\phi^*}_{\pi[\Delta]} V^{\phi^*}_{\pi[\Delta,\lambda^*]}$, so that

$$\begin{aligned}
\lambda^* &= \left(T^{\phi^*}_{\pi[\Delta]} V^{\phi^*}_{\pi[\Delta,\lambda^*]}\right)(1) \\
&\geq \left(T^{\hat{\phi}}_{\pi[\Delta]} V^{\hat{\phi}}_{\pi[\Delta,\lambda^*]}\right)(1) \\
&\geq \left(T^{\hat{\phi}}_{\pi[\Delta]} V^{\hat{\phi}}_{\pi[\Delta,\lambda_{\hat{\phi}}]}\right)(1) \\
&= \lambda_{\hat{\phi}},
\end{aligned}$$

where the final inequality follows from the fact that $R_{\lambda_1}(x,a) \leq R_{\lambda_2}(x,a)$ over all $(x,a) \in \mathcal{K}$ if $\lambda_1 \geq \lambda_2$. Therefore, the assumption that $\lambda_{\hat{\phi}} > \lambda^*$ leads to a contradiction, and we conclude that a policy yielding optimal average cost λ^* can be replaced by a policy optimizing the shortest path cost for model $\pi[\Delta, \lambda^*]$. We already know how to determine the optimal policy for this problem, namely, by solving the fixed point equation,

$$\begin{aligned}
\bar{V}_{\pi[\Delta,\lambda^*]} &= \bar{T}_{\pi[\Delta,\lambda^*]} \bar{V}_{\pi[\Delta,\lambda^*]}, \text{ or equivalently} \\
\bar{V}_{\pi[\Delta,\lambda^*]} + \lambda^* &= \bar{T}_{\pi[\Delta]} \bar{V}_{\pi[\Delta,\lambda^*]}.
\end{aligned} \tag{12.40}$$

The development of a VIA for the optimal average cost MDP will be taken up in Section 13.9.

Chapter 13

Markov decision processes – value iteration

If we are given a Banach space $\mathcal{F}(\mathcal{X}, \|\cdot\|_w)$ on which a policy operator T_π^ϕ or DPO $\bar{T}_\pi$ is contractive then a VIA is simply the sequence of iterations $V_{k+1} = T_\pi^\phi V_k$ or $V_{k+1} = \bar{T}_\pi V_k$ for any suitable starting point $V_0 \in \mathcal{F}(\mathcal{X}, \|\cdot\|_w)$. The sequence V_k converges uniformly to $\bar{V}_\pi$, but since a VIA is exact only in the limit, the remaining component is a stopping rule N which permits the claim $\| V_N - \bar{V}_\pi \|_w \leq \epsilon$ for some fixed tolerance $\epsilon > 0$. The stopping rules discussed in Section 6.2.1 apply to any contractive operator on a metric space (including multistage or pseudocontractive), and will be used to develop stopping rules for VIAs.

Regarding the starting point, to fix ideas, assume that costs are bounded. We may always select $V_0 = \vec{0}$. If so, the first DPO iteration yields $V_1 = R^{inf}$, and subsequent iterations add an amount to each state between R^{inf} and R^{sup}. For discounted costs we expect $(1-\beta)^{-1} R^{inf} \leq \bar{V}_\pi \leq (1-\beta)^{-1} R^{sup}$, so the number of iterations needed is of order $O\left((1-\beta)^{-1}\right)$.

The preceding inequality suggests an alternative starting point, namely $V_0 = (1-\beta)^{-1} R^{inf}$, since we must have $\|(1-\beta)^{-1} R^{inf}\|_w < \infty$.

We may also use the comparison approach discussed in Section 12.6.3. If for some policy ϕ and integer m we have weight equivalence $\bar{R}[m] \equiv R^\phi[m] \equiv w$ and T_π^ϕ is contractive *wrt* norm $\|\cdot\|_w$, then $\| \bar{V}_\pi \|_w \leq \| V_\pi^\phi \|_w < \infty$, and we may use starting point $V_0 = V_\pi^\phi$.

If V_π^ϕ is to be calculated by VI, starting point $V_0 = (1-\beta)^{-1} R^\phi$ may be used, since under the given conditions $\| R^\phi \|_w < \infty$, and in fact from the discussion in Section 7.5.2 we have $\| V_\pi^\phi \|_w \leq \| (1-\beta)^{-1} R^\phi \|_w$. Although the evaluation of V_π^ϕ can be expected to have the same linear convergence rate as a VIA based on $\bar{T}_\pi$, the computation of T_π^ϕ may be considerably more efficient, since minimization over the action space is not needed. Alternatively, since $V_\pi^\phi = T_\pi^\phi V_\pi^\phi$ is a linear system, more efficient specialized methods may be available. Methods for *policy evaluation* (the calculation of V_π^ϕ) are discussed in detail in, for example, Puterman (1994) or Bertsekas and Tsitsiklis (1996). Possibly, it will be easier to verify the contractive property for multistep policies $\tilde{\phi} \in \mathcal{K}_f^m$, in which case, the preceding discussion also applies.

13.1 VALUE ITERATION ON QUOTIENT SPACES

Suppose T is an operator (not necessarily linear) on Banach space $\mathcal{V} = \mathcal{F}(\mathcal{X}, \|\cdot\|_w)$ which possesses an eigenpair (β, ν), in particular, for any scalar c and $V \in \mathcal{V}$

$$T(V + c\nu) = TV + \beta c\nu \tag{13.1}$$

(at this point, the reader may wish to review the material in Chapters 6 and 7).

As a technical matter, although T is constructed as a mapping between single elements of $\mathcal{V}$ we may also regard it as a mapping between sets of functions, so that if $E \subset \mathcal{V}$, we take $TE \subset \mathcal{V}$ to be a mapping of subsets, evaluated as the image of E under T.

Next, note that $\mathcal{N}_\nu = \{c\nu \mid c \in \mathbb{R}\}$ is a subspace of $\mathcal{V}$. Recall that the quotient space $\mathcal{V}/\mathcal{N}_\nu$ consists of all cosets $[V] = \{V + n \mid n \in \mathcal{N}_\nu\}$ over all $V \in \mathcal{V}$. These form a vector space of equivalence classes, with additive identity $\mathcal{N}_\nu$ (see Section 6.3.1). In this case, we may say that T is an operator on $\mathcal{V}/\mathcal{N}_\nu$ if $TE \in \mathcal{V}/\mathcal{N}_\nu$ whenever $E \in \mathcal{V}/\mathcal{N}_\nu$. In terms of the original meaning of the operator, this is equivalent to the claim that for any $E_1 \in \mathcal{V}/\mathcal{N}_\nu$ there is another $E_2 \in \mathcal{V}/\mathcal{N}_\nu$ such that $T : E_1 \to E_2$ is a bijective mapping. The condition (13.1) suffices for this to hold.

Next, we may define for any coset $[V] \in \mathcal{V}/\mathcal{N}_\nu$

$$\|[V]\|_\alpha = \inf_{n \in \mathcal{N}_\nu} \|V - \nu\|_w,$$

noting that this definition is consistent, since V may be replaced by any other element of $[V]$. Then by Theorem 6.18 of Section 6.3.1, if $\mathcal{V}$ is a Banach space, then so is $(\mathcal{V}/\mathcal{N}_\nu, \|\cdot\|_\alpha)$. Finally, by Theorem 6.25 of Section 6.8.4, if we may set $w = \nu$ (which we can if $w \sim \nu$) we have

$$\|[V]\|_\alpha = \frac{1}{2} \|V\|_{SP(\nu)},$$

which again is consistent in the sense that $\|V_1\|_{SP(\nu)} = \|V_2\|_{SP(\nu)}$ whenever V_1, V_2 are in the same coset.

At this point we refer to the Banach space $(\mathcal{V}/\mathcal{N}_\nu, \|\cdot\|_{SP(\nu)})$ on which $\|\cdot\|_{SP(\nu)}$ is a true norm which defines convergence on $\mathcal{V}/\mathcal{N}_\nu$. Furthermore, contractive properties for $T : \mathcal{V}/\mathcal{N}_\nu \to \mathcal{V}/\mathcal{N}_\nu$ may be analyzed as for the original Banach space $\mathcal{V}$ and we may be able to conclude that T possesses fixed point $[V^*] = T[V^*] \in \mathcal{V}/\mathcal{N}_\nu$. This does not (necessarily) mean that there exists $V^* \in \mathcal{V}$ for which $V^* = TV^*$. It does mean that T is a bijective mapping on $[V^*]$, so that $TV \in [V^*]$ whenever $V \in [V^*]$, and this turns out to be a precise enough statement to resolve the optimality problem. In particular, for the average cost MDP a fixed point in $\mathcal{V}$ generally does not exist, but, under suitable regularity conditions a fixed point in $\mathcal{V}/\mathcal{N}_\nu$ does exist in this sense.

A similar issue arises in the actual implementation of value iteration. Whether we define the algorithm on $\mathcal{V}$ or $\mathcal{V}/\mathcal{N}_\nu$ the iterates will still take the form $V_{k+1} = TV_k$, where V_k is a specific function in $\mathcal{V}$. If we may guarantee convergence in $\mathcal{V}$ (for example, T is contractive on $\mathcal{V}$), we also have convergence in the span seminorm (Theorem 6.25). However, if we may only claim convergence in the span seminorm then the iterates $V_{k+1} = TV_k$ need not converge, and a modification is needed to yield a numerically stable algorithm. Fortunately, we need only rely on the fact that a coset may be represented by any of its elements. Therefore, define a canonical member $V_0 \in [V]$, and use evaluations $T[V] = TV_0$. For example, if $1 \in \mathcal{X}$, we may take V_0 to be the unique element of $[V]$ for which $V(1) = 0$. The iterations then assume the form

$$V_{k+1} = T(V_k - (V_k(1)/\nu(1))\nu).$$

Essentially, we have devised a modified operator $T_0 V = T(V - (V(1)/\nu(1))\nu)$, which yields iterations $V_{k+1} = T_0 V_k$.

The value of constructing a VIA on $\mathcal{V}/\mathcal{N}_\nu$ is clear if T is not contractive on $\mathcal{V}$. In this case, the essential step is to determine the contractivity properties *wrt* $\|\cdot\|_{SP(\nu)}$. However, even when T is contractive on $\mathcal{V}$, and we can achieve our goal by allowing the iterates $V_{k+1} = TV_k$ to converge to the fixed point, there can be considerable advantage to conceiving of the algorithm as defined on $\mathcal{V}/\mathcal{N}_\nu$. In this chapter, we consider both cases.

13.2 CONTRACTION IN THE SPAN SEMINORM

Since T_π^ϕ is a linear operator based on a stochastic kernel, Theorem 6.28 of Section 6.9.2 gives the Lipschitz constant directly:

$$\left\|\left| Q^\phi \right\|\right|_{SP(\nu)} = \frac{1}{2} \sup_{x,y \in \mathcal{X}} \|(I_\nu^{-1} Q^\phi)(x) - (I_\nu^{-1} Q^\phi)(y)\|_{TV(\nu)} \leq \left\|\left| Q^\phi \right\|\right|_\nu, \tag{13.2}$$

provided ν is an eigevector of Q^ϕ. As discusses in Section 7.5.2 the quantity $\left\|\left| Q^\phi \right\|\right|_{SP(\nu)}$ is equivalent to Dobrushin's ergodicity coefficient (for example, Section 6.7 of Brémaud (1999) when Q^ϕ is a proper stochastic kernel.

To fix ideas, suppose $Q^\phi = \beta Q_0^\phi$ where $\beta < 1$ and Q_0^ϕ is a proper stochastic kernel, so that T_π^ϕ is β-contractive, and $(1, \vec{1})$ is an eigenpair of Q_0^ϕ. Then $\left\|\left| Q^\phi \right\|\right|_\nu = \left\|\left| \beta Q_0^\phi \right\|\right|_\nu = \beta$, so by (13.2) $\left\|\left| Q^\phi \right\|\right|_{SP(\nu)} \leq \beta$, and we know that T_π^ϕ is contractive on both $\mathcal{V}$ and $\mathcal{V}/\mathcal{N}_\nu$.

So far, all contractive properties in the span seminorm are attributable to the discount factor β. However, since $\left\|\left| Q^\phi \right\|\right|_{SP(\nu)} = \beta \left\|\left| Q_0^\phi \right\|\right|_{SP(\nu)}$ the contraction rate is reduced further if $\left\|\left| Q_0^\phi \right\|\right|_{SP(\nu)} < 1$, or more generally if $\left\|\left| (Q_0^\phi)^J \right\|\right|_{SP(\nu)} < 1$.

13.2.1 Contraction properties of the DPO

In the context of $\mathcal{V}$ we have seen that if the Lipschitz constants of all linear operators T_π^ϕ over $\phi \in \mathcal{K}_f$ are bounded by β then β is also a Lipschitz constant of the DPO $\bar{T}_\pi$. This is stated in Theorem 12.15, but is largely a consequence of a technical theorem (Theorem 12.13) proposed in Hinderer (1970), which under the stated hypothesis asserts that

$$|\inf_x f_1(x) - \inf_x f_2(x)| \leq \sup_x |f_1(x) - f_2(x)|$$

for real valued functions on a general domain.

We can extend this approach to the span seminorm, up to a point, but we don't have quite as strong a result. First, we present the extension of Theorem 12.13:

Theorem 13.1 *Suppose two functions f, g map $\mathcal{K}$ to $\mathbb{R}$, and that*

$$|f(x,a)| < \infty \quad \text{and} \quad |g(x,a)| < \infty.$$

In addition, assume that for each x we have

$$\inf_{a \in \mathcal{K}_x} f(x,a) > -\infty \quad \textit{and} \quad \inf_{a \in \mathcal{K}_x} g(x,a) > -\infty.$$

Then

$$\left\| \inf_a f(x,a) - \inf_a g(x,a) \right\|_{SP} \leq \sup_{x,a \in \mathcal{K}} (f(x,a) - g(x,a)) - \inf_{x,a \in \mathcal{K}} (f(x,a) - g(x,a))$$

Proof From the proof of Theorem 12.13 we have

$$\sup_x (\inf_a f(x,a) - \inf_a g(x,a)) \leq \sup_x \sup_a (f(x,a) - g(x,a)),$$
$$\sup_x (\inf_a g(x,a) - \inf_a f(x,a)) \leq \sup_x \sup_a (g(x,a) - f(x,a)).$$

Then

$$\begin{aligned}
&\left\| \inf_a f(x,a) - \inf_a g(x,a) \right\|_{SP} \\
&\quad = \sup_x (\inf_a f(x,a) - \inf_a g(x,a)) - \inf_x (\inf_a f(x,a) - \inf_a g(x,a)) \\
&\quad = \sup_x (\inf_a f(x,a) - \inf_a g(x,a)) + \sup_x (\inf_a g(x,a) - \inf_a f(x,a)) \\
&\quad \leq \sup_x \sup_a (f(x,a) - g(x,a)) - \inf_x \inf_a (f(x,a) - g(x,a)),
\end{aligned}$$

which completes the proof. ///

Thus, by Theorem 4.22

$$\begin{aligned}
&\| \bar{T}_\pi V_1 - \bar{T}_\pi V_2 \|_{SP(\nu)} \\
&\quad = \sup_x \sup_a Q(x,a)(V_1 - V_2) - \inf_x \inf_a Q(x,a)(V_1 - V_2) \\
&\quad \leq \sup_{(x,a),(x',a') \in \mathcal{K}^2} \| Q(x,a) - Q(x',a') \|_{TV(\nu)} \| V_1 - V_2 \|_{SP(\nu)}
\end{aligned}$$

so that the span seminorm operation is simply extended from $\mathcal{X}$ to $\mathcal{K}$, and a Lipschitz constant of $\bar{T}_\pi$ *wrt* $\|\cdot\|_{SP(\nu)}$ is given by

$$L = \sup_{(x,a),(x',a') \in \mathcal{K}^2} \| Q(x,a) - Q(x',a') \|_{TV(\nu)}.$$

However, it is not sufficient to consider $\left|\!\left|\!\left| Q^\phi \right|\!\right|\!\right|_{SP(\nu)}$ over all policy functions. To see this, by decomposing into the cases $x = x'$ and $x \neq x'$ we obtain

$$\begin{aligned}
\| \bar{T}_\pi V_1 - \bar{T}_\pi V_2 \|_{SP(\nu)} &= \sup_x \sup_a Q(x,a)(V_1 - V_2) - \inf_x \inf_a Q(x,a)(V_1 - V_2) \\
&= \max \{A, B\},
\end{aligned}$$

where

$$A = \sup_{\phi \in \mathcal{K}} \sup_{x,x' \in \mathcal{X} \times \mathcal{X}} \left\| Q(x, \phi(x)) - Q(x', \phi(x')) \right\|_{TV(\nu)} \| V_1 - V_2 \|_{SP(\nu)}$$
$$= \sup_{\phi \in \mathcal{K}} \left\| Q^{\phi} \right\|_{SP(\nu)} \| V_1 - V_2 \|_{SP(\nu)},$$

and

$$B = \sup_{x \in \mathcal{X}} \sup_{a,a' \in \mathcal{K}_x \times \mathcal{K}_x} \left\| Q(x, a) - Q(x, a') \right\|_{TV(\nu)} \| V_1 - V_2 \|_{SP(\nu)}.$$

Therefore, the uniform contraction structure of Theorem 12.15 does not necessarily extend to the span seminorm, unless we can assert that $A \geq B$.

13.3 STOPPING RULES FOR VALUE ITERATION

Suppose we are given any fixed point algorithm $V_{k+1} = TV_k$, with starting point V_0 for some operator T, and assume T^J possesses Lipschitz constant ρ_J. From Section 7.6 the stopping rules

$$N_J^{\epsilon} = \min\{N \mid \rho_J (1 - \rho_J)^{-1} \left\| V_N - V_{N-J} \right\| \leq \epsilon\},$$
$$N_{\infty}^{\epsilon} = \min\{N \mid \rho_N (1 - \rho_N)^{-1} \| V_N - V_0 \| \leq \epsilon\},$$

guarantee absolute approximation bound

$$\left\| V_N - V^* \right\| \leq \epsilon,$$

where N is any of the stopping times. For this example we consider $J = 1$ and ∞. The stopping rules may be expressed as stopping bounds:

$$\left\| V_N - V^* \right\| \leq \frac{\rho_1}{1 - \rho_1} \| V_N - V_{N-1} \|, \quad (J = 1)$$
$$\left\| V_N - V^* \right\| \leq \frac{\rho_N}{1 - \rho_N} \| V_N - V_0 \|, \quad (J = \infty).$$

This applies to any Banach space.

13.4 VALUE ITERATION IN THE SPAN SEMINORM

Next, suppose (β, ν) is an eigenpair for operator T on $\mathcal{V} = \mathcal{F}(\mathcal{X}, \|\cdot\|_\nu)$, with $\beta < 1$, so that T is β-contractive. Then T is also β-contractive on the quotient space $(\mathcal{V}/\mathcal{N}_\nu, \|\cdot\|_{SP(\nu)})$. Of course, as we have seen, the contraction constant *wrt* $\|\cdot\|_{SP(\nu)}$ may be considerable smaller, and convergence to the fixed point may be considerably faster on $\mathcal{V}/\mathcal{N}_\nu$ than on $\mathcal{V}$. The price we pay for this is that convergence is to a equivalence class, and not to a specific solution.

This turns out not to be too high a price. To see this, the discussion in Section 7.5 regarding solutions to linear fixed point equations may be applied here. Because T is contractive on $\mathcal{V}$, we know there is a unique fixed point satisfying $V^* = TV^*$. Suppose we are given an equivalence class solution $E = TE$. Following the discussion in Section 7.5 this means that if $\hat{V} \in E$ we may obtain the fixed point of T in $\mathcal{V}$ by the formula

$$V^* = \hat{V} + (1-\beta)^{-1}(T\hat{V} - \hat{V}). \tag{13.3}$$

This holds for any $\hat{V} \in E$, so that a fixed point solution in $\mathcal{V}/\mathcal{N}_\nu$ immediately yields the fixed point solution in $\mathcal{V}$. It will therefore be useful to construct a composite operator based on (13.3), which we may refer to as *span adjustment*

$$HV = V + (1-\beta)^{-1}(TV - V). \tag{13.4}$$

Then $V^* = HV_0$ is an exact solution to the fixed point equation $V^* = TV^*$ in $\mathcal{V}$ for any $V_0 \in E$ where E is a solution to the fixed point equation in $\mathcal{V}/\mathcal{N}_\nu$.

However, we are also interested in the case in which we have a bound $\left\| \hat{V} - V^* \right\|_{SP(\nu)} \leq \epsilon$ where $\hat{V}$ is an approximation of a fixed point $V^* = TV^*$ in $\mathcal{V}$ measured in the span seminorm, presumably some iterate $\hat{V} = V_N$ of a VIA. By Theorem 7.11 of Section 7.5.1 we have

$$\left\| H\hat{V} - V^* \right\|_\nu \leq \frac{1+\beta/2}{(1-\beta)} \left\| \hat{V} - V^* \right\|_{SP(\nu)}. \tag{13.5}$$

In this way we construct both an exact fixed point, or its appproximation, in $\mathcal{V}$ from an exact fixed point, or its approximation, in $\mathcal{V}/\mathcal{N}_\nu$.

13.5 EXAMPLE: *M/D/1/K* QUEUEING SYSTEM

Consider a $M/D/1/K$ queueing system with arrival rate λ, a deterministic service time of 1 unit, and system capacity K (see Section 5.4 for a discussion of queueing models). The control model will be based on the imbedded service model, so that decision epochs occur at the beginning of service periods. The control variable consists of an integer service capacity d, meaning that d customers can be served at once. If fewer than d customers are in the queue at the beginning of a service period, all are serviced, but the capacity may remain d.

Costs are calculated on a per service period basis, and there will be a number of cost determinants. We assume there is a time loss cost, so that a unit cost is assumed for each customer in the system. The service cost is given by $C_s(x,d)$ where x is the system occupancy and d is the service capacity. There may also be a policy cost $C_p(d_x, d)$ where d_x is the capacity of the previous stage and d is the capacity selected for the current stage. Note that if $C_p(d_x, d)$ depends on d_x then the previous capacity must be incorporated into the state space. A discount factor of β will be used to calculate total cost.

The transition kernel $Q(x,d)$ may be calculated using the methods of Section 5.4.

Contraction properties

We first explore some of the contraction properties of $\bar{T}_\pi$. First assume $C_p(d_x, d) \equiv C_p(d)$ so that the state space becomes $\mathcal{X} = \{1, \ldots, K\}$ (an empty queue transitions to an occupancy of 1, assuming no cost, so will be combined with the state representing an occupancy of 1). For the moment, consider a single policy $\phi \equiv N_s$ for some fixed N_s.

The principal eigenpair of $Q = Q^\phi$ is taken to be $(1, \vec{1})$. The linear operator used in $\bar{T}_\pi$ will be βQ, with a principal eigenpair of $(\beta, \vec{1})$. In the supremum norm the contraction constant is $|||\beta Q|||_{sup} = \beta$. Of course, the contractive properties of Q in the span seminorm are usefully examined independently. Then the contraction properties of βQ attributable to β and those attributable to its spectral properties are easily compounded with the formula $|||(\beta Q)^J|||_{SP} = \beta^J \, |||Q^J|||_{SP}$.

For an example, take $K = 100$, $\beta = 0.999$ and parameter pairs $(\lambda, d) = (2, 5)$ and $(0.4, 1)$, which we denote models π_1 and π_2 respectively. Each queue has the same utilization factor 0.4, but the λ_{SLEM} values differ considerably, with values $\lambda_{SLEM} = 0.3231, 0.7725$ (abusing notation somewhat, we will take $\lambda_{SLEM} = |\lambda_{SLEM}|$). Recall that λ_{SLEM} determines the rate of convergence of a Markov chain to its steady state (Section 2.3.4) and that it is related to the span operator norm of a stochastic kernel (Section 7.5.2). The values of λ_{SLEM} are clearly dependent on the model parameters, but are both considerably smaller than typical values used for the discount factor β.

Figure 13.1 displays values of $|||Q^J|||_{SP}$, standardized as rates using two devices, geometric standardization $|||Q^J|||_{SP}^{1/J}$ and geometric rate $|||Q^J|||_{SP} \, / \, |||Q^{J-1}|||_{SP}$ for models π_1, π_2 (left column of plot array). The values β and λ_{SLEM} are superimposed for comparison. A number of trends are apparent. Recall that $|||Q|||_{SP}$ is the maximum L^1 distance of all pairs of distributions of the kernel Q. We expect the distributions $Q(1)$ and $Q(K)$ to be close to singular, and that this effect will persist for some number of iterates Q^J, as can be seen in Figure 13.1. Of course, as discussed in Section 5.2, Q^J approaches the steady state transition matrix Q^∞, in which all rows are identically π_Q, the steady state distribution of Q (this is the fact responsible for the contraction property). Furthermore, progress towards the steady state eventually occurs at the rate λ_{SLEM}. Thus, the geometric average of the contraction rate of Q^J in the the span seminorm eventually becomes considerably smaller than the discount rate β. The right columns of Figure 13.1 compare the total cumulative compounded contraction rates of the operator βQ in the supremum norm β^J to that in the span seminorm $\beta^J \, |||Q^J|||_{SP}$.

Policy evaluation

We next consider the evaluation of a value function V_π^ϕ for the simple policy ϕ used in the previous example, with cost function $R(x) = 1 + x$. As a first demonstration, the approximation error for the iterates $V_i - T_\pi^\phi[i]v_0$ in the supremum norm, setting $v_0 = \vec{0}$ are shown in Figure 13.2 (left plots) for model π_2, using both the unadjusted iterates, and the span adjusted iterates, as defined in (13.4). In addition, values of $\beta = 0.999, 0.9999$ were considered.

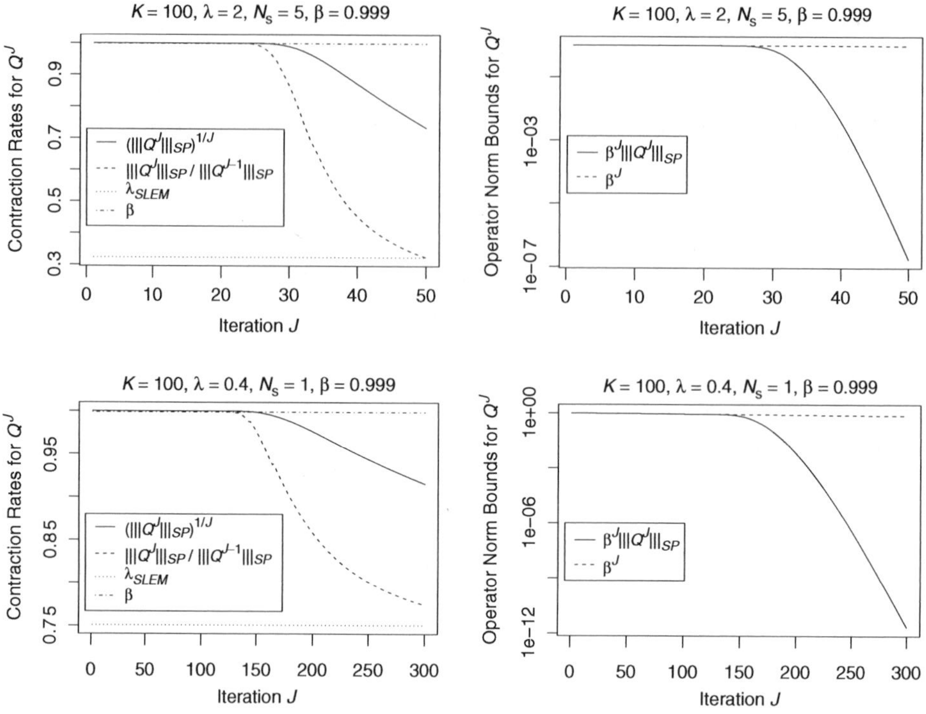

Figure 13.1 Representations of operator norms $\left|\left|\left|Q^J\right|\right|\right|_{SP}$, discount factor β and λ_{SLEM}, for $M/D/1/K$ model of Section 13.5. Parameters $K=100$, $(\lambda, N_s)=(2,5)$, $(0.4,1)$, $\beta=0.999$ are represented.

Clearly, convergence of the span adjusted iterates is order of magnitudes faster than the unadjusted iterates in the supremum norm itself. The relatively large discount factor β makes this a challenging problem for standard VI, but has little effect on the span adjusted method. Furthermore, increasing β from 0.999 to 0.9999 has little effect on the convergence of the span adjusted iterates, but has, predictably considerable effect on the unadjusted iterates.

Next, consider stopping rules based on the span seminorm. From Section 7.6 the stopping rules

$$N_J^{\epsilon} = \min\{N \mid \rho_J(1-\rho_J)^{-1} \left\| V_N - V_{N-J} \right\| \leq \epsilon\},$$
$$N_{\infty}^{\epsilon} = \min\{N \mid \rho_N(1-\rho_N)^{-1} \left\| V_N - V_0 \right\| \leq \epsilon\},$$

guarantee absolute approximation bound

$$\left\| V_N - V^* \right\|_{SP} \leq \epsilon,$$

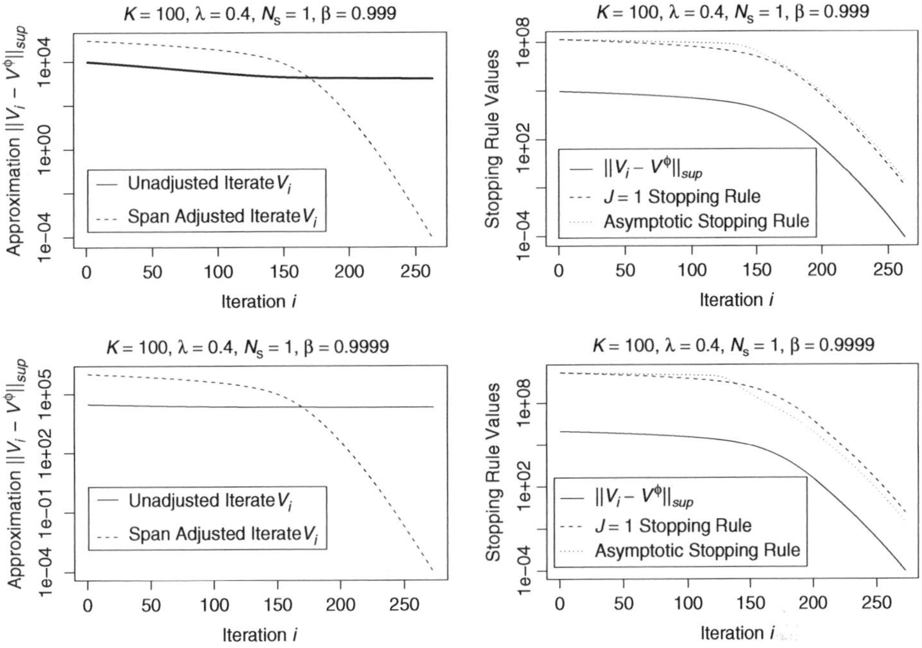

Figure 13.2 Approximation error for VI iterates V_i (unadjusted and span adjusted) for $M/D/1/K$ model of Section 13.5. Parameters $K = 100$, $d = 1$, $\beta = 0.999$, $\lambda = 0.4$. [Left Plot]. Bounds for $J = 1, \infty$ stopping rules [Right Plot].

where N is any of the stopping times. For this example we consider $J = 1, \infty$. The stopping rules may be expressed as stopping bounds:

$$\left\| V_N - V^* \right\|_{SP} \leq \frac{\rho_1}{1 - \rho_1} \| V_N - V_{N-1} \|_{SP}, \quad (J = 1)$$

$$\left\| V_N - V^* \right\|_{SP} \leq \frac{\rho_N}{1 - \rho_N} \| V_N - V_0 \|_{SP}, \quad (J = \infty).$$

The choice of stopping time involves a trade-off between smaller increment sizes and smaller factor. When $\beta = 1$, some J larger than 1 will have to be chosen in many applications. These stopping bounds are shown in Figure 13.2 (right plots). The relative advantage depends on the value of β, but both methods are effective in these examples. As suggested in Section 6.2.1, the relative advantage of the asymptotic stopping rule will increase as $\beta \uparrow 1$, but requires the calculation of ρ_N. In principal, this may be obtained from a spectral decomposition of Q, but it is well known that accurate calcuations of eigenvectors for large matrices may be difficult to obtain. The natural alternative is to update $Q^i = Q^{i-1}Q$, and calculate $\left\|\left| Q^i \right|\right\|_{SP}$ directly. Of course, absent any additional structure to the problem, this is an order $O(K^3)$ calculation, as is the calculation of $\left\|\left| Q^i \right|\right\|_{SP}$, while the value iteration step has a calculation cost of only $O(K^2)$. Therefore, if the cost of an iteration is dominated by K, a comparison of the $J = \infty$

stopping rule must take this into account. If as a simple device we convert the iteration scale to a computational cost scale, then the iteration scale for the $J = \infty$ stopping rule should be rescaled by a factor of $K = 100$ in comparison to the $J = 1$ stopping rule, and it can be seen by inspection of Figure 13.2 that the $J = \infty$ stopping rule is much less efficient from this point of view. The $J = 1$ stopping rule then becomes recommendable, and its reliance on the discount factor β is largely offest by the rapid convergence of the increments $\| V_N - V_{N-1} \|_{SP}$, which clearly occurs much more quickly than $\Omega(\beta^N)$.

If Q has N_Q nonzero elements, then the update $Q^i = Q^{i-1}Q$ may be implemented as an order $O(KN_Q)$ computation, while the value iteration step is an order $O(N_Q)$ computation. The worst case is that $N_Q = O(K^2)$ but if Q is sparse we would have $N_Q = O(K)$. The update of Q^i is still larger than the value iteration step by a factor of K, but in this scenario the state look-up cost may no longer be dominant. Of course, the evaluation of $\||Q^J\||_{SP}$ remains $O(K^3)$, but we will next consider a number of modifications.

13.6 EFFICIENT CALCULATION OF $\||Q^J\||_{SP}$

We have seen that more detailed knowledge of $\||Q^J\||_{SP}$ will result in sharper approximation bounds, but also that in the worst case the calculation of $\||Q^J\||_{SP}$ may exceed that of the value iteration algorithm itself by a factor of $O(K)$ or even $O(K^2)$. However, we show that we may reduce the order of complexity by exploiting the anticipated behavior of Q^J.

First note that from the spectral decomposition of Q, we can expect that $\||Q^{J+1}\||_{SP} \approx \lambda_{SLEM} \||Q^J\||_{SP}$ for all large enough J. For the examples of Figure 13.1, this occurs at approximately $J = 50, 300$ for models π_1, π_2. Furthermore, the value of λ_{SLEM} may be calculated accurately using a variety of numercial packages. However, it is also clear that accepting an approximation of the form $\||Q^J\||_{SP} \approx \lambda^J_{SLEM}$ will result in considerable underestimation, so it will be worth considering how to efficiently calculate $\||Q^J\||_{SP}$ before this point is reached.

Recall that for discrete state spaces

$$\||Q^J\||_{SP} = \frac{1}{2} \max_{i,j} \|Q^J(i) - Q^J(j)\|_{TV},$$

so that the complexity of the calculation is primarily driven by the maximization operation. Thus, it is worth considering if the search over pairs i, j may be reduced.

In our example, the distributions of kernel $Q(i)$ change gradually as i increases, so we would expect $\|Q^J(i) - Q^J(j)\|_{TV}$ to increase with index distance $|i - j|$. It will generally be difficult, other than for certain special cases, to verify such relations precisely, but it would be possible to design an algorithm which would work well if such a conjecture held.

To see this, set $D_{i,j} = \|Q^J(i) - Q^J(j)\|_{TV}$. By the triangle inequality, for $i < j$

$$D_{i,j} \leq \sum_{k=i+1}^{j} D_{k-1,k}. \tag{13.6}$$

Then construct

$$F_1 = 0, \quad \text{and} \quad F_j = \sum_{k=2}^{j} D_{k-1,k} \text{ for } j = 2, \ldots, K, \tag{13.7}$$

so that we have bound, for $i < j$,

$$D_{i,j} \leq F_j - F_i.$$

Then, suppose we anticipate that the index pair (i^*, j^*) will be close to maximizing. Set $D^* = D_{i^*,j^*}$. Then enumerate all remaining pairs (i,j). If $F_j - F_i < D^*$ then i,j can be excluded from further consideration (therefore eliminating the order $O(K)$ calculation of $D_{i,j}$). The construction of F_j is an order $O(K^2)$ calculation, so if the number of nonexcluded pairs is small compared to the full enumeration order $K(K-1)/2$, then the computation of $\|\!|Q^J|\!\|_{SP}$ may be reduced by an order of magnitude, so that the computation is dominated by the calculation of F_j.

The choice of index pair (i^*, j^*) may be based on the anticipated behavior of Q^J, or if $\|\!|Q^J|\!\|_{SP}$ is being calculated sequentially, it may be the maximizing pair from the calculation of $\|\!|Q^{J-1}|\!\|_{SP}$, which would be known using this algorithm.

It should also be noted that $\|\!|Q^J|\!\|_{SP}$ is nondecreasing, with a maximum of 1. Therefore, if we find $D_{i^*,j^*} = 1$ or $D_{i^*,j^*} = 2\,\|\!|Q^{J-1}|\!\|_{SP}$ within a suitable tolerance, then we set $\|\!|Q^J|\!\|_{SP} = D_{i^*,j^*}$ without the need to calculate F_j.

We summarize the algorithm, using the notation of (13.6)–(13.7):

Algorithm 1 *Given kernel Q on state space $i = 1, \ldots, K$, and maximum iteration N_J:*

1 *Set $J = 1$.*
2 *Determine maximum $D_{i,j} = \|Q^J(i) - Q^J(j)\|_{TV}$ over all pairs $i < j$, as well as the maximizing pair (i^*, j^*).*
3 *Output $\|\!|Q^J|\!\|_{SP} = \frac{1}{2} D_{i^*,j^*}$.*
4 *If $J = N_J$ stop algorithm otherwise $J = J + 1$.*
5 *Calculate $D^* = D_{i^*,j^*}$.*
6 *If $D^* \approx 2\,\|\!|Q^{J-1}|\!\|_{SP}$ then output $\|\!|Q^J|\!\|_{SP} = \|\!|Q^{J-1}|\!\|_{SP}$ and go to Step 4.*
7 *Calculate F_j, $j = 1, \ldots, K$, where $D_{i,j} = \|Q^J(i) - Q^J(j)\|_{TV}$.*
8 *For each pair $i < j$:*

   (*a*) *If $F_j - F_i < D^*$ go to next pair,*
   (*b*) *otherwise if $D_{i,j} > D^*$ then assign $D^* = D_{i,j}$ and $(i^*, j^*) = (i, j)$.*

9 *Output $\|\!|Q^J|\!\|_{SP} = \frac{1}{2} D_{i^*,j^*}$.*
10 *Go to Step 4.*

Algorithm 1 would be suitable for any model in which the kernel Q can be ordered so that proximate distributions $Q^J(i), Q^J(i+1)$ can be expected to have small variational distance.

Algorithm 1 was implemented for the example of Figure 13.2. The computational cost is given as the number of pairs (i,j) for which the comparison $\|Q^J(i) - Q^J(j)\|_{TV}$ is evaluated, which is itself an order $O(K)$ computation. The maximum complexity is

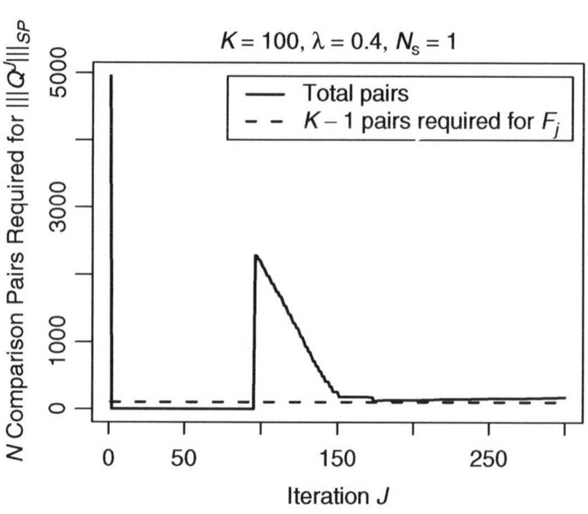

Figure 13.3 Number of comparison pairs needed by Algorithm 1 (Step 7(b)) for the calculation of $|||Q^J|||_{SP}$, for the $M/D/1/K$ model of Section 13.5. Parameters $K = 100$, $d = 1$, $\lambda = 0.4$ are represented.

therefore $\binom{100}{2} = 4950$, and the minimum is 1, which occurs if the condition in Step 6 is met. If it is not, then the calculation of F_j in Step 7 requires $K - 1$ pair comparisons.

The complexity achieved by Algorithm 1 for model π_2 for the evaluation $|||Q^J|||_{SP}$ is shown in Figure 13.3 up to $J = 300$. The number of comaprisons $K - 1$ required to evaluate F_j is shown separately. The maximum complexity is required for $|||Q|||_{SP}$, although this number could be significantly reduced by an initial guess of, for example, $(i^*, j^*) = (1, K)$. As suggested by Figure 13.1, we have $|||Q^J|||_{SP} \approx 1$ for a large number of iterations. This is easily detected by Algorithm 1, so that only one comparison pair is required for $2 \leq J \leq 95$. The complexity rises to 2,284 at $J = 96$, then decreases rapidly, remaining below 200 for $151 \leq J \leq 300$. At $J = 300$, $|||Q^J|||_{SP}$ decreases at the rate λ_{SLEM}, so that no further evaluations are needed, and we may use the updating rule $|||Q^{J+1}|||_{SP} \approx \lambda_{SLEM} |||Q^J|||_{SP}$.

The total complexity for the first 300 iterations was 95,285 comparison pairs, or 317.6 per iteration. Evaluation of $\|Q^J(i) - Q^J(j)\|_{TV}$ requires $2K$ state lookups, yielding approximetely 63,500 state lookups per iteration, which compares with $K^2 = 10,000$ state lookups required for a single VI iteration.

Of course, this still leaves the order $O(K^3)$ computation for the update $Q^J = Q^{J-1}Q$. We will see in Chapter 16 that truncation of the distributions defining kernel Q can be an effective way of reducing computation time with a well controlled approximation error. When the distributions $Q(i)$ are dominated by small tail probabilities, the resulting approximate kernel $\hat{Q}$ will typically have $N_{\hat{Q}} = O(K)$ nonzero elements. Taken together, the calculation of $Q^J = Q^{J-1}Q$ and $|||Q^J|||_{SP}$ will then each have computation cost $O(K^2)$ per iteration, while each VI iteration will have computation cost of $O(K)$. We will generally find that the $J = 1$ stopping rule will work effectively for the values of β used in these examples. For the average cost criterion (that is, $\beta = 1$) the developement of an effective stopping rule will require finding a small enough value of $|||Q^J|||_{SP}$. This will often require allowing J to approach a value comparable in magnitude to $|\mathcal{X}|$, so an effective exploitation of the spectral properties of Q may become important in the design of the most efficient possible algorithms. Of course, this holds also when especially small values of $1 - \beta$ are anticipated.

13.7 EXAMPLE: *M/D/1/K* SYSTEM WITH OPTIMAL CONTROL OF SERVICE CAPACITY

The example of the preceding section will now be expanded to incorporate control structure. Control epochs occur at the beginning of service periods. The state variable is (x, d_x) where x is the system occupancy at the control epoch (customers serviced during the previous service period have left the system) and d_x is the service capacity in effect during the previous service period. The available action from state (x, d_x) is the service capacity of the imminent service period, with admissible actions $d \in \{1, 2, \ldots, N_s\}$, where N_s is the maximum service capacity. The state space is therefore of size $|\mathcal{X}| = KN_s$ and the action space is of size $|\mathcal{A}| = N_s$. Nominally, the computation cost of single VI iteration is $|\mathcal{X}|^2|\mathcal{A}| = K^2N_s^3$. However, the transition kernel is sparse, since transitions from state (x, d_x) under action d must be to (y, d). Therefore, the computation cost in state lookups is actually $K^2N_s^2$.

For demonstration, we have $\lambda = 2$, $K = 100$, $N_s = 10$, $\beta = 0.999$. The service cost will be $C_s(x, d) = d^{1.5}$ and the policy cost will be $C_p(d_x, d) = 50|d_x - d|$. Top row plots in Figure 13.4 show progress in terms of $\| V_i - V^* \|_{sup}$ for both unadjusted iterates [left plot] and span adjusted iterates [right plot]. As in the previous examples, convergence in the span seminorm is orders of magnitude faster. In addition, full and accurate solutions to the optimality equation are obtainable within 300 iterations using the span adjustment procedure, whereas the unadjusted iterates have not yielded a comparable accuracy after 15,000 iterations.

Value iteration can be seen to consist of two essentially independent calculations, the first consisting of the 'shape' of V^*, which is reducible to the centered V^*, representable as, for example, $V^* - \inf V^*$. This gives any relative value of $V^*(x) - V^*(y)$, but not an absolute value $V^*(x)$. The second calculation gives the 'location' of V^*, and is expressible as a single constant which, when added to a centered estimate of V^* gives the complete value function.

Figure 13.4 (bottom row) shows sequences of VI iterates, both in absolute form (left plot) and centered form (right plot). The problem of estimating the 'shape' of V^*, that is, V^* in centered form, is accomplished very quickly, within 100 iterations. At this point, using span adjustment yields an absolute solution within approximately 300 iterations (Figure 13.4, top right plot). Otherwise, as can be seen in Figure 13.4 (bottom left plot), unadjusted iterates will not be close to V^* until approximately 10,000 iterations.

This type of decomposition reveals an important point. Once V^* is solved in a centered form, the remaining problem reduces to the calculation of a single number, which represents the multiple of the principal eigenvector required to complete the solution. This number may be calculated directly, making further iterations unnecessary.

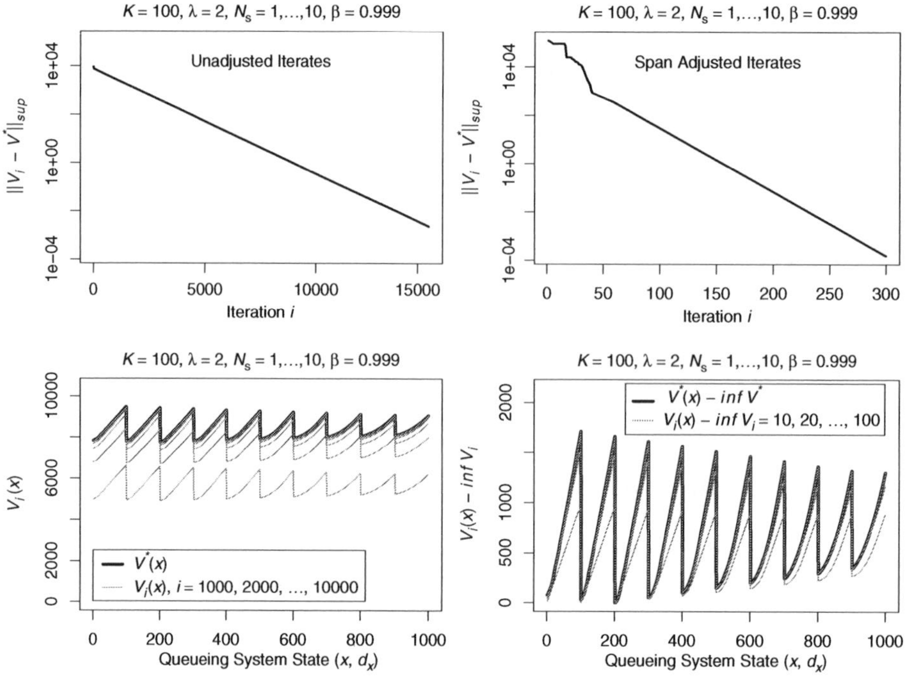

Figure 13.4 Summary of value iteration for control problem for $M/D/1/K$ model of Section 13.5 with parameters $\lambda = 2$, $K = 100$, $N_s = 10$, $\beta = 0.999$. Top row plots show progress in terms of $\|V_i - V^*\|_{sup}$ for both unadjusted iterates [left plot] and span adjusted iterates [right plot]. Bottom row plots show the full (unadjusted) iterates for various sequences of iterates V_i, as well as the optimal value function V^*. In the right plot the value functions are centered as $V^*(x) - \inf V^*$ and $V_i(x) - \inf V_i$. The state space is represented sequentially by index $j = x + K(d_x - 1)$.

13.8 POLICY ITERATION

Policy iteration is an alternative to value iteration which generates a sequence of stationary policies $\phi_1, \phi_2, \ldots$ in the following manner:

Algorithm 2

1 *Begin with any policy ϕ_1, set counter $n = 1$.*
2 ***Policy Evaluation:** For step n calculate value function $V_\pi^{\phi_n}$.*
3 ***Policy Improvement:** Generate new policy $\phi_{n+1}(x) = argmin_{a \in K_x} T_\pi^{(x,a)} V_\pi^{\phi_n}$ for each $x \in \mathcal{X}$. If possible, set $\phi_{n+1}(x) = \phi_n(x)$.*
4 *If $\phi_{n+1} = \phi_n$ then stop, otherwise increment n, then go to Step 2.*

In the policy improvement step of Algorithm 2 it is important to resolve ties in the minimization operation in favor of the current policy because of the stopping rule in Step 4. Clearly, we have $V_\pi^{\phi_{n+1}} \leq V_\pi^{\phi_n}$, so that if the number of policies is finite policy iteration will converge to the optimal policy in a finite number of iterations.

In the literature, policy iteration is reported to be less efficient than value iteration (an extensive treatment of this topic can be found in Chapter 6 of Puterman (1994)). Each value iteration step $V_{k+1} = \bar{T}_\pi V_k$ implicitly defines a control policy (at least under suitable continuity assumptions) but as a practical matter there is no need to capture this until the final iterate. If we do choose to capture these policies, we can expect to find a process similar to policy iteration, at least in the later iterations, with the final iterations essentially representing a policy evaluation for the optimal control. The difference is that until the final optimal policy is reached, what appears to be an interim policy evaluation is not carried out to convergence.

It is also worth noting that the conditions for convergence of policy iteration are in one important sense less stringent than for value iteration. It is necessary that the policy evaluation step yield the correct policy value function, possibly by verifying the contractive property for each T_π^ϕ. However, the DPO $\bar{T}_\pi$ plays no role, so there is no need to verify its contractive properties. We have seen in Chapter 12 that contractivity in the weighted supremum norm of a rich enough class of operators T_π^ϕ implies contractivity of $\bar{T}_\pi$, but the same is not true *wrt* the span seminorm (Section 13.2.1). In such cases, policy iteration at least guarantees the existence of a method of determining the optimal control.

13.9 VALUE ITERATION FOR THE AVERAGE COST OPTIMIZATION

In this section we consider a VIA for the average cost MDP defined in Section 12.7. Recall that the average cost model was reformulated as a shortest path MDP. One particular state was taken to be the initial state, with the process terminating upon the return to that state. The introduction of a kill state induces the contraction property.

This formulation replaces model π with $\pi[\Delta]$, and we may deduce contraction properties for operators $T_{\pi[\Delta]}^\phi$ and $\bar{T}_{\pi[\Delta]}$. However, we may still consider the fixed point equation

$$V + \lambda = T_\pi^\phi V. \tag{13.8}$$

If (β, ν) is an eigenpair for Q^ϕ then T_π^ϕ is a mapping on the quotient space $\mathcal{V}/\mathcal{N}_\nu$. If T_π^ϕ is not contractive on the underlying Banach space, it may be on the quotient space, in which case it has a unique fixed point. At this point we must be clear as to what this means. It does not mean that there is some $V \in \mathcal{V}$ for which $V = T_\pi^\phi V$. It does mean that there is a unique coset $[V] \in \mathcal{V}/\mathcal{N}_\nu$ on which T_π^ϕ is a bijective mapping. This means for any $V' \in [V]$ we also have $T_\pi^\phi V' \in [V]$, equivalently $V' + \lambda' \nu = T_\pi^\phi V'$ for some scalar λ'. If V'' is any other element in $[V]$ then we similarly have $V'' + \lambda'' \nu = T_\pi^\phi V''$ for some scalar λ''. But there exists scalar c for which $V'' = V' + c\nu$. This means

$$\begin{aligned}
\lambda'' &= T_\pi^\phi V'' - V'' \\
&= T_\pi^\phi (V' + c\nu) - V' - c\nu \\
&= T_\pi^\phi V' + c\nu - V' - c\nu \\
&= T_\pi^\phi V' - V' \\
&= \lambda',
\end{aligned}$$

that is, among the class of solutions (λ, V) to equation (13.8), λ is unique, and all V comprise a single equivalence class in $\mathcal{V}/\mathcal{N}_\nu$. Thus, (13.8) is the type of invariant equation defined in part (*iii*) of Theorem 7.8 (the reader can refer to the discussion in Sections 7.4 and 7.5).

This equivalence structure is indispensible to our model, and allows us to equate (12.39) with (13.8). The eigenpair is $(1, \vec{1})$, and there is exactly one solution (λ, V) to (13.8) for which $V(1) = 0$. Similarly, by construction the operators $T^{\phi}_{\pi[\Delta]}$ and T^{ϕ}_{π} of equations (12.39) and (13.8) satisfy $T^{\phi}_{\pi[\Delta]}V = T^{\phi}_{\pi}V$ whenever $V(1) = 0$. Therefore, we conclude that there is a unique solution (λ, V) to (13.8) for which $V(1) = 0$, and this solution is precisely $\lambda = \lambda_\phi$ and $V = V^{\phi}_{\pi[\Delta,\lambda]}$. Of course, there is no special reason to associate Δ with any particular state, and formulating the problem in terms of (13.8) removes the need to do so. Given any solution (λ, V) we may conclude that λ is the average cost per stage.

Essentially the same argument holds for the dynamic programming equation (12.40). If (β, ν) is an eigenpair for each Q^{ϕ}, then $\bar{T}_\pi$ is an operator in the quotient space $\mathcal{V}/\mathcal{N}_\nu$, since = $\bar{T}_\pi(V + c\nu) = \bar{T}_\pi v + c\nu$. We then have optimality equation

$$V + \lambda = \bar{T}_\pi V, \tag{13.9}$$

which behaves essentially the same way as (13.8). If $\bar{T}_\pi$ is contractive on the quotient space, there exists a unique fixed point in the form of an equivalence class, which yields a unique value of λ, which we have already argued equals the optimal average stage cost.

To summarize, the optimal MDP by the average cost criterion can be determined by value iteration in the span seminorm quotient space, as described in this chapter, provided T^{ϕ}_{π} or $\bar{T}_\pi$ can be shown to be contractive in $\|\cdot\|_{SP(\nu)}$ (which does not require discount factor $\beta < 1$). An excellent discussion of the contraction properties of the DPO for the average cost MDP can be found in Chapter 3 of Hernández-Lerma (1989b).

Chapter 14

Model approximation in dynamic programming – general theory

In this chapter we consider the problem of determining approximation bounds on attained costs which result from replacing a model with either a single approximation, or a sequence of approximations. We are given model $\pi = (R, Q)$, satisfying definitions (M1)–(M6) of Section 12.1, where Q may be a single or multiple kernel, and so both policy evaluation and dynamic programming are of interest. If the model were known, a VIA $V_{k+1} = \bar{T}_\pi V_k$ would be employed.

Possibly, the model π is unknown but can be estimated by a single approximation $\hat{\pi} = (\hat{R}, \hat{Q})$, or sequence of approximations $\hat{\pi}_k = (\hat{R}_k, \hat{Q}_k)$. It may also be that π is known, but the computation of $\bar{T}_\pi$ can be simplified by replacing π with a simpler model $\hat{\pi}$.

The former case will typically be of interest when π is unknown and is to be estimated by $\hat{\pi}$, and a suitable error bound for this estimate is available, for example, in the form of a statistical estimate. Interest is in $\bar{V}_\pi$, but only $\bar{V}_{\hat{\pi}}$ is available. If Q is a single kernel, it will usually suffice to determine an approximation bound of the form $\|\bar{V}_\pi - \bar{V}_{\hat{\pi}}\|$. If Q is a multiple kernel, then $\|\bar{V}_\pi - \bar{V}_{\hat{\pi}}\|$ will similarly serve as an approximation bound for the optimal achievable costs, but a complete assessement of the effect of the approximation on a control policy will usually require further analysis. Applying dynamic programming to the approximate model $\hat{\pi}$ will yield control $\phi_{\hat{\pi}}$, and it cannot be expected that this control will be the same as the optimal control ϕ_π based on the true model. For this reason, there may be more interest in the quantity $V_\pi^{\phi_{\hat{\pi}}}$, which is the actual cost that would be obtained by applying the approximate control $\phi_{\hat{\pi}}$ to the true model π, as well as the regret $V_\pi^{\phi_{\hat{\pi}}} - \bar{V}_\pi$. Of course, a small error bound for the model approximation will generally mean that the various value functions $\bar{V}_{\hat{\pi}}$, $V_\pi^{\phi_{\hat{\pi}}}$ and $\bar{V}_\pi$ will be within a common approximation bound of each other.

14.1 THE GENERAL INEQUALITY FOR MDPs

Essentially, we have four classes of problems, according to whether Q is a single or multiple kernel, and whether we have a single approximation or a sequence. It turns out that all classes can be unified under the most complex case.

Suppose we are given an EIA based either on DPO $\bar{T}_\pi$ or policy operator T_π^ϕ. For convenience we will consider $\bar{T}_\pi$ only, recognizing that T_π^ϕ is a special case of the optimization problem with a single admissible policy. Thus, the EIA is based on a single operator $\bar{T}_\pi$. If $\bar{T}_\pi$ is β-contractive the Lipschitz kernel for the EIA is of product form with $\bar{\lambda}^{k,i} = \beta^i$. There may be interest in multistage Lipschitz constants, either because $\bar{T}_\pi$ is not single stage contractive, or because the asymptotic contraction rate is significantly lower than the single stage contraction rate. In this case, we let β_J be the J-stage Lipschitz constant, and set $\bar{\lambda}^{k,i} = \beta_i$. We also assume that the DPO is nonexpansive, which will generally hold by Theorems 7.7 and 12.15.

We then have a sequence of model approximations $\hat{\pi}_k$, $k \geq 1$. This generates an AIA based on operator sequence $\bar{T}_{\hat{\pi}_k}$, $k \geq 1$, with iterates

$$\begin{aligned} V_k &= \bar{T}_{\hat{\pi}_k} V_{k-1}, \quad k \geq 1, \\ V_0 &= v_0. \end{aligned}$$

We then have approximation terms

$$\begin{aligned} V_k &= \bar{T}_\pi V_{k-1} + U_k, \quad k \geq 1, \text{ where} \\ U_k &= \bar{T}_{\hat{\pi}_k} V_{k-1} - \bar{T}_\pi V_{k-1}. \end{aligned}$$

Recall the general inequality of (10.30). We will assume $p = 1$. The norm may be any weighted supremum norm, or a span seminorm. The general inequality then becomes:

$$\| V_k - \bar{V}_\pi \| = \bar{\lambda}^{k,k} \| V_0 - \bar{V}_\pi \| + \mathcal{I}_k(\tilde{d}, \bar{\lambda}) \quad k \geq 1, \tag{14.1}$$

where $\tilde{d} = (d_1, d_2, \ldots)$, with d_k serving as a bound on $\| U_k \|$.

In general, we must expect that U_n is proportional in magnitude to V_{n-1}, as will be the case for approximation methods introduced in subsequent chapters. We therefore assume that the relative error model, defined by (ARE) in Section 10.4.1, holds:

$$\| U_k \| \leq a_k + b_k \| V_{k-1} \|_0$$

where $\| \cdot \|_0$ is a seminorm satisfying $\| V \|_0 \leq \kappa_0 \| V \|$ for some $\kappa_0 < \infty$. In this case, the error bound of the general inequality (14.1) depends explicitly on the magnitude of the iterates V_n, so we must either impose a bound on V_n, or confirm that the AIA is convergent without benefit of a bound. The latter holds if $a_k \to 0$ and $b_k \to 0$. In fact, under the hypothesis of Theorem 10.4, it suffices that a_k is bounded and $\limsup_k b_k < \delta$ for a small enough but positive δ, but the constraint on this number is proportional to $(1 - \beta)^{-1}$, so little is to be gained from this fact.

However, in many applications the relative error model holds, but it is not anticipated that $b_k \to 0$, and it would be restrictive to impose a bound on b_k. In this case, the boundedness of the AIA can follow from the boundedness of each approximate operator $\bar{T}_{\hat{\pi}}$. The following theorem summarizes this approach.

Theorem 14.1 *Suppose for constants $\beta < 1$, $M < \infty$ each approximate operator $\bar{T}_{\hat{\pi}}$ is β-pseudocontractive with respect to unique fixed point $\bar{V}_{\hat{\pi}}$ for which $\| \bar{V}_{\hat{\pi}} \| \leq M$, then the AIA is bounded in $\| \cdot \|$.*

Proof We may write

$$\begin{aligned}
\| V_k - \bar{V}_{\hat{\pi}_k} \| &\leq \| \bar{T}_{\hat{\pi}_k} V_{k-1} - \bar{V}_{\hat{\pi}_k} \| \\
&\leq \beta \| V_{k-1} - \bar{V}_{\hat{\pi}_k} \| \\
&= \beta \| V_{k-1} - \bar{V}_{\hat{\pi}_{k-1}} \| + \beta \| \bar{V}_{\hat{\pi}_{k-1}} - \bar{V}_{\hat{\pi}_k} \| \\
&= \beta \| V_{k-1} - \bar{V}_{\hat{\pi}_{k-1}} \| + \beta 2M.
\end{aligned}$$

If the preceding inequality is applied iteratively we have

$$\| V_k - \bar{V}_{\hat{\pi}_k} \| \leq \beta^k \| \bar{T}_{\hat{\pi}_1} V_0 - \bar{V}_{\hat{\pi}_1} \| + (1 - \beta)^{-1} 2M,$$

which completes the proof. ///

If each approximate operator $\bar{T}_{\hat{\pi}}$ is β-pseudocontractive and the cost functions $\hat{R} \leq M_R < \infty$ are uniformly bounded over all approximate models $\hat{\pi}$ then the hypothesis of Theorem 14.1 holds, since $\| \bar{V}_{\hat{\pi}} \| \leq (1 - \beta)^{-1} M_R$.

First, for $\rho \in (0,1)$ the following families of sequences were defined in Section 10.11:

$$\begin{aligned}
\mathcal{F}_\rho^L &= \left\{ \{d_k\} \in \mathcal{S} : \sum_{k=1}^{\infty} \rho^{-k} d_k < \infty \right\}, \\
\mathcal{F}_\rho &= \left\{ \{d_k\} \in \mathcal{S} : \hat{\lambda}\{d_k\} = \rho \right\}, \text{ and} \\
\mathcal{F}_\rho^U &= \left\{ \{d_k\} \in \mathcal{S} : \lambda^l\{d_k\} > \rho \right\}.
\end{aligned} \tag{14.2}$$

The following two theorems presents a summary of the relevant convergence rates, based on Theorems 10.10, 10.19 and Theorem 2.12 of Almudevar (2008).

Theorem 14.2 *Suppose the J-stage Lipshitz constant of $\bar{T}_\pi$ is β_J, setting $\beta_0 = 1$. Then*

$$\limsup_{n \to \infty} \| V_n - \bar{V}_\pi \|_w \leq \left[\sum_{i=0} \beta_i \right] \limsup_{n \to \infty} d_n. \tag{14.3}$$

In particular, if $\bar{T}_\pi$ is nonexpansive then

$$\limsup_{n \to \infty} \| V_n - \bar{V}_\pi \|_w \leq J(1 - \beta_J)^{-1} \limsup_{n \to \infty} d_n. \tag{14.4}$$

Theorem 14.3 *Suppose $\bar{T}_\pi$ is nonexpansive and possesses J-stage contraction rate ρ. Then*

(*i*) *If $\tilde{d} \in \mathcal{F}_\rho^L$ then*

$$\limsup_{n\to\infty} \rho^{-n} \| V_n - \bar{V}_\pi \|_w < \infty \tag{14.5}$$

(*ii*) *If $\tilde{d} \in \mathcal{F}_\rho$ then*

$$\limsup_{n\to\infty} n^{-1} \log(\| V_n - \bar{V}_\pi \|_w \leq \log(\rho) \tag{14.6}$$

(*iii*) *If $\tilde{d} \in \mathcal{F}_\rho^U$ then*

$$\limsup_{n\to\infty} d_n^{-1} \| V_n - \bar{V}_\pi \|_w \leq K(1 - \rho/r_l)^{-1} \tag{14.7}$$

where $r_l = \lambda^l\{d_k\}$, and K is a finite constant dependent on J and r_l alone.

14.2 MODEL DISTANCE

The crucial quantity in our approximation theory is the operator tolerance $\|\bar{T}_\pi V - \bar{T}_{\hat{\pi}} V\|$, allowing V to range over $\mathcal{V}$. For single kernel linear models, this quantity is easily expressible as

$$\begin{aligned}\|\bar{T}_\pi V - \bar{T}_{\hat{\pi}} V\| &= \|R + QV - \hat{R} - \hat{Q}V\| \\ &\leq \|R - \hat{R}\| + \|\!|Q - \hat{Q}|\!\| \, \|V\|.\end{aligned}$$

If Q and $\hat{Q}$ possess a common principal eigenvector ν, we may write

$$\begin{aligned}\|\bar{T}_\pi V - \bar{T}_{\hat{\pi}} V\|_\nu &\leq \|R - \hat{R}\|_\nu + \|QV - \hat{Q}V\|_\nu \\ &\leq \|R - \hat{R}\|_\nu + \|\!|Q - \hat{Q}|\!\|_\nu \|V\|_{SP(\nu)},\end{aligned}$$

which will generally be preferable. The evaluation of $\|\!|Q - \hat{Q}|\!\|$ will usually be based on L^1 distances of density functions on a common measure.

The principal is much the same for multiple kernels, and we have

$$D_Q^w(Q_1, Q_2) = \sup_{(x,a)\in\mathcal{K}} w^{-1}(x) \|Q_1(x,a) - Q_2(x,a)\|_{TV(w)}.$$

Similarly, the distance between two cost functions R_1, R_2 will be defined

$$D_R^w(R_1, R_2) = \sup_{(x,a)\in\mathcal{K}} w(x)^{-1} |R_1(x,a) - R_2(x,a)| .$$

Theorem 14.4 *The operator tolerance* $\|\bar{T}_\pi V - \bar{T}_{\hat{\pi}} V\|$ *is bounded by*

$$\|\bar{T}_{\hat{\pi}} V - \bar{T}_\pi V\|_w \leq D_R^w(R, \hat{R}) + D_Q^w(Q, \hat{Q})\|V\|_w. \tag{14.8}$$

Furthermore, suppose (ρ, ν) *is a positive eigenpair for* Q*, in the sense that*

$$Q(x,a)\nu = \rho\nu(x) \tag{14.9}$$

for all $(x,a) \in \mathcal{K}$*. Then*

$$\|\bar{T}_{\hat{\pi}} V - \bar{T}_\pi V\|_\nu \leq D_R^\nu(R, \hat{R}) + \frac{1}{2} D_Q^\nu(Q, \hat{Q})\|V\|_{SP(\nu)}. \tag{14.10}$$

Proof From Theorem 12.13, for each $x \in \mathcal{X}$ and any $V \in \mathcal{V}$ we have

$$\begin{aligned}\left|\inf_{a\in\mathcal{K}_x} T_\pi^a V - \inf_{a\in\mathcal{K}_x} T_{\hat{\pi}}^a V\right| &\leq \sup_{a\in\mathcal{K}_x} \left|T_\pi^a V - T_{\hat{\pi}}^a V\right| \\ &= \sup_{a\in\mathcal{K}_x} \left|R(x,a) + Q(x,a)V - \hat{R} + \hat{Q}(x,a)V\right|.\end{aligned}$$

It follows that

$$\begin{aligned}\|\bar{T}_\pi V - \bar{T}_{\hat{\pi}} V\|_w &\leq \sup_{x\in\mathcal{X}} w(x)^{-1} \sup_{a\in\mathcal{K}_x} \left|R(x,a) + Q(x,a)V - \hat{R} + \hat{Q}(x,a)V\right| \\ &\leq D_R^w(R, \hat{R}) + D_Q^w(Q, \hat{Q})\|V\|_w,\end{aligned} \tag{14.11}$$

so that (14.8) holds. If (14.9) holds, then V in the upper bound of (14.11) may be replaced by $V + a\nu$ for any scalar a. This leads to

$$\begin{aligned}&\|\bar{T}_\pi V - \bar{T}_{\hat{\pi}} V\|_\nu \\ &\quad\leq \sup_{x\in\mathcal{X}} \nu(x)^{-1} \sup_{a\in\mathcal{K}_x} \left|R(x,a) + Q(x,a)(V + a\nu) - \hat{R} + \hat{Q}(x,a)(V + a\nu)\right| \\ &\quad\leq D_R^\nu(R, \hat{R}) + D_Q^\nu(Q, \hat{Q})\|V + a\|_\nu.\end{aligned}$$

Using Theorem 6.25 to minimize over a gives (14.10). ///

Suppose for model π, $\bar{T}_\pi$ is defined on Banach space $\mathcal{F}(\mathcal{X}, \|\cdot\|_w)$. Then let $\hat{\pi}_k = (R_k, Q_k)$ be a sequence of approximations of π. It follows from Theorem 14.4 that if $D_Q^w(Q, Q_k) < \infty$ and $D_R^w(R, R_k) < \infty$ for all $k \geq 1$, then the resulting AIA is defined on $\mathcal{F}(\mathcal{X}, \|\cdot\|_w)$.

The relative error model follows directly from Theorem 14.4. Using inequality (14.8) we obtain

$$\|U_k\|_w \leq D_R^w(R, R_k) + D_Q^w(Q, Q_k)\|V_{k-1}\|_w, \tag{14.12}$$

and using inequality (14.10) we obtain

$$\|U_k\|_\nu \le D^\nu_R(R, R_k) + \frac{1}{2} D^\nu_Q(Q, Q_k) \| V_{k-1} \|_{SP(\nu)} \tag{14.13}$$

for eigenvector ν, noting that $\|V\|_{SP(\nu)} \le 2\|V\|_\nu$, so we may equate $a_k = D^w_R(R, R_k)$ and $b_k = D^w_Q(Q, Q_k)$ in (14.12), with a similar derivation for (14.13).

14.3 REGRET

Suppose we are given a single model approximation $\hat{\pi}$ of π. A reasonable strategy would be to accept $\hat{\pi}$ as the true model, then to apply the control $\hat{\phi}$ that would be optimal under $\hat{\pi}$ (that is, the *certainty equivalence control*). This could be done using a VIA based on DPO operator $\bar{T}_{\hat{\pi}}$, and the algorithm need not be considered an AIA.

Of course, we wish to estimate the cost of using what we should anticipate to be a suboptimal control, specifically, the quantity introduced as regret in Chapter 12. The first step can be to regard the certainty equivalence EIA as an AIA with constant model approximations $\hat{\pi}_n = \hat{\pi}$, $n \ge 1$. The operator tolerance can be bounded by Theorem 14.4, which together with Theorem 14.2 gives

$$\limsup_k \| V_k - \bar{V}_\pi \|_w \le \frac{D^w_R(R, \hat{R}) + D^w_Q(Q, \hat{Q}) \limsup_k \| V_k \|_\alpha}{1 - \beta},$$

where $\|\cdot\|_\alpha$ is either $\|\cdot\|_w$ or $(1/2)\,\|\cdot\|_{SP(w)}$ as specified in Theorem 14.4. Since the AIA is equivalently an EIA with respect to the DPO $\bar{T}_{\hat{\pi}}$ we have

$$\| \bar{V}_{\hat{\pi}} - \bar{V}_\pi \|_w \le \frac{D^w_R(R, \hat{R}) + D^w_Q(Q, \hat{Q}) \| \bar{V}_{\hat{\pi}} \|_\alpha}{1 - \beta}, \tag{14.14}$$

which provides a direct method of comparison between the value functions between models $\pi, \hat{\pi}$. One notable feature of (14.14) is that when $\|\cdot\|_\alpha = (1/2)\,\|\cdot\|_{SP(w)}$ the order of magnitude of $\| \bar{V}_{\hat{\pi}} - \bar{V}_\pi \|_w$ is $O(\delta(1-\beta)^{-1})$, where $\max(D^w_R(R, \hat{R}), D^w_Q(Q, \hat{Q})) = O(\delta)$, which is appropriate since $\| \bar{V}_\pi \|_w = O((1-\beta)^{-1})$. In this case, the objective is to set $\delta \ll 1$. In contrast, if $\|\cdot\|_\alpha = \|\cdot\|_w$ the corresponding approximation error is $O(\delta(1-\beta)^{-2})$, and therefore to achieve a comparable tolerance we would need $\delta \ll (1-\beta)$, which would be a considerably more burdensome requirement. The tightness of approximation bounds such as (14.14) is discussed in some detail in Chapter 6 of Bertsekas and Tsitsiklis (1996); see also Tsitsiklis and Roy (1996).

We present two types of bounds for regret. The first is simpler, and is concerned with the regret accruing from a stationary suboptimal policy $\Phi = \phi_{\hat{\pi}}$, where $\phi_{\hat{\pi}}$ is the certainty equivalence policy based on approximate model $\hat{\pi}$. Although we will make use of bound (14.14), it is not the quantity we are ultimately interested in. The true model is π, and so the costs must be calculated using this model, whereas $\bar{V}_{\hat{\pi}}$ is calculated assuming $\hat{\pi}$ holds. The actual cost realized is $V^{\phi_{\hat{\pi}}}_\pi$, and so the relevant approximation bound is $\| V^{\phi_{\hat{\pi}}}_\pi - \bar{V}_\pi \|_w$, which bounds the total discounted regret.

The second involves bounds on $\lambda_\pi(x,a)$, defined in (12.24) of Section 12.5, and permits a more detailed analysis. In particular, it will permit models for regret in adaptive systems which assume cost in parallel with model refinement, which will be discussed in Chapter 18. However, even in the case of the single model estimate a comparison between the two will be instructive, and both are worth considering.

Given models π we can denote the regret under policy $\phi_{\hat{\pi}}$

$$Z_\pi^{\phi_{\hat{\pi}}} = V_\pi^{\phi_{\hat{\pi}}} - \bar{V}_\pi \geq 0,$$

which is positive by the optimality of $\bar{V}_\pi$. Appropriately, we are assigning a special role for model π by calling it the 'true' model. However, it will occasionally be more convenient to consider π and $\hat{\pi}$ simply as two alternative models between which certain relationships can be established. To suggest this structure more explicitly, we will assume that stationary policies $\phi_{\hat{\pi}}$ and ϕ_π are optimal for models π and $\hat{\pi}$ respectively. An interesting relationship exists between $\| Z_\pi^{\phi_{\hat{\pi}}} + Z_{\hat{\pi}}^{\phi_\pi} \|_\alpha$.

Theorem 14.5 *Suppose ϕ_π and $\phi_{\hat{\pi}}$ are optimal policies for models π and $\hat{\pi}$. Suppose for seminorm $\|\cdot\|_\alpha$ the following bounds hold.*

$$\| \bar{V}_\pi - V_{\hat{\pi}}^{\phi_\pi} \|_\alpha \leq \delta \quad \textit{and} \quad \| V_\pi^{\phi_{\hat{\pi}}} - \bar{V}_{\hat{\pi}} \|_\alpha \leq \delta. \tag{14.15}$$

Then

$$\| Z_\pi^{\phi_{\hat{\pi}}} + Z_{\hat{\pi}}^{\phi_\pi} \|_\alpha \leq 2\delta. \tag{14.16}$$

Proof By the triangle inequality

$$\begin{aligned} \| Z_\pi^{\phi_{\hat{\pi}}} + Z_{\hat{\pi}}^{\phi_\pi} \|_\alpha &= \| V_\pi^{\phi_{\hat{\pi}}} - \bar{V}_\pi + V_{\hat{\pi}}^{\phi_\pi} - \bar{V}_{\hat{\pi}} \|_\alpha \\ &\leq \| V_\pi^{\phi_{\hat{\pi}}} - \bar{V}_{\hat{\pi}} \|_\alpha + \| \bar{V}_\pi - V_{\hat{\pi}}^{\phi_\pi} \|_\alpha \end{aligned}$$

which completes the proof following (14.16). ///

We next present an approximation bound for $\lambda_\pi(x,a)$.

Theorem 14.6 *Suppose ϕ_π and $\phi_{\hat{\pi}}$ are optimal policies for models π and $\hat{\pi}$. Then*

$$\lambda_\pi(x, \phi_{\hat{\pi}}(x)) \leq V_\pi^{\phi_{\hat{\pi}}}(x) - \bar{V}_\pi(x) \tag{14.17}$$

for all $x \in \mathcal{X}$.

In addition, suppose the following bounds hold:

$$\begin{aligned} \| \bar{V}_\pi - \bar{V}_{\hat{\pi}} \|_\nu &\leq \delta_\nu \\ \| \bar{V}_\pi - \bar{V}_{\hat{\pi}} \|_{SP(\nu)} &\leq \delta_{SP(\nu)}, \end{aligned}$$

and that (β, ν) *is a principal eigenpair for* $\bar{T}_\pi$ *and* $\bar{T}_{\hat{\pi}}$*. Then*

$$\nu^{-1}\lambda_\pi(x, \phi_{\hat{\pi}}(x)) \leq \| T_\pi^{\hat{\phi}} \bar{V}_\pi - \bar{T}_\pi \bar{V}_\pi \|_\nu \tag{14.18}$$
$$\leq D_R^\nu(R, \hat{R}) + \frac{1}{2} D_Q^\nu(Q, \hat{Q}) \| \bar{V}_\pi \|_{SP(\nu)} + (1-\beta)\delta_\nu + \beta\delta_{SP(\nu)}.$$

Proof We obtain (14.17) from the optimality of ϕ_π, which implies

$$T_\pi^{\phi_{\hat{\pi}}} \bar{V}_\pi - \bar{T}_\pi \bar{V}_\pi \leq T_\pi^{\phi_{\hat{\pi}}} V_\pi^{\phi_{\hat{\pi}}} - \bar{T}_\pi \bar{V}_\pi = V_\pi^{\phi_{\hat{\pi}}} - \bar{V}_\pi.$$

Next, suppose the remainder of the hypothesis holds. Then there exists scalar a for which

$$\| \bar{V}_\pi - (\bar{V}_{\hat{\pi}} + a\nu) \|_\nu \leq \frac{1}{2}\delta_{SP(\nu)}.$$

This gives

$$\begin{aligned}
T_\pi^{\phi_{\hat{\pi}}} \bar{V}_\pi - \bar{T}_\pi \bar{V}_\pi &= \left[T_\pi^{\phi_{\hat{\pi}}} \bar{V}_\pi - \bar{T}_{\hat{\pi}} \bar{V}_\pi \right] + \bar{T}_{\hat{\pi}} \bar{V}_\pi - \bar{V}_\pi \\
&\leq \left[T_\pi^{\phi_{\hat{\pi}}} \bar{V}_\pi - \bar{T}_{\hat{\pi}} \bar{V}_\pi \right] + \bar{T}_{\hat{\pi}} \left(\bar{V}_{\hat{\pi}} + a\nu + \frac{1}{2}\delta_{SP(\nu)}\nu \right) - \bar{V}_\pi \\
&= \left[T_\pi^{\phi_{\hat{\pi}}} \bar{V}_\pi - \bar{T}_{\hat{\pi}} \bar{V}_\pi \right] + \bar{V}_{\hat{\pi}} + \beta a\nu + \beta\frac{1}{2}\delta_{SP(\nu)}\nu - \bar{V}_\pi \\
&= \left[T_\pi^{\phi_{\hat{\pi}}} \bar{V}_\pi - \bar{T}_{\hat{\pi}} \bar{V}_\pi \right] + (1-\beta)\left(\bar{V}_{\hat{\pi}} - \bar{V}_\pi \right) \\
&\quad + \beta \left(\bar{V}_{\hat{\pi}} - \bar{V}_\pi + a\nu + \frac{1}{2}\delta_{SP(\nu)}\nu \right) \\
&\leq \left[T_\pi^{\phi_{\hat{\pi}}} \bar{V}_\pi - \bar{T}_{\hat{\pi}} \bar{V}_\pi \right] + (1-\beta)\delta_\nu\nu + \beta\delta_{SP(\nu)}\nu,
\end{aligned}$$

from which (14.18) follows after applying Theorem 14.4. ///

14.4 A COMMENT ON THE APPROXIMATION OF REGRET

If we are given estimate $\hat{\pi}$ of model π, the problem of estimating $\bar{V}_\pi$ with $\bar{V}_{\hat{\pi}}$ is straightforward, but the quantity of interest is more likely to be $V_\pi^{\phi_{\hat{\pi}}}$, which is the realized cost under the true model using control $\phi_{\hat{\pi}}$, which would be optimal for model $\hat{\pi}$. If we are given $D_R^w(R, \hat{R}), D_Q^w(Q, \hat{Q})$, we may use Theorem 14.5 to obtain a bound for $\| V_\pi^{\phi_{\hat{\pi}}} - \bar{V}_\pi \|_w$ based on bounds for $\| V_\pi^{\phi_{\hat{\pi}}} - V_{\hat{\pi}}^{\phi_{\hat{\pi}}} \|_w$ and $\| V_\pi^{\phi_\pi} - V_{\hat{\pi}}^{\phi_\pi} \|_w$. We can therefore bound regret using the model approximation bounds $D_R^w(R, \hat{R})$ and $D_Q^w(Q, \hat{Q})$.

However, some care is needed in determing a suitable bound for regret. Clearly, it should be expressed as a proportion of some cost related quantity, and a number of possibilities present themselves. An obvious choice is to compare regret to the optimal cost $\bar{V}_\pi$. While this is natural mathematically, from an economic point of view a number of complications arise. Recall that the optimization problem does not change when a constant is added uniformly to $R(x, a)$. In particular, regret does not change by replacing R with $R + cw$, but the magnitude of the value function does. Thus, making c arbitrarily large makes regret arbitrarily small as a proportion of $\| \bar{V}_\pi \|_w$. Furthermore, in the example we will consider next regret is only marginally sensitive to factors which otherwise greatly influence the total optimal cost.

An alternative approach is to regard the cost $\bar{V}_\pi$ as given, and regret as a cost to be considered independently. In this case, the relationship between regret and the bounds $D^w_Q(R, \hat{R})$ and $D^w_Q(Q, \hat{Q})$ are emphasized. Viewed this way, there is no particular reason why regret should be compared to $\bar{V}_\pi$, any more than to any other cost (a reason to do so might exist, but this would be independent of our analysis). The important comparison is between regret and the cost of reducing $D^w_R(R, \hat{R})$ and $D^w_Q(Q, \hat{Q})$.

If this point of view is accepted, the question remains as to what the appropriate comparison for regret would be. If a cost can be assigned to the reduction of $D^w_R(R, \hat{R})$ and $D^w_Q(Q, \hat{Q})$, then regret can be compared to this cost directly. This situation might arise when data from a active system is to be used. If data is acquired from a computer simulated model, it is more likely that the data acquisition cost would be small compared to an indefinite accrual of regret, and the problem is to determine the running time for the simulation needed to guarantee negligible regret.

A detailed analysis of regret can be based on the quantities $\lambda_\pi(x,a)$, which we have seen is directly interpretable as the regret accrued by using action a from state x. We have seen that $\lambda_\pi(x,a)$ can be bounded using $D^w_R(R, \hat{R})$ and $D^w_Q(Q, \hat{Q})$, and these quantities approach 0 at a rate of $O(n^{-1/2})$, that normally associated with statistical estimation. A number of other methods, based on adaptive sampling methods such as bandit processes, report bounds on regret of order $O(\log(n)/n)$ (Chang et al. (2007)). A consideration of the quantity $\lambda_\pi(x,a)$ yields a direct connection between these two rates. Suppose $\mathcal{K}$ is finite. For each x let $\lambda^s_\pi(x) = \inf_{a:\lambda_\pi(x,a)>0} \lambda_\pi(x,a)$, and then $\lambda^s_\pi = \inf_{x \in \mathcal{X}} \lambda^s_\pi(x)$.

Next, suppose we may claim for some certainty equivalence policy $\phi_{\hat{\pi}}$ the bound $\lambda_\pi(x, \phi_{\hat{\pi}}(x)) \leq \delta$ (by Theorem 14.6 such a bound will follow from $D^w_R(R, \hat{R})$ and $D^w_Q(Q, \hat{Q})$). If $\delta < \lambda^s_\pi$ we must have $\lambda_\pi(x, \phi_{\hat{\pi}}(x)) \equiv 0$, so that $\phi_{\hat{\pi}}$ is optimal for model π, that is, regret is zero. Then let δ_n be a sequence of bounds based on a sample size of n. The distributional properties of δ_n vary, but we generally expect that $E[\delta^p_n]^{1/p} \approx n^{-1/2} C_p$ where C_p increases with p but does not depend on n. In most cases $C_p < \infty$ for all p (although exceptions should be anticipated). In addition, the tail probability $P(\delta_n \geq t)$ will almost alway approach zero much more quickly than $n^{-1/2}$ (for the normal distribution convergence is faster than exponential). Therefore, the regret may be modeled as

$$Z_n \leq \delta_n I\{\delta_n \geq \lambda^s_\pi\}, \tag{14.19}$$

since regret is zero if $\delta_n < \lambda^s_\pi$. Applying Hölder's inequality, we have

$$E_\pi[Z_n] \leq n^{-1/2} C_p^{1/p} P\left(\delta_n \geq \lambda^s_\pi\right)^{1/q}, \tag{14.20}$$

for any conjugate pair $p^{-1} + q^{-1} = 1$. If $p = 1$ the original rate $n^{-1/2}$ holds, while if we may set a maximum value Z_{max} for regret, then $q = 1$ implies a bound $E_\pi[Z_n] \leq Z_{max} P\left(\delta_n \geq \lambda^s_\pi\right)$. Markov's inequality yields $E_\pi[Z_n] \leq n^{-k/2} Z_{max}(C_k/\lambda^s_\pi)^k$, so that as long as $C_k < \infty$, we may report convergence at rate $O(n^{-k/2})$ for any k. Of course, in order for the tail probability $P\left(\delta_n \geq \lambda_\pi^s\right)$ to play an important role in the bound, it will be necessary for δ_n to first approach a neighborhood of λ_π^s. Thus, convergence to zero regret occurs in two stages. First, δ_n approaches λ_π^s in magnitude, the rate of which is determined by the convergence rate of the mean and standard deviation of δ_n, usually $O(n^{-1/2})$. At this point the quantity λ_π^s becomes decisive, and convergence to zero regret occurs at a rate determined by the tail probability $P\left(\delta_n \geq \lambda_\pi^s\right)$, which can be expected to be considerably faster than $n^{-1/2}$. In fact, the assumption that $\lambda_\pi^s > 0$ is explicitly required to obtain a convergence rate for regret of $O(\log(n)/n)$ in Chang et al. (2007) (for example, Theorem 2.3, Chapter 2, for a finite horizon model). This condition naturally holds for finite $\mathcal{K}$.

This phenomenon can be interpreted in terms of functional analysis. Given model π we may define $\bar{V}_\pi(x,a) = (T_\pi^a \bar{V}_\pi)(x,a)$, so that the optimal policy is given by $\phi_\pi(x) = \text{argmin}_{a\in\mathcal{K}_x} \bar{V}_\pi(x,a)$. We then define the same quantity $\bar{V}_{\hat{\pi}}(x,a) = (T_{\hat{\pi}}^a \bar{V}_{\hat{\pi}})(x,a)$ for model $\hat{\pi}$. We have established that if $\pi \approx \hat{\pi}$ then $\bar{V}_\pi(x,a) \approx \bar{V}_{\hat{\pi}}(x,a)$, and it can be seen that in order for an approximate model $\hat{\pi}$ to yield the policy optimal for π it is necessarily only that $\hat{\pi}$ be in a neighborhood of π, which is definable by the approximation theory we have discussed.

When viewed this way, a problem emerges regarding the role played by λ_π^s. Suppose we may define a family of models M_π to which the true model π and any estimate $\hat{\pi}$ belong. It will be possible to partition M_π into subsets of models with a common optimal policy (but different value functions). If an approximation $\hat{\pi}$ of π is sufficiently accurate then it will be in the same partition subset as π and will therefore yield zero regret, assuming general continuity conditions hold.

Of course, we need to take into account the possibility that π is near or on a partition boundary, so that $\hat{\pi}$ could be close to π while remaining in a separate partition. Suppose we are given models π and $\hat{\pi}$ in different partition subsets which share a boundary, and that for some state $x \in \mathcal{X}$, $a \neq \hat{a}$ are the respective (unique) optimal actions. This means $\bar{V}_\pi(x,a) - \bar{V}_\pi(x) = 0$ and $\bar{V}_\pi(x,\hat{a}) - \bar{V}_\pi(x) > 0$, and that $\bar{V}_{\hat{\pi}}(x,a) - \bar{V}_{\hat{\pi}}(x) > 0$ and $\bar{V}_{\hat{\pi}}(x,\hat{a}) - \bar{V}_{\hat{\pi}}(x) = 0$. Of course, if π and $\hat{\pi}$ are arbitrarily close, we would expect $\bar{V}_\pi(x,\hat{a}) - \bar{V}_\pi(x)$ to be close to $\bar{V}_{\hat{\pi}}(x,\hat{a}) - \bar{V}_{\hat{\pi}}(x) = 0$, which would mean that λ_π^s could itself be made arbitrarily small, although the optimization problem itself is otherwise a standard one.

This possibility poses no problem, provided the target of zero regret is replaced by an acceptable maximum regret of λ_{max}, which is guaranteed by the bound $\lambda_\pi(x, \phi_{\hat{\pi}}(x)) \leq \lambda_{max}$, similarly expressible in terms of $D_R^w(R, \hat{R})$ and $D_Q^w(Q, \hat{Q})$. Then (14.19) becomes

$$Z_n \leq \lambda_{max} + \delta_n I\{\delta_n \geq \lambda_{max}\}, \tag{14.21}$$

and so the rate of approach to zero regret implied by (14.20) becomes the rate of approach to λ_{max}.

14.5 EXAMPLE

We will illustrate these issues with an example of the variable capacity queueing model. We assume a policy cost $C_p \equiv 0$, so that the state space is $\mathcal{X} = \{1, \ldots, K\}$, and retain the previous service cost will be $C_s(x,d) = d^{1.5}$. We will use $K = 10$, $N_s = 5$, and vary

$\lambda = 1, 3$, $\beta = 0.99, 0.9999$. Table 14.1 gives for each model the complete values of $\lambda_\pi(x, d)$, as well as the value function $\bar{V}_\pi$, optimal control ϕ_π, the cost function under the optimal control $R(x, \phi_\pi(x))$, the steady state distribution $\bar{Q}(x)$, as well as the value for $\lambda^s_\pi(x)$. A number of features of these quantities are worth noting.

Table 14.1 Calculations or regret for example of Section 14.5.

	$\lambda_\pi(x,d)$									
x	$d=1$	$d=2$	$d=3$	$d=4$	$d=5$	$\bar{V}(x)$	$\phi_\pi(x)$	$R(x,\phi_\pi(x))$	$\bar{Q}(x)$	$\lambda^s_\pi(x)$
					$\lambda = 1, \beta = 0.99$					
1	0	1.83	4.2	7	10.18	304.19	1	2	0.58	1.83
2	0	0.01	2.38	5.18	8.36	307.01	1	3	0.26	0.01
3	1.15	0	0.55	3.35	6.53	309.84	2	5.83	0.11	0.55
4	2.05	0.61	0	0.99	4.17	313.20	3	9.2	0.035	0.61
5	2.92	1.07	0.17	0	1.36	317.01	4	13	0.0083	0.17
6	3.88	1.77	0.47	0	0.2	320.98	4	14	0.0016	0.2
7	4.81	2.53	0.96	0.09	0	325.16	5	18.18	0.00026	0.09
8	5.77	3.37	1.63	0.49	0	329.43	5	19.18	3.70E–05	0.49
9	6.25	3.92	2.06	0.76	0	334.10	5	20.18	4.60E–06	0.76
10	5.81	4.14	2.35	0.93	0	339.04	5	21.18	5.60E–07	0.93
					$\lambda = 1, \beta = 0.9999$					
1	0	1.83	4.2	7	10.18	30,562.19	1	2	0.7	1.83
2	0.01	0	2.37	5.17	8.35	30,565.01	2	4.83	0.2	0.01
3	1.19	0	0.52	3.33	6.51	30,567.86	2	5.83	0.074	0.52
4	2.13	0.65	0	0.96	4.14	30,571.22	3	9.2	0.02	0.65
5	3.04	1.15	0.21	0	1.34	30,575.03	4	13	0.0043	0.21
6	4.02	1.85	0.51	0	0.16	30,579.04	4	14	0.00078	0.16
7	5.02	2.67	1.04	0.13	0	30,583.22	5	18.18	0.00012	0.13
8	6	3.53	1.73	0.54	0	30,587.53	5	19.18	1.60E–05	0.54
9	6.53	4.12	2.19	0.82	0	30,592.25	5	20.18	1.90E–06	0.82
10	6.1	4.36	2.49	0.99	0	30,597.25	5	21.18	2.20E–07	0.99
					$\lambda = 3, \beta = 0.99$					
1	0	1.83	4.2	7	10.18	890.32	1	2	0.18	1.83
2	1.77	0	2.37	5.17	8.35	893.15	2	4.83	0.21	1.77
3	3.53	1.23	0	2.8	5.98	896.52	3	8.2	0.22	1.23
4	5.21	2.55	0.79	0	3.18	900.32	4	12	0.17	0.79
5	6.77	3.86	1.74	0.42	0	904.50	5	16.18	0.11	0.42
6	7.94	5	2.63	0.95	0	909.10	5	17.18	0.06	0.95
7	8.24	5.64	3.24	1.31	0	914.23	5	18.18	0.029	1.31
8	7.37	5.59	3.52	1.56	0	919.72	5	19.18	0.013	1.56
9	5.04	4.47	3.22	1.59	0	925.45	5	20.18	0.0052	1.59
10	1.4	2.1	2.07	1.25	0	931.22	5	21.18	0.003	1.25
					$\lambda = 3, \beta = 0.9999$					
1	0	1.83	4.2	7	10.18	89,797.84	1	2	0.18	1.83
2	1.82	0	2.37	5.17	8.35	89,800.67	2	4.83	0.21	1.82
3	3.64	1.28	0	2.8	5.98	89,804.03	3	8.2	0.22	1.28
4	5.4	2.66	0.84	0	3.18	89,807.84	4	12	0.17	0.84
5	7.06	4.05	1.85	0.46	0	89,812.02	5	16.18	0.11	0.46
6	8.3	5.24	2.77	1.01	0	89,816.66	5	17.18	0.06	1.01
7	8.66	5.94	3.42	1.39	0	89,821.85	5	18.18	0.029	1.39
8	7.81	5.92	3.74	1.66	0	89,827.42	5	19.18	0.013	1.66
9	5.44	4.8	3.45	1.71	0	89,833.26	5	20.18	0.0052	1.71
10	1.72	2.38	2.28	1.37	0	89,839.14	5	21.18	0.003	1.37

The magnitude of the value function $\bar{V}_\pi$ is highly dependent on the parameters. This is obviously true of β, but the variation of λ changes the achieved cost by a factor of almost 3, since larger values of λ force the system to spend a greater proportion of time in higher cost states. Suppose we chose to express regret as a proportion of optimal cost. Since $\lambda_\pi(x,d)$ gives regret on a per stage basis, it would be reasonable to standardize the optimal cost using the factor $(1-\beta)$. If we, conservatively, use $\bar{V}_\pi(1)$ as a reference, we obtain

$$\begin{aligned}
(1-\beta)\bar{V}_\pi(1) &= 3.04, \quad \text{for} \quad \lambda=1,\beta=0.99,\\
(1-\beta)\bar{V}_\pi(1) &= 3.06, \quad \text{for} \quad \lambda=1,\beta=0.9999,\\
(1-\beta)\bar{V}_\pi(1) &= 8.90, \quad \text{for} \quad \lambda=3,\beta=0.99,\\
(1-\beta)\bar{V}_\pi(1) &= 8.98, \quad \text{for} \quad \lambda=3,\beta=0.9999.
\end{aligned}$$

Thus, a comparison between regret and achieved cost can be standardized with respect to β. However, a remarkable feature of Table 14.1 is that the values of $\lambda_\pi(x,d)$ do not change greatly with changes in either β or λ, even though they have considerable effect on $\bar{V}_\pi$. If β is held constant, the problem of bounding regret would be quite similar for either value $\lambda=1,3$, requiring similar values of $D_R^w(R,\hat{R})$ and $D_Q^w(Q,\hat{Q})$ for a similar achieved regret. Therefore, reporting regret as a proportion of the value function would yield a spurious advantage for the $\lambda=3$ model, even after standardization for β.

We next note the values $\lambda_\pi^s=0.01,0.01,0.42,0.46$ for the models $(\beta,\lambda)=(1,0.99),(1,0.9999),(3,0.99),(3,0.9999)$. Clearly, we cannot rely on the assumption $\lambda_\pi^s>0$ to obtain a practical convergence rate. We may instead select a suitable value of λ_{max} for the bound (14.21). If estimates of $\lambda_\pi(x,a)$ are available, possibly $\lambda_{\hat{\pi}}(x,a)$, a more refined choice for λ_{max} may be made. Values analagous to $\lambda_\pi^s(x)$ may be defined, for example

$$\lambda_\pi^s(x\mid\epsilon)=\inf_{a:\lambda_\pi(x,a)>\epsilon}\lambda_\pi(x,a), \quad \text{and} \quad \lambda_\pi^s(\epsilon)=\inf_{x\in\mathcal{X}}\lambda_\pi^s(x\mid\epsilon),$$

so that a bound on per stage regret of ϵ is obtainable by the bound $\lambda_\pi(x,a)<\lambda_\pi^s(\epsilon)$. We may also derive the converse, the regret attained by enforcing bound $\lambda_\pi(x,a)<\epsilon$, which may be expressed

$$Z^{reg}(x\mid\epsilon)=\sup_{a:\lambda_\pi(x,a)<\epsilon}\lambda_\pi(x,a), \quad \text{and} \quad Z^{reg}(\epsilon)=\sup_{x\in\mathcal{X}}Z^{reg}(x\mid\epsilon).$$

Clearly, for fixed ϵ we have $Z^{reg}(x\mid\epsilon)\le\lambda_\pi^s(x\mid\epsilon)$ and $Z^{reg}(\epsilon)\le\lambda_\pi^s(\epsilon)$, so a distinction must be made between using ϵ as a bound for $\lambda_\pi(x,a)$, and as a target for attainable regret.

Table 14.2 gives values of $Z^{reg}(x\mid 0.5)$ and $\lambda_\pi^s(x\mid 0.5)$ for model $(\beta,\lambda)=(1,0.99)$. We obtain $Z^{reg}(0.5)=0.49$ and $\lambda_\pi^s(0.5)=0.55$, which suggests little difference between using ϵ as a bound for $\lambda_\pi(x,a)$ and as a target for regret. However, note that the combined steady state probabilities for states $1,\ldots,4$ is approximately 0.99 ($\bar{Q}(x)$ is included in Table 14.2 for reference). If bound $\lambda_\pi(x,a)<0.05$ holds, the achieved regret for three of these states is 0, and is within 0.01 for the remaining state 2, so that while the nominal regret bound is $\epsilon=0.5$ the achieved regret will be much

Table 14.2 Calculations of regret for example of Section 14.5.

x	$Z^{reg}(x \mid 0.5)$	$\lambda_{\pi}^{s}(x \mid 0.5)$	$\bar{Q}(x)$
1	0	1.83	0.58
2	0.01	2.38	0.26
3	0	0.55	0.11
4	0	0.61	0.035
5	0.17	1.07	0.0083
6	0.47	1.77	0.0016
7	0.09	0.96	0.00026
8	0.49	1.63	3.70E–05
9	0	0.76	4.60E–06
10	0	0.93	5.60E–07
	$Z^{reg}(0.5) = 0.49$	$\lambda_{\pi}^{s}(0.5) = 0.55$	

smaller. Similarly, the bound $\lambda_{\pi}(x,a) < \lambda_{\pi}^{s}(0.5) = 0.55$ guarantees an attained bound of 0.5. If we examine the values of $\lambda_{\pi}(x,a)$ for states $x = 1, 2$, with a combined steady state probability of 0.84, we find that for any bound $\lambda_{\pi}(x,a) < \epsilon < 1.83$ we have $Z^{reg}(1 \mid \epsilon) = 0$ and $Z^{reg}(2 \mid \epsilon) = 0.01$, attaining negligible regret for the two highest frequency states with a much higher approximation bound for $\lambda_{\pi}(x,a)$. Similar comments hold for the remaining models.

Chapter 15

Sampling based approximation methods

As a first application of the model based approximation methods of Chapter 14 we consider the problem of estimating the model elements (R, Q) by sampling. This may be based on histories of an active system or on computer simulations. The main problem we consider is the determination of a sample size sufficient to achieve a fixed algorithm tolerance. We assume the concern is entirely with model estimation, but will later consider the problem of determining the optimal rate of exploratory behavior in an active system also subject to optimal cost control (Chapter 18).

All model based approaches are based on developing an estimate $(\hat{R}, \hat{Q})$ of (R, Q) (both elements may be estimated). The *certainty equivalence* value function $\hat{V}^*$, that obtained by accepting $(\hat{R}, \hat{Q})$ as the true model, will then have an error $\| \hat{V}^* - V^* \|_w$ determined directly by $D_Q^w(\hat{Q}, Q)$ and $D_R^w(\hat{R}, R)$, so these quantities are the crucial ones. We have already seen that we may obtain an approximation bound satisfying

$$\| \hat{V}^* - V^* \|_w \leq (1-\beta)^{-1} \left(D_R^w(\hat{R}, R) + D_Q^w(\hat{Q}, Q) \, \| V^* \|_{SP(w)} \right),$$

and that the advantage of the span seminorm is that the approximation bound is of order $O((1-\beta)^{-1}\delta)$, where $\delta = \max\left(D_R^w(\hat{R}, R), D_Q^w(\hat{Q}, Q) \right)$, rather than $O((1-\beta)^{-2}\delta)$, which would be the case if $\| V^* \|_{SP(w)}$ were replaced with $\| V^* \|_w$. Therefore, it suffices for δ to be small, rather than $(1-\beta)^{-1}\delta$.

We will assume for the moment that $\mathcal{K}$ is finite. The quantities $D_Q^w(\hat{Q}, Q)$ and $D_R^w(\hat{R}, R)$ are then maxima of a finite, but possibly quite large set of estimation errors. The first step is to charactize what is unknown. It will often be the case that the model π is known up to a relatively small number of parameters (compared to $|\mathcal{K}|$), as would be the case for a queueing system in which transitions are governed by an arrival process. In this case the parametric estimation approach of Section 15.3 could be used.

The most difficult case, on which we will focus primarily, occurs when no analytical relationships are available. This means the estimation of elements $R(x, a)$ by $\hat{R}(x, a)$ and of $Q(x, a)$ by $\hat{Q}(x, a)$ for each $(x, a) \in \mathcal{K}$ are essentially separate estimation problems. In statistical estimation theory, this distinction is governed by the *degrees of freedom*, which is the number of parameters minus the number of constraints on the parameters. If the transition law is determined by K Poisson arrival processes, the degrees of freedom will be the number of distinct arrival rates (that is, 1 if all rates are constrained to be equal, or K if they are allowed to vary independently).

A nonparametric estimate of a distribution on $\mathcal{S}=\{1,\ldots,N\}$ requires estimates of N parameters p_i, but we have a single constraint $p_1+\cdots+p_N=1$, so the degrees of freedom is $N-1$ (see Section 4.15). Although the vector $(p_1,\ldots,p_N)$ may be regarded as a parameter, the term *parametric* is usually used to denote the dependence of a distribution on a relatively small number of parameters, the number of which is unrelated to the cardinality of the sample space $\mathcal{S}$. Thus, in nonparametric estimation, all distributions on $\mathcal{S}$ are considered while the range of distributions considered in parametric estimation is considerable smaller. Here, the importance of the degrees of freedom is that it determines the complexity of the estimation problem, and required sample sizes may be usefully expressed on a per degrees of freedom basis.

15.1 MODELING MAXIMA

In Chapter 8 the problem of estimating the maximum of a set of random variables was considered in some detail. We are given a set of random variables $X_1, X_2, \ldots, X_m$ and define the maximum and expected maximum

$$M=\max_{i\leq m} X_i \quad \text{and} \quad \bar{M}=E[M].$$

Table 15.1 summarizes a number of the bounds on $\bar{M}$ we have already introduced. Some of these bounds may be refined, but generally without altering the orders of magnitude. None of the inequalities require the assumption of independence, which plays a surprisingly small role in this problem.

The approximation problem is reducible to bounding the maximum of m RVs $X_1,\ldots,X_m$, where X_i represents the statistical error of the ith estimation problem. We can generally expect that each element of $\mathcal{X}$ or $\mathcal{K}$ require one or several estimates.

To analyze the problem, it is helpful to assume there is a fixed sample size n from which n_i is to be allocated to the ith estimaton problem, so that $n_1+\cdots+n_m=n$, and we denote the vector of sample sizes $\tilde{n}=(n_1,\ldots,n_m)$. The allocation can also be conceived as a probability distribution $\tilde{q}=(q_1,\ldots,q_m)$ with q_i representing the proportion of the total sample allocated to estimation problem i, so that $n_i\approx q_i n$ (we will ignore rounding error from this point, which will be negligible for large enough n). We will refer to the *uniform allocation* $\tilde{q}_{unif}=(1/m,\ldots,1/m)$.

Table 15.1 Summary of maximum inequalities. All inequalities assume general dependence structure.

Theorem	Result
(a) Lemma 8.1	$\bar{M}\leq\sum_{i\leq m} E[\lvert X_i\rvert^p]^{1/p}$
(b) Theorem 8.7 *(i)*	$\bar{M}\leq\mu+(m-1)^{1/2}(\bar{\sigma}^2+\nu)^{1/2}$
(c) Theorem 8.7 *(ii)*	$\bar{M}\leq\mu_{max}+(m-1)^{1/2}(\bar{\sigma}^2)^{1/2}$
(d) Theorem 8.4	$\bar{M}\leq\mu_{max}+\sigma_{max}\sqrt{2\log(m)}$

In most statistical estimates, the deviation of an estimate based on sample size n is of order $O(n^{-1/2})$ and its variance is of order $O(n^{-1})$. Therefore, it will serve our purpose to assume

$$\mu_i = \frac{\mu}{n_i^{1/2}} = \frac{\mu}{(q_i n)^{1/2}} \quad \text{and} \quad \sigma_i^2 = \frac{\sigma^2}{n_i} = \frac{\sigma^2}{q_i n} \tag{15.1}$$

for some finite μ and σ. The inequalities are monotone in μ_i and σ_i^2, so these may be exact values or upper bounds. Then denote power transformations $\tilde{q}^p = (q_1^p, \ldots, q_m^p)$ and $\tilde{n}^p = (n_1^p, \ldots, n_m^p)$. This gives mean and variance vectors

$$\tilde{\mu} = \mu\tilde{n}^{-1/2} = n^{-1/2}\mu\tilde{q}^{-1/2} \quad \text{and} \quad \tilde{\sigma^2} = \sigma^2\tilde{n}^{-1} = n\sigma^2\tilde{q}^{-1}$$

expressed in terms of the allocation distribution $\tilde{q}$.

We may then express the various quantities appearing in Table 15.1 in terms of the parameters μ, σ^2, sample size n, order of maximum m and allocation distribution $\tilde{q}$, where $\tilde{q}$ is represented through power means (Section 2.1.17):

$$\begin{aligned}
\bar{\mu} &= \frac{\mu}{(nM_{-1/2}[\tilde{q}])^{1/2}} \\
\bar{\mu^2} &= \frac{\mu^2}{nM_{-1}[\tilde{q}]} \\
\nu = \bar{\mu^2} - \bar{\mu}^2 &= \frac{\mu^2}{n}\left(\frac{1}{M_{-1}[\tilde{q}]} - \frac{1}{M_{-1/2}[\tilde{q}]}\right) \\
\mu_{max} &= \frac{\mu}{(nM_{-\infty}[\tilde{q}])^{1/2}} \\
\bar{\sigma^2} &= \frac{\sigma^2}{nM_{-1}[\tilde{q}]} \\
\sigma^2_{max} &= \frac{\sigma^2}{nM_{-\infty}[\tilde{q}]}.
\end{aligned} \tag{15.2}$$

It will also be convenient to use the notation

$$\nu[\tilde{q}] = \frac{1}{M_{-1}[\tilde{q}]} - \frac{1}{M_{-1/2}[\tilde{q}]}.$$

We expect the magnitude of the estimation problem to be driven by m, and so it will be reasonable to express the sample size requirements in terms of the quantity n/m, that is, the number of samples available per estimation problem. Ideally, the upper bound depends on m and n only through n/m, but the maximization operation forces an additional dependence on m independently. Therefore, we express the bounds in the form

$$\bar{M} \le \left(\frac{m}{n}\right)^{1/2} C_I(\tilde{q}, m) \tag{15.3}$$

for the inequality indexed I. Applied to inequalities (b), (c) and (d) of Table 15.1, this yields, respectively,

$$C_b(\tilde{q},m) = \frac{\mu}{\left(mM_{-1/2}[\tilde{q}]\right)^{1/2}} + [m-1]^{1/2}\left[\frac{\sigma^2}{mM_{-1}[\tilde{q}]} + \frac{\mu^2}{mM_{-1}[\tilde{q}]} - \frac{\mu^2}{mM_{-1/2}[\tilde{q}]}\right]^{1/2},$$

$$C_c(\tilde{q},m) = \frac{\mu}{\left(mM_{-\infty}[\tilde{q}]\right)^{1/2}} + [m-1]^{1/2}\frac{\sigma}{\left(mM_{-1}[\tilde{q}]\right)^{1/2}},$$

$$C_d(\tilde{q},m) = \frac{\mu}{\left(mM_{-\infty}[\tilde{q}]\right)^{1/2}} + [2\log(m)]^{1/2}\frac{\sigma}{\left(mM_{-\infty}[\tilde{q}]\right)^{1/2}}. \tag{15.4}$$

All terms in (15.4) are of the form $K/mM_p[\tilde{q}]^s$, where $p=-\infty, -1$ or $-1/2$ and $k=1/2$ or 1. In all but one case $K>0$. The single exception occurs in $C_b(\tilde{q},m)$. However, noting that $0 \le \nu[\tilde{q}] \le M_{-1}[\tilde{q}]^{-1}$, we can see that this type reciprocal relationship extends to that term as well.

Dependence on maximum order m

We first consider how the coefficients $C_I(\tilde{q},m)$ scale with m. As already discussed, the bound depends stongly on n/m, but the order m also affects the bound independently. Here, the important fact is that $C_b(\tilde{q},m)$ and $C_c(\tilde{q},m)$ are of order $O(m^{1/2})$ while $C_d(\tilde{q},m)$ is of order $O(\log(m)^{1/2})$ (we will see below that it is reasonable to regard $mM_p[\tilde{q}]$ as fixed in this context).

Suppose m increases by a factor of r to m'. We consider the problem of determing the sample size n' which keeps the upper bound constant. Clearly (assuming that the m estimation problems are independent), we must have at least linear scaling $n' \ge rn$, so the interesting quantity is n'/m'. For bounds $C_b(\tilde{q},m)$, $C_c(\tilde{q},m)$, in order to maintain a fixed bound n'/m' must increase from n/m by a factor of approximately r, whereas for bound $C_d(\tilde{q},m)$, n'/m' need only increase by a factor of $1+log(r)/log(m)$. For example, if m doubles from 1000 to 2000, then $r=2$, and $n'/m' \approx 2(n/m)$, meaning that the new sample size must be $n' \approx 4n$, whereas using bound C_d the new sample size scales almost linearly as $n' \approx 2.2n$. In addition, if m increases exponentially as K^N, for example, as in a queueing system with N queues of capacity K, the sample size per state required to maintain a fixed approximation increases proportionally with N.

These forms are directly comparable, and after any necessary modifications, inequality (d) will be most efficient for all large enough m.

Uniform sample size allocation

Clearly, absent any further structure, the best choice of allocation is the uniform case $\tilde{q}_{unif}$, and the coefficients of (15.4) can give a quantitative measure of the consequences of any deviation from the optimal. We first introduce the following lemma:

Lemma 15.1 *If $\tilde{q}=(q_1,\ldots,q_m)$ is a probability distribution on $\mathcal{S}=\{1,\ldots,m\}$ then for any $p<1$ we have*

$$M_p[\tilde{q}] \le 1/m, \tag{15.5}$$

with equality if and only if $\tilde{q} = \tilde{q}_{unif}$.

Proof The bound (15.5) follows from Theorem 2.4 and the fact that $M_1[\tilde{q}] = 1/m$ (since $\tilde{q}$ is a probability distribution). ///

Lemma 15.1 then provides a sharp lower bound for coefficients $C_b(\tilde{q}, m)$, $C_c(\tilde{q}, m)$ and $C_d(\tilde{q}, m)$.

Theorem 15.1 *For any distribution* $\tilde{q}$ *on* $S = \{1, \ldots, m\}$ *the coefficients* $C_b(\tilde{q}, m)$, $C_c(\tilde{q}, m)$ *and* $C_d(\tilde{q}, m)$ *defined in (15.4) satisfy the inequalities:*

$$\begin{aligned} C_b(\tilde{q}, m) &\geq \mu + (m-1)^{1/2}\sigma, \\ C_c(\tilde{q}, m) &\geq \mu + (m-1)^{1/2}\sigma, \\ C_d(\tilde{q}, m) &\geq \mu + [2\log(m)]^{1/2}\sigma, \end{aligned} \tag{15.6}$$

with equality if and only if $\tilde{q} = \tilde{q}_{unif}$.

Proof For fixed μ, σ and m, each term of $C_c(\tilde{q}, m)$ and $C_d(\tilde{q}, m)$ is a strictly decreasing function of $mM_p[\tilde{q}]$ for some $p = -\infty, -1, -1/2$, so that the theorem follows directly from Lemma 15.1. In reference to $C_b(\tilde{q}, m)$, it may be verified that $\nu[\tilde{q}]$ is always nonnegative, and is zero if and only if $\tilde{q} = \tilde{q}_{unif}$, which also uniquely minimizes the remaining terms. ///

The lower bounds of (15.6) then provide an optimal baseline against which any other allocation can be compared.

15.1.1 Nonuniform sample allocation: Dependence on q_{min}, and the 'Curse of the Supremum Norm'

The use of the supremum norm to quantify approximation error forces the need for a uniform bound on all estimation problems, therefore an approximation bound can be no better than the worst case among all estimation problems considered separately. In the sample allocation model this must mean that an approximation measured in the supremum norm will be especially sensitive to the value q_{min}, which defines the worst case.

More precisely, as discussed in Chapter 8, we always have $\bar{M} \geq \mu_{max} = \mu/(nq_{min})^{1/2}$, and so the influence of an especially large variate, say X_1, cannot be 'averaged out'. Suppose X_1 has especially large values of μ_1 and σ_1^2. We would then have $\bar{\mu} \approx \mu_1/m$, $\bar{\sigma^2} \approx \sigma_1^2/m$, $\bar{\mu^2} \approx \mu_1^2/m$. Bounds (b) and (c) of Table 15.1 are then approximately $(\sigma_1^2 + \mu_1^2)^{1/2}$ and $\mu_1 + \sigma_2$ respectively (within this approximation, bound (b) is smaller). In this case, bounds (b) and (c) are sharper than (d), since for the latter dependence on m is not affected by the variation in distributions.

We have already seen, by Theorem 15.1, that for the uniform allocation case $C_d(\tilde{q}, m)$ gives the best scaling with respect to m, and this advantage will be maintained for $\tilde{q}$ which do not deviate greatly from the uniform. Furthermore, this deviation can be quantified by the value of $(mM_{-\infty}[\tilde{q}])^{-1/2}$. To do this, note, that the uniform allocation provides the smallest bound for all coefficients in (15.4), and for all $m \geq 4$ the optimal choice will be $C_d(\tilde{q}, m)$. For this reason, it will useful to define (based on Theorem 15.1)

$$C_{unif}(m) = \mu + [2\log(m)]^{1/2}\sigma,$$

as the best possible bound. Clearly,

$$C_d(\tilde{q}, m) = (mq_{min})^{-1/2} C_{unif}(m),$$

so that $(mq_{min})^{-1/2}$ measures directly the effect of deviation from uniform allocation.

Thus, if $(mq_{min})^{-1/2}$ does not greatly exceed 1, we will be close to the optimal efficiency of uniform allocation. Alternatively, when $q_{min} \ll 1/m$ the remaining bounds $C_b(\tilde{q}, m)$ or $C_c(\tilde{q}, m)$ will become more efficient. For very small q_{min} the bound is likely to be very large for any of the coefficients, but the interaction of the effect of small q_{min} with m tends to be smaller for $C_b(\tilde{q}, m)$ and $C_c(\tilde{q}, m)$ than for $C_d(\tilde{q}, m)$.

15.1.2 Some queueing system examples

Figure 15.1 shows values of $C_{unif}(m)$ for the range $1 \leq m \leq 1000$, for simplicity taking $\mu = \sigma = 1$. Recalling (15.3), the best bound is given by $\bar{M} \leq (n/m)^{-1/2} C_{unif}(m)$. The value of $C_{unif}(m)$ rises quickly for small values of m, reaching $C_{unif}(8) \approx 3.04$ at $m = 8$. However, the remaining increase is smaller, so that at $m = 1000$ we have $C_{unif}(1000) \approx 4.72$. Thus, using uniform allocation, for large enough m the required total sample size n for a fixed approximation bound will scale almost linearly with m. Thus, a uniform allocation design can be quite effective for large state-space models even when the approximation tolerance is measured using the supremum norm.

It will be important to consider as an alternative to uniform allocation the online model, in which data is collected as the system is allowed to operate according to its stochastic kernel. In this case we also rely on the concept of the proportion of allocation q_i of a total sample size n to estimation problem indexed by i. Here, i may refer to a single state, or to a state/action pair. In the context of MDPs, if the object is to estimate the model for a fixed policy ϕ, i would represent a state, and the stochastic kernel would define a specific Markov chain with transition law $Q^{\phi}(x) = Q(x, \phi(x))$. Then q_i would simply be given by the steady state frequency of state i, and the sampling is of order $m = |\mathcal{X}|$.

The situation is less precise when the object is to estimate (R, Q) over the entire state action space $\mathcal{K}$, and a number of strategies are available. The simplest is to use a stationary randomized exploration control which selects an action a from $\mathcal{K}_x$ whenever the process enters state x, at which point cost $R(x, a)$ is captured as data, and the process selects the next state from distribution $Q(x, a)$. In this case, the estimation problem i is associated with a state/action pair, and the sampling is of order $m = |\mathcal{K}|$. In this case, the observed process is also a Markov chain, but the transition law depends on both Q and the randomized control.

Under a reasonable conjecture, we may predict that sampling over $\mathcal{K}$ under an exploratory randomized control need not be significantly less efficient that sampling under a fixed control over $\mathcal{X}$, on a per estimate basis (assuming $m = |\mathcal{K}|$ and $m = |\mathcal{X}|$ required estimates respectively). Given model (R, Q) the transition law on state space

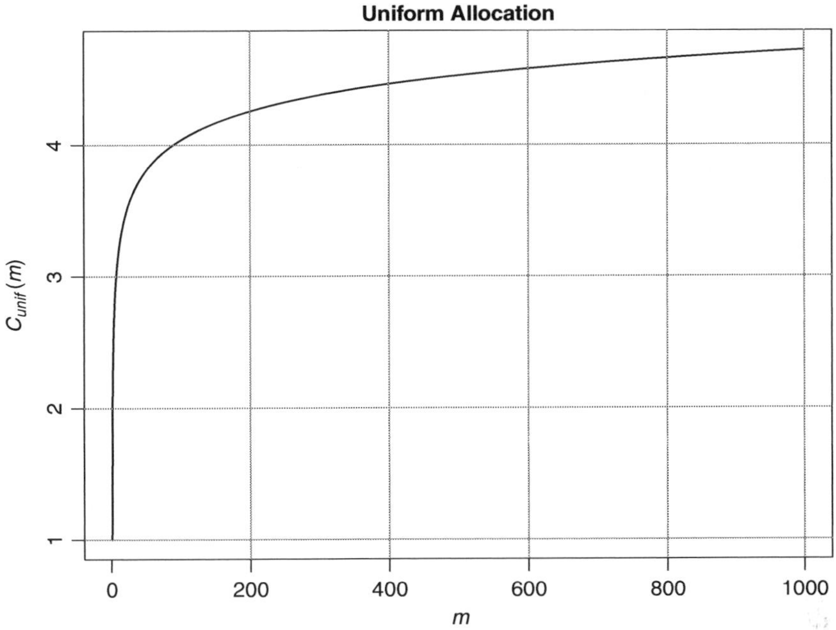

Figure 15.1 Values of $C_{unif}(m)$ for the range $1 \leq m \leq 1000$, for $\mu = \sigma = 1$. See Section 15.1.2.

$\mathcal{X}$ for a stationary determinstic policy ϕ is $Q^\phi = Q(x, \phi(x))$. Suppose we may specify a set of frequencies $\tilde{q}$ on $\mathcal{X}$ such that the steady state frequencies following from Q^ϕ remain within some neighborhood for each ϕ (in the sense that the relevant power means $M_p[\tilde{q}]$ would not vary greatly). Then, a randomized policy Φ defines a Markov chain on $\mathcal{K}$. However, we may still define state occupancy frequencies q'_i as the sum of those steady-state frequencies on $\mathcal{K}$ associated with state i. We then extend our conjecture to assume $q'_i \approx q_i$. To fix ideas, suppose $n_a = |\mathcal{K}_x|$ is constant for each $x \in \mathcal{X}$, and Φ selects each available action with equal probability. This yields steady state frequencies for $z \in \mathcal{K}$ of value $q''_z = q'_i / n_a$ for any $z = (i, a) \in \mathcal{K}$. It is easily verified that for any power mean $M_p[\tilde{q}''] = n_a^{-1} M_p[\tilde{q}']$, and therefore $M_p[\tilde{q}''] \approx n_a^{-1} M_p[\tilde{q}]$. If the order of the sampling problem on $\mathcal{X}$ is m, then it is of order $m'' = n_a m$ on $\mathcal{K}$, and so the crucial quantities $(m'' M_p[\tilde{q}''])^{1/2} \approx (m M_p[\tilde{q}])^{1/2}$ remains approximately the same, so that the affect on the bound will be expressed through the factors $[m-1]^{1/2}$ or $[2\log(m)]^{1/2}$ appearing in (15.4).

Given this simple extension, we will confine attention to sampling over a state space $\mathcal{X}$, with the control problem playing no role.

15.1.3 Truncated geometric model

It will generally be necessary to make some conjecture regarding $\tilde{q}$ in order to estimate the required sample size n. In our discussion of queueing models (Section 5.4) it was pointed out that even in such well defined models (for which the control is fixed) a closed form solution for the steady state frequencies may not be obtainable, and this appears to be the rule, rather than the exception. However, it has also been observed (see, for example, Kleinrock (1975)) that there is a tendency for the steady state frequencies of commonly used queuieng system models to resemble the geometric distribution, or more generally, for steady state frequencies to decrease geometrically with respect to some naturaly state ordering. It will therefore be useful to rely on the truncated geometric distribution defined by

$$q_i = \beta^i \frac{1-\beta}{\beta(1-\beta^m)}, \quad i = 1, \ldots, m \tag{15.7}$$

for positive integer m and constant $\beta \in (0,1)$. First, we may directly evaluate the quantity

$$q_{min} = \frac{\beta^m}{1-\beta^m} \frac{1-\beta}{\beta}, \tag{15.8}$$

to determine if steady state allocation is feasible. For example, if $r = 0.9$ and $m = 100$, then $q_{min} \approx 2.95 \times 10^{-6}$, and $(mq_{min})^{1/2} \approx 0.017 \ll 1$, so we can expect the small value of q_{min} to dominate the bound. Alternatively, if $r = 0.9$ and $m = 25$ we have $q_{min} \approx 0.00859$ and $(mq_{min})^{1/2} \approx 0.464$, and so we may predict that steady state sampling will be at least of the same order of magnitude of efficiency as uniform sampling.

Figure 15.2 shows the approximation bounds $C_b(\tilde{q}, m)$, $C_c(\tilde{q}, m)$ and $C_d(\tilde{q}, m)$ relative to $C_{unif}(m)$ for the truncated geometric distribution for values $\beta = 0.25, 0.5, 0.8, 0.9, 0.95, 0.99$, and for values of m for which the relative value does not exceed 100. We expect that small values of q_{min} dominate for smaller values of β, in which case $C_b(\tilde{q}, m)$ or $C_c(\tilde{q}, m)$ will be the smallest coefficient. On the other hand, the distribution approaches the uniform allocation case as β approaches 1, so that $C_d(\tilde{q}, m)$ will be smallest for all large enough β. This tendency can be seen in Figure 15.2, with values of $\beta < 0.9$ favoring $C_b(\tilde{q}, m)$ and values of $\beta > 0.9$ favoring $C_d(\tilde{q}, m)$ (all coefficients are nearly equal for $\beta = 0.9$).

15.1.4 *M/G/1/K* queueing model

As discussed in Section 5.4, closed forms of steady-state distributions for queueing system models are available only for a small set of special cases. Suppose for the $M/G/1/K$ model we are given service rate μ, arrival rate λ and queue capacity K. This defines a system for which the state is the system occupancy, which ranges over $x = 0, 1, \ldots, K$ (in some models states 0 and 1 are consolidated into state 1). Then μ and λ are interpretable as the rates of decrease and of increase of the state index, except possibly at the boundaries. Under well defined distributional conditions these rates may be used in balance equations such as (5.5) or (5.15) to precisely obtain steady-state occupancy rates (whether in the time domain, or relative to the embedded Markov chain). These conditions are met by the $M/M/1/K$ queuing model, and the balance equations yield steady-state occupancy frequencies which decrease geometrically at rate $\rho = \lambda/\mu$ for either the time-domain system or the embedded Markov chain, except possibly at the boundaries. Here, ρ equals the utilization factor defined in Section

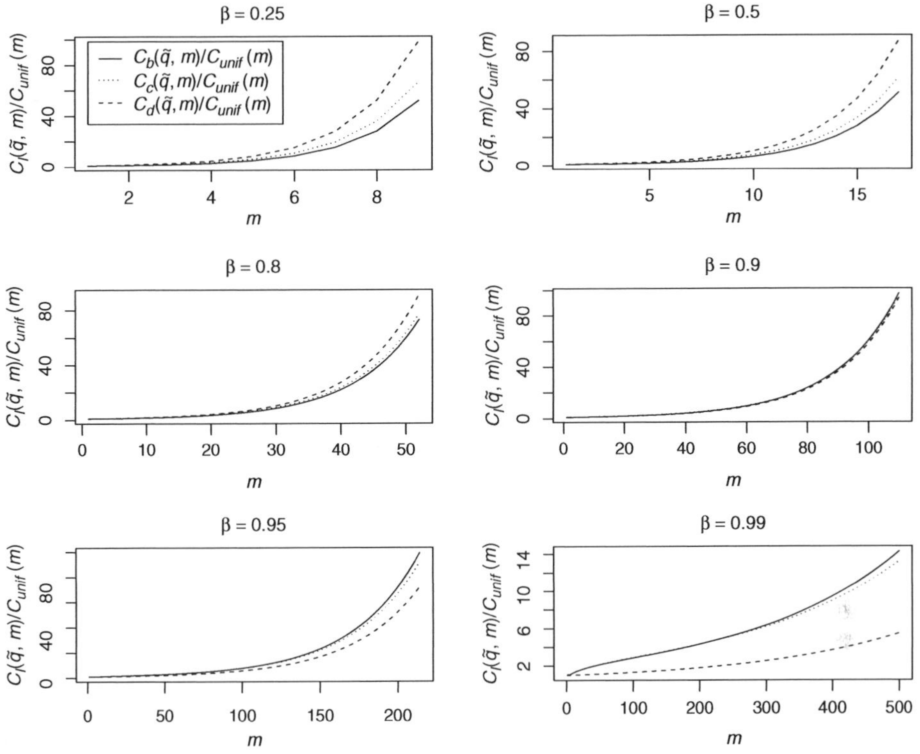

Figure 15.2 Approximation bounds $C_b(\tilde{q}, m)$, $C_c(\tilde{q}, m)$ and $C_d(\tilde{q}, m)$ relative to $C_{unif}(m)$ for the truncated geometric distribution for values $\beta = 0.25, 0.5, 0.8, 0.9, 0.95, 0.99$. Only values of m for which the relative value does not exceed 100 are shown. See Section 15.1.3.

5.4.2. Thus, the truncated geometric model (15.7) for steady-state sample allocation will approximately hold (the occupancy of the embedded Markov chain will be of more relevance to this problem).

For the more general $M/G/1/K$ model, μ and λ retain the same interpretation as rates of transitions, while the utilization factor $\rho = \lambda/\mu$ still decisively determines the burden placed on the queue. Thus, while balance equations (5.5) or (5.15) do not hold strictly, the fact that transitions from state i to $i+j$ or $i-j$ occur at rates independent of i, at least far enough from the boundaries, and that, aggregately, the rates of decrease are higher than the rates of increase, together suggest that steady-state occupancy frequencies π_i should decrease in a regular multiplicative manner, with scalar increments close to ρ, resulting in a distribution close to the geometric.

As an example, consider the $M/D/1/K$ model with unit service time, arrival rate λ and system capacity K. Here we analyze the embedded Markov chain, and consolidate states 0 and 1 into 1. We present an analysis of the same form as that of Section 15.1.3, with Figure 15.3 similarly displaying values of $C_b(\tilde{q}, m)$, $C_c(\tilde{q}, m)$ and $C_d(\tilde{q}, m)$ relative

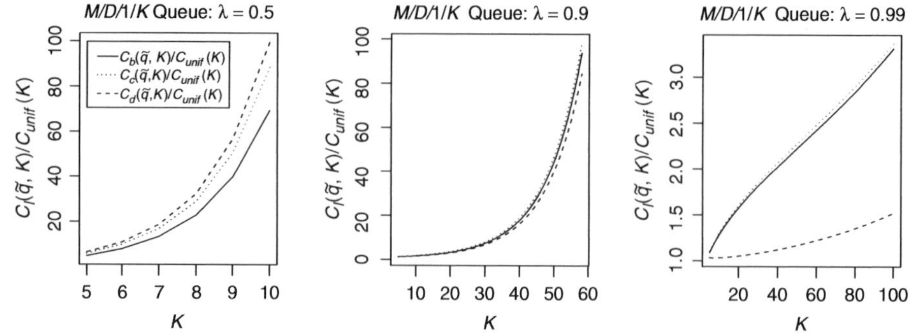

Figure 15.3 M/D/1/K model with unit service time, arrival rate λ and system capacity K. Values of $C_b(\tilde{q}, m)$, $C_c(\tilde{q}, m)$ and $C_d(\tilde{q}, m)$ relative to $C_{unif}(m)$ are given for parameters $\lambda = 0.5$, 0.9, 0.99. Only values of K for which the relative value does not exceed 100 are shown. See Section 15.1.4.

to $C_{unif}(m)$ for parameters $\lambda = 0.5, 0.9, 0.99$, and values of K giving ratios not greater than 100, up to $K = 100$. The utilization factors for the three models are given directly by $\rho = \lambda/1$, and we find the results comparable, if not identical, to those of Figure 15.2 for which $\beta = \rho$.

An most important example is the $M/M/1/K$ queue with service rate μ, arrival rate λ and queue capacity K. The state is the system occupancy, which ranges over $x = 0, 1, \ldots, K$, so that the order of the sampling process is $m = K + 1$. In this case, the steady state frequencies π_i of the embedded Markov chain can be calculated analytically, and closely conform to the truncated geometric distribution, with the relationship $\pi_{i+1} = \rho\pi_i$ holding $i = 1, \ldots, K - 2$, but not $i = 0, K - 1$, for $\rho = \lambda/\mu$.

15.1.5 Restarting schemes

The issue of *mixing* is an important one in the study of algorithms based on simulated Markov chains, which refers to progress to its steady state distribution. The simple device of *restarting* the simulated process at a new initial state can sometimes improve efficiency by improving mixing properties (see, for example, Shonkwiler and Mendivil (2009)). Frequent enough restarting will alter the steady-state attained by the sampler, which may interfere with the objectives of some applications, but will help attain the objectives stated here.

Suppose we propose the following simple algorithm, in which a simulation is restarted at regular intervals of M transitions.

Algorithm 3 *We are given a Markov chain transition kernel Q on state space $\mathcal{X} = \{1, \ldots, m\}$. Define array $n[i]$ on index set $\mathcal{X}$ and initialize to 0. Fix transition count N and interval count M. Fix current state i^*.*

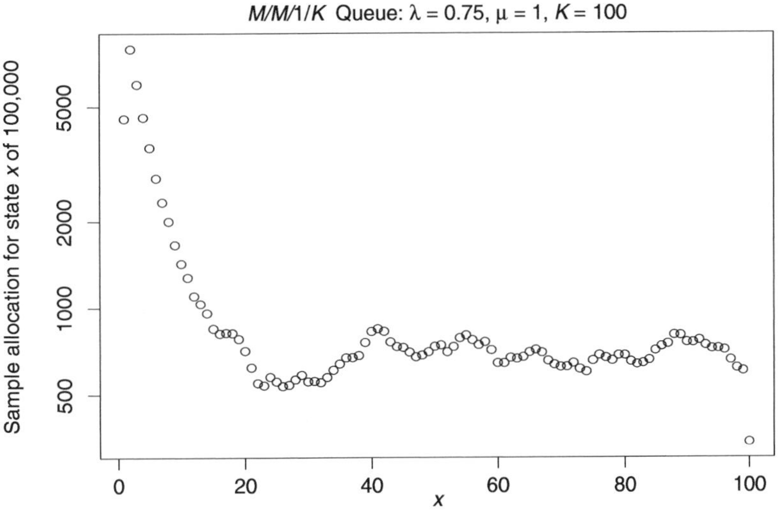

Figure 15.4 Example of Algorithm 3 discussed in Section 15.1.5.

for J = *1* to N begin loop
  if J mod $M = 0$ then
    select i^* = $argmin_{i=\mathcal{X}}\ n[i]$
  else
    sample i^* from distribution $Q(i^*)$
  end if
  visit i^*
  set $n[i^*] = n[i^*] + 1$
end loop

Clearly, setting restart interval $M = 1$ in Algorithm 3 yields uniform allocation, which might be more easily accomplished by repeated enumeration of $\mathcal{X}$. However, depending on the simulation platform, it may be considerably more convenient to set M much larger.

We give a simple demonstration that this can yield close to uniform allocation. Consider an $M/M/1/K$ queue with service rate $\mu = 1$, arrival rate $\lambda = 0.75$, and queue capacity $K = 100$. The state is the system occupancy, which ranges over $x = 0, 1, \ldots, K$, so that the order of the sampling process is $m = K + 1$. In this case, the steady state frequencies π_i of the embedded Markov chain can be calculated analytically, and closely conform to the truncated geometric distribution, with the relationship $\pi_{i+1} = \rho\pi_i$ holding for $i = 1, \ldots, K - 2$, but not $i = 0, K - 1$, for $\rho = \lambda/\mu = 0.75$. Based on Figure 15.2 we may predict that steady-state allocation will not be feasible.

Figure 15.4 shows the resulting sample allocation resulting from a single simulation of Algorithm 3 for this model, using sampling parameters $N = 100,000$ and $M = 1000$. Ties in the selection of the lowest visit frequency state were resolved by selecting the highest index (the rationale for this being that π_i is decreasing in i above $i = 0$). It is clearly apparent that the algorithm is close to achieving uniform allocation, with some clearly apparent exceptions. A set of states between $i=0$ and $i=20$ are over represented, in the extreme case by a factor of about 10. In addition, a boundary effect can be seen at $i=0$ and $i=K$, due to the fact that the steady-state equations force the values of π_0 and π_K to be smaller than those of neighboring states by a factor of $\mu/(\mu+\lambda)$ and $\lambda/(\mu+\lambda)$ respectively. Of course, deviations from uniformity such as this may be anticipated and corrected by suitable modifications of Algorithm 3.

15.2 CONTINUOUS STATE/ACTION SPACES

The case of discrete $\mathcal{K}$ receives most attention in the literature, although the principle of dynamic programming extends also to general state/action spaces. We consider here the case in which $\mathcal{X}$ and $\mathcal{A}$ are continuous subsets of Euclidean space. The problem of estimating model (R, Q) involves a suitable choice of model approximation technique. The theory of Chapter 14 requires that the model be estimated explicitly in terms of densities, hence nonparametric density estimation techniques are especially suitable. We introduce one scheme for convergent approximation of the value function in a continuous state/action space context. The purpose is to establish the existence of such approximation schemes with deducible convergence rates.

Suppose the spaces $\mathcal{X}$ and $\mathcal{A}$ are continuous subsets of $\mathbb{R}^p$, $\mathbb{R}^q$. Let $f_K(x,a)$ be a continuous density on $\mathcal{K}$. We assume for the moment that R is known. For simplicity, we assume that the unweighted supremum norm is contractive ($w\equiv 1$) and that $\mathcal{X}$ and $\mathcal{A}$ are bounded. Suppose a sequence of vectors $H_j=(X_j, A_j, Y_j), j\geq 1$ may be simulated such that (X_j, A_j) has density f_K, and Y_j has conditional density $f(y|X_j, A_j)$ (the density of $Q(\,\cdot\,|X_j, A_j)$). The density f_K is known, and selected to be suitable for use in a simulation algorithm. The vector H_j therefore has density $f_H(x,a,y)=f_y(y|x,a)f_K(x,a)$ on $\mathcal{K}\times\mathcal{X}$. If $\hat{f}_{H,n}(x,a,y)$ is an estimate of $f_H(x,a,y)$ based on n random vectors, then the transition kernel is estimated by

$$\hat{f}_{y,n}(y|x,a)=\hat{f}_{H,n}(x,a,y)/f_K(x,a).$$

We then have the model error of Chapter 14

$$\begin{aligned} D_Q^w(Q,\hat{Q}) &= \sup_{(x,a)\in\mathcal{K}} \int_{y\in\mathcal{X}} |\hat{f}_{y,n}(y|x,a)-f_y(y|x,a)|dy \\ &= \sup_{(x,a)\in\mathcal{K}} f_K(x,a)^{-1}\int_{y\in\mathcal{X}} |\hat{f}_{H,n}(x,a,y)-f_H(x,a,y)|dy. \end{aligned}$$

If we assume $f_K(x,a)\geq c>0$ then

$$D_Q^w(Q,\hat{Q})\leq c^{-1}vol(\mathcal{X}) \sup_{(x,a,y)\in\mathcal{K}\times\mathcal{X}} |\hat{f}_{H,n}(x,a,y)-f_H(x,a,y)|.$$

We then have approximate algorithm

$$V_n(x) = \hat{T}_n V_{n-1}(x) = \min_{a \in \mathcal{K}_x} R(x,a) + \rho \int V_{n-1}(y) \hat{f}_{y,n}(y|x,a)dy \quad (15.9)$$

$n \geq 1$. Hence uniform convergence of $\hat{f}_{H,n}(x,a,y)$ suffices for the convergence of V_n, with convergence rate bounded by the convergence rate of $\hat{f}_{H,n}(x,a,y)$.

Many scenarios exist under which consistent estimation of $f_H(x,a,y)$ based on $H_1,\ldots,H_n$ is possible. In particular the vectors H_j need not be independent. Convergence rates for kernel density estimates on subsets of $\mathbb{R}^d$ based on sequences of vectors satisfying a geometric α-mixing condition are given in Kim and Cox (1996). It is shown that kernel density estimates exist which $wp1$ converge in supremum norm at a rate $o(n^{-\delta})$ when $\delta < 1/(2+d)$ (here $d = 2p + q$). Thus we have convergence $wp1$ at a rate $o(n^{-\delta})$. We note that $R(x,a)$, if unknown, may be similarly estimated using nonparametric kernel regression estimation, with similar convergent behavior (see Györfi et al. (2002)).

15.3 PARAMETRIC ESTIMATION OF MDP MODELS

It will often be the case that the model $\pi = (R, Q)$ is indexed by a parameter $\theta \in \Theta \subset \mathbb{R}^k$, given a metric *parameter space* (Θ, d). This generates a family of models $\Pi_\Theta = \{(R^\theta, Q^\theta) \mid \theta \in \Theta\}$. A typical example is a queueing system in which all model quantities are known up to a Poisson arrival rate λ. Of course, any model with finite $\mathcal{K}$ is reducible to this type of parametric model, but there will be some advantage to considering separately the case in which $k \ll |\mathcal{K}|$, since a more refined estimate than that permitted by the high dimensional sampling model will usually be possible. We will assume throughout that for some measure μ, usually Lebesgue or counting measure, all distributions $Q^\theta(x,a) \ll \mu$, and so can be represented by density kernel $g^\theta(\cdot \mid x,a)$, with the expectation operator denoted $E^\theta_{x,a}$ (or E^θ when $\mathcal{K}$ is unspecified).

The theory of parametric estimation usually employs a metric defined on a parameter space $\Theta \subset \mathbb{R}^k$, whereas approximation bounds for operators are naturally expressed using L^1 distance between density functions. We therefore need to consider how to express bounds on D^w_R and D^w_Q in terms of parametric estimates. Ideally, we have the Lipschitz continuity condition:

$$D^w_R(R^\theta, R^{\hat{\theta}}) \leq M_R d(\theta, \hat{\theta})$$
$$D^w_Q(Q^\theta, Q^{\hat{\theta}}) \leq M_Q d(\theta, \hat{\theta}). \quad (15.10)$$

The main technical issue is that this Lipschitz bound will usually not hold uniformly over the entire parameter space. The problem is then to identify conditions under which a local Lipschitz constant will suffice. It should also be noted that the magnitude of this Lipschitz constant may have considerable impact on the computational cost of an approximation. Even if a uniform Lipschitz constant did exist, it may be worth considering whether or not a tighter bound could be obtained by considering the relationship between $\hat{\theta}$ and θ in more detail.

The precise objective will be to find subset $B \subset \Theta$ and constants M_R, M_Q for which

$$\sup_{(x,a)\in\mathcal{K}} \sup_{\theta,\hat{\theta}\in B} w(x)^{-1} \left| R^{\theta}(x,a) - R^{\hat{\theta}}(x,a) \right| \leq M_R d(\theta,\hat{\theta})$$

$$\sup_{(x,a)\in\mathcal{K}} \sup_{\theta,\hat{\theta}\in B} w(x)^{-1} L^1_w(g^{\theta} - g^{\hat{\theta}}) \leq M_Q d(\theta,\hat{\theta}). \tag{15.11}$$

The key is to determine a *bounding function* $t : \mathcal{X} \times \Theta \times \Theta \to \mathbb{R}_+$ for which

$$|g^{\theta}(x) - g^{\hat{\theta}}(x)| \leq t(x,\theta,\hat{\theta}) \left(g^{\theta}(x) + g^{\hat{\theta}}(x) \right), \quad \forall x \in \mathcal{X}, \text{ and } \theta, \hat{\theta} \in \Theta, \tag{15.12}$$

and from which an approximation bound can be conveniently formed. This leads to

$$\begin{aligned} L^1_w(g^{\theta} - g^{\hat{\theta}}) &= \int_{x\in\mathcal{X}} w(x)|g^{\theta}(x) - g^{\hat{\theta}}(x)| d\mu(x) \\ &\leq \int_{x\in\mathcal{X}} w(x)t(x,\theta,\hat{\theta}) \left(g^{\theta}(x) + g^{\hat{\theta}}(x) \right) d\mu(x) \\ &= E^{\theta}\left[w(X)t(X,\theta,\hat{\theta}) \right] + E^{\hat{\theta}}\left[w(X)t(X,\theta,\hat{\theta}) \right]. \end{aligned} \tag{15.13}$$

The form of the inequality (15.12) will imply a natural continuity the exponential family type of density commonly employed in statistical inference. This form of density usually provides a natural relationship between a parameter space and a probability measure. See, for example, Lehmann and Casella (1998).

Definition 15.1 *An* exponential family of densities *on $\mathcal{X}$ with parameter space $\Theta \subset \Re^k$ takes the form*

$$g^{\theta}(x) = \exp\left(v(x,\theta) - b(\theta) + h(x) \right), \quad x \in \mathcal{X} \tag{15.14}$$

where

$$v(x,\theta) = \sum_{i=1}^{k} \theta_i v_i(x), \tag{15.15}$$

for real valued functions $v_1, \ldots, v_k, h$ defined on $\mathcal{X}$ and b defined on Θ.

Any density may be transformed to the form (15.14), the distinguishing condition being given by (15.15), and it permits the development of a general theory of estimation which is applicable to any exponential family. It is important to note that the definition makes a distinction between fixed and unknown parameters, the parameter space Θ including only unknown parameters. The normal distribution $N(\mu, \sigma^2)$ is an exponential family distribution whether μ, σ^2 or both parameters are unknown. On the other hand, the Weibull distribution $g(x \mid \kappa, \lambda) = (\kappa\lambda)(\lambda x)^{\kappa-1} \exp(-(\lambda x)^{\kappa})$, $x \geq 0$ is an exponential family distribution only if the shape parameter κ is fixed. This is typical of densities used to model positive random lifetimes.

The Poisson density will be an important example, and may be written

$$g(x \mid \lambda) = \frac{\lambda^x}{x!} \exp(-\lambda) = \exp(x \log(\lambda) - \lambda - \log(x!)), \quad x = 0, 1, \ldots,$$

and so is an exponential family density, with $k = 1$, $v_1(x) = x$, $\theta = \log(\lambda)$, $b(\theta) = \lambda$ and $h(x) = -\log(x!)$ (this representation often transforms parameters from commonly used forms). Exponential family densities also include the gamma, geometric and χ^2 densities, but not the uniform or t densities.

The quantities which will define continuity may be given as

$$\begin{aligned} \Delta_v(x, \theta, \hat{\theta}) &= v(x, \hat{\theta}) - v(x, \theta) \\ \Delta_b(\theta, \hat{\theta}) &= b(\hat{\theta}) - b(\theta), \end{aligned} \tag{15.16}$$

and lead to a convenient form for a bounding function. This is given in the following theorem:

Theorem 15.2 *Suppose we have given a parametric family of densities g^θ on $\mathcal{X}$, with $\theta \in \Theta \subset \mathbb{R}^k$. Using the notion of Defintion 15.1 and (15.16),*

$$t(x, \theta, \hat{\theta}) = \left| \Delta_v(x, \theta, \hat{\theta}) - \Delta_b(\theta, \hat{\theta}) \right| \tag{15.17}$$

is a bounding function as defined in (15.12). In addition, if (15.15) holds (that is, g^θ is an exponential family of densities) then for any conjugate pair $p^{-1} + q^{-1} = 1$.

$$t(x, \theta, \hat{\theta}) = \|\tilde{v}(x)\|_p \left\| \theta - \hat{\theta} \right\|_q + \left| b(\theta) - b(\hat{\theta}) \right| \tag{15.18}$$

is a bounding function as defined in (15.12), where $\tilde{v}(x) = (v_1(x), \ldots, v_k(x))$.

Proof We will make use of the inequality

$$\begin{aligned} |e^x - 1| &\leq |x| e^{\max(0,x)} \\ &\leq |x|(1 + e^x), \quad x \in \mathbb{R}. \end{aligned} \tag{15.19}$$

Using the representation (15.14) and the definitions (15.16) we may write directly

$$\left| g^{\hat{\theta}}(x) - g^\theta(x) \right| = \left| \exp(\Delta_v(x, \theta, \hat{\theta}) - \Delta_b(\theta, \hat{\theta})) - 1 \right| |g^\theta(x),$$

then using (15.19) gives

$$\begin{aligned} &|g^{\hat{\theta}}(u) - g^\theta(u)| \\ &\leq \left| \Delta_v(x, \theta, \hat{\theta}) - \Delta_b(\theta, \hat{\theta}) \right| \left(1 + \exp(\Delta_v(x, \theta, \hat{\theta}) - \Delta_b(\theta, \hat{\theta})) \right) g^\theta(x), \\ &= \left| \Delta_v(x, \theta, \hat{\theta}) - \Delta_b(\theta, \hat{\theta}) \right| \left(g^\theta(x) + g^{\hat{\theta}}(x) \right), \end{aligned}$$

so that (15.2) is a bounding function. If (15.15) holds then $\Delta_v(x,\theta,\hat{\theta}) = \sum_{i=1}^k v_i(x)(\hat{\theta}_i - \theta_i)$, so that (15.18) is a bounding function, following Hölder's inequality. ///

Continuing the example of the Poisson distribution, from Theorem 15.2 we have bounding function:

$$t(x,\lambda,\hat{\lambda}) = x\left|\log(\hat{\lambda}) - \log(\lambda)\right| + \left|\hat{\lambda} - \lambda\right|$$

so that by (15.13) we have, for $w \equiv 1$,

$$L^1(g^\theta - g^{\hat{\theta}}) \leq (\hat{\lambda} + \lambda)\left|\log(\hat{\lambda}) - \log(\lambda)\right| + 2\left|\hat{\lambda} - \lambda\right|. \tag{15.20}$$

We will generally have a locally, but not uniformly, Lipschitz bound for $L^1(g^\theta - g^{\hat{\theta}})$. It will usually be reasonable to assume that $\hat{\lambda}$, if interpreted as an estimate of λ, will be within a bounded neighborhood of λ. Using Taylor's approximation theorem, (15.20) may be approximated by

$$L^1(g^\theta - g^{\hat{\theta}}) \leq \left(\frac{\hat{\lambda} + \lambda}{\lambda} + 2\right)\left|\hat{\lambda} - \lambda\right| + O\left(\left|\hat{\lambda} - \lambda\right|^2\right). \tag{15.21}$$

Note that the local Lipschitz constant given in (15.21) may be uniformly bounded if a scalar proximity condition such as $\hat{\lambda}/\lambda \in (r^{-1}, r)$, $r > 1$ holds, or if λ is considered fixed and $\hat{\lambda} - \lambda \in (-\epsilon, \epsilon)$.

Theorem 15.3 *Suppose we have given an exponential family of densities g^θ on $\mathcal{X}$ (Defintion 15.1), and for a subset $B \subset \Theta$ and conjugate pair $p^{-1} + q^{-1} = 1$ the following hold:*

(*i*) *$b(\theta)$ possesses Lipschitz constant M_b wrt $\|\cdot\|_q$ on B,*
(*ii*) $\sup_{\theta \in B} E^\theta\left[w(X)\,\|v(X)\|_p\right] = M_v < \infty$,
(*iii*) $\sup_{\theta \in B} E^\theta\left[w(X)\right] = M_w < \infty$.

Then

$$L^1_w(g^\theta - g^{\hat{\theta}}) \leq 2(M_v + M_v M_b)\left\|\theta - \hat{\theta}\right\|_q \tag{15.22}$$

for $\theta, \hat{\theta} \in B$.

Proof The theorem is a direct application of Theorem 15.2 and (15.13). ///

A bound such as (15.11) could be obtained by applying Theorem 15.3 uniformly over multiple kernel $g^\theta(\cdot \mid x,a)$, so that conditions (*i*)–(*iii*) would be replaced by:

(*i'*) $b(\theta \mid x,a)$ possesses Lipschitz constant M_b *wrt* $\|\cdot\|_q$ on B for each $(x,a) \in \mathcal{K}$,
(*ii'*) $\sup_{(x,a)\in\mathcal{K}} \sup_{\theta \in B} w(x)^{-1} E^\theta_{x,a}\left[w(X)\,\|v(X \mid x,a)\|_p\right] = M_v < \infty$,
(*iii'*) $\sup_{(x,a)\in\mathcal{K}} \sup_{\theta \in B} w(x)^{-1} E^\theta_{x,a}\left[w(X)\right] = M_w < \infty$.

Of course, the elements $b(\theta \mid x,a)$ and $v(\cdot \mid x,a)$ used in the exponential family representation of $g^{\theta}(\cdot \mid x,a)$ would usually depend on $(x,a) \in \mathcal{K}$ and would have to be deduced. A simpler approach would be to use the *disturbance model*, in which the transition distribution for each (x,a) is modeled as a mapping of a single, relatively simple disturbance process (see, for example, Hernández-Lerma and Lasserre (1996)). An obvious example is a embedded queueing system, in which random transitions follow from a service time S and the number of arrivals Y in time intervals $(0,S]$.

We present one version of this model, which can be generalized when needed. Suppose there is an *iid* sequence of random quantities $U_1, U_2, \ldots$, and the transition law of a MDP may be written

$$
\begin{aligned}
X_0 &= x \\
X_n &= H(X_{n-1}, A_{n-1}, U_n), \quad n \geq 1,
\end{aligned}
\tag{15.23}
$$

for some fixed mapping H. In the queueing example, $U_n = (Y_n, S_n)$ where Y_n and S_n are the nth stage arrivals and service time. In the simplest $M/G/1$ case we have $H(X_{n-1}, A_{n-1}, U_n) = H(X_{n-1}, A_{n-1}, Y_n) = \max(X_{n-1} + Y_n - 1, 1)$ (assuming the empty queue state is consolidated into the single occupancy state). Straightforward modifications could take into account finite capacity, or controllable service capacity and service time distribution.

We then suppose the density for the disturbance process $U \in \mathcal{U}$ belongs to a parametric family g^{θ}_U, $\theta \in \Theta$. The kernel must then satisfy

$$
Q^{\theta}(E_x \mid x,a) = \int_{u \in \mathcal{U}} I\{H(x,a,u) \in E_x\} g^{\theta}(u)_U d\mu(u)
$$

over all $(x,a) \in \mathcal{K}$, so that $Q^{\theta}(x,a)$ is the distribution of the random quantity $H(x,a,U)$, and the density $g^{\theta}(\cdot \mid x,a)$ of $Q^{\theta}(x,a)$ is obtainable from g^{θ}_U using a standard measure transformation method.

Theorem 15.4 *For the disturbance model (15.23) the following inequality holds:*

$$
D^{w}_{Q}(\theta, \hat{\theta}) \leq \sup_{x,a \in \mathcal{K}} w(x)^{-1} \int_{u \in \mathcal{U}} w(H(x,a,u)) \left| g^{\theta}(u) - g^{\hat{\theta}}(u) \right| d\mu_u(u).
$$

Proof Define

$$
\chi(y \mid x, a, \theta, \hat{\theta}) = \begin{cases} 1; & g^{\theta}(y \mid x,a) \geq g^{\hat{\theta}}(y \mid x,a) \\ 0; & \text{otherwise} \end{cases}.
$$

Fix $(x,a)\in\mathcal{K}$ and for convenience set $\chi(y)=\chi(y\,|\,x,a,\theta,\hat{\theta})$ and $\chi(H)=\chi(H(x,a,u)\,|\,x,a,\theta,\hat{\theta})$. Then we may write

$$\begin{aligned}
&\int_{y\in\mathcal{X}} w(y)\left|g^{\theta}(y|x,a)-g^{\hat{\theta}}(y\,|\,x,a)\right| d\mu_{\Theta}(y)\\
&=\int_{y\in\mathcal{X}} w(y)\chi(y)\left(g^{\theta}(y\,|\,x,a)-g^{\hat{\theta}}(y\,|\,x,a)\right) d\mu_{\Theta}(y)\\
&+\int_{y\in\mathcal{X}} w(y)(1-\chi(y))\left(g^{\hat{\theta}}(y\,|\,x,a)-g^{\theta}(y\,|\,x,a)\right) d\mu_{\Theta}(y)\\
&=\int_{u\in\mathcal{U}} w(H(x,a,u))\chi(H)\left(g^{\theta}(u)-g^{\hat{\theta}}(u)\right) d\mu_{u}(u)\\
&+\int_{u\in\mathcal{U}} w(H(x,a,u))(1-\chi(H))\left(g^{\hat{\theta}}(u)-g^{\theta}(u)\right) d\mu_{u}(u)\\
&\leq\int_{u\in\mathcal{U}} w(H(x,a,u))|g^{\theta}(u)-g^{\hat{\theta}}(u)|d\mu_{u}(u),
\end{aligned}$$

from which the result follows. ///

Chapter 16

Approximate value iteration by truncation

One simple approximation scheme is to truncate evaluation of the integral $\int_{y\in\mathcal{X}} V(y)f(y\,|\,x,a)d\mu$. This would be especially advantageous if the support of density $f(y\,|\,x,a)$ was large, hence computationally burdensome, while concentrating most of the probability mass onto a relatively small subset $E\subset\mathcal{X}$. It would then be expected that a good approximation could be obtained by restricting evaluation of $\int_{y\in\mathcal{X}} V(y)f(y\,|\,x,a)d\mu$ to E, while significantly reducing the computation time. The material in this chapter is based on Almudevar and de Arruda (2012) and Arruda et al. (2013).

In such a case, we may simply regard such a scheme as a type of model approximation:

$$\int_{y\in E} V(y)f(y\,|\,x,a)d\mu = \int_{y\in\mathcal{X}} V(y)f_E(y\,|\,x,a)d\mu,$$

where $f_E(y\,|\,x,a)=f(y\,|\,x,a)I\{y\in E\}$ is a truncated density. Of course, f_E is not a proper density, but may be normalized to a proper density by evaluating

$$\bar{f}_E(y\,|\,x,a) = \frac{f_E(y\,|\,x,a)}{\int_{y\in E} f(y\,|\,x,a)d\mu}$$

The L^1_w distances between an arbitrary density f and its truncation f_E or $\bar{f}_E$ can be evaluated as

$$\|f - f_E\|_{TV(w)} = \int_{y\in\mathcal{X}} w\left|f(y) - f_E(y)\right| d\mu = \int_{y\in E^c} wf(y)d\mu,$$

$$\|f - \bar{f}_E\|_{TV(w)} = \int_{y\in E} w(P(E)^{-1} - 1)f(y)d\mu + \int_{y\in E^c} wf(y)d\mu.$$

There are two terms of interest. The integral $\int_{y\in E^c} wf(y)d\mu$ appears in both evaluations. If weight function $w\equiv 1$, then we have

$$\begin{aligned} \|f - f_E\|_{TV(w)} &= P(E^c) \\ \|f - \bar{f}_E\|_{TV(w)} &= 2P(E^c). \end{aligned} \tag{16.1}$$

We will develop a model for bounded costs on a finite state space, setting $\|\cdot\|_w = \|\cdot\|_{sup}$.

16.1 TRUNCATION ALGORITHM

Once we have established that kernel truncation is a form of model approximation yielding a tractable tolerance, we may develop an AIA directly. We must first associate a truncation support for each element of the kernel. Accordingly, define a class of state space subsets $E' = \{E_z : z \in \mathcal{K}\}$, then define approximate operators

$$T_{E'}V(x) = \inf_{a \in \mathcal{K}_x} R(x,a) + \beta \int_{E_{(x,a)}} V(y)Q(dy \mid x,a), \tag{16.2}$$

and

$$\tilde{T}_{E'}V(x) = \inf_{a \in \mathcal{K}_x} R(x,a) + \beta \frac{\int_{E_{(x,a)}} V(y)Q(dy \mid x,a)}{Q(E_{(x,a)} \mid x,a)}. \tag{16.3}$$

for each $x \in \mathcal{X}$. It will be convenient to define the complementary class $\bar{E}' = \{\mathcal{X} - E_z : z \in \mathcal{K}\}$, and to define the quantity

$$P^*(E') = \sup_{z \in \mathcal{K}} Q(E_z \mid z)$$

for any such class. We then conclude from (16.1) and Theorem 14.4 that the operator tolerance is of order $O(P^*(\bar{E}'))$. The cost of evaluating $T_{E'}$ or $\tilde{T}_{E'}$ may be given directly as $\sum_z |E_z|$, the sum of the cardinalities of the sets in E'. Clearly, the computation cost can be minimized for a fixed operator tolerance by selecting E_z based on the largest probabilities.

For convenience set $\mathcal{X} = \{1, \ldots, J\}$, and denote probabilities $p_j(z) = Q(j \mid z)$. For any state/action pair $z \in \mathcal{K}$ let $\tilde{J}^z = (j_1^z, j_2^z, \ldots)$ be an ordering of the states in decreasing order of transition probabilities, that is $p_{j_k^z}(z) \geq p_{j_{k+1}^z}(z)$. This leads to the distribution functions $F(m,z) = \sum_{k=1}^m p_{j_k^z}(z)$, with $\bar{F}(m,z) = 1 - F(m,z)$. Let $n_{max}(z,\tau) = \min\{m : \bar{F}(m,z) \leq \tau\}$, $\tau \in [0,1)$, so thet τ becomes an approximation parameter. We may then define a class of approximate operators in terms of the approximation parameter τ:

$$T_\tau V(i) = \inf_{a \in \mathcal{K}_i} R(i,a) + \beta \sum_{k=1}^{n_{max}(i,a,\tau)} V\left(j_k^{(i,a)}\right) p_{j_k^{(i,a)}}(i,a),$$

$$\tilde{T}_\tau V(i) = \inf_{a \in \mathcal{K}_i} R(i,a) + \frac{\beta \sum_{k=1}^{n_{max}(i,a,\tau)} V\left(j_k^{(i,a)}\right) p_{j_k^{(i,a)}}(i,a)}{F(n_{max}(i,a,\tau),i,a)}.$$

We then have operator tolerance

$$\|T_\tau V - TV\|_{sup} \leq (1-\beta)^{-1} \|V\|_{sup} \tau,$$
$$\|\tilde{T}_\tau V - TV\|_{sup} \leq (1-\beta)^{-1} \|V\|_{sup} 2\tau. \tag{16.4}$$

16.2 REGULARITY CONDITIONS FOR TOLERANCE-COST MODEL

Next, assume we have an AIA with an approximation schedule τ_k, $k \geq 1$, based on approximate operators $\hat{T}_k = T_{\tau_k}$ (the analysis for operator sequence $\tilde{T}_{\tau_k}$ is identical). Evaluation of T_{τ_k} requries summation of $n_{max}(z, \tau_k)$ terms for each $z \in \mathcal{K}$, so that the computation cost of an iteration, as a function of τ, may be taken as

$$G^{app}(\tau) = \sum_{z \in \mathcal{K}} n_{max}(z, \tau).$$

We will make a simplifying assumption that all transition distributions have similar enough structure to be represented by a single distribution $F(m)$, in the sense that $F(m, z) = F(t(m, z))$ for a suitable transformation $m = t(m, z)$ (this transformation would then generate the orderings $\tilde{J}^z$). We may expect such a structure in disturbance models, for example, when all transition measures are derived from a common arrival process. We then have

$$G^{app}(\tau) = |\mathcal{K}|\bar{F}^{-1}(\tau).$$

Following (16.4) and the discussion of Section 11.1.1 we have approximation tolerance $u_k \approx C^{-1}\tau_k$, and so the cost function becomes

$$G(u) = G^{app}(Cu) = |\mathcal{K}|\bar{F}^{-1}(Cu).$$

Of course, the behavior of $\bar{F}(m)$ may vary considerably, however, the relationship of the cost function to regularity conditions (A), (B1) or (B2) of Section 11.1.2 may be easily deduced from the form of the tail probabilities. In particular, for power law tail probabilities $\bar{F}(m) \propto m^{-d}$, by Theorem 11.4, (A) holds, and (B1) holds for schedules with $\lambda_l\{u_k\} > 0$. For exponential tail probabilities $\bar{F}(m) \propto \rho^m$, by Theorem 11.5, (A) holds, and (B2) holds for schedules with $\lambda_l\{u_k\} > 0$. Applying, as appropriate, Theorems 11.1, 11.2 and 11.3, and noting operator tolerance (16.4), we adopt the simple recommendation that an AIA be based on tolerance schedule $\tau_k \approx \beta^k$. Note that the cost function G is used only to verify the regularity conditions (A), (B1) or (B2), and otherwise plays no role in the selection of the tolerance schedule.

16.2.1 Suboptimal orderings

It is important to note that the truncation algorithms do not depend on an availability of the perfect orderings of the transition probabilities for each state/action pair z. While a perfect ordering will be most efficient, the convergence properties reported here are still valid when a suboptimal ordering is used.

Consider the following general example. We may take the previously defined orderings $\tilde{J}^z = (j_1^z, j_2^z, \ldots)$ for each $z \in \mathcal{K}$ as the optimal. Suppose we have a suboptimal ordering $\hat{j}_1^z, \hat{j}_2^z, \ldots$ for which $\hat{j}_{k'}^z = j_k^z$ implies $k' \leq Bk$ for some number B and all $k \leq C$, for some large number C (the original ordering of the larger probabilities is maintained with regular insertions). We then must have $n_{max}(z, \tau_k) \leq \hat{n}_{max}(z, \tau_k) \leq Bn_{max}(z, \tau_k)$, where $\hat{n}_{max}(z, \tau_k)$ is analagously defined for the suboptimal orderings. Note that the truncation algorithms will have approximately the same convergence properties with respect to iteration count for both orderings. The difference is that the suboptimal ordering yields a higher cost per iteration, but by a factor no greater than B.

16.3 EXAMPLE

We will continue the example ontroduced in Section 13.7 (see also Figure 13.4). Since convergence in the span quotient space has been shown to be considerably faster than convergence *wrt* the supremum norm, the trunctation scheme will be developed in this context.

The first point to note is that we have a choice of two approximate operators, the unnormalized and normalized truncation operators $T_{E'}V$ and $\tilde{T}_{E'}V$ defined in (16.2) and (16.3), respectively. The operator $\tilde{T}_{E'}V$ is based on a strictly stochastic kernel, due to the normalization, and is therefore continuous on the span quotient space. Since the same cannot in general be guaranteed of the unnormlized approximate operator we will use $\tilde{T}_{E'}$.

The second point to note that the approximation schedule is determined by the contraction rate of the operator, which is not necessarily equivalent to the discount factor β, and in the span quotient space generally won't be. We have seen that this contraction rate may be difficult to evaluate. In addition, the asymptotic contraction rate may be significantly smaller than any single J-stage contraction rate.

Formally, the problem is to determine the tolerance α_k of the EIA in the tolerance model (11.1) of Chapter 11. With respect to the supremum norm this would be $\alpha_k = \beta^k$,

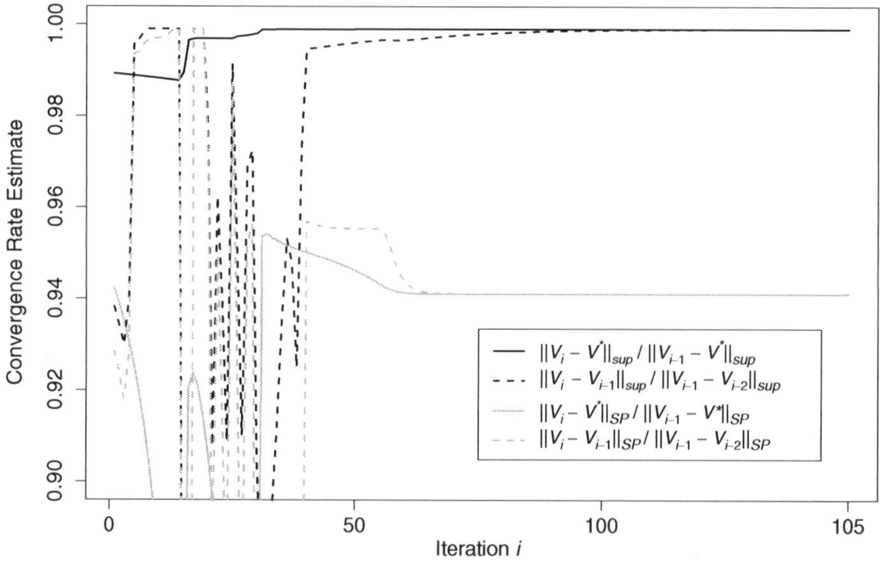

Figure 16.1 Various estimates of linear convergence rates for the example of Section 16.3.

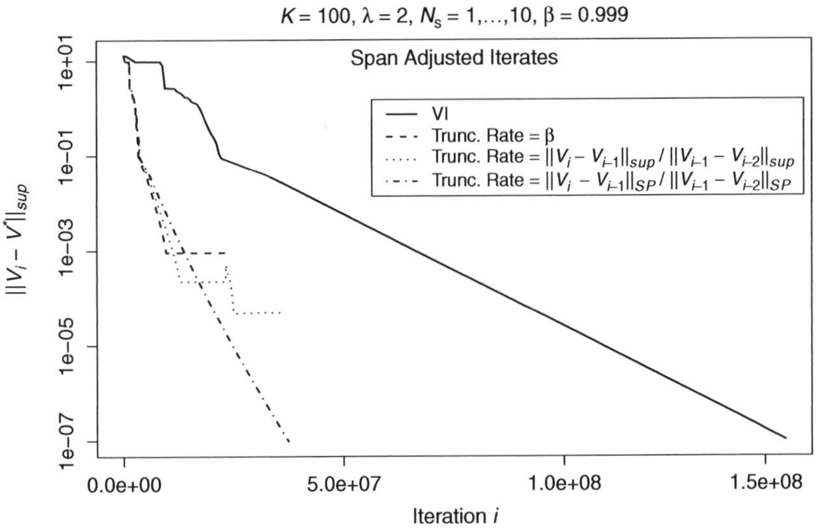

Figure 16.2 Various estimates of linear convergence rates for the example of Section 16.3.

but is more difficult to determine *wrt* the span seminorm. The rate can be estimate empirically, however, by using the ratio updating rule

$$\alpha_{k+1} \approx \frac{\| V_k - V_{k-1} \|}{\| V_{k-1} - V_{k-2} \|} \alpha_k \tag{16.5}$$

using the appropriate norm.

Figure 16.1 plots the ratios

$$\frac{\| V_k - V^* \|_{sup}}{\| V_{k-1} - V^* \|_{sup}}, \quad \frac{\| V_k - V_{k-1} \|_{sup}}{\| V_{k-1} - V_{k-2} \|_{sup}}, \quad \frac{\| V_k - V^* \|_{SP}}{\| V_{k-1} - V^* \|_{SP}}, \quad \frac{\| V_k - V_{k-1} \|_{SP}}{\| V_{k-1} - V_{k-2} \|_{SP}},$$

for our example. The linear convergence rate is given directly by $\| V_k - V^* \| / \| V_{k-1} - V^* \|$ in the appropriate norm. As seen in Figure 16.1 this rate can be accurately estimated also by the ratios of the value iteration increments, which are available for use by the algorithm.

Figure 16.2 shows the progressive accuracy of the truncation algorithm using three methods for estimating α_k. The first sets $\alpha_k \propto \beta^k$, while the other two use the updating rule (16.5) based on norms $\| \cdot \|_{sup}$ and $\| \cdot \|_{SP}$. The progress is measured, appropriately, not by the iteration index but by the cumulative computation effort. The nontruncated VI algorithm is also included for comparison (apart from the change of units in the horizontal axis, this plot is equivalent to the upper right plot of Figure 13.4). The truncation method clearly results in considerable savings in computation. Furthermore, we find the updating rule (16.5), based on the span seminorm, is the appropriate choice for this application.

Chapter 17

Grid approximations of MDPs with continuous state/action spaces

Much of the theory of MDPs and their approximation does not rely on the assumption that $\mathcal{K}$ is finite or countable, but this question is of significant practical consequence, arising from the need to first compute T_π^a, then to optimize over the control variable $a \in \mathcal{K}_x$. The performance of these operations enumeratively when $\mathcal{K}$ is finite leads to issues of complexity, but also permits the development of general algorithms with a well defined performance, and so the approximation of continuous MDP models using discrete models seems a natural solution.

The problem can be conceived in two ways. A type of certainty equivalence approach would be to accept a finite approximation as a substitute for the true model. Alternatively, we may accept discretization as a method of developing approximate operators which are tractable, but which also converge to the true operator as defined on the original continuous state/action space, so that all iterates V_k are formally defined on the original state space $\mathcal{X}$. We will adopt this approach, which can be see to contain within itself the certainty equivalence approach as a special case.

The material of this chapter largely follows the seminal methods described in Chow and Tsitsiklis (1989) and Chow and Tsitsiklis (1991). We will find that the conclusions reached in that work also follow from the functional analytic approach considered in this volume.

17.1 DISCRETIZATION METHODS

Suppose f is a function on $\mathcal{X}$. Suppose $\mathcal{E} = \{E_1, \ldots, E_n\}$ is a partition of $\mathcal{X}$. Define $E(x) = \{y \in E_j \mid x \in E_j\}$, leading to

$$f_{\mathcal{E}}(x) = 2^{-1}\left(\inf_{y \in E(x)} f(y) + \sup_{y \in E(x)} f(y)\right).$$

If we may claim the bound

$$\sup_{E_j \in \mathcal{E}} \sup_{x,y \in E_j} f(y) - f(x) \leq \tau,$$

then it follows that

$$\|f_{\mathcal{E}} - f\|_{sup} \leq \tau/2. \tag{17.1}$$

No assumption of continuity for f has been made, and τ may be made arbitrarily small by defining $\mathcal{E}$ using inverse images of a grid partition of the range of f. For example, if we have $f \in [0,1]$, we may define for $\tau \leq 1$ partition $\mathcal{E}_\tau = \{E_1, \ldots, E_n\}$, $n = \lceil 1/\tau \rceil$, by

$$E_j = f^{-1}((\tau(j-1), \tau j] \cap [0,1]), \quad j = 2, \ldots, n, \text{ and } E_1 = f^{-1}([0,\tau]),$$

then (17.1) holds for $f_{\mathcal{E}_\tau}$. Furthermore, it is not required that $\mathcal{X}$ be bounded.

The model elements of π to be discretized are cost function $R(x,a)$ defined on $\mathcal{K}$ and density kernel $f(y \mid x, a)$ (*wrt* Lebesgue measure) defined on $\mathcal{K} \times \mathcal{X}$. A number of partition schemes may be considered, but we will follow the approach used in Chow and Tsitsiklis (1991), in which the state and action space are assumed to be subsets of unit cubes $\mathcal{X} \subset [0,1]^n$ and $\mathcal{A} \subset [0,1]^m$, with the dimensions given explicitly in order to capture their effect on computational complexity. An n dimensional grid partition on $[0,1]^n$ is defined by the product $\mathcal{E}_\tau^n = \times^n \mathcal{E}_\tau$, consisting of all cubes in $[0,1]^n$ with sides of length τ defined by the grid $\mathcal{E}_\tau$, with the possible exception of cubes bordering any upper bound 1, which have sides no greater than τ.

Formally, we need to identify elements of a partition $I \in \mathcal{E}_\tau^n$ for which $I \cap \mathcal{X} \neq \emptyset$, and similarly, $I \in \mathcal{E}_\tau^{n+m}$ for which $I \cap \mathcal{K} \neq \emptyset$. However we lose little generality in assuming $\mathcal{K} = [0,1]^{n+m}$, although careful attention to this point would be needed in any implementation.

For each element $I' \in \mathcal{E}_\tau^{n+m}$ select a representative element $(x', a') \in I'$, and let $(x_\tau, a_\tau) = (x', a')$ for all $(x,a) \in I'$. This leads to approximations

$$\begin{aligned} \hat{R}_\tau(x,a) &= R(x_\tau, a_\tau), \\ \hat{f}_\tau(y \mid x, a) &= \frac{f(y_\tau \mid x_\tau, a_\tau)}{\int_{\mathcal{X}} f(y_\tau \mid x_\tau, a_\tau) dy}. \end{aligned} \tag{17.2}$$

We have normalized $\hat{f}_\tau(y \mid x, a)$ to be a proper density function. As in the case of the truncation approximation method, this is not formally needed to obtain convergence, but will be needed to exploit contraction in the span seminorm, as indicated in Chow and Tsitsiklis (1991). The next objective is to define conditions under which the approximation (17.2) leads to a suitable bound on the operator tolerance. A Lipschitz type condition is imposed on R and $f(y \mid x, a)$ over the spaces $\mathcal{X}$ and $\mathcal{K} \times \mathcal{X}$ respectively in Chow and Tsitsiklis (1991). We give similar conditions directly in terms of the quantities $D_r(R, \hat{R}_\tau)$ and $D_q(Q, \hat{Q}_\tau)$, where $\hat{Q}_\tau$ is the kernel defined by (17.2).

Theorem 17.1 *Suppose for model $\pi = (R, Q)$ there exists a finite constant K for which:*

(*i*) $|R(x,a) - R(x',a')| \leq K \left\| (x,a) - (x',a') \right\|_{sup}$,
(*ii*) *For each $(x,a) \in \mathcal{K}$ we have $|f(y \mid x,a) - f(y' \mid x,a)| \leq K|y - y'|$,*
(*iii*) $\left\| f(\cdot \mid x,a) - f(\cdot \mid x',a') \right\|_{TV} \leq K \left\| (x,a) - (x',a') \right\|_{sup}$.

Then

$$D_r(R, \hat{R}_\tau) \leq K\tau, \quad \text{and} \tag{17.3}$$

$$D_q(Q, \hat{Q}_\tau) \leq 3K\tau + O(\tau^2). \tag{17.4}$$

Proof Suppose $I \in \mathcal{E}_\tau^{n+m}$. By (i), for any $(x,a),(x',a') \in I$ we have $|R(x,a) - R(x',a')| \leq K\tau$, and therefore $D_r(R, \hat{R}_\tau) \leq K\tau$.

Suppose $I \in \mathcal{E}_\tau^n$. If (ii) holds, then for any $(x,a) \in \mathcal{K}$ we have

$$|f(y \mid x,a) - f(y' \mid x,a)| \leq K\tau$$

when $y, y' \in I$, from which it follows that

$$|f(y \mid x,a) - f(y_\tau \mid x,a)| \leq K\tau$$

for all $y \in \mathcal{X}$. This further implies that

$$\int_{\mathcal{X}} f(y_\tau \mid x,a) dy \in [1 - K\tau, 1 + K\tau],$$

so that

$$\left\| \hat{f}_\tau(y \mid x,a) - f(y_\tau \mid x,a) \right\|_{TV} \leq 2K\tau + O(\tau^2).$$

Note that, using assumption (iii),

$$\begin{aligned} |f(y \mid x,a) - f_\tau(y \mid x,a)| &= |f(y \mid x,a) - f(y_\tau \mid x_\tau, a_\tau)| \\ &\leq |f(y \mid x,a) - f(y \mid x_\tau, a_\tau)| + |f(y \mid x_\tau, a_\tau) - f(y_\tau \mid x_\tau, a_\tau)| \\ &\leq 3K\tau + O(\tau^2), \end{aligned}$$

from which (17.4) follows. ///

The approximate operator is defined simply as T_π^τ, with $\pi_\tau = (\hat{R}_\tau, \hat{Q}_\tau)$, and from Theorem 14.4 we have operator tolerance

$$\left\| \bar{T}_\pi V - T_\pi^\tau V \right\|_{sup} \leq \tau K(1 + 3 \left\| V \right\|_{sup}) + O(\tau^2), \tag{17.5}$$

and given principal eigenvector ν we also have

$$\left\| \bar{T}_\pi V - T_\pi^\tau V \right\|_\nu \leq \tau K \left(1 + \frac{3}{2} \left\| V \right\|_{SP(\nu)} \right) + O(\tau^2). \tag{17.6}$$

17.2 COMPLEXITY ANALYSIS

We may develop a cost tolerance model as described in Chapter 11. We may interpret T_π^τ as the DPO for a discrete model with a state space and action space defined by a τ-spaced regular grids in $[0,1]^n$ and $[0,1]^m$, in which case the cardinalities may be taken as $|\mathcal{X}| \approx \tau^{-n}$ and $|\mathcal{K}| \approx \tau^{-n-m}$. Thus, the computational effort required to evaluate T_π^τ may be given as a function of τ by

$$G^{grid}(\tau) = K_3 \tau^{-2n-m}, \tag{17.7}$$

where K_3 depends only on π, and the cumulative complexity of the algorithm is then summed over the iterations.

A number of important results exist concerning the complexity of this problem. Suppose we wish to estimate value function V^* within an error of ϵ, using discount factor α. For any algorithm $\mathcal{A}$ we define $C(\mathcal{A};\alpha,\epsilon,n,m)$ to be the worst case computation time over all MDPs $\pi \in \mathcal{P}$ with fixed state/action space dimension (n,m), assuming discount factor α, and assuming $\mathcal{A}$ outputs an ϵ-optimal solution V^ϵ (that is, $\|V^* - V^\epsilon\| \leq \epsilon$). In Chow and Tsitsiklis (1989) it is shown that a lower bound on $C(\mathcal{A};\alpha,\epsilon,n,m)$ exists of order $C(\alpha,\epsilon,n,m) = \Omega([(1-\alpha)^2\epsilon]^{-2n-m})$.

It turns out that the multigrid algorithm proposed in Chow and Tsitsiklis (1991) achieves complexity of order $(1-\alpha)^{-1}C(\alpha,\epsilon,n,m)$, that is, it is optimal with respect to ϵ and dimensions n,m but not with respect to $1-\alpha$. In comparison, the single grid algorithm achieves complexity of order $-\log((1-\alpha)\epsilon)(1-\alpha)^{-1}C(\alpha,\epsilon,n,m)$.

Reductions in complexity are possible for subsets of $\mathcal{P}$. In Chow and Tsitsiklis (1991) an ergodicity condition (Definition 4.1) imposed on $\mathcal{P}$ permits a reduction of the minimum worst-case complexity to order $\Omega([(1-\alpha)\epsilon]^{-2n-m})$, which is the minimum achievable (Chow and Tsitsiklis (1989)). For finite action sets randomized algorithms can achieve further reductions in complexity (Rust (1997); Kearns et al. (2000); Szepesvari (2001); Kearns et al. (2002)). We will confine attention to the general case $\mathcal{P}$.

17.3 APPLICATION OF APPROXIMATION SCHEDULES

An approximation schedule as defined in Chapter 11 will be generated by considering a sequence of grid sizes $\tau_1, \tau_2, \ldots$ leading to AIA

$$V_k = T_\pi^{\tau_k} V_{k-1}, \quad k \geq 1 \tag{17.8}$$

given starting point V_0, and assuming $\{\tau_k\} \in \mathcal{S}^-$. Recall that T_π^τ remains formally an operator on the original Banach space $\mathcal{F}(\mathcal{X}, \|\cdot\|_{sup})$. If τ is held constant, then the approximate model π_τ has been constructed so that TV is measurable *wrt* $\mathcal{E}_\tau^n$ if V is. If the schedule τ_k is nonincreasing, then the simplest way to maintain consistent measurability is to use only reciprocals of powers of 2, setting $\tau_k = 2^{i_k}$ for some nondecreasing sequence of integers i_k.

For any approximation schedule $\{\tau_k\}$ we have operator tolerance

$$\left\| \bar{T}_\pi V_{k-1} - T_\pi^{\tau_k} V_{k-1} \right\|_{sup} = O(\tau_k). \tag{17.9}$$

If the parameter τ is held fixed, we attain an approximation bound of order $O(\tau)$. Using (17.9) and Theorems 14.2–14.3 we have, for some finite constant L,

$$\limsup_{k\to\infty} \alpha^{-k} \|V_k - V^*\|_{sup} < \infty, \quad \text{for } \{\tau_k\} \in \mathcal{F}_\alpha^L, \tag{17.10}$$

$$\limsup_{k\to\infty} \tau_k^{-1} \|V_k - V^*\|_{sup} \leq L\left(1 - \alpha/\lambda^l\{\tau_k\}\right)^{-1}, \quad \text{for } \{\tau_k\} \in \mathcal{F}_\alpha^U, \tag{17.11}$$

and we may in general conclude that V_k converges to the true fixed point in the supremum norm for any sequence $\tau_k \to 0$.

Our strategy is then to verify that the tolerance model of Definition 11.1 holds for pairs of algorithms, following the discussion in Chapter 11. First suppose two algorithms $\{\tau_k\}$ and $\{\tau'_k\}$belong to $\mathcal{F}^L_\alpha$, with $\tau'_k = o(\tau_k)$. Define respective approximation schedules S, S' by $u_k = O(\tau_k)$, $u'_k = O(\tau'_k)$, with computation function $G(u) = G^{grid}(u)$ defined in (17.7). Given (17.10) it is easily verified that this pair of algorithms satisfy Definition 11.1 for tolerance model (α, G).

Furthermore, from Theorem 11.4, G satisfies condition (A). If $\{\tau_k\}$ is linearly convergent then condition (C) holds, and by Theorem 11.1 algorithm $\{\tau_k\}$ has strictly better computational convergence. Otherwise, again by Theorem 11.1, the computational convergence of $\{\tau_k\}$ is no worse than that of $\{\tau'_k\}$.

A similar argument can be made for pairs of algorithms $\{\tau_k\}$, $\{\tau'_k\}$ in $\mathcal{F}^U_\alpha$. Here we assume $\tau_k = o_\ell(\tau'_k)$. Using (17.11) we may select a finite constant b such that Definition 11.1 holds for schedules $u_k = b\tau_k$, $u'_k = b\tau'_k$ and computation function $G(u) = G^{grid}(b^{-1}u)$. We may use Theorem 11.4 to verify that G satisfies condition (B1) with each schedule. Our conditions constrain $\{\tau_k\}$ to be linearly convergent. Applying Theorem 11.2 we may conclude that if $\{\tau'_k\}$ is also linearly convergent then $\{\tau_k\}$ has no worse than the same order computational efficiency, whereas if $\{\tau'_k\}$ is sublinearly convergent then $\{\tau_k\}$ is strictly more efficient.

In general, the theory of approximation schedules permits us to rule out as optimal algorithms any sublinearly or superlinearly convergent approximation schedules. To simplify the analysis, we consider schedules of the form

$$\tau_k = Cr^k, \quad k \geq 1 \tag{17.12}$$

where $r \in (0, 1)$, and $C > 0$. Suppose we have algorithm tolerance

$$\| V_k - V^* \|_{sup} \leq \eta_k.$$

A final algorithm tolerance of v is achieved after t_v iterations, where $v = \eta_{t_v}$. From the computational effort given in (17.7) for iteration k, the cumulative complexity becomes

$$\bar{G}_k = K_3 \sum_{i=1}^{k} (Cr^i)^{-m-2n} = K_3 C^{-m-2n} \frac{(r^k)^{-m-2n} - 1}{1 - r^{m+2n}}, \tag{17.13}$$

and an algorithm tolerance of v can be achieved using a total computational effort of

$$\bar{G}_\eta(v) = \bar{G}_{t_v}. \tag{17.14}$$

Next, suppose the cost function is bounded by $0 \leq h \leq h^{max}$. The iterative relationship $\| V_k \|_{sup} \leq h^{max} + \alpha \| V_{k-1} \|_{sup}$ is easily verified, giving $\| V_k \|_{sup} \leq (1 - \alpha)^{-1} h^{max} + \alpha^k \| V_0 \|_{sup}$. We may set $V_0 \equiv 0$, leading to bound $\| V_k \|_{sup} \leq h^{max}(1 - \alpha)^{-1}$. In general, for the span seminorm we have $\| V \|_{SP} \leq 2 \| V \|_{sup}$, so we may make use of the inequalities

$$\| V_k \|_{sup} \leq (1 - \alpha)^{-1} h^{max} \tag{17.15}$$

$$\| V_k \|_{SP} \leq 2(1 - \alpha)^{-1} h^{max}. \tag{17.16}$$

From (17.15) and (17.16), we may substitute into (17.12), yielding

$$\| V_k - V^* \|_{sup} \leq \begin{cases} K_4^\alpha \alpha^k + K_5^\alpha C\zeta(r,\alpha)(r^k - \alpha^k); & r \neq \alpha, \\ K_4^\alpha \alpha^k + K_5^\alpha Ckr^k & ; \quad r = \alpha \end{cases} \tag{17.17}$$

where

$$\begin{aligned} K_4^\alpha &= (1-\alpha)^{-1} b^{max}, \\ K_5^\alpha &= (K_1 + 2K_2(1-\alpha)^{-1} b^{max}), \\ \zeta(r,\alpha) &= r(r-\alpha)^{-1}. \end{aligned}$$

We first consider the choice of constant C in (17.12). We need not make τ greater than 1 (assuming $\mathcal{S}$ is a unit cube). Thus, if C is made large enough that $\tau_k > 1$ for the first several iterations, such iterations may be considered void in that they do not contribute to model refinement and assume no cost. The following lemma verifies that the computational cost required to attain a given algorithm tolerance increases unboundedly as C approaches 0.

Lemma 17.1 *For fixed r, α, as $C \to 0$ we have $G_\eta(v) \to \infty$ for any v.*

Proof From (17.17), for a fixed algorithm tolerance v we have

$$v \geq K_4^\alpha \alpha^{t_v} \tag{17.18}$$

for all $C > 0$, where $v = \eta_{t_v}$. From (17.13) and (17.14) we have

$$\begin{aligned} \bar{G}_\eta(v) &= C^{-m-2n} K_3 \sum_{i=1}^{t_v} (r^i)^{-m-2n} \\ &\geq C^{-m-2n} K_3 \sum_{i=1}^{\log(v/K_4^\alpha)/\log(\alpha)} (r^i)^{-m-2n} \end{aligned}$$

where the inequality follows from (17.18). The lemma follows by letting $C \to 0$. ///

Following the above discussion, in the remaining development we will set $C = 1$.

The immediate objective is to verify that the same complexity obtained by the multigrid algorithm proposed in Chow and Tsitsiklis (1991) can be attained simply by varying the grid size according to (17.12).

We will consider $r \in (\alpha, 1)$. From (17.17) the algorithm tolerance may be bounded as follows:

$$\eta_k \leq \left(K_4^\alpha + K_5^\alpha \zeta(r,\alpha)\right) r^k \tag{17.19}$$

We solve $v = \eta_{t_v}$ then substitute inequality (17.19) into (17.14) to give

$$\bar{G}_\eta(v) \leq \frac{K_3}{1 - r^{m+2n}} \left(\left[\frac{K_4^\alpha + K_5^\alpha \zeta(r,\alpha)}{v} \right]^{m+2n} - 1 \right). \tag{17.20}$$

Then note that $K_4^\alpha = \Omega((1-\alpha)^{-1})$ and $K_5^\alpha = \Omega((1-\alpha)^{-1})$, while $\zeta(r,\alpha) \geq \alpha/(1-\alpha)$. In this case the dominant term of (17.20) becomes

$$\bar{G}_\eta^1(v) = \frac{K_3}{1-r^{m+2n}} \left[\frac{2K_2 h^{max} \zeta(r,\alpha)}{v(1-\alpha)} \right]^{m+2n}. \tag{17.21}$$

Since an investigation of the complexity involves allowing $\alpha \to 1$, and we are assuming $r > \alpha$, we must allow r to depend on α, which will be denoted r_α. We may write

$$\bar{G}_\eta^1(v) = \frac{K_6}{[v(1-\alpha)]^{m+2n}} \times \frac{r^{m+2n}}{(1-r^{m+2n})(r-\alpha)^{m+2n}} \tag{17.22}$$

where K_6 does not depend on r,α or v. It is easily verified that over $[\alpha, 1]$ this quantity is minimized with respect to r by setting $r_\alpha = \alpha^{1/(m+2n+1)}$. Substituting into (17.21) gives

$$\bar{G}_\eta^{1,\alpha}(v) = \frac{K_6}{[v(1-\alpha)]^{m+2n}} \times \frac{1}{\left(1-\alpha^{\frac{m+2n}{m+2n+1}}\right)^{m+2n+1}}.$$

It can be verified that the following limits hold:

$$\lim_{m+2n\to\infty} \frac{\left(1-\alpha^{\frac{m+2n}{m+2n+1}}\right)^{m+2n+1}}{(1-\alpha)^{m+2n+1}} = \exp\left(\frac{\alpha}{1-\alpha}\log(\alpha)\right),$$

$$\lim_{\alpha\to 1} \frac{\left(1-\alpha^{\frac{m+2n}{m+2n+1}}\right)^{m+2n+1}}{(1-\alpha)^{m+2n+1}} = \left(\frac{m+2n}{m+2n+1}\right)^{m+2n+1}. \tag{17.23}$$

The limits of (17.23) are both approximately e^{-1} for α close to 1 and large $m+2n$, so we have the uniform approximation

$$\begin{aligned} \bar{G}_\eta^{1,\alpha}(v) &\approx \frac{K_3}{1-\alpha}\left[\frac{2K_2 h^{max}}{v(1-\alpha)^2}\right]^{m+2n} e^1 \\ &= \Omega([(1-\alpha)^2\epsilon]^{-2n-m}) \end{aligned} \tag{17.24}$$

which is equivalent to the complexity reported in equation (7.11) of Chow and Tsitsiklis (1991).

As a final note, the selection of r affects the order of $\bar{G}_\eta(v)$ significantly with respect to $m+2n$. To see this, when r is close to 1 we may write (17.22)

$$\bar{G}_\eta^{1,\alpha}(v) \geq \frac{K_3}{1-\alpha}\left[\frac{2K_2 h^{max}}{v(1-\alpha)^2}\right]^{m+2n} (m+2n)^{-1}\left[\frac{r-r\alpha}{r-\alpha}\right]^{m+2n},$$

noting that $r > \alpha$. Since $r - r\alpha > r - \alpha$ this introduces an exponentially ordered factor into $\bar{G}_\eta^1(v)$ which is not present in when r_α is substituted for r, as in (17.24).

Chapter 18

Adaptive control of MDPs

As a final case study, we consider the problem of adaptive control. We have already considered the problem of estimating regret when the MCM π is unknown but can be estimated by $\hat{\pi}$. Specifically, we have considered the problem of replacing π with a single statistical estimate $\hat{\pi}$, and the problem of employing successively refined numerical estimates of a known model π.

We return to the problem of statistical model identification. We are able to translate statistical model error directly into an estimate of regret, in effect, estimating the cost of statistical error. However, if the resulting certainty equivalence policy $\phi_{\hat{\pi}}$ is applied indefinitely to an MDP governed by model π, a constant amount of regret will also be accrued indefinitely. Of course, while the MDP is on-line new data is being collected with which the model can be refined. With a refined model, a new certainty equivalence policy can be calculated, reducing the regret in a predictable way.

In principle, there is no reason not to continue this process indefinitely. If we do, we have an *adaptive control policy*, which is formally a sequence of control functions $\phi_1, \phi_2, \ldots$, such that ϕ_n is applied at stage n, and is allowed to depend on history H_n^x. A *certainty equivalence adaptive policy* simply sets $\phi_n = \phi_{\hat{\pi}_n}$, the policy optimal for model $\hat{\pi}_n$, the best model estimate based on history data H_n^x (which may include any other auxiliary data or information).

This is an intuitively appealing approach, but we have no guarantee that the sequence of policies ϕ_n converges to the optimal, or equivalently, that regret approaches 0. In fact, there really is no reason to think it should. In the absence of special structure, in order to calculate the optimal control for π, we must have the entire model π. To fix ideas, suppose $\mathcal{K}$ is finite. For each $(x,a) \in \mathcal{K}$ we must estimate $R(x,a)$ and $Q(\cdot \mid x,a)$. When an online system visits state/action pair (X_n, A_n), it transitions to state X_{n+1} according to distribution $Q(\cdot \mid X_n, A_n)$. As for cost, if the realized cost R_n is deterministic, we have identified without error $R(X_n, A_n)$ with a single visit. We consider instead the more general problem in which it is assumed only that $E[R_n \mid H_n^a] = R(X_n, A_n)$. Clearly, we can only expect a sequence of models $\hat{\pi}_n$ to converge to the true model π if each state/action pair (X_n, A_n) is visited infinitely often. This might reasonably be guaranteed for the state variable, but not the action variable.

To see this, suppose for state x we have $\mathcal{K}_x = \{0,1\}$. For the sake of argument, assume $Q(\cdot \mid x,0) = Q(\cdot \mid x,1)$ and that $R(x,0) < R(x,1)$. We can conclude on this basis that $\phi_\pi(x) = 0$. This particular cost inequality will be correctly infered with probability approaching 1 as the number of visits to $(x,0)$ and $(x,1)$ approach ∞. Next, suppose that for some interim model $\hat{\pi}_n$ the distributions $Q(\cdot \mid x, 0)$ and $Q(\cdot \mid x, 1)$ are accurately estimated, as is $R(x, 1)$, but that the estimate of $R(x, 0)$ (incorrectly) exceeds that of $R(x, 1)$ by some large enough amount. On this basis we would probably have $\phi_{\hat{\pi}_n}(x) = 1$. This means that the online process will choose action 1 for the next visit to state x. If the estimation process is close to convergence, we would expect little further change in subsequent estimates of $R(x, 1)$ and $Q(\cdot \mid x, 1)$, and no further change in the estimate of $R(x, 0)$ unless state/action pair $(x, 0)$ is visited again. In order for this to happen, there must be a certainty equivalence policy at some stage $n' > n$, for which $\phi_{\hat{\pi}_{n'}}(x) = 0$, which may not happen under the given scenario. In this case, the certainty equivalence adaptive policy never identifies the action $a = 0$ as optimal from stage x, and regret does not approach 0. The issue of identifiabilty of certainty equivalence adaptive controls is discussed in detail, with some interesting counterexamples, in Bertsekas (1995a) and Kumar and Varaiya (1986).

A number of approaches to this problem can be taken. If the objective is to ensure that each state/action pair is visited infinitely often, then this can be easily achieved using a suitable randomized control policy. If the purpose of this policy is to ensure that all state/action pairs are visited, we may refer to it as an *exploratory control policy*. Of course, the object is to attain minimum regret, which generates conflicting goals. Minimum regret cannot be achieved by a certainty equivalence adaptive policy unless all state/action pairs are visited infinitely often. In the absence of additional structure, we assume this can only be achieved by enforcing an exploratory control. We also assume that exploratory control results in regret bounded away from zero. We therefore have two sources of regret, that due to suboptimal control policies resulting from imperfect model identification, and that due to exploration.

Of course, these two forms of regret can be quantified and balanced. At any stage of an online MDP we have access to model estimate $\hat{\pi}_n$, and therefore to certainty equivalence control policy $\phi_{\hat{\pi}_n}$. In addition, suppose that at any stage we may choose between applying the current certainty equivalence control and an exploratory control. Next, we may define an *exploration rate* α_n, so that at stage n, in some sense, the probability that the applied control is exploratory is α_n. If the exploratory control satisfies certain assumptions, the number of visits to any state/action pair will be proportional to $\sum_{i=1}^{n} \alpha_i$ at stage n. We may always choose α_n so that $\alpha_n \rightarrow_n 0$ and $\sum_{i=1}^{n} \alpha_i \rightarrow_n \infty$, so that each state/action pair is visited infintely often, but also that the stage frequency of exploratory behavior approaches 0. In this case we can expect the certainty equivalence policy to approach the optimal, while being applied with a stage frequency approaching 1.

18.1 REGRET BOUNDS FOR ADAPTIVE POLICIES

Theorem 12.8 gives a direct method of bounding regret through the formula

$$\Lambda_n^{\Phi}(H_n^x) = \bar{V}_\pi(X_n) + E_x^{\Phi}\left[\sum_{i=0}^{\infty} \beta^i \lambda_\pi(X_{n+i}, A_{n+i}) \mid H_n^x\right].$$

Under an adaptive policy $\Phi = (\phi_1, \phi_2, \dots)$ the regret accrued from state X_n at stage n is

$$\Lambda_n^{\Phi}(H_n^x) - \bar{V}_\pi(X_n) = E_x^{\Phi}\left[\sum_{i=0}^{\infty} \beta^i \lambda_\pi(X_{n+i}, \phi_n(X_{n+i})) \mid H_n^x\right]. \tag{18.1}$$

In addition, by Theorem 14.6 we have bound

$$\lambda_\pi(x, \phi_{\hat{\pi}}(x)) \leq K_1 \max(D_R^w(R, \hat{R}), D_Q^w(Q, \hat{Q})), \tag{18.2}$$

for some constant K_1, where $\hat{\pi} = (\hat{R}, \hat{Q})$. Immediately, we may reach some conclusions for the simplest forms of adaptive controls. In some cases, the certainty equivalence adaptive policy will suffice. For example, for a queueing system the model may be known up to a small number of parameters, for example, an arrival rate λ. Suppose cost function R is known. In most models, each stage will generate a datum which can contribute to an estimate of λ. Suppose the estimate of λ at stage n is $\hat{\lambda}_n$. Section 15.3 shows how to construct a bound of the form $D_Q^w(Q, \hat{Q}_n) \leq K_2|\hat{\lambda}_n - \lambda|$ for some constant K_2. We can generally expect $|\hat{\lambda}_n - \lambda| = O(n^{-1/2})$. We need to next make an argument such as

$$E_x^\Phi\left[|\hat{\lambda}_{n+m} - \lambda| \mid H_n^x\right] \leq |\hat{\lambda}_{n-1} - \lambda| + o(n^{-1/2}). \tag{18.3}$$

Combining (18.1), (18.2) and (18.3) immediately yields

$$\Lambda_n^\Phi(H_n^x) - \bar{V}_\pi(X_n) = O(n^{-1/2}), \tag{18.4}$$

that is, if the certainty equivalence adaptive policy does yield a consistent estimate of the model π, we can generally expect regret to conform to the usually statistical error of $O(n^{-1/2})$.

18.2 DEFINITION OF AN ADAPTIVE MDP

The point was made earlier, in Section 4.4 for example, that statistical procedures which rely on assumptions of independence may not be appropriate as part of an adaptive system with a well defined history process H_n, $n \geq 1$. This history defines a filtration (Definition 4.2), with respect to which all processes we study are adapted. It will usually be the case that properties of adapted statistical procedures are easier to define conditioned on the history processes. This sometimes necessitates some cumbersome notation, and the relationship between conditional events and the history process always needs to be precisely defined. However, this leads to procedures of considerably more flexibility than would be possible by relying on assumptions of independence.

The first step is to expand on the definition of a MDP to include exploratory behavior and auxiliary data. For example, we have defined $R(x, a)$ as an expected cost. The actual cost may be random, and this is not formally modeled by the MDP measure P_x^Φ defined in Section 12.1 (this distinction is explicitly characterized in Bertsekas and Shreve (1978), Chapters 8–9). Now that the stochastic propreties of the cost are of relevance, a probability measure must be defined which can do this.

As before, we have MCM $\pi = (\mathcal{K}, Q, R, \beta)$, and we add two new elements to definitions $(M1)$–$(M6)$ of Section 12.1.

(M7) A Borel space $\mathcal{O}$, called the *observation space*, and a stochastic kernel $Q^{o,x}$: $\mathcal{K} \to \mathcal{M}(\mathcal{O} \times \mathcal{X})$ for which $Q^{o,x}(\mathcal{O} \times E_x \mid x, a) = Q(E_x \mid x, a)$ for all $E_x \in \mathcal{B}(\mathcal{X})$, $(x, a) \in \mathcal{K}$. In addition, $E_{x,a}^{Q^{o,x}}$ is the expectation operator associated with $Q^{o,x}(\cdot \mid x, a)$.

(M8) A binary outcome $\mathcal{Z} = \{0, 1\}$ and a sequence of measurable mappings $p_n^e : (\mathcal{XZAO})^{n-1} \times \mathcal{X} \to [0, 1], n \geq 1$.

Two new quantities will be associated with stage n. First, we have O_n, defined on $\mathcal{O}$, with distribution calculable from $Q^{o,x}(\cdot \mid X_n, A_n)$ according to (M7). This represents information available to the controller following the realization of the state/action pair (X_n, A_n), including that pertaining to the realized stage n cost, and the transition from X_n to X_{n+1}. This information is assumed to be available in time to influence the control applied at the $n + 1$st stage.

Second, given state X_n, a binary randomization quantity $Z_n \in \mathcal{Z} = \{0, 1\}$ is observed. The action A_n is permitted to depend on Z_n as well as history H_n^a. The role of Z_n is to select between exploratory $(Z_n = 1)$ and certainty equivalence control $(Z_n = 0)$, as described in the introduction to this chapter. The distributional properties of the sequence $Z_1, Z_2, \ldots$ are determined by the sequence of mappings p_n^e, as described below. It will be helpful to think of the order of realization of the stage quantities as $X_n \to Z_n \to A_n \to O_n \to X_{n+1} \to Z_{n+1} \to \ldots$. We accordingly define the Borel space $\mathcal{S} \subset \mathcal{XZAO}$ to be all elements $(x, w, a, o) \in \mathcal{XZAO}$ for which $(x, a) \in \mathcal{K}$. The history vectors are expanded accordingly:

$$
\begin{aligned}
H_n^a &= (X_1, Z_1, A_1, O_1, \ldots, X_n, Z_n, A_n) \\
H_n^z &= (X_1, Z_1, A_1, O_1, \ldots, X_n, Z_n) \\
H_n^x &= (X_1, Z_1, A_1, O_1, \ldots, X_n) \qquad (18.5)
\end{aligned}
$$

The adaptive control will be a mixture of two policies, $\Phi^e = (\Phi_1^e, \Phi_2^e, \ldots)$ and $\Phi^o = (\Phi_1^o, \Phi_2^o, \ldots)$, where Φ_n^e and Φ_n^o are measurable mappings of H_n^x to $\mathcal{M}(\mathcal{A})$. The randomization variable Z_n is used to select the policy according to the form

$$
\Phi_n(E_a \mid H_n^z) = (1 - Z_n)\Phi_n^o(E_a \mid H_n^x) + Z_n \Phi_n^e(E_a \mid H_n^x), \quad E_a \in \mathcal{B}(\mathcal{A}). \qquad (18.6)
$$

The intention is that Φ^e is used to explore. We accept that while this policy is used an amount of regret bounded away from 0 is accrued, so that no attempt to minimize regret is made under Φ^e.

On the other hand Φ^o is intended to be the best control available with respect to the minimization of regret. In our example, this will be the current certainty equivalence policy.

Although we have explictly defined two new stage quantities, from the point of view of measure construction we can incorporate Z_n into the action space and O_n into the state space, retaining the original definition of a MCM given in Section 12.1. Thus, given elements $(M1)$–$(M8)$, for any admissible starting state $X_1 = x$ a unique measure P_x^Φ exists on the Borel space $\mathcal{S}^\infty$ satisfying

$$
\begin{aligned}
P_x^\Phi(X_1 = x) &= 1, \\
P_x^\Phi((O_n, X_{n+1}) \in E_{ox} \mid H_n^a) &= Q^{o,x}(E_{ox} \mid X_n, A_n), \quad E_{ox} \in \mathcal{B}(\mathcal{OX}), \\
P_x^\Phi(X_{n+1} \in E_x \mid H_n^a) &= Q(E_x \mid X_n, A_n), \quad E_x \in \mathcal{B}(\mathcal{X}), \\
P_x^\Phi(Z_n = 1 \mid H_n^x) &= p_n^e(H_n^x), \\
P_x^\Phi(A_n \in E_a \mid H_n^z) &= \Phi_n(E_a \mid H_n^z), \quad E_a \in \mathcal{B}(\mathcal{A}), \qquad (18.7)
\end{aligned}
$$

for $n \geq 1$ and each admissible history H_n^x, H_n^z, H_n^a. As above, we let E_x^Φ be the expectation operator of P_x^Φ, and x may be any initial state.

18.3 ONLINE PARAMETER ESTIMATION

Assume that model π is fully defined by a parameter vector $\theta = (\theta_1, \ldots, \theta_k)$, and that a Lipschitz relationship exists between θ and the model π, in the sense that for any estimate $\hat{\theta} = (\hat{\theta}_1, \ldots, \hat{\theta}_k)$ of θ there is a constant K_θ for which

$$\max(D_R^w(R_\pi, R_{\hat{\pi}}), D_Q^w(Q_\pi, \hat{Q}_{\hat{\pi}})) \leq K_\theta d(\theta, \hat{\theta}),$$

for a suitable metric d. We have alreay seen that this will be possible under general conditions. We may also assert that if $\phi_{\hat{\pi}}$ is the certainty equivalence policy for model $\hat{\pi}$ we have

$$\lambda_\pi(x, \phi_{\hat{\pi}}(x)) \leq K_\lambda d(\theta, \hat{\theta}).$$

This will permit us to construct a bound on regret directly from the statistical error of the model estimates.

The procedure we present will be illustrated using estimators formed from sample averages, but the essential requirements, of which there are two, will be stated explicitly.

The first problem which arises concerns the amount of information in the history process regarding any specific parameter θ_j. Possibly, each parameter is associated with a specific state/action pair, so that the properties of the estimation process is closely dependent on the exploration process, so the two must be considered together.

Accordingly, we offer the following defintion:

Definition 18.1 Suppose we have a MCM $\pi = (\mathcal{K}, Q, R, \beta)$, observation space $\mathcal{O}$ and kernel $Q^{o,x}$ defined in (M7). The *informative subset* of $\mathcal{K}$ for component θ_j of $\theta = (\theta_1, \ldots, \theta_k)$, which we denote $\mathcal{K}(\theta_j)$, consists of all $(x, a) \in \mathcal{K}$ for which a measurable estimator $\bar{\theta}_j(O)$, $O \in \mathcal{O}$ exists satisfying

$$E_{x,a}^{Q^{o,x}}[\bar{\theta}_j(O)] = \theta'_j \quad \text{and} \quad E_{x,a}^{Q^{o,x}}[(\bar{\theta}_j(O) - \theta'_j)^2] \leq \nu \tag{18.8}$$

for some constant $0 \leq \nu < \infty$, where $\theta' = (\theta'_1, \ldots, \theta'_k)$ is the true parameter. ///

For convenience set $I_n(\theta_j) = I\{(X_n, A_n) \in \mathcal{K}(\theta_j)\}$ and

$$M_n(\theta_j) = \sum_{i=1}^{n} I_i(\theta_j), \quad n \geq 1.$$

Next, define the sequence

$$W_n(\theta_j) = \sum_{i=1}^{n} (\bar{\theta}_j(O_i) - \theta'_j) I_i(\theta_j), \quad n \geq 1. \tag{18.9}$$

It is easily verified that under Definition 18.1 the process defined by (18.9) is a martingale on filtration $(\sigma(H_2^a), \sigma(H_3^a), \ldots)$, since

$$\begin{aligned} E_x^{\Phi}[W_n(\theta_j) \mid H_n^a] &= E_x^{\Phi}[(\bar{\theta}_j(O_n) - \theta'_j)I_n(\theta_j) \mid H_n^a] + W_{n-1}(\theta_j) \\ &= E_x^{\Phi}[(\bar{\theta}_j(O_n) - \theta'_j)I_n(\theta_j) \mid X_n, A_n] + W_{n-1}(\theta_j) \\ &= W_{n-1}(\theta_j), \quad n \geq 1. \end{aligned}$$

For each m the quantity $\tau_m = \min\{n \geq 1 \mid M_n(\theta_j) = m\}$ represents the stage at which $I_n(\theta_j) = 1$ for exactly the mth time. As discussed in Section 4.4.1, τ_m defines an increasing sequence of stopping times, so that $W_{\tau_m}(\theta_j)$, $m \geq 1$ is also a martingale, by the optional sampling theorem (Theorem 4.7). Under Definition 18.1 the martingale differences of both $W_n(\theta_j)$ and $W_{\tau_m}(\theta_j)$ are square integrable, so by the martingale SLLN (Theorem 4.34) we have

$$\left| \frac{W_{\tau_m}(\theta_j)}{m} \right| = o\left(m^{-1/2+\epsilon}\right), \quad wp1,$$

which is equivalent to

$$\left| \frac{W_n(\theta_j)}{M_n(\theta_j)} \right| = o\left(M_n(\theta_j)^{-1/2+\epsilon}\right), \quad wp1, \tag{18.10}$$

for any small $\epsilon > 0$. This leads to component estimates

$$\hat{\theta}_{n,j} = \begin{cases} M_n(\theta_j)^{-1} \sum_{i=1}^n \bar{\theta}_j(O_i) I_i(\theta_j); & M_n(\theta_j) \geq 1 \\ \hat{\theta}_{0,j} & ; \; M_n(\theta_j) = 0 \end{cases} \tag{18.11}$$

for $n \geq 1, j = 1, \ldots, k$, where $\hat{\theta}_0 = (\hat{\theta}_{0,1}, \ldots, \hat{\theta}_{0,k})$ is a suitably chosen starting value. The parameter estimate sequence is then $\hat{\theta}_n = (\hat{\theta}_{n,1}, \ldots, \hat{\theta}_{n,k})$. From (18.10) we have

$$\begin{aligned} |\hat{\theta}_{n,j} - \theta'_j| &= o\big(M_n(\theta_j)^{-1/2+\epsilon}\big), \\ d(\hat{\theta}_n, \theta') &= o\big(M_n(\theta)^{-1/2+\epsilon}\big) \end{aligned} \tag{18.12}$$

for any $\epsilon > 0$ where

$$M_n(\theta) = \min_{1 \leq j \leq k} M_n(\theta_j), \quad n \geq 1.$$

Hence, convergence of $\hat{\theta}_n$ to θ' follows from $M_n(\theta) \to \infty$, at a rate implied by $M_n(\theta)$.

We have established the first requirement of an online estimation scheme, that a rate of convergence of $d(\hat{\theta}_n, \theta')$ to 0 can be established based on the rate at which information for the parameters is collected, relying only on minimal conditional properties.

This suffices to bound regret on a per stage basis. We also wish to bound expected future regret. While we expect that $d(\hat{\theta}_n, \theta')$ decreases in the long run as $n \to \infty$

we will need to bound short term variation of $\hat{\theta}_n$. This is done in the following theorem:

Theorem 18.1 *Under Definition 18.1 the following inequality holds:*

$$E_x^\Phi\left[|\hat{\theta}_{n+m-1,j} - \theta_j'| \mid H_n^a\right] \le |\hat{\theta}_{n-1,j} - \theta_j'| + \frac{m\nu^{1/2}}{M_n(\theta_j)} \tag{18.13}$$

for $m \ge 0$, $n \ge 1$.

Proof First note that $\hat{\theta}_{n-1,j}$ is $\sigma(H_n^a)$-measurable, so that (18.13) holds for $m=0$. Next assume $m \ge 1$. For $n \ge 1$, if $M_n(\theta_j) \ge 1$ we may write

$$\begin{aligned}
|\hat{\theta}_{n+m-1,j} - \theta_j'| &= \frac{\left|\sum_{i=1}^{n+m-1} (\bar{\theta}_j(O_i) - \theta_j')I_i(\theta_j)\right|}{M_{n+m-1}(\theta_j)} \\
&\le \frac{\sum_{i=n}^{n+m-1} \left|(\bar{\theta}_j(O_n) - \theta_j')I_n(\theta_j)\right|}{M_{n+m-1}(\theta_j)} + \frac{\left|\sum_{i=1}^{n-1} (\bar{\theta}_j(O_i) - \theta_j')I_i(\theta_j)\right|}{M_{n+m-1}(\theta_j)} \\
&\le \frac{\sum_{i=n}^{n+m-1} \left|(\bar{\theta}_j(O_n) - \theta_j')I_n(\theta_j)\right|}{M_n(\theta_j)} + |\hat{\theta}_{n-1,j} - \theta_j'|,
\end{aligned} \tag{18.14}$$

since $M_n(\theta_j)$ is nondecreasing. We then note that under Definition 18.1

$$E_x^\Phi\left[|(\bar{\theta}_j(O_n) - \theta_j')I_n(\theta_j)| \mid H_n^a\right] \le \nu^{1/2},$$

and consequently for any $m \ge 0$

$$\begin{aligned}
&E_x^\Phi\left[|(\bar{\theta}_j(O_{n+m}) - \theta_j')I_{n+m}(\theta_j)| \mid H_n^a\right] \\
&= E_x^\Phi\left[E_x^\Phi\left[|(\bar{\theta}_j(O_{n+m}) - \theta_j')I_{n+m}(\theta_j)| \mid H_{n+m}^a\right] \mid H_n^a\right] \\
&\le E_x^\Phi\left[\nu^{1/2} \mid H_n^a\right] = \nu^{1/2}.
\end{aligned}$$

The proof is completed by taking the expectation of (18.14) conditional on H_n^a, applying the preceding inequality, and noting that $\hat{\theta}_{n-1,j}$ and $M_n(\theta_j)$ are measurable *wrt* H_n^a. ///

18.4 EXPLORATION SCHEDULE

When the online certainty equivalence policies can be shown to converge to the optimal (so that exploration is not needed) we have seen that regret will generally approach 0 at a rate of order $O(n^{-1/2})$. We have also argued that this cannot generally be expected. We introduced earlier the concept of an exploration rate α_n, roughly, the probability that the control is exploratory at stage n. In this section we show that the optimal exploration rate will be $\alpha_n = O(n^{-1/3+\epsilon})$, for which regret converges to 0 at a rate of $O(n^{-1/3+\epsilon})$, for any $\epsilon > 0$.

Under an estimation model such as Definition 18.1 the goal of an exploratory policy is to ensure sufficient visits to each informative subset $\mathcal{K}(\theta_j)$ to allow $M_n(\theta_j) \to_n \infty$. Returning to the problem posed in the introduction, we can conceive of an exploration rate α_n, comparable to an arrival rate, which describes the proportion of stages in the neighborhood of n at which exploratory control is applied. Since a regret bounded away from zero is accrued under exploratory control, and the object is to allow regret to approach 0, the exploration rate must also approach 0. However, it must do so at a slow enough rate to allow $M_n(\theta_j) \to_n \infty$, so that the certainty equivalence policy approaches the true optimal policy, and our objective is achieved.

Of course, we may take the analysis one step further. We have a model which permits us to determine in terms of exploration rate α_n the rate at which regret due to exploration and regret due to suboptimal certainty equivalence control is accrued. Therefore, analysis permitting, we may determine the optimal exploration rate, that is, the rate minimizing the combined regret.

In our model, exploratory behavior is defined by Definition (M8) of Section 18.2. We refer to Z_n, $n \geq 1$ as the *exploration schedule*. Then assume that $\hat{\pi}_n$ is the model estimate available from history H_n^o. Note that at the time at which control is to be applied at stage n, only model $\hat{\pi}_{n-1}$ is available. The control policy is defined in (18.6). According to the certainty equivalence principle, set $\Phi_n^o(E_a \mid H_n^x) = \phi_{\hat{\pi}_{n-1}}$.

It remains to construct $p_n^e(H_n^x)$ as defined in (M8), which is the subject of Section 5.5. This can be done from two points of view. The first step, clearly, is to establish the existence of an exploration schedule which achieves convergence to zero of *total* regret, and, if possible, the optimal rate. This is largely a mathematical problem, so that the schedule may be designed only with this in mind. We will see below that defining Z_n as a two state nonhomogenous Markov chain with transition matrices

$$Q_n = \begin{bmatrix} 1-\alpha_n & \alpha_n \\ 1-\gamma & \gamma \end{bmatrix},$$

with certain additional constraints on α_n and γ, will suffice (see definition in (5.21) for more detail). The resulting exploration schedule exhibits the block structure underlying the methods of Section 5.5. If at stage n the system is not under exploratory control ($Z_n = 0$) then it transfers in the next stage to exploratory control with probability α_n, otherwise ($Z_n = 1$) it remains in exploratory control with probability γ for stage $n+1$. In both cases, the selection is made independently of the current state and any process history. This defines *exploration blocks*, that is, maximal blocks of consecutive stages in exploratory control. We then have a well defined block length distribution which remains the same indefinitely, in this case given by the geometric distribution with parameter γ. In addition, these blocks occur at a rate determined by α_n. If we can assert that each informative subset $\mathcal{K}(\theta_j)$ is visited within any block with a minimum probability $\delta > 0$, then data is accumulated at a rate $M_n(\theta) = O(\xi_n)$ where $\xi_n = \sum_{i=1}^n \alpha_i$, the regret due to suboptimal certainty equivalence control is of order $O\left(\xi_n^{-1/2+\epsilon}\right)$ and the regret due to exploration is of order $O(\alpha_n)$.

For the sake of argument, suppose $\alpha_n \propto n^{-r}$ for $0 < r \leq 1$. Then $\xi_n \propto n^{1-r}$ for $r < 1$ and $\xi_n \propto \log(n)$ for $r = 1$. This gives $\xi_n^{-1/2+\epsilon} = o(n^{(r-1)/2+\epsilon})$. The remaining step is to minimize the maximum of the two rates over r, which, within ϵ, is attained simply by setting $-r=(r-1)/2$, yielding $r=1/3$. On this basis, the optimal exploration rate is $\alpha_n \approx n^{-1/3}$.

The remaining step is to formalize this argument.

We accept the model of an adapted counting process Z_n, $n \geq 1$ discussed in Section 5.5, and use the notation introduced in (5.22). In addition, following (5.23) we define a sequence of measurable mappings $\alpha_n(H_n^x) \in [0,1]$ for which

$$P(B_n = 1 \mid H_n^x) = \alpha_n(H_n^x) I\{Z_{n-1} = 0\}, \quad n \geq 1, \tag{18.15}$$

where $B_n = 1$ is the event that a block starts at stage n (see (5.22)). Here, Z_{n-1} is $\sigma(H_n^x)$-measurable.

We will assume Theorem 5.14 holds for model (18.15), and that the assumptions of Theorem 5.15 hold for each informative subset $\mathcal{K}(\theta_j)$ for some common $\delta > 0$ (Theorem 5.16 may also be used to introduce contraints into the exploration schedule). This suffices to conclude that $M_n(\theta) = O(\xi_n)$.

Finally, we will make use of the following lemma.

Lemma 18.1 *If for model* (18.15) $\alpha_n(H_n^x) \leq \alpha_n$, $n \geq 1$ *for some nonincreasing sequence of constants* α_n *then for* $n \geq 1$, $m \geq 0$,

$$E_x^{\Phi}[I\{Z_{n+m} = 1\} \mid H_n^x] \leq I\{Z_{n-1} = 1\} + \alpha_n(m+1). \tag{18.16}$$

Proof We have

$$\{Z_{n+m} = 1\} \subset \{Z_{n-1} = 1\} \cup \left(\cup_{j=0}^{m} \{B_{n+j} = 1\} \right),$$

which implies

$$E_x^{\Phi}[I\{Z_{n+m} = 1\} \mid H_n^x] \leq I\{Z_{n-1} = 1\} + \sum_{j=0}^{m} E_x^{\Phi}[I\{B_{n+j} = 1\} \mid H_n^x]. \tag{18.17}$$

To analyze the terms in (18.17), we write, for $j \geq 0$

$$\begin{aligned} E_x^{\Phi}[I\{B_{n+j} = 1\} \mid H_n^x] &= E_x^{\Phi}[E_x^{\Phi}[I\{B_{n+j} = 1\} \mid H_{n+j}^x] \mid H_n^x] \\ &\leq \alpha_{n+j} \end{aligned} \tag{18.18}$$

which completes the proof. ///

The following theorem completes the argument.

Theorem 18.2 *If for positive constants* b_π *and* K_λ

$$\sup_{x,a \in \mathcal{K}} \lambda_\pi(x,a) \leq b_\pi,$$

$$\lambda_\pi(x, \phi_{\hat{\pi}_n}(x)) \leq K_\lambda d(\hat{\theta}_{n-1}, \theta'),$$

then under the conditions of Theorem 18.1 and Lemma 18.1 the following bound on regret holds:

$$\Lambda_x^{\Phi}(H_n^x) - \bar{V}_{\pi}(X_n) \tag{18.19}$$
$$\leq \frac{K_{\lambda}d(\hat{\theta}_{n-1}, \theta') + b_{\pi}I\{Z_{n-1}=1\} + (1-\beta)^{-1}\left[b_{\pi}\alpha_n + K_{\lambda}kv^{1/2}/M_n(\theta)\right]}{1-\beta}.$$

Proof For fixed $n, m \geq 0$ consider a term of the form

$$\lambda_{\pi}(X_{n+m}, A_{n+m}) \leq \lambda_{\pi}(X_{n+m}, A_{n+m})I\{Z_{n+m}=1\} + \lambda_{\pi}(X_{n+m}, A_{n+m})I\{Z_{n+m}=0\}$$
$$= B_{n+m}^1 + B_{n+m}^2, \tag{18.20}$$

and consider the problem of estimating $E_x^{\Phi}[\lambda_{\pi}(X_{n+m}, A_{n+m}) \mid H_n^x]$. For term B_{n+m}^1, by Lemma 18.1 we may write

$$E_x^{\Phi}[B_{n+m}^1 \mid H_n^x] \leq b_{\pi}E_x^{\Phi}[I\{Z_{n+m}=1\} \mid H_n^x]$$
$$\leq b_{\pi}(I\{Z_{n-1}=1\} + (m+1)\alpha_n). \tag{18.21}$$

For term B_{n+m}^2 note that $Z_{n+m}=0$ implies $A_{n+m}=\hat{\phi}_{n+m}(X_{n+m})$, so that

$$B_{n+m}^2 \leq K_{\lambda}d(\hat{\theta}_{n+m-1}, \theta'). \tag{18.22}$$

We similarly have, by Theorem 18.1

$$E_x^{\Phi}[B_{n+m}^2 \mid H_n^x] \leq K_{\lambda}\left(d(\hat{\theta}_{n-1}, \theta') + kmv^{1/2}M_n(\theta)^{-1}\right) \tag{18.23}$$

Then (18.19) follows from a direct application of Theorem 12.8. ///

Bibliography

D. Aldous. *Probability Approximations via the Poisson Clumping Heuristic*, volume 77 of *Applied Mathematical Sciences*. Springer-Verlag, New York, NY, 1989.

A. Almudevar. A dynamic programming algorithm for the optimal control of piecewise deterministic Markov processes. *SIAM J. Control Optim.*, 4(1):525–539, 2001.

A. Almudevar. Approximate fixed point iteration with an application to infinite horizon Markov decision processes. *SIAM J. Control Optim.*, 47(5):2303–2347, 2008.

A. Almudevar and E.F. de Arruda. Optimal approximation schedules for a class of iterative algorithms, with an application to multigrid value iteration. *IEEE Transactions on Automatic Control*, 57(12):3132–3146, 2012.

B. C. Arnold and R. A. Groeneveld. Bounds on expectations of linear systematic statistics based on dependent samples. *Annals of Statistics*, 7:220–223, 1979.

E. F. Arruda, F. Ourique, J. Lacombe and A. Almudevar. Accelerating the convergence of value iteration by using partial transition functions. *European Journal of Operational Research*, 229(1):190–198, 2013.

R. B. Ash. *Real Analysis and Probability*. Academic Press, Orlando, Florida, first edition, 1972.

K. Atkinson and W. Han. *Theoretical Numerical Analysis: A Functional Analysis Framework*, volume 39 of *Texts in Applied Mathematics*. Springer-Verlag, New York, NY, 2001.

T. Aven. Upper (lower) bounds on the mean of the maximum (minimum) of a number of random variables. *Journal of Applied Probability*, 22(3):723–728, 1985.

V. Berinde. *Iterative Approximation of Fixed Points*. Springer, New York, NY, second edition, 2007.

D. P. Bertsekas. *Dynamic Programming and Optimal Control, Volume* 1. Athena Scientific, Belmont, MA, second edition, 1995a.

D. P. Bertsekas. *Dynamic Programming and Optimal Control, Volume* 2. Athena Scientific, Belmon, MA, second edition, 1995b.

D. P. Bertsekas and S. E. Shreve. *Stochastic Optimal Control: The Discrete-Time Case*. Academic Press, New York, NY, 1978.

D. P. Bertsekas and J. N. Tsitsiklis. *Neuro-dynamic Programming*. Athena Scientific, Belmont, MA, 1996.

D. Bertsimas, K. Natarajan and C. Teo. Tight bounds on expected order statistics. *Probab. Eng. Inf. Sci.*, 20(4):667–686, 2006.

P. Billingsley. *Probability and Measure*. John Wiley and Sons, New York, NY, third edition, 1995.

C. R. Blyth. Expected absolute error of the usual estimator of the binomial parameter. *The American Statistician*, 34(3):155–157, 1980.

P. Brémaud. *Markov Chains: Gibbs Fields, Monte Carlo Simulation and Queues*. Springer, New York, NY, 1999.

L. Buşoniu, R. Babuška, B. De Schutter and D. Ernst. *Reinforcement Learning and Dynamic Programming Using Function Approximators*. CRC Press, Boca Raton, FL, 2010.

G. Casella and R. L. Berger. *Statistical Inference*. Duxbury, Pacific Grove, CA, second edition, 2002.

K. S. Chan. A note on the geometric ergodicity of a Markov chain. *Advances in Applied Probability*, 21(3): pp. 702–704, 1989.

H. S. Chang, M. C. Fu, J. Hu and S. I. Marcus. *Simulation-based Algorithms for Markov Decision Processes*. Springer, New York, NY, 2007.

C. Chow and J. N. Tsitsiklis. The complexity of dynamic programming. *Journal of Complexity*, 5:466–488, 1989.

C. Chow and J. N. Tsitsiklis. An optimal one-way multigrid algorithm for discrete-time stochastic control. *IEEE Transactions on Automatic Control*, 36(8):898–914, 1991.

H. A. David and H. N. Nagaraja. *Order Statistics*. John Wiley and Sons, Hoboken, NJ, third edition, 2003.

M. H. A. Davis. *Markov Models and Optimization*. Chapman and Hall, London, 1993.

M. H. A. Davis, M. A. H. Dempster, S. P. Sethi and D. Vermes. Optimal capacity expansion under uncertainty. *Advances in Applied Probability*, 19:156–176, 1987.

L. Deng and S. Li. Ishikawa iteration process with errors for nonexpansive mappings in uniformly convex Banach spaces. *International Journal of Mathematics and Mathematical Sciences*, 24 (1):49–53, 2000.

L. Devroye. Exponential inequalities in nonparametric estimation. In G. Roussas, editor, *Nonparametric Functional Estimation and Related Topics*, volume 335 of *NATO Adv. Sci. Inst. Ser. C Math. Phys. Sci.*, 31–44. Kluwer Academic Publishers, Dordrecht, The Netherlands, 1991.

P. Diaconis and S. Zabell. Closed form summation for classical distributions: variations on a theme of de Moivre. *Statistical Science*, 6(3):284–302, 1991.

J. L. Doob. *Stochastic Processes*. John Wiley and Sons, New York, NY, 1953.

L.E. Dubins and D.A. Freedman. A sharper form of the Borel-Cantelli lemma and the strong law. *The Annals of Mathematical Statistics*, 36:800–807, 1965.

L.E. Dubins and L.J. Savage. *Inequalities for Stochastic Processes: How to Gamble if You Must*. Dover Publications, New York, NY, second edition, 1976.

R. Durrett. *Probability: Theory and Examples*. Cambridge University Press, New York, NY, fourth edition, 2010.

N. Etemadi. An elementary proof of the strong law of large numbers. *Z. Warsch. verw. Gebiete.*, 55:119–122, 1981.

W. Feller. *Probability Theory and Its Applications, Volume 1*. John Wiley and Sons, New York, NY, third edition, 1968.

W. Feller. *Probability Theory and Its Applications, Volume 2*. John Wiley and Sons, New York, NY, second edition, 1971.

E. Fischer. *Intermediate Real Analysis*. Springer-Verlag, New York, NY, 1983.

P. W. Glynn. Upper bounds on poisson tail probabilities. *Operations Research Letters*, 6(1): 9–14, 1987.

E. J. Gumbel. The maxima of the mean largest value and of the range. *Annals of Mathematical Statistics*, 25(1):76–84, 1954.

L. Györfi, M. Kohler, A. Krzyzk and H. Walk. *A Distribution-Free Theory of Nonparametric Regression*. Springer, New York, NY, 2002.

P. Hall and C. C. Heyde. *Martingale Limit Theory and Its Application*. Academic Press, New York, NY, 1980.

P. Hall. On the rate of convergence of normal extremes. *Journal of Applied Probability*, 16(2):433–439, 1979.

T.E. Harris. The existence of stationary measures for certain Markov processes. In *Proc. 3rd Berkeley Sympos. Math. Statist. Probability* 2, 113–124, 1956.

H. O. Hartley and H. A. David. Universal bounds for mean range and extreme observation. *Annals of Mathematical Statistics*, 25(1):85–89, 1954.

O. Hernández-Lerma. *Adaptive Markov Control Processes.* Springer-Verlag, New York, NY, 1989.

O. Hernández-Lerma and J.B. Lasserre. *Discrete-Time Markov Control Processes: Basic Optimality*. Springer, New York, NY, 1996.

O. Hernández-Lerma and J.B. Lasserre. *Further Topics on Discrete-Time Markov Control Processes*. Springer, New York, NY, 1999.

O. Hernández-Lerma and J.B. Lasserre. Further criteria for positive Harris recurrence of Markov chains. *Proceedings of the American Mathematical Society*, 129(5):1521–1524, 2001.

K. Hinderer. *Foundations of Non-stationary Dynamic Programming with Discrete Time Parameter*, volume 33 of *Lecture Notes Oper. Res.* Springer-Verlag, New York, NY, 1970.

W. Hoeffding. Probability inequalities for sums of bounded random variables. *Journal of the American Statistical Association*, 58(301):13–30, 1963.

R. A. Horn and C. R. Johnson. *Matrix Analysis.* Cambridge University Press, Cambridge, UK, 1985.

R. Isaac. A general version of Doeblin's condition. *Ann. Math. Stat.*, 34:668–671, 1963.

E. Isaacson and H. B. Keller. *Analysis of Numerical Methods.* John Wiley & Sons, New York, NY, 1966.

S. Ishikawa. Fixed points by a new iteration method. *Proceedings of the American Mathematical Society*, 44:147–150, 1974.

S. Karlin and H. M. Taylor. *A First Course in Stochastic Processes.* Academic Press, San Diego, CA, second edition, 1975.

S. Karlin and H. M. Taylor. *A Second Course in Stochastic Processes.* Academic Press, San Diego, CA, 1981.

M. Kearns, Y. Mansour and A. Ng. Approximate planning in large POMDPs via reusable trajectories. In *Advances in Neural Information Processing Systems*, volume 12. MIT Press, Cambridge, MA, 2000.

M. Kearns, Y. Mansour and A. Ng. A sparse sampling algorithm for near-optimal planning in large Markov decision processes. *Machine Learning*, 49(2–3):193–208, 2002.

A.S. Kechris. *Classical Descriptive Set Theory.* Graduate Texts in Mathematics. Springer, New York, NY, 1995.

D. G. Kendall. Stochastic processes occurring in the theory of queues and their analysis by the method of the imbedded Markov chain. *Ann. Math. Statist.*, 24(3):338–354, 1953.

J. Kiefer and J. Wolfowitz. Stochastic estimation of the maximum of a regression function. *The Annals of Mathematical Statistics*, 23(3):462–466, 1952.

T. Y. Kim and D. D. Cox. Uniform strong consistency of kernel density estimators under dependence. *Statistics & Probability Letters*, 26:179–185, 1996.

L. Kleinrock. *Queueing Systems (Volume 1: Theory).* John Wiley and Sons, New York, NY, 1975.

A. N. Kolmogorov. *Grundbegriffe der Wahrscheinlichkeitrechnung, Ergebnisse Der Mathematik; translated as Foundations of Probability (1950).* Chelsea Publishing Company, New York, NY, 1933.

A. N. Kolmogorov and S. V. Fomin. *Introductory Real Analysis.* Dover, Mineola, New York, 1970.

M. Kuczma. *An Introduction to the Theory of Functional Equations and Inequalities: Cauchy's Equation and Jensen's Inequality.* Birkhäuser, Basel, second edition, 2009.

P.R. Kumar and P. Varaiya. *Stochastic Systems: Estimation, Identification and Adaptive Control.* Prentice-Hall, Englewood Cliffs, New Jersey, 1986.

H.J. Kushner and G. Yin. *Stochastic Approximation and Recursive Algorithms and Applications.* Applications of Mathematics. Springer, New York, NY, second edition, 2003.

E.L. Lehmann and G. Casella. *Theory of Point Estimation.* Springer, New York, NY, second edition, 1998.

X. Li. Some properties of ageing notions based on the moment-generating-function order. *J. Appl. Probab.*, 41(3):927–934, 2004.

S.A. Lippman. On dynamic programming with unbounded rewards. *Management Science*, 21: 1225–1233, 1975.

L.S. Liu. Ishikawa and Mann iterative process with errors for nonlinear strongly accretive mappings in Banach spaces. *Journal of Mathematical Analysis and Applications*, 194(1):114–125, 1995.

L.S. Liu. Ishikawa iteration process with errors for nonexpansive mappings. *International Journal of Mathematics and Mathematical Sciences*, 27(7):413–417, 2001.

L. Ljung. Strong convergence of a stochastic approximation algorithm. *The Annals of Statistics*, 6(3):680–696, 1978.

P. McCullagh. Tensor notation and cumulants of polynomials. *Biometrika*, 71:461–476, 1984.

C. McDiarmid. On the method of bounded differences. In J. Siemons, editor, *Surveys in Combinatorics*, volume 141 of *LMS Lecture Note Series*, pages 148–188. Morgan Kaufmann Publishers, San Mateo, CA, 1989.

E. Nummelin and P. Tuominen. Geometric ergodicity of Harris recurrent Markov chains with applications to renewal theory. *Stochastic Processes and their Applications*, 12(2):187–202, 1982.

J. M. Ortega and W. C. Rheinboldt. On a class of approximate iterative processes. *Archive for Rational Mechanics and Analysis*, 23:352–365, 1967.

M.O. Osilike. Ishikawa and Mann iteration methods with errors for nonlinear equations of the accretive type. *Journal of Mathematical Analysis and Applications*, 213(1):91–105, 1997.

A. Ostrowski. The rounding-off stability of iterations. *Basel Math. Notes*, BMN-12, 1964.

E. Parzen. *Stochastic Processes.* Holden-Day, San Fransisco, CA, 1962.

N. N. Popov. Conditions for geometric ergodicity of countable Markov chains. *Soviet Math. Dokl.*, 18(3):676–679, 1977.

W. B. Powell. *Approximate Dynamic Programming: Solving the Curses of Dimensionality.* John Wiley and Sons, Hoboken, NJ, second edition, 2011.

M. L. Puterman. *Markov Decision Processes: Discrete Stochastic Dynamic Programming.* John Wiley and Sons, New York, NY, 1994.

H. Robbins and S. Monro. A stochastic approximation method. *The Annals of Mathematical Statistics*, 22(3):400–407, 1951.

S. M. Ross. *Stochastic Processes.* John Wiley and Sons, New York, NY, second edition, 1996.

H. L. Royden. *Real Analysis.* MacMillan Publishing, New York, NY, second edition, 1968.

J. Rust. Using randomization to break the curse of dimensionality. *Econometrica*, 65(3):487–516, 1997.

M. Schäl. Estimation and control in discounted stochastic dynamic programming. *Stochastics*, 20:51–71, 1987.

R. W. Shonkwiler and F. Mendivil. *Explorations in Monte Carlo Methods.* Springer, New York, NY, 2009.

J. Si, A. Barto, W. Powell and D. Wunsch (editors). *Handbook of Learning and Approximate Dynamic Programming.* John Wiley & Sons-IEEE Press, Piscataway, NJ, 2004.

C. Szepesvári. Efficient approximate planning in continuous space Markovian decision problems. *AI Communications*, 14(3):163–176, 2001.

J. N. Tsitsiklis and B. Roy. Feature-based methods for large scale dynamic programming. *Machine Learning*, 22(1–3):59–94, 1996.

J.A.E.E. Van Nunen and J. Wessels. A note on dynamic programming with unbounded rewards. *Management Science*, 24:576–580, 1978.

W. Wang and Y. Ma. Stochastic orders and aging notions based upon the moment generating function order: Theory. *Journal of the Korean Statistical Society*, 38(1):87–94, 2009.

P. Whittle. *Probability via Expectation*. Springer-Verlag, New York, NY, fourth edition, 2000.

Subject index